WDM TECHNOLOGIES: ACTIVE OPTICAL COMPONENTS

Edited by

Achyut K. Dutta
Fujitsu Compound Semiconductors, Inc.
San Jose, California, USA

Niloy K. Dutta
University of Connecticut
Storrs, Connecticut, USA

Masahiko Fujiwara
Networking Research Laboratories
NEC Corporation
Tsukuba, Ibaraki, Japan

ACADEMIC PRESS

An imprint of Elsevier Science

Amsterdam   Boston   London   New York   Oxford   Paris
San Diego   San Francisco   Singapore   Sydney   Tokyo

Academic Press
An imprint of Elsevier Science
525 B Street, Suite 1900, San Diego, California 92101-4495, USA
http://www.academicpress.com

Academic Press
An imprint of Elsevier Science
84 Theobald's Road, London WC1X 8RR, UK
http://www.academicpress.com

Library of Congress Control Number: 2002107431

International Standard Book Number: 0-12-225261-6

PRINTED IN THE UNITED STATES OF AMERICA
02   03   04   05   06   MB   9   8   7   6   5   4   3   2   1

Dedicated to our parents,
Harish Chandra and Kalpana Rani Dutta,
Debakar and Madabor Datta,
and to our families,
Keiko, Jayoshree, Jaydeep,
Sudeep Hiroshi, and Cristine Dutta

Contents

Chapter **3**   High Power Semiconductor Lasers for EDFA Pumping   59

Akihiko Kasukawa

Chapter **4**   Tunable Laser Diodes   105

Gert Sarlet, Jens Buus, and Pierre-Jean Rigole

Chapter 5   Vertical-Cavity Surface-Emitting Laser Diodes   167

Kenichi Iga and Fumio Koyama

Chapter **11**   Dry Etching Technology for Optical Devices   533

Stella W. Pang

Contributors

Jens Buus (Chapter 4), Gayton Photonics Ltd., 6 Baker Street, Gayton, Nothants, NN7 3EZ, United Kingdom.

Achyut K. Dutta (Chapters 1, 12, & 13), Fujitsu Compound Semiconductors, Inc, 2355 Zanker Road, San Jose, CA 95131, USA.

Niloy K. Dutta (Chapters 1 & 2), Department of Physics and Photonics Research Center, University of Connecticut, Storrs, CT 06269-3046, USA.

Masahiko Fujiwara (Chapter 1), Networking Research Laboratories, NEC Corporation, 34, Miyukigaoka, Tsukuba, Ibaraki, 305-8501, Japan.

Kenichi Iga (Chapter 5), The Japan Society for the Promotion of Science, 6 Ichibancho, Chiyodaku, Tokyo 102-8471, Japan.

Akihiko Kasukawa (Chapter 3), Yokohama R&D Laboratories, The Furukawa Electric Co., Ltd., 2-4-3 Okano, Nishi-ku, Yokohama 220-0073, Japan.

Masahiro Kobayashi (Chapters 9, 12 & 13), Fujitsu Quantum Devices Limited, Kokubo Kogyo Danchi, Showa-Cho, Nakakoma-Gun, Yamanashi-Ken 409-3883, Japan.

Fumio Koyama (Chapter 5), Precision & Intelligence Lab., Tokyo Institute of Technology, 4259 Nagatsuta, Midoriku, Yokohama 226-8503, Japan.

Koji Kudo (Chapter 10), Photonic and Wireless Devices Research Labs., System Devices and Fundamental Research, NEC Corporation, 2-9-1 Seiran, Ohtsu-shi, Shiga 520-0833, Japan.

Rangaraj Madabhushi (Chapter 6), Optoelectronics Center, Room 3I-153, Agere Systems Inc., 9999 Hamilton Blvd, Breinigsville, PA 18031, USA.

T.G. Beck Mason (Chapter 7), Optoelectronics Center, Agere Systems Inc., 9999 Hamilton Blvd., Breinigsville, PA 18031, USA.

Takashi Mikawa (Chapter 9), Fujitsu Quantum Devices Limited, Kokubo Kogyo Danchi, Showa-Cho, Nakakoma-Gun, Yamanashi-Ken 409-3883, Japan.

Stella W. Pang (Chapter 11), Dept. of Electrical Engineering & Computer Science, 304, EECS Bldg., University of Michigan, 1301 Beal Ave., Ann Arbor, MI 48109-2122, USA.

Pierre-Jean Rigole (Chapter 4), ADC–Sweden, Bruttov. 7, SE-175 43 Järfälla-Stockholm, Järfälla, Sweden.

Gert Sarlet (Chapter 4), Orkanvägen 35, 17771 Järfälla, Sweden.

Tatsuya Sasaki (Chapter 10), Photonic and Wireless Devices Research Labs., System Devices and Fundamental Research, NEC Corporation, 2-9-1 Seiran, Ohtsu-shi, Shiga 520-0833, Japan.

Kenko Taguchi (Chapter 8), Development Department, Optoelectronic Industry and Technology Development Association, Sumitomo Edogawabashiekimae Bldg., 7F, 20-10 Sekiguchi 1-Chome, Bunkyo-ku, Tokyo, 112-0014, Japan.

Foreword

The WDM Revolution

This book is the first of four about wavelength division multiplexing (WDM), the most recent technology innovation in optical fiber communications. In the past two decades, optical communications has totally changed the way we communicate. It is a revolution that has fundamentally transformed the core of telecommunications, its basic science, its enabling technology, and its industry. The WDM innovation represents a revolution inside the optical communications revolution and it is allowing the latter to continue its exponential growth.

The existence and advance of optical fiber communications is based on the invention of the laser, particularly the semiconductor junction laser, the invention of low-loss optical fibers, and on related disciplines such as integrated optics. We should never forget that it took more than 25 years from the early pioneering ideas to the first large-scale commercial deployment of optical communications, the Northeast Corridor system linking Washington and New York in 1983 and New York with Boston in 1984. This is when the revolution got started in the marketplace, and when optical fiber communications began to seriously impact the way information is transmitted. The market demand for higher capacity transmission was helped by the fact that computers continued to become more powerful and needed to be interconnected. This is one of the key reasons why the explosive growth of optical fiber transmission technology parallels that of computer processing and other key information technologies. These technologies have combined to meet the explosive global demand for new information services including data, internet, and broadband services—and, most likely, their rapid advance has helped fuel this demand. We know that this demand is continuing its strong growth as internet traffic, even by reasonably conservative estimates, keeps doubling every year. Today, we optical scientists and engineers are naturally puzzling the question why this traffic growth does not appear to be matched by a corresponding growth in revenue. Another milestone in the optical communications revolution we remember with pride is the deployment of the first transatlantic fiber system, TAT8, in 1988 (today, of course, the map of undersea systems deployed in the oceans of the globe looks like a dense spider web). It was around this time that researchers began exploring the next step forward, optical fiber amplifiers and WDM transmission.

WDM technology has an interesting parallel in computer architecture. Computers have a similar problem as lightwave systems: both systems trends—pulled by demand and pushed by technology advances—show their key technological figure of merit (computer processing power in one case, and fiber transmission capacity in the other) increasing by a factor 100 or more every ten years. However, the raw speed of the IC technologies computers and fiber transmission rely on increases by about a factor of 10 only in the same time frame. The answer of computer designers is the use of parallel architectures. The answer of the designers of advanced lightwave system is similar: the use of many parallel high-speed channels carried by different wavelengths. This is WDM or "dense WDM." The use of WDM has other advantages such as the tolerance of WDM systems of the high dispersion present in the low loss window of embedded fibers, the fact that WDM can grow the capacity incrementally, and that WDM provides great simplicity and flexibility in the network.

WDM required the development of many new enabling technologies, including broadband optical amplifiers of high gain, integrated guided-wave wavelength filters and multiplexers, WDM laser sources such as distributed-feedback (DFB) lasers providing spectral control, high-speed modulators, etc. It also required new systems and fiber techniques to compensate fiber dispersion and to counteract nonlinear effects caused by the large optical power due to the presence of many channels in the fiber. The dispersion management techniques invented for this purpose use system designs that avoid zero dispersion locally, but provide near-zero dispersion globally.

Vigorous R&D in WDM technologies led to another milestone in the history of optical communications, the first large-scale deployment of a commercial WDM system in 1995, the deployment of the NGLN system in the long-distance network of AT&T.

In the years that followed, WDM led the explosive growth of optical communications. In early 1996, three research laboratories reported prototype transmission systems breaking through the Terabit/second barrier for the information capacity carried by a single fiber. This breakthrough launched lightwave transmission technology into the "tera-era." All three approaches used WDM techniques. Five years later, in 2001 and exactly on schedule for the factor-100-per-decade growth rate, a WDM research transmission experiment demonstrated a capacity of 10 Tb/s per fiber. This is an incredible capacity: recall that, at the terabit/sec rate, the hair-thin fiber can support a staggering 40 million 28-K baud data connections, transmit 20 million digital voice telephony channels, or a half million compressed digital TV channels. Even more importantly, we should recall that the dramatic increase in lightwave systems capacity has a very strong impact on lowering the cost of long-distance transmission. The Dixon-Clapp rule projects that the cost per voice channel reduces with the square root of the systems capacity. This allows one to estimate that the above technology growth rate reduces the technology cost of transmitting one voice channel by a factor of ten every ten years. As a consequence of this trend, one finds that the distance of transmission plays a smaller and smaller role in the equation of telecom economics: An internet user, for example, will click a web site regardless of its geographical distance.

WDM technology is progressing at a vigorous pace. Enabled by new high-speed electronics the potential bit-rate per WDM channel has increased to 40 Gb/s and higher, broadband Raman fiber amplifiers are being employed in addition to the early erbium-doped fiber amplifiers, and there are new fibers and new techniques for broadband dispersion compensation and broadband dispersion management, etc.. The dramatic decrease in transmission cost, combined with the unprecedented capacities appearing at a network node as well as the new traffic statistics imposed by the internet and data transmission have caused a rethinking of long-haul and ultra-long-haul network architectures. New designs are being explored that take advantage of the fact that WDM has opened up a new dimension in networking: it has added the dimension of wavelength to the classical networking dimensions of space and time. New architectures are under exploration that are transparent to bit-rate, modulation format, and protocol. A recent example for this are the recent demonstrations of bit-rate transparent fiber cross-connects based on photonic MEMS fabrics, arrays of micromirrors fabricated like integrated silicon integrated circuits.

Exactly because of this rapid pace of progress, these volumes will make a particularly important contribution. They will provide a solid assessment and teaching of the current state of the WDM art serving as a valuable basis for further progress.

Herwig Kogelnik
Bell Labs
Lucent Technologies
Crawford Hill Laboratory
Holmdel, NJ 07733-0400

Acknowledgments

Future communication networks will require total transmission capacities of few Tb/s. Such capacities could be achieved by wavelength division multiplexing (WDM). This has resulted in increasing demand of WDM technology in communication. With increase in demand, many students and engineers are migrating from other engineering fields to this area. Based on our many years of experience, we felt that it is necessary to have a set of books which could help all engineers wishing to work or already working in this field. Covering a fast-growing subject such as WDM technology is a very daunting task. This work would not have been possible without the support and help from all chapter contributors. We are indebted to our current and previous employers, NEC Research Labs, Fujitsu, Bell Laboratories, and the University of Connecticut for providing the environment, which enabled and provided the intellectual stimulation for our research and development in the field of optical communication and their applications. We are grateful to our collaborators over the years. We would also like to convey our appreciation to our colleagues with whom we have worked for many years. Thank you also to the author of our foreword, H. Kogelnik, for his kindness in providing his gracious remarks on *The WDM Revolution* for our four books on WDM Technologies. Last but not least, many thanks also go to our family members for their patience and support, without which this book could not have been completed.

Achyut K. Dutta
Niloy K. Dutta
Masahiko Fujiwara

Chapter 1 | Overview

Achyut K. Dutta

Fujitsu Compound Semiconductors Inc., 2355 Zanker Road,
San Jose, CA 95131, USA

Niloy K. Dutta

Department of Physics and Photonics Research Center
University of Connecticut, Storrs, CT 06269-3046, USA

Masahiko Fujiwara

Networking Research Laboratories, NEC Corporation,
34, Miyukigaoka, Tsukuba, Ibaraki, 305-8501, Japan

1.1. Prospectus

With the recent exponential growth of Internet users and the simultaneous proliferation of new Internet protocol applications such as web browsing, e-commerce, Java applications, and video conferencing, there is an acute need for increasing the bandwidth of the communications infrastructure all over the world. The bandwidth of the existing SONET and ATM networks is pervasively limited by electronic bottlenecks, and only recently was this limitation removed by the first introduction of wavelength-division multiplexing (WDM) systems in the highest capacity backbone links. The capacity increase realized by the first WDM systems was quickly exhausted/utilized, and both fueled and accommodated the creation of new Internet services. This, in turn, is now creating a new demand for bandwidth in more distant parts of the network. The communication industries are thus at the onset of a new expansion of WDM technology necessary to meet the new and unanticipated demand for bandwidth in elements of the telephony and cable TV infrastructure previously unconsidered for WDM deployment. The initial deployments of WDM were highly localized in parts of the communications infrastructure and supported by a relatively small group of experts. The new applications in different parts of the network must be implemented by a much larger group of workers from a tremendous diversity of technical backgrounds. To serve this community

1

WDM TECHNOLOGIES: ACTIVE
OPTICAL COMPONENTS
$35.00

involved with the optical networking, a series of volumes covering all WDM technologies (from the optical components to networks) is introduced.

Many companies and new start-ups are trying to make the WDM-based products as quickly as possible, hoping to become leaders in that area. As the WDM-based products need wide knowledge, ranging from components to network architecture, it is difficult for the engineers to grasp all the related areas quickly. Today, engineers working specifically in one area are always lacking in the other areas, which impedes the development of the WDM products. The main objective of these volumes will be to give details on the WDM technology varying from the components (all types) to network architecture. We expect that this book and series will not only be useful for graduate students specifically in electrical engineering, electronic engineering, and computer engineering, but that instructors could consider it for their courses either as the textbook or a reference book.

Because the major developments in optical communication networks have started to capture the imagination of the computing, telecommunications, and opto-electronics industries, we expect that industry professionals will find this book useful as a well-rounded reference. Through our wide experience in industries on the optical networking and optical components, we know that there are many engineers who are expert in the physical layer, but still must learn the optical system and networks, and corresponding engineering problems in order to design new state-of-the-art optical networking products. We had all these groups of people in mind while we prepared these books.

1.2.   Organization and Features of the Volumes

Covering this broad an area is not an easy task, as the volumes will need to cover everything from optical components (to be/already deployed) to the network. WDM includes areas of expertise from electrical engineering to computer engineering and beyond, and the field itself is still evolving. This volume is not intended to include any details about the basics of the related topics; readers will need to search out the reference material on more basic issues, especially the undergraduate-level books for such materials. These references together with this series of books can provide a systematic in-depth understanding of multidisplinary fields to graduate students, engineers, and scientists who would like to increase their knowledge in order to potentially contribute more to these WDM technologies.

An important organizing principle that we attempted while preparing the contents was that research, development, and education on WDM technologies should allow tight coupling between the network architectures and device capabilities. Research on WDM has taught us that, without sound knowledge of device or component capabilities and limitations, one can produce architecture that would be completely unrealizable; new devices developed without the concept of the useful system can lead to sophisticated technology with limited or no usefulness. This idea motivated us to prepare this series of books, which will be helpful to professional and academic personnel, working in different area of WDM technologies.

This series on various areas of WDM technologies is divided into four volumes, each of which is divided into a few parts to provide a clear concept among the readers or educators of the possibilities of their technologies in particular networks of interest to them. The series starts with two complete volumes on optical components. Because many of the chapters relate to components, we decided to publish one volume for active and one volume for passive components. This format should prove more manageable and convenient for the reader. Other volumes are on optical systems and optical networks. Volume I gives a clear view on the WDM components, especially all kinds of active optical components. Volume II, covering key passive optical components, follows this. Volume III covers WDM networks and their architecture possibly implementable in near-future networks. Finally, Volume IV will describe the WDM system, especially including a system aspects chapter implementable in the WDM equipment. All of these volumes cover not only recent technologies, but also future technologies. Chapter 1 of that volume's contents, each volume will explain to accommodate users who choose to buy just one volume. This chapter contains survey of this volume.

1.3. Survey of Volume I

WDM TECHNOLOGIES: ACTIVE OPTICAL COMPONENTS

Unlike most of the available textbooks on optical fiber communication, our Volume I covers several key active optical components and their key technologies from the standpoint of WDM-based application. Based on our own hands-on experience in this area for the past 25 years, we tend to cover only those components and technologies that could be practically used in the most WDM communication. This volume is divided into five parts; Part I: Laser sources, Part II: Optical Modulators, Part III: Photodetectors, Part IV: Fabrication Technologies, and Part V: Optical Packaging Technologies. Next, we briefly survey the chapters of each part to attempt to put the elements of the book into context.

Part I: Laser Sources

Ever since the invention of the semiconductor laser in 1962 [1], development has been on going to improve performance and functionality for optical communication application. This part covers several kinds of laser sources being used in the optical networks from the edge to core networks. Each chapter provides the current network application and future direction.

Chapter 2: Long-Wavelength Laser Source

Semiconductor lasers, especially 1.3 μm and 1.55 μm wavelengths, have been widely used as the transmitter source in optical communication since their invention. Now, in each transmission system, whether a short- or long-haul application, long-wavelength semiconductor lasers fabricated on InP substrate are being used, and their performance has been improved tremendously. The fabrication technologies, performance characteristics, current state-of-the-art, and research direction of long-wavelength laser diodes are examined in Chapter 2 by Niloy K. Dutta, a pioneer of the laser diode.

Chapter 3: High-Power Semiconductor Lasers for EDFA Pumping

The introduction of two technologies, WDM and optical amplifier, makes it possible to increase the capacity and transmission distances, respectively, helpful in extending the optical domains from core to edge. The realization of the optical amplifier, especially using the Er-doped fiber base and later the Raman amplifier, is possible because of tremendous improvement of high-power laser diodes of wavelengths 1.4 μm and 0.98 μm for use in pumping. The trends of high-power semiconductor laser along with the design, fabrication, characteristics, reliability, and packaging, are described in Chapter 3 by A. Kasukawa, a pioneer in the pump laser.

Chapter 4: Tunable Laser Diodes

In a WDM transmission system whether in long- or short-haul applications, optical sources capable of generating a number of wavelengths are required. From the viewpoint of system complexities and cost, especially with WDM applications, it is very unrealistic to use an optical source for each wavelength. This drives the development of the tunable laser diodes, with tunability ranges from a few nanometers to whole c-band wavelengths. Tunable lasers offer many compelling advantages over fixed wavelength solutions in optical networks in that they simplify the planning, reduce inventories, allow dynamic wavelength provisioning, and simplify network control software. This is also expected to be a feature in optical network developments spanning nearly all application segments, from access/enterprise through metropolitan and long-haul networks, which has lead to a variety of desired specifications and approaches. Gert Sarlet, Jens Buus and Pierre-Jean Rigole describe design and performances of different kinds of tunable semiconductor laser diodes in Chapter 4.

Chapter 5: Vertical Cavity Surface-Emitting Laser Diodes (VCSELs)

A cornerstone of the optical network revolution is the semiconductor laser, the component that literally sheds light on the whole industry. The most prevalent semiconductor laser in telecommunication has been the edge-emitting laser, which has enabled many facets of today's optical revolution in the long-haul application. Its improvement, along with other optical components, has increased the data rate from OC 3 to OC 192, and very soon to OC 768, and distances from a few kilometers to thousands of kilometers. The dense WDM (DWDM) application is also possible due to the semiconductor laser's improvements.

The edge-emitting laser enabled the first wave of optical networking. The next wave will be enabled by laser technology that substantially reduces costs and improves performance. That technology is the vertical cavity surface-emitting laser (VCSELs). After its invention in 1979 [2], 850-nm VCSEL development quickly evolved into successful commercial components for data communications in the mid 1990s. The benefits are so compelling in the application that 850-nm VCSELs completely replaced the edge-emitting lasers as the technology of choice. The benefits and success of 850-nm VCSELs are now driving its development to apply to telecommunication applications where more expensive edge-emitting lasers are currently used, and in 1999 VCSEL entered into the third generation of development. In Chapter 5, K. Iga, an inventor of VCSELs, and Fumio Koyama describe the progress of VCSELs in a wide range of optical spectra based on GaInAsP, AlGaInAs, GaInNAs, GaInAs, AlGaAsSb, GaAlAs, AlGaInP, ZnSe, GaInN, and some other materials.

Part II: Optical Modulators

The presence of chirp in direct modulation laser diodes limits the transmission distance, and the effect is more pronounced as the bit rate increases. This limitation can be overcome by using the external modulation technique. This part covers two kinds of key external modulators frequently used in telecommunications.

Chapter 6: Lithium Niobate Optical Modulators

More than 25 years have passed since the invention of the titanium-diffused waveguides in titanium niobate [3], and the associated integrated optic waveguide electrooptic modulator [4]. In the beginning, while the data rate was low, electrooptic mechanisms had to compete with the direct modulation technique. Later, with an increase of the data rate, electrooptic modulators using lithium niobate (LN) have been considered to be the best technique for long-distance transmission. In Chapter 6, Raj Madabhushi of Agere Systems describes the design and progress of LN modulators. Raj has lengthy experience with LN modulators in University and in different industries in North America and Japan.

Chapter 7: Electroabsorption Modulators

The EA modulator is another external modulator that can be fabricated using semiconductor laser technology. The main advantage of the EA modulator over the LN modulator is that EA can be monolithically integrated with a laser diode and semiconductor amplifier on the single substrate for higher functionality Beck Mason of Agere Systems explains the basic principle design, fabrication, and characterization of the EA modulator, including its progress, in Chapter 7. Current developments on the 40G EA modulator are also included in this chapter.

Part III: Photodetectors

The heart of a receiver for any optical transmission system is the optoelectronics component that is used as the photodetector. This part covers two kinds of key photodetectors frequently used in optical communication.

Chapter 8: P-I-N Photodiodes

K. Taguchi has many years of experience in designing various photodetectors for optical communication. In Chapter 8, Taguchi describes basic concepts, details, design, and fabrication of PIN-type photodiodes composed mainly of InGaAs as a light absorption layer with no internal gain. The photonic integrated circuit including the photodetector is also included in this chapter.

Chapter 9: Avalanche Photodiodes

The first avalanche photodiode (APD) made commercially available for long-wavelength optical communication (1.3-mm wavelength window) and frequently useful in the 1980s was Germanium APD (Ge-APD). Limitations of Ge-APD performances are dark current, multiplication noise, and sensitivity at longer wavelength window at 1.55 μm—these are material-induced parameters. To respond to higher sensitivity APDs at both 1.3 μm and 1.55 μm, InGaAs-based APDs are introduced. Chapter 9, by M. Kobayashi, and T. Mikawa, pioneers in APD, describes the design, fabrication, and reliability of avalanche photodiodes with an internal gain for optical communication. This chapter also includes various APDs from Ge-APD. Recent progress and the future direction of APD are also included in this chapter.

Part IV: Fabrication Technologies

Some of the great advances in semiconductor laser performances in recent years can be traced to advanced fabrication technology. This part provides the advanced fabrication technology of the semiconductor photonics devices.

Chapter 10: Selective Growth Techniques and Their Application in WDM Device Fabrication

The recent trend of DWDM application necessitates the cost-effective photonics device. Device fabrication strongly affects the device performance and production yield, particularly for the complicated integrated photonics devices. Recent development in fabrication technology make it possible to reduce the cost and improve performance of the photonics devices. In Chapter 10, J. Sasaki and K. Kudo describe the selective area growth for multiwavelength laser diode and EA modulator integrated LD fabrication. Details of growth mechanism for controlling the band energy are also included in this chapter.

Chapter 11: Dry-Etching Technology for Optical Devices

Today's advanced dry-etching technology enables the high-performance and low-cost photonic devices. Their development is also underway in different research organizations and academia, focusing on the future monolithic integration of high functional photonics devices on the single wafer. In Chapter 11, S. Pang, pioneer in dry etching, describes the dry-etching technologies for the fabrication of high-performance photonics devices.

Part V: Optical Packaging Technologies

Today more than 50% of the total cost in optical module is accounted for by the packaging and assembly technologies. The main reason is that packaging technology is not yet matured and all industries are using their respective proprietary technology. No design guideline has been published for designing the photonic device. This part, comprising two chapters, covers the packaging technologies for optical components.

Chapter 12: Optical Packaging/Module Technologies: Design Methodology

Chapter 12 by A. K. Dutta and M. Kobayashi describes the design methodologies as required systematically for optical package/module design. Different kinds of optical packages are also included for giving insight about the optical packages. For the most part, emphasis is on different design considerations, necessary for high-performance and cost-effective optical package. Related examples are also included.

Chapter 13: Packaging Technologies for Optical Components: Integrated Module

Integrating multiple optical functions monolithically into the single device is a key step to lowering the costs of the optical networks. Integrating multiple functions into the single device can reduce the cost of labor, packaging, and testing. The primary challenges to monolithic integration are finding a material that can perform multiple functions and understanding the impact that concatenating functions has on fabrication yields. The integration technology is not matured enough to apply to the field-implementable optical devices. Prior to available monolithic integration technology, the path to integration will take the sequential steps, from packaging the discrete optical devices together in the modules, eventually leading to monolithic integration. In Chapter 13, A. K. Dutta and M. Kobayashi review the technologies available for integrating multifunctional devices into the modules. Future directions on various optical module technologies are also included in this chapter.

References

1. H. Kressel and J. K. Butler, Semiconductor Lasers and Heterojunctions LEDs, (Academic Press, NY, 1977).
2. H. Soda, K. Iga, C. Kitahara, and Y. Suematsu, "GaInAsP/InP surface emitting injection lasers," Jpn. J. Appl. Phys., 18 (1979) 2329–2330.
3. I. P. Kaminow, L. W. Stulz, and E. H. Turner, "Efficient strip-waveguide modulator," Appl. Phys. Lett., 27 (1975) 555–557.
4. R. V. Schmidt and I. P. Kaminow, "Metal-diffused optical waveguides in LiNbO3," Appl. Phys. Lett., 25 (1974) 458–460.

Part 1 | Laser Sources

Chapter 2 | Long-Wavelength Laser Source

Niloy K. Dutta

Department of Physics and Photonics Research Center
University of Connecticut, Storrs, CT 06269-3046, USA

2.1. Introduction

Phenomenal advances in research results, and development and application of optical sources have occurred over the last decade. The two primary optical sources used in telecommunications are the semiconductor laser and the light-emitting diode (LED). The LEDs are used as sources for low data rate (<200 Mb/s) and short-distance applications, and lasers are used for high data rate and long-distance applications. The fiber optic revolution in telecommunications, which provided several orders of magnitude improvement in transmission capacity at low cost, would not have been possible without the development of reliable semiconductor lasers. Today, semiconductor lasers are used not only for fiber optic transmission but also in optical reading and recording (e.g., CD players), printers, Fax machines, and in numerous applications as a high-power laser source. Semiconductor injection lasers continue to be the laser of choice for various system applications, primarily because of their small size, simplicity of operation, and reliable performance. For most transmission system applications the laser output is encoded with data by modulating the current. However, for some high data rate applications, which require long-distance transmission, external modulators are used to encode the data.

This chapter describes the fabrication, performance characteristics, current state of the art, and research directions for semiconductor lasers and,

13

WDM TECHNOLOGIES: ACTIVE
OPTICAL COMPONENTS
$35.00

integrated laser with modulators. The focus of this chapter is laser sources needed for fiber optic transmission systems. These devices are fabricated using the InP material system. For early work and thorough discussion of semiconductor lasers, see Refs. [1–4].

The semiconductor injection laser was invented in 1962 [5–7]. With the development of epitaxial growth techniques and the subsequent fabrication of double heterojunction, the laser technology advanced rapidly in the 1970s and 1980s [1–4]. The demonstration of CW operation of the semiconductor laser in the early 1970s [8] was followed by an increase in development activity in several industrial laboratories. This intense development activity in the 1970s was aimed at improving the performance characteristics and reliability of lasers fabricated using the AlGaAs material system [1]. These lasers emit near 0.8 μm and were deployed in early optical fiber transmission systems (in the late 1970s and early 1980s).

The optical fiber has zero dispersion near 1.3 μm wavelength and has lowest loss near 1.55 μm wavelength. Thus semiconductor lasers emitting near 1.3 μm and 1.55 μm are of interest for fiber optic transmission application. Lasers emitting at these wavelengths are fabricated using the InGaAsP/InP materials system, and were first fabricated in 1976 [9]. Much of the fiber optic transmission systems around the world that are in use or are currently being deployed utilize lasers emitting near 1.3 μm or 1.55 μm.

Initially these lasers were fabricated using liquid phase epitaxy (LPE) growth technique. The development of metal-organic chemical vapor deposition (MOCVD) and gas source molecular beam epitaxy (GSMBE) growth techniques in the 1980s, not only improved the reproducibility of the fabrication process but also led to advances in laser designs such as quantum well lasers and very high speed lasers using semi-insulating Fe doped InP current blocking layers [10].

2.2. Laser Designs

A schematic of a typical double heterostructure used for laser fabrication is shown in Fig. 2.1. It consists of n-InP, undoped $In_{1-x}Ga_xP_yAs_{1-y}$, p-InP and p-InGaAsP grown over (100) oriented n-InP substrate. The undoped $In_{1-x}Ga_xP_yAs_{1-y}$ layer is the light-emitting layer (active layer). It is lattice matched to InP for $x \sim 0.45y$. The band gap of the $In_{1-x}Ga_xP_yAs_{1-y}$ material (lattice matched to InP), which determines the laser wavelength, is given by [11]

$$Eg(eV) = 1.35 - 0.72y + 0.12y^2.$$

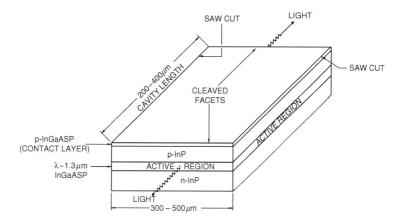

Fig. 2.1 Schematic of a double heterostructure laser.

For lasers emitting near $1.3\,\mu m$ $y \sim 0.6$. The double heterostructure material can be grown by LPE, GSMBE, or MOCVD growth technique. The double heterostructure material can be processed to produce lasers in several ways. Perhaps the simplest is the broad area laser (Fig. 2.1), which involves putting contacts on the p- and n-side and then cleaving. Such lasers do not have transverse mode confinement or current confinement, which leads to high threshold and nonlinearities in light vs. current characteristics. Several laser designs have been developed to address these problems. Among them are the gain guided laser, weakly index guided laser, and buried heterostructure (strongly index guided) laser. A typical version of these laser structures is shown in Fig. 2.2. The gain guided structure uses a dielectric layer for current confinement. The current is injected in the opening in the dielectric (typically 6 to $12\,\mu m$ wide), which produces gain in that region and hence the lasing mode is confined to that region. The weakly index guided structure has a ridge etched on the wafer, a dielectric layer surrounds the ridge. The current is injected in the region of the ridge, and the optical mode overlaps the dielectric (which has a low index) in the ridge. This results in weak index guiding.

The buried heterostructure design shown in Fig. 2.2 has the active region surrounded (buried) by lower index layers. The fabrication process of DCPBH (double channel planar buried heterostructure) laser involves growing a double heterostructure, etching a mesa using a dielectric mask, and then regrowing the layer surrounding the active region using a second epitaxial growth step. The second growth can be a single Fe doped InP (Fe:InP) semi-insulating layer or a combination of p-InP, n-InP, and

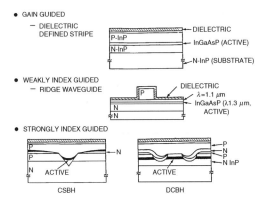

InGaAsP LASER STRUCTURES

Fig. 2.2 Schematic of a gain guided, weakly index guided, and strongly index guided buried heterostructure laser.

Fe:InP layer. Generally MOCVD growth process is used for the growth of the regrown layer. Researchers have often given different names to the particular buried heterostructure laser design that they discovered. These are described in detail in Ref. 12. For the structure of Fig. 2.2, the Fe doped InP layer provides both optical confinement to the lasing mode and current confinement to the active region. Buried heterostructure lasers are generally used in communication system applications because a properly designed strongly index guided buried heterostructure design has superior mode stability, higher bandwidth, and superior linearity in light vs. current (L vs I) characteristics compared to the gain guided and weakly index guided designs. Early recognition of these important requirements of communication-grade lasers led to intensive research on InGaAsP BH laser designs all over the world in the 1980s. It is worth mentioning that BH lasers are more complex and difficult to fabricate compared to the gain guided and weakly index guided lasers. Scanning electron micrograph of a capped mesa buried heterostructure (CMBH) laser along with the laser structure is shown in Fig. 2.3. The current blocking layers in this structure consist of i-InP (Fe doped InP), n-InP, i-InP, and n-InP layers. These sets of blocking layers have the lowest capacitance and are therefore needed for high-speed operation. An optimization of the thickness of these layers is needed for highest speed performance. The laser fabrication involves the following steps. One-micron-wide mesas are etched on the wafer and the current blocking layers consisting of i-InP, n-InP, i-InP, and n-InP layers

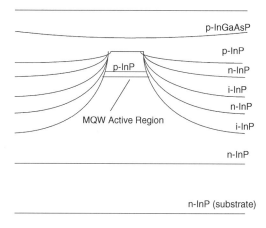

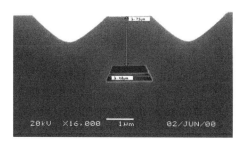

Fig. 2.3 Schematic of a BH laser and scanning electron photomicrograph of the same laser.

are grown on the wafer with an oxide layer on top of the mesa in place. The oxide layer is then removed and a third growth of p-InP cladding layer and p-InGaAs contact layer is carried out. The wafer is then processed using standard lithography, metallization, and cleaving techniques to produce the lasers.

The light vs. current characteristics at different temperatures of an InGaAsP BH laser emitting at 1.3 μm are shown in Fig. 2.4. Typical threshold current of a BH laser at room temperature is in the 5 to 10 mA range. For gain guided and weakly index guided lasers, typical room temperature threshold currents are in the 25–50 mA and 50–100 mA range, respectively. The external differential quantum efficiency defined as the derivative of the L vs I characteristics above threshold is ~0.25 mW/mA/facet for a cleaved uncoated laser emitting near 1.3 μm.

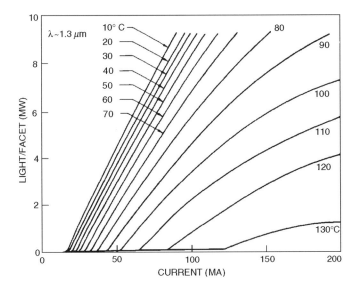

Fig. 2.4 Light vs. current characteristics of an InGaAsP buried heterostructure laser emitting at 1.3 μm.

An important characteristic of the semiconductor laser is that its output can be modulated easily and simply by modulating the injection current. The relative magnitude of the modulated light output is plotted as a function of the modulation frequency of the current in Fig. 2.5 at different optical output powers. The laser is of the BH type (shown in Fig. 2.3), has a cavity length of 250 μm, and the modulation current amplitude was 5 mA. Note that the 3-dB frequency to which the laser can be modulated increases with increasing output power and the modulation response is maximum at a certain frequency (ω_r). The resonance frequency ω_r is proportional to the square root of the optical power. The modulation response determines the data transmission rate capability of the laser, for example, for 10 Gb/s data transmission, the 3-dB bandwidth of the laser must exceed 10 GHz. However, other system level considerations, such as allowable error-rate penalty, often introduce much more stringent requirements on the exact modulation response of the laser.

A semiconductor laser with cleaved facets generally emits in a few longitudinal modes of the cavity. Typical spectrum of a laser with cleaved facets is shown in Fig. 2.6. The discrete emission wavelengths are separated by the longitudinal cavity mode spacing, which is ~10 A for a laser ($\lambda \sim 1.3$ μm) with 250-μm cavity length. Lasers can be made to emit in a

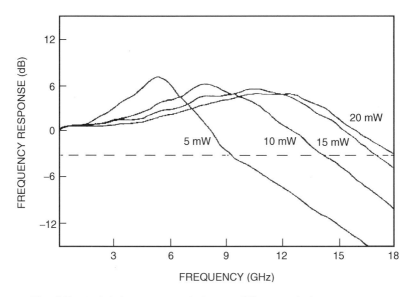

Fig. 2.5 Modulation response of a laser at different optical output powers.

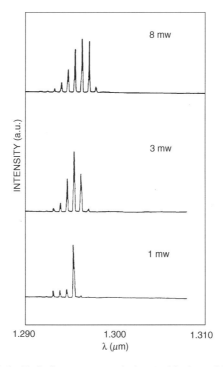

Fig. 2.6 Emission spectrum of a laser with cleaved facets.

single frequency using frequency selective feedback, for example, using a grating internal to the laser cavity as described in Section 2.3.

2.3. Quantum Well Lasers

So far we have described the fabrication and performance characteristics of regular double heterostructure (DH) laser that has an active region ~0.1 to 0.2 μm thick. Beginning in the 1980s, lasers with very thin active regions, quantum well lasers, were being developed in many research laboratories [13–22]. Quantum well (QW) lasers have active regions ~100 A thick, which restricts the motion of the carriers (electrons and holes) in a direction normal to the well. This results in a set of discrete energy levels and the density of states is modified to a "two-dimensional-like" density of states. This modification of the density of states results in several improvements in laser characteristics such as lower threshold current, higher efficiency, higher modulation bandwidth, and lower CW and dynamic spectral width. All of these improvements were first predicted theoretically and then demonstrated experimentally [23–32].

The development of InGaAsP QW lasers was made possible by the development of MOCVD and GSMBE growth techniques. The transmission electron micrograph (TEM) of a multiple QW laser structure is shown in Fig. 2.7. Shown are four InGaAs quantum wells grown over n-InP substrate. The well thickness is 70 A and they are separated by barrier layers of InGaAsP ($\lambda \sim 1.1 \, \mu m$). Multiquantum well (MQW) lasers with threshold current densities of 600 A/cm^2 have been fabricated [33]. The schematic of a MQW BH laser is shown in Fig. 2.8. The composition of the InGaAsP material from the barrier layers to the cladding layer (InP) is gradually varied in this structure over a thickness of ~0.1 μm. This produces a graded variation in index (GRIN structure), which results in a higher optical confinement of the fundamental mode than that for an abrupt interface design. Larger mode confinement factor results in lower threshold current. The laser has a MQW active region and it utilizes Fe doped semi-insulating (SI) InP layers for current confinement and optical confinement. The light vs. current characteristics of a MQW BH laser is shown in Fig. 2.9. The laser emits near 1.5 μm. The MQW lasers have lower threshold currents than regular DH lasers. Also, the two-dimensional-like density of states of the QW lasers makes the transparency current density of these lasers significantly lower than that for regular DH lasers [30]. This allows the fabrication of very low threshold lasers using high reflectivity coatings.

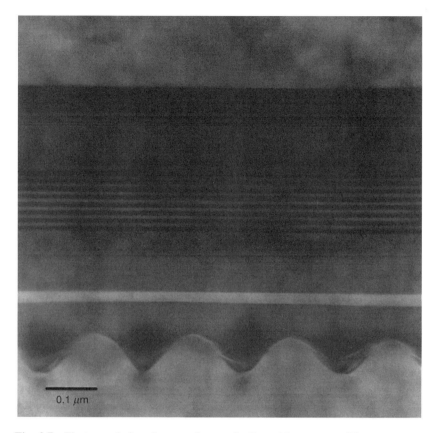

Fig. 2.7 The transmission electron micrograph of a multiquantum well laser structure.

Fig. 2.8 Schematic of multiquantum well buried heterostructure laser.

The optical gain (g) of a laser at a current density J is given by

$$g = a(J - J_0), \tag{2.1}$$

where a is the gain constant and J_0 is the transparency current density. Although a logarithmic dependence of gain on current density [23] is often

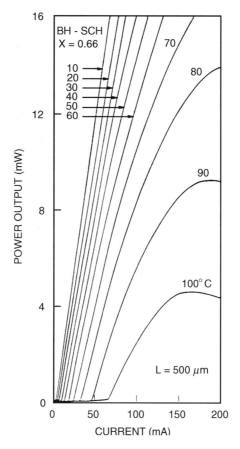

Fig. 2.9 Light vs. current characteristics of a multiquantum well buried heterostructure laser at different temperatures.

used in order to account for gain saturation, a linear dependence is used here for simplicity. The cavity loss α is given by

$$\alpha = \alpha_c + (1/L)\ln(1/R_1 R_2), \tag{2.2}$$

where α_c is the free carrier loss, L is the length of the optical cavity, and R_1, R_2 are the reflectivity of the two facets. At threshold, gain equals loss, hence it follows from (2.1) and (2.2) that the threshold current density (J_{th}) is given by

$$J_{th} = \alpha_c/a + (1/La)\ln(1/R_1 R_2) + J_0 \tag{2.3}$$

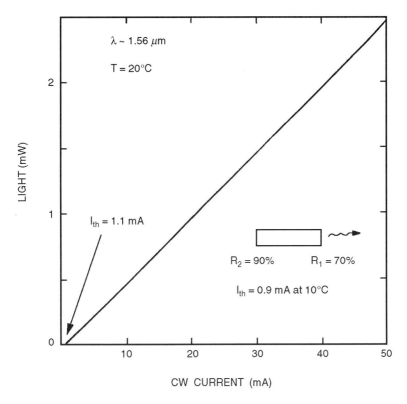

Fig. 2.10   Light vs. current of a quantum well laser with high reflectivity coatings on both facets.

Thus for a laser with high reflectivity facet coatings (R_1, $R_2 \sim 1$) and with low loss ($\alpha_c \sim 0$), $J_{th} \sim J_0$. For a QW laser, $J_0 \sim 50$ A/cm^2 and for a DH laser, $J_0 \sim 700$ A/cm^2, hence it is possible to get much lower threshold current using QW as the active region.

The light vs. current characteristics of a QW laser with high reflectivity coatings on both facets is shown in Fig. 2.10 [33]. The threshold current at room temperature is $\sim$1.1 mA. The laser is 170 μm long and has 90% and 70% reflective coating at the facets. This laser has a compressively strained MQW active region. For lattice matched MQW active region, a threshold current of 2 mA has been reported [34]. Such low-threshold lasers are important for array applications. Recently, QW lasers were fabricated which have higher modulation bandwidth than regular DH lasers. The current confinement and optical confinement in this laser is carried out using MOCVD grown Fe doped InP lasers similar to that shown in Fig. 2.2.

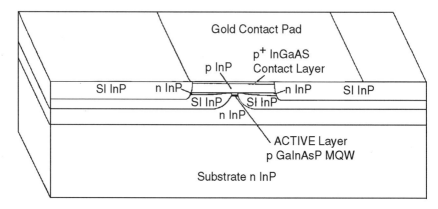

Fig. 2.11   Schematic of a laser designed for high speed. (Morton *et al.* [35])

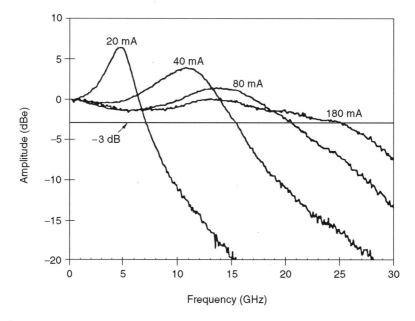

Fig. 2.12   Modulation response of multiquantum well high speed lasers. (Morton *et al.* [35])

The laser structure is then further modified by using a small contact pad and etching channels around the active region mesa (Fig. 2.11). These modifications are designed to reduce the capacitance of the laser structure. The modulation response of the laser is shown in Fig. 2.12. A 3-dB bandwidth of 25 GHz is obtained [35].

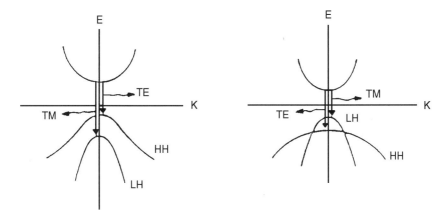

Fig. 2.13   Band structures under stress. The figures on the left and right represent situations under compressive and tensile strain respectively.

2.3.1.   STRAINED QUANTUM WELL LASERS

Quantum well lasers have also been fabricated using an active layer whose lattice constant differs slightly from that of the substrate and cladding layers. Such lasers are known as strained quantum well lasers. Over the last few years, strained quantum well lasers have been extensively investigated all over the world [36–43]. They show many desirable properties such as (i) a very low threshold current density and (ii) a lower linewidth than regular MQW lasers both under CW operation and under modulation. The origin of the improved device performance lies in the band-structure changes induced by the mismatch-induced strain [44, 45]. Figure 2.13 shows the band structure of a semiconductor under tensile and compressive strains. Strain splits the heavy-hole and the light-hole valence bands at the Γ point of the Brillouin zone where the bandgap is minimum in direct bandgap semiconductors.

Two material systems have been widely used for strained quantum well lasers: (i) InGaAs grown over InP by the MOCVD or the CBE growth technique [36–40] and (ii) InGaAs grown over GaAs by the MOCVD or the MBE growth technique [41–43]. The former material system is of importance for low-chirp semiconductor laser for lightwave system applications, while the latter material system has been used to fabricate high-power lasers emitting near 0.98 µm, a wavelength of interest for pumping erbium-doped fiber amplifiers.

The alloy $In_{0.53}Ga_{0.47}As$ has the same lattice constant as InP. Semiconductor lasers with an $In_{0.53}Ga_{0.47}As$ active region have been grown on InP by the MOCVD growth technique. Excellent material quality is also obtained for $In_{1-x}Ga_xAs$ alloys grown over InP by MOCVD for nonlattice-matched compositions. In this case the laser structure generally consists of one or many $In_{1-x}Ga_xAs$ quantum well layers with InGaAsP barrier layers whose composition is lattice matched to that of InP. For $x < 0.53$ the active layer in these lasers is under tensile stress, while for $x > 0.53$ the active layer is under compressive stress.

Superlattice structures of InGaAs/InGaAsP with tensile and compressive stress have been grown by both MOCVD and CBE growth techniques over an n-type InP substrate. Figure 2.14 shows the broad-area threshold current density as a function of cavity length for strained MQW lasers with four $In_{0.65}Ga_{0.35}As$ [39] quantum wells with InGaAsP ($\lambda \sim 1.25\,\mu m$) barrier layers. The active region in this laser is under 0.8% compressive strain. Also shown for comparison is the threshold current density as a function of cavity length of MQW lattice-matched lasers with $In_{0.53}Ga_{0.47}As$ wells. The entire laser structure, apart from the quantum well composition, is identical for the two cases. The threshold current density is lower for the compressively strained MQW structure than for the lattice-matched MQW structure.

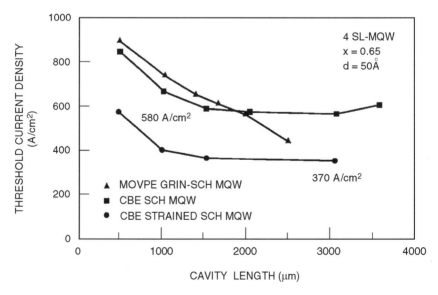

Fig. 2.14 Broad-area threshold current density as a function of cavity length for strained and lattice-matched InGaAs/InP MQW lasers. (Tsang *et al.* [39])

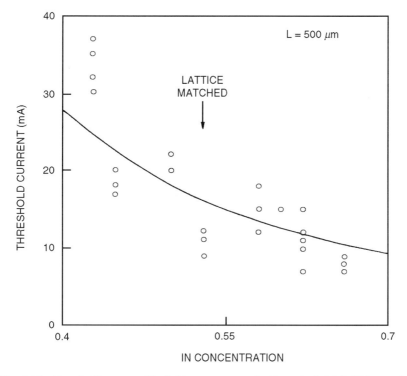

Fig. 2.15   Threshold current of buried heterostructure $In_x Ga_{1-x} As/InP$ MQW lasers plotted as a function of In concentration x. (Temkin *et al.* [46])

Buried heterostructure (BH) lasers have been fabricated using compressive and tensile strained MQW lasers. The threshold current of these lasers as a function of the In concentration is shown in Fig. 2.15 [46]. Lasers with compressive strain have a lower threshold current than do lasers with tensile strain. This can be explained by splitting of the light-hole and heavy-hole bands under stress [47, 48]. However, more recent studies have shown that it is possible to design tensile strained lasers with lower threshold [37, 42].

Strained quantum well lasers fabricated using $In_{1-x} Ga_x$ As layers grown over a GaAs substrate have been extensively studied [41, 43, 49–54]. The lattice constant of InAs is 6.06 A and that of GaAs is 5.654 A. The $In_{1-x} Ga_x As$ alloy has a lattice constant between these two values, and to a first approximation it can be assumed to vary linearly with x. Thus, an increase in the In mole fraction x increases the lattice mismatch relative to the GaAs substrate and therefore produces larger compressive strain on the active region.

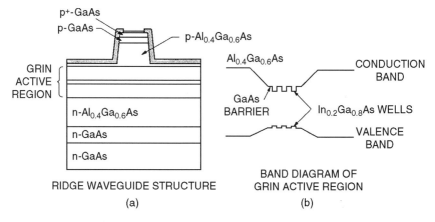

Fig. 2.16   Typical $In_{1-x}Ga_xAs/GaAs$ MQW laser structure.

A typical laser structure grown over the n-type GaAs substrate is shown in Fig. 2.16 [41] for this material system. It consists of a MQW active region with one to four $In_{1-x}Ga_xAs$ wells separated by GaAs barrier layers. The entire MQW structure is sandwiched between n- and p-type $Al_{0.3}Ga_{0.7}As$ cladding layers, and the P-cladding layer is followed by a p-type GaAs contact layer. Variations of the structure with different cladding layers or large optical cavity designs have been reported. Emission wavelength depends on the In composition, x. As x increases, the emission wavelength increases and for x larger than a certain value (typically ~ 0.25), the strain is too large to yield high-quality material. For $x \sim 0.2$, the emission wavelength is near 0.98 μm, a wavelength region of interest for pumping fiber amplifiers [49]. Threshold current density as low as 47 A/cm^2 has been reported for $In_{0.2}Ga_{0.8}As/GaAs$ strained MQW lasers [52]. High-power lasers have been fabricated using $In_{0.2}Ga_{0.8}As/GaAs$ MQW active region. Single-mode output powers of greater than 200 mW have been demonstrated using a ridge-waveguide-type laser structure.

Frequency chirp of strained and unstrained QW lasers has been investigated. Strained QW lasers (InGaAs/GaAs) exhibit the lowest chirp (or dynamic linewidth) under modulation. The lower chirp of strained QW lasers is consistent with a small linewidth enhancement factor (α-factor) measured in such devices. The α-factor is the ratio of the real and imaginary part of the refractive index. A correlation between the measured chirp and linewidth enhancement factor for regular double-heterostructure, strained and unstrained QW lasers is shown in Table 2.1. The high efficiency, high

Table 2.1   **Linewidth Enhancement Factor and Chirp of Lasers.**
FWHM = Full Width at Half Maximum

Laser Type	Linewidth Enhancement Factor	FWHM Chirp at 50 mA and 1 Gb/s (A)
DH Laser	5.5	1.2
MQW Laser	3.5	0.6
Strained MQW Laser InGaAs/GaAs, $\lambda \sim 1\,\mu m$	1.0	0.2
Strained MQW Laser InGaAsP/InP, $\lambda \sim 1.55\,\mu m$	2.0	0.4

power and low chirp of strained and unstrained QW lasers make these devices attractive candidates for lightwave transmission applications.

2.3.2.   OTHER MATERIAL SYSTEMS

A few other material systems have been reported for lasers in the 1.3-μm wavelength range. These are the AlGaInAs/InP and InAsP/InP materials grown over InP substrates and more recently the InGaAsN material grown over GaAs substrates.

The AlGaInAsP/InP system has been investigated with the aim of producing lasers with better high-temperature performance for uncooled transmitters [55]. This material system has a larger conduction band offset than the InGaAsP/InP material system, which may result in lower electron leakage over the heterobarrier and thus better high-temperature performance. The energy band diagram of a GRINSCH (graded index separate confinement heterostructure) laser design is shown in the Fig. 2.17. The laser has five compressively strained quantum wells in the active region. The 300-μm-long ridge waveguide lasers typically have a threshold current of 20 mA. The measured light vs. current characteristics of a laser with 70% high reflectivity coating at the rear facet is shown in Fig. 2.18. These lasers have somewhat better high-temperature performance than InGaAsP/InP lasers.

The InAsP/InP material system has also been investigated for 1.3-μm lasers [56]. InAsP with an arsenic composition of 0.55 is under 1.7% compressive strain when grown over InP. Using MOCVD growth technique buried heterostructure lasers with InAsP quantum well, InGaAsP ($\lambda \sim 1.1\,\mu m$) barrier layers, and InP cladding layers have been reported.

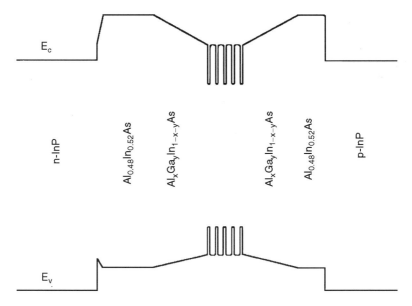

Fig. 2.17 Band diagram of a AlGaInAs GRINSCH with five quantum wells. (Zah *et al.* [55])

The schematic of the laser structure is shown in Fig. 2.19. Typical threshold current of the BH laser diodes are ~20 mA for 300-μm cavity length.

The material InGaNAs when grown over GaAs can have very large (~300 meV) conduction band offset, which can lead to much better high-temperature performance than the InGaAsP/InP material system [57]. The temperature dependence of threshold is characterized by $I_{th}(T) = I_o \exp(T/T_o)$, where T_o is generally called the characteristic temperature. Typical T_o values for InGaAsP/InP laser are ~60–70 K in the temperature range of 300–350 K. The predicted T_o value for the InGaNAs/GaAs system is ~150 K and recently $T_o = 126$ K has been reported for a InGaNAs laser emitting near 1.2 μm [57].

2.4. Distributed Feedback Lasers

Semiconductor lasers fabricated using the InGaAsP material system are widely used as sources in many lightwave transmission systems. One measure of the transmission capacity of a system is the data rate. Thus the drive toward higher capacity pushes the systems to higher data rates where the

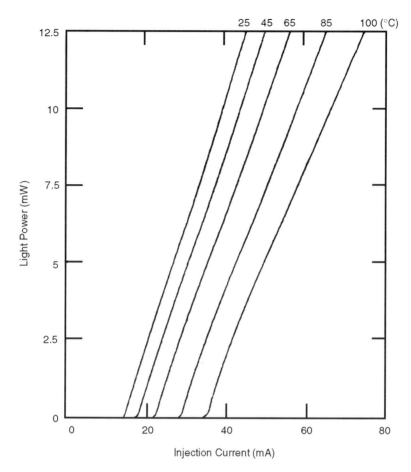

Fig. 2.18 Light vs. current characteristics of a AlGaInAs quantum well laser with five wells. (Zah. *et al.* [55])

chromatic dispersion of the fiber plays an important role in limiting the distance between regenerators. Sources emitting in a single wavelength help reduce the effects of chromatic dispersion and are therefore used in most systems operating at high data rates (>1.5 Gb/s).

The single wavelength laser source used in most commercial transmission systems is the distributed feedback (DFB) laser where a diffraction grating etched on the substrate close to the active region provides frequency selective feedback which makes the laser emit in a single wavelength. This section reports the fabrication, performance characteristics, and reliability of DFB lasers [58].

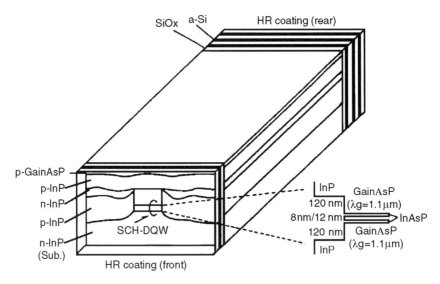

Fig. 2.19 Schematic of a buried heterostructure InAsP/InGaAsP quantum well laser. (Kusukawa *et al.* [56])

The schematic of our DFB laser structure is shown in Fig. 2.20. The fabrication of the device involves the following steps. First, a grating with a periodicity of 2400 A is fabricated on a (100) oriented n-InP substrate using optical holography and wet chemical etching. Four layers are then grown over the substrate. These layers are (i) n-InGaAsP ($\lambda \sim 1.3\,\mu m$) waveguide layer, (ii) undoped InGaAsP ($\lambda \sim 1.55\,\mu m$) active layer, (iii) p-InP cladding layer, and (iv) p-InGaAsP ($\lambda \sim 1.3\,\mu m$) contact layer. Mesas are then etched on the wafer using a SiO_2 mask and wet chemical etching. Fe doped InP semi-insulating layers are grown around the mesas using the MOCVD growth technique. The semi-insulating layers help confine the current to the active region and also provide index guiding to the optical mode. The SiO_2 stripe is then removed and the p-InP cladding layer and a p-InGaAsP contact layer are grown on the wafer using the vapor phase epitaxy growth technique. The wafer is then processed to produce 250-μm-long laser chips using standard metallization and cleaving procedures. The final laser chips have antireflection coating (<1%) at one facet and high reflection coating (~65%) at the back facet. The asymmetric facet coatings help remove the degeneracy between the two modes in the stop band.

The CW light vs. current characteristics of a laser are shown in Fig. 2.21. Also shown is the measured spectrum at different output powers.

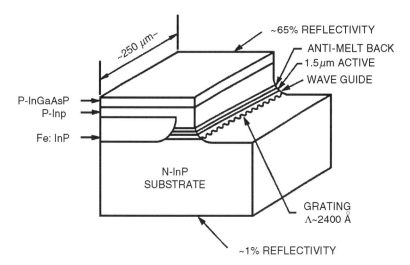

Fig. 2.20   Schematic of a capped mesa buried heterostructure (CMBH) distributed feedback laser.

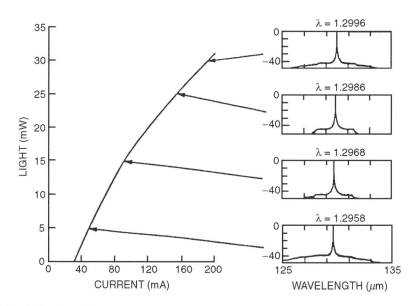

Fig. 2.21   CW light vs. current characteristics and measured spectrum at different output powers. Temperature = 30°C.

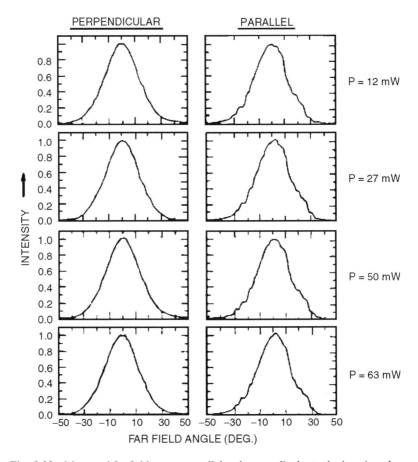

Fig. 2.22 Measured far field pattern parallel and perpendicular to the junction plane.

The threshold current of these lasers is in the 15 to 20 mA range. For high fiber coupling efficiency, it is important that the laser emit in the fundamental transverse mode. The measured far field pattern parallel and perpendicular to the junction plane at different output powers of a device is shown in Fig. 2.22. The figure shows that the laser operates in the fundamental transverse mode in the entire operating power range from threshold to 60 mW. The full width at half maximum of the beam divergences parallel and normal to the junction plane are 40° and 30° respectively.

The dynamic spectrum of the laser under modulation is an important parameter when the laser is used as a source for transmission. The measured 20 dB full width is shown in Fig. 2.23 at two different data rates as a function

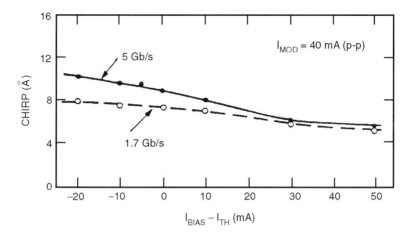

Fig. 2.23 Measured chirp as a function of bias.

of bias level. Note that for a laser biased above threshold, the chirp width is nearly independent of the modulation rate.

2.4.1. TUNABLE LASERS

Tunable semiconductor lasers are needed for many applications. Examples of applications in lightwave transmission systems are (i) wavelength-division multiplexing where signals at many distinct wavelengths are simultaneously modulated and transmitted through a fiber and (ii) coherent transmission systems where the wavelength of the transmitted signal must match that of the local oscillator. Several types of tunable laser structures have been reported in the literature [59–64]. Two principle schemes are (i) multisection DFB laser and (ii) multisection distributed Bragg reflector (DBR) laser. The multisection DBR lasers generally exhibit higher tunability than do the multisection DFB lasers. The design of a multisection DBR laser is shown schematically in Fig. 2.24 [59]. The three sections of this device are (i) the active region section that provides the gain, (ii) the grating section that provides the tunability, and (iii) the phase-tuning section that is needed to access all wavelengths continuously. The current through each of these sections can be varied independently. The tuning mechanism can be understood by noting that the emission wavelength λ of a DBR laser is given by $\lambda = 2n\Lambda$ where Λ is the grating period and n is the effective refractive index of the optical mode in the grating section. The latter can be changed simply by varying the current in the grating section.

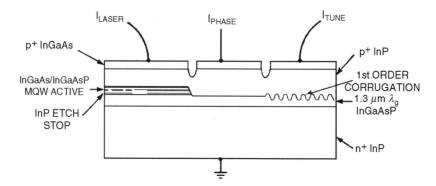

Fig. 2.24 Schematic of a multisection DBR laser. The laser has a MQW active region. The three sections are optically coupled by the thick waveguide layer below the MQW active region. (Koch *et al.* [59])

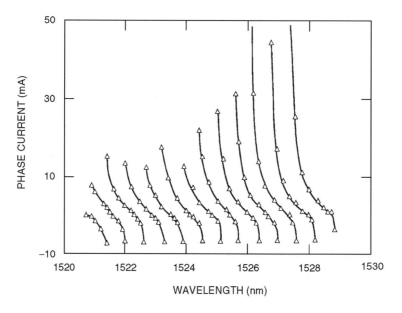

Fig. 2.25 Frequency tuning characteristics of a three-section MQW DBR laser. (Koch *et al.* [59])

The extent of wavelength tunability of a three-section DBR laser is shown in Fig. 2.25 [61]. Measured wavelengths are plotted as a function of phase-section current for different currents in the tuning section. A tuning range in excess of 6 nm can be obtained by controlling currents in the grating and phase-tuning sections.

An important characteristic of lasers for applications requiring a high
degree of coherence is the spectral width (linewidth) under CW operation.
The CW linewidth depends on the rate of spontaneous emission in the laser
cavity. For coherent transmission applications, the CW linewidth must be
quite small. The minimum linewidth allowed depends on the modulation
format used. For differential phase-shift keying (DPSK) transmission, the
minimum linewidth is approximately given by $B/300$ where B is the bit
rate. Thus, for 1-Gb/s transmission rate, the minimum linewidth is 3 MHz.
The CW linewidth of a laser decreases with increasing length and varies
as α^2, where α is the linewidth enhancement factor. Because α is smaller
for a multiquantum well (MQW) laser, the linewidth of DFB or DBR
lasers utilizing MQW active region is smaller than that for lasers with
regular DH active region. The linewidth varies inversely with the output
power at low powers (<10 mW) and shows saturation at high powers.
The measured linewidth as a function of output power of a 850-µm-long
DFB laser with MQW active region is shown in Fig. 2.26. The minimum
linewidth of 350 kHz was observed for this device at an operating power
of 25 mW. For multisection DBR lasers of the type shown in Fig. 2.24,
the linewidth varies with changes in currents in the phase-tuning and the
grating sections. The measured data for a MQW three-section DBR laser

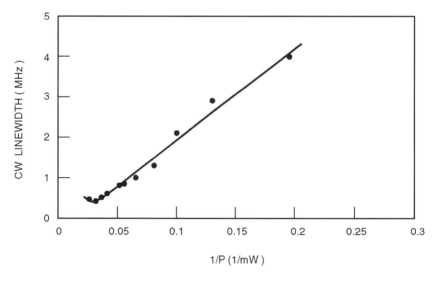

Fig. 2.26 Measured CW linewidth plotted as a function of the inverse of the output power
for a MQW DFB laser with a cavity length of 850 µm.

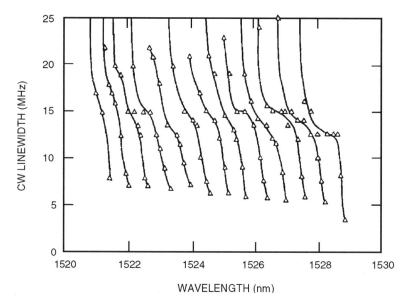

Fig. 2.27 Measured CW linewidth as a function of wavelength for a 3-section MQW DBR laser. (Koch *et al.* [61])

is shown in Fig. 2.27. Measured linewidths are plotted as a function of phase-section current for different currents in the tuning section.

2.5. Surface-Emitting Lasers

Semiconductor lasers described in the previous chapters have cleaved facets that form the optical cavity. The facets are perpendicular to the surface of the wafer and light is emitted parallel to the surface of the wafer. For many applications requiring a two-dimensional laser array or monolithic integration of lasers with electronic components (e.g., optical interconnects), it is desirable to have the laser output normal to the surface of the wafer. Such lasers are known as surface-emitting lasers (SEL). A class of surface-emitting lasers also have optical cavity normal to the surface of the wafer [65–72]. These devices are known as vertical-cavity surface-emitting lasers (VCSEL) in order to distinguish them from other surface emitters.

A generic SEL structure utilizing multiple semiconductor layers to form a Bragg reflector is shown in Fig. 2.28. The active region is sandwiched between n- and p-type cladding layers, which are themselves sandwiched between the two n- and p-type Bragg mirrors. This structure is shown using

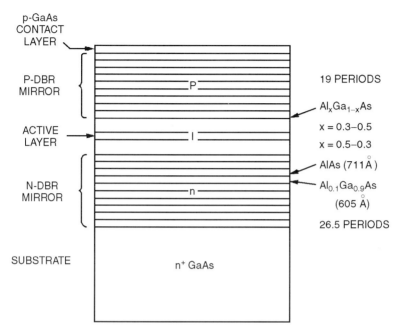

Fig. 2.28 Schematic illustration of a generic SEL structure utilizing distributed Bragg mirrors formed by using multiple semiconductor layers. DBR pairs consist of AlAs (711 A thick) and $Al_{0.1}Ga_{0.9}As$ (605 A thick) alternate layers. Active layer could be either a quantum well or similar to a regular double heterostructure laser.

the AlGaAs/GaAs material system, which has been very successful in the fabrication of SELs. The Bragg mirrors consist of alternating layers of low-index and high-index materials. The thicknesses of each layer is one-quarter of the wavelength of light in the medium. Such periodic quarter-wave-thick layers can have very high reflectivity. For normal incidence, the reflectivity is given by [73]

$$R = \frac{(1 - n_4/n_1(n_2/n_3)^{2N})^2}{(1 + n_4/n_1(n_2/n_3)^{2N})^2}, \tag{2.4}$$

where n_2, n_3 are the refractive indices of the alternating layer pairs, n_4, n_1 are the refractive indices of the medium on the transmitted and incident sides of the DBR mirror, and N is the number of pairs. As N increases, R increases. Also for a given N, R is larger if the ratio of n_2/n_3 is smaller. For a AlAs/$Al_{0.1}Ga_{0.9}$ As set of quarter-wave layers, typically 20 pairs are needed for a reflectivity of $\sim$99.5%. Various types of AlGaAs/GaAs SELs have been reported [74–82].

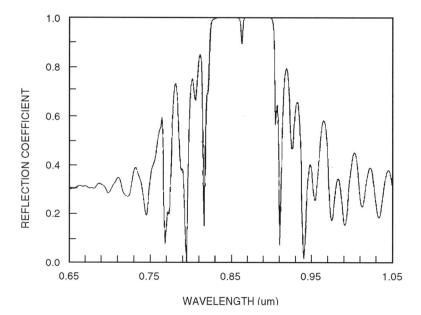

Fig. 2.29 Typical reflectivity spectrum of a SEL stack.

For a SEL to have a threshold current density comparable to that of an edge-emitting laser, the threshold gains must be comparable for the two devices. The threshold gain of an edge-emitting laser is ~100 cm^{-1}. For a SEL with an active-layer thickness of $0.1\,\mu$m, this value corresponds to a single-pass gain of $\sim1\%$. Thus, for the SEL device to lase with a threshold current density comparable to that of an edge emitter, the mirror reflectivities must be $>99\%$.

The reflectivity spectrum of a SEL structure is shown in Fig. 2.29. The reflectivity is $>99\%$ over a 10 nm band. The drop in reflectivity in the middle of the band is due to the Fabry-Perot mode.

The number of pairs needed to fabricate a high-reflectivity mirror depends on the refractive index of layers in the pair. For large index differences fewer pairs are needed. For example, in the case of CaF$_2$ and ZnS, for which the index difference is 0.9, only 6 pairs are needed for a reflectivity of 99%. By contrast for a InP/InGaAsP ($\lambda \sim 1.3\,\mu$m) layer pair, for which the index difference is 0.3, more than 40 pairs are needed to achieve a reflectivity of 99%.

Five principal structures (Fig. 2.30) used in SEL fabrication are (i) etched mesa structure, (ii) ion-implanted structure, (iii) dielectric isolated structure, (iv) buried heterostructure, and (v) metallic reflector structure.

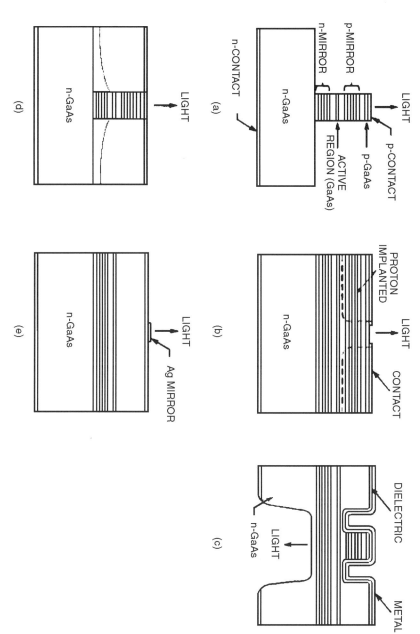

Fig. 2.30   Schematic of several SEL designs.

Zone Laser (Z-Laser)

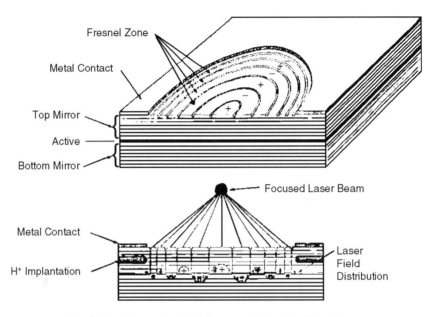

Fig. 2.31   Schematic of a high-power SEL with a Fresnel zone.

Threshold current of ∼0.3 mA has been reported for InGaAs/GaAs SEL devices. A SEL design has been demonstrated whose output can be focussed to a single spot [80]. The laser has a large area (∼100 μm dia) and it has a Fresnel-zone-like structure etched on the top mirror (Fig. 2.31). The lasing mode with the lowest loss has π phase shift in the near field as it traverses each zone. The laser emits 500 mW in a single mode.

An important SEL structure for the AlGaAs/GaAs material system is the oxide aperture device [81] (Fig. 2.32). AlAs has the property that it oxidizes rapidly in the presence of oxygen to aluminum oxide, which forms an insulating layer. Thus by introducing a thin AlAs layer in the device structure it is possible to confine the current to a very small area. This allows the fabrication of very low threshold (<0.2 mA) and high bandwidth (∼14 GHz) lasers.

Central to the fabrication of low-threshold SELs is the ability to fabricate high-reflectivity mirrors. In the late 1970s, Soda *et al.* [83] reported on a SEL fabricated using the InP material system. The surfaces of the wafer

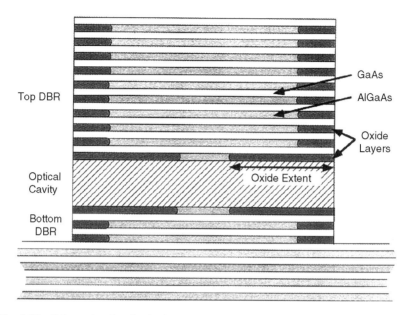

Fig. 2.32 Schematic of a selectively oxidized SEL consisting of AlGaAs/GaAs multilayers and buried aluminum oxide layers. AlGaAs layers with higher Al content are oxidized more.

form the Fabry-Perot cavity of the laser. Fabrication of the device involves the growth of a double heterostructure on an n-InP substrate. A circular contact is made on the p-side using an SiO_2 mask. The substrate side is polished making sure that it is parallel to the epitaxial layer, and ring electrodes (using an alloy of Au-Sn) are deposited on the n-side. The laser had a threshold current density of $\sim$11 kA/cm^2 at 77 K and operated at output powers of several milliwatts.

InGaAsP/InP SELs have been investigated by many researchers over the last few years [83–89]. Many of the schemes utilize alternating layers of InP and InGaAsP to produce Bragg mirrors [86] (Fig. 2.33). The refractive index difference between InGaAsP ($\lambda \sim 1.3\,\mu$m) and InP layers is smaller than that in GaAs SELs, hence InGaAsP/InP SELs utilize more pairs (typically 40 to 50) to produce a high-reflectivity ($>$99%) mirror. Such mirror stacks have been grown by both chemical beam epitaxy (CBE) and MOCVD growth techniques and have been used to fabricate InGaAsP SELs. Room-temperature pulsed operation of InGaAsP/InP SELs using these mirror stacks and emitting near 1.5 μm have been reported [88].

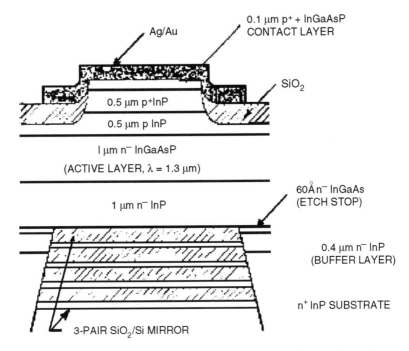

Fig. 2.33   Schematic of a InGaAsP SEL fabricated using multilayer mirors. (Yang *et al.* [86])

An alternative approach is using the technique of wafer fusion [89, 90]. In this technique the Bragg mirrors are formed using GaAs/AlGaAs system grown by MBE and the active region of InGaAsP bounded by thin InP layers is formed by MOCVD. The post type structure of a 1.5-μm wavelength SEL formed using this technique is shown in Fig. 2.34 [89]. The optical cavity is formed by wafer fusion of the InGaAsP quantum well between the Bragg mirrors. Room-temperature CW threshold current of 2.3 mA has been reported for the 8-μm diameter post device [89].

2.6.   Laser Reliability

The performance characteristics of injection lasers used in lightwave systems can degrade during their operation. The dominant mechanism responsible for the degradation is determined by any or all of the several fabrication processes including epitaxial growth, wafer quality, device

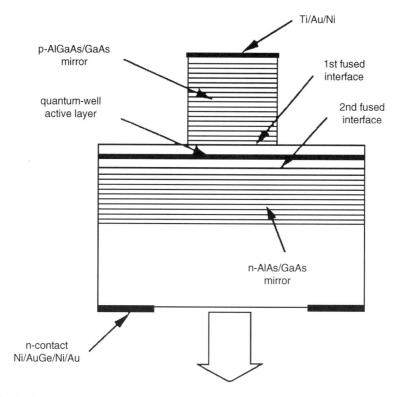

Fig. 2.34 Schematic of a InGaAsP SEL fabricated using wafer fusion. (Babic *et al.* [89])

processing and bonding [91–101]. In addition, the degradation rate of devices processed from a given wafer depends on the operating conditions, viz., the operating temperature and the injection current. Although many of the degradation mechanisms are not fully understood, extensive amounts of empirical observations exist in the literature, which have allowed the fabrication of InGaAsP laser diodes with extrapolated median lifetimes in excess of 25 years at an operating temperature of 20°C [93].

The detailed studies of degradation mechanisms of optical components used in lightwave systems have been motivated by the desire to have a reasonably accurate estimate of the operating lifetime before they are used in practical systems. For many applications, the components are expected to operate reliably over a period in excess of 10 years, so an appropriate reliability assurance procedure becomes necessary, especially for applications such as an undersea lightwave transmission system where the replacement cost is very high. The reliability assurance is usually carried out by operating

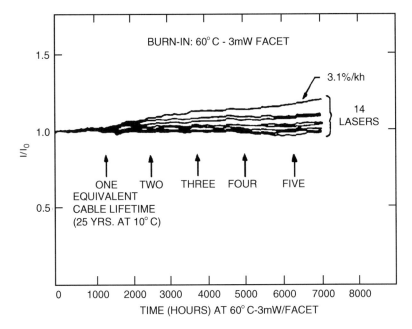

Fig. 2.35 Operating current for 3 mW output at 60°C as a function of operating time. These data were generated for 1.3 μm InGaAsP lasers used in the first submarine fiber optic cable. (Nash *et al.* [91])

the devices under a high stress (e.g., high temperature) which enhances the degradation rate so that a measurable value can be obtained in an operating time of a few hundred hours. The degradation rate under normal operating conditions can then be obtained from the measured high-temperature degradation rate using the concept of an activation energy [93].

The light output vs. current characteristics of a laser change after stress aging. There is generally a small increase in threshold current and a decrease in external differential quantum efficiency following the stress aging. Aging data for 1.3 μm InGaAsP lasers used in the first submarine fiber optic cable is shown in Figure 2.35 [91].

Some lasers exhibit an initial rapid degradation after which the operating characteristics of the lasers are very stable. Given a population of lasers, it is possible to quickly identify the "stable" lasers by a high stress test (also known as the purge test) [91, 92, 102, 103]. The stress test implies that operating the laser under a set of high stress conditions (e.g., high current, high temperature, high power) would cause the weak lasers to fail and stabilize the possible winners. Observations on the operating current after stress aging have been reported by Nash *et al.* [91]. It is important to point out that the determination of the duration and the specific conditions for stress aging are critical to the success of this screening procedure.

The expected operating lifetime of a semiconductor laser is generally determined by accelerated aging at high temperatures and using an activation energy. The lifetime (t) at a temperature T is experimentally found to vary as $\exp(-E/kT)$, where E is the activation energy and k is the Boltzmann constant [104, 105]. The operating current of good lasers increases at a rate of less than 1%/khr of aging time at 60°C operating temperature. Assuming a 50% change in operating current as the useful lifetime of the device and an activation energy of 0.7 eV, this aging rate corresponds to a light-emitting lifetime of greater than 100 years at 20°C.

A parameter that determines the performance of the DFB laser is the side mode suppression ratio (SMSR), that is, the ratio of the intensity of the dominant lasing mode to that of the next most intense mode [105]. The SMSR of good DFB lasers does not change significantly after aging, which confirms the spectral stability of the emission.

For some applications such as coherent transmission systems, the absolute wavelength stability of the laser is important. The measured change in emission wavelength at 100 mA before and after aging of several devices is shown in Fig. 2.36. Note that most of the devices do not exhibit any

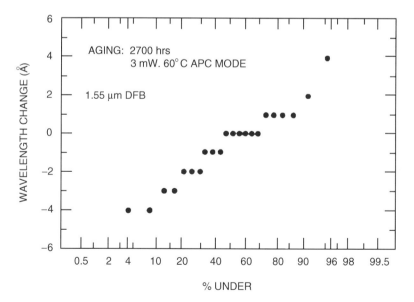

Fig. 2.36 Change in emission wavelength after aging is plotted in the form of a normal probability distribution.

change in wavelength and the standard deviation of the change is less than
2 A. This suggests that the absolute wavelength stability of the devices is
adequate for coherent transmission applications.

2.7. Integrated Laser Devices

There have been a significant number of developments in the technology
of optical integration of semiconductor lasers and other related devices on
the same chip. These chips allow higher levels of functionality than that
achieved using single devices. For example, laser and optical modulators
have been integrated, serving as simple monolithic transmitters.

2.7.1. LASER ARRAYS

The simplest of all integrated laser devices are one-dimensional arrays
of lasers, LEDs, or photodetectors. These devices are fabricated exactly
the same way as individual devices except the wafers are not scribed to
make single-device chips but left in the form of a bar. The main required
characteristics of a laser array are low threshold current and good electrical
isolation between the individual elements of the array. The schematic of
two adjacent devices in a 10-channel low threshold laser array is shown in
Fig. 2.37 [106]. These lasers emit near 1.3 μm and are grown by MOCVD
on p-InP substrate. The lasers have multiquantum well active region with
five 7-nm-thick wells as shown in the insert of Fig. 2.37. The light vs.
current characteristics of all the lasers in a 10-element array are uniform.
The average threshold current and quantum efficiency are 3.2 mA and
0.27 W/A respectively. The cavity length was 200 μm and the facets of

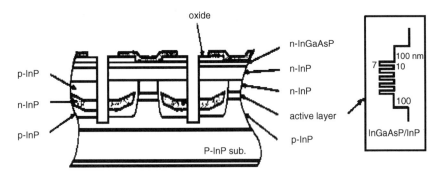

Fig. 2.37 Schematic of two adjacent devices in a laser array.

the lasers were coated with dielectric to produce 65% and 90% reflectivity respectively [106].

The vertical-cavity surface-emitting laser (SEL) design is more suitable for the fabrication of two-dimensional arrays than the edge-emitting laser design. Several researchers have reported two-dimensional arrays of SELs. Among the individual laser design used are the proton implanted design and the oxide confined design. These SELs and SEL arrays have been fabricated so far using the GaAs/AlGaAs material system for the emission near 0.85 μm.

2.7.2. *INTEGRATED LASER MODULATOR*

Externally modulated lasers are important for applications where low spectral width under modulation is needed [107–110]. The two types of integrated laser modulator structures that have been investigated are the integrated electroabsorption modulated laser (EML) and the integrated electrorefraction modulated laser. The electrorefraction property is used in a Mach-Zehnder configuration to fabricate a low-chirp modulated light source.

For some applications it is desirable to have the laser and the modulator integrated on the same chip. Such devices, known as electroabsorption modulated lasers (EMLs), are used for high data rate transmission systems with large regenerator spacing. The schematic of an EML is shown in Fig. 2.38 [111]. In this device, the light from the DFB laser is coupled directly to the modulator. The modulator region has a slightly higher bandgap than that of the laser region, which results in very low absorption of the laser light in the absence of bias. However, with reverse bias, the effective bandgap decreases, which results in reduced transmission through the modulator. For very high-speed operation, the modulator region capacitance must be sufficiently small, which makes the modulator length small, resulting in low on/off ratio. Recently, very high-speed EMLs have been reported using a growth technique where the laser and modulator active regions are fabricated using two separate growths, thus allowing independent optimization of the modulator bandgap and length for high on/off ratio and speed. Operation at 40 Gb/s has been demonstrated using this device [111].

EML devices have also been fabricated using the selective area epitaxy growth process [112]. In this process the laser and the modulator active region are grown simultaneously over a patterned substrate. The patterning allows the materials grown to have slightly different bandgaps, resulting

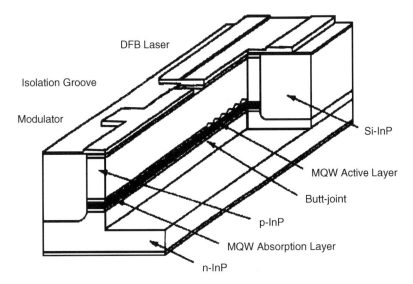

Fig. 2.38   Schematic of a electroabsorption modulated laser structure. (Takeuchi *et al.* [111])

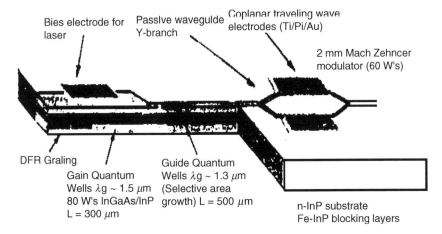

Fig. 2.39   Schematic of a integrated laser and Mach-Zehnder modulator.

in separate laser and modulator regions. EMLs have been fabricated with bandwidths of 15 GHz and have operated error free over 600 km at 2.5 Gb/s data rate.

An integrated laser Mach-Zehnder device is shown in Fig. 2.39. This device has a ridge-waveguide-type DFB laser integrated with a Mach-Zehnder modulator, which also has a lateral guiding provided by a ridge structure. The Mach-Zehnder traveling wave phase modulator is designed so that the microwave and optical velocities in the structure are identical. This allows good coupling of the electrical and optical signal.

2.7.3.   MULTICHANNEL WDM SOURCES

An alternative to single channel very high speed (>20 Gb/s) data transmission for increasing transmission capacity is multichannel transmission using wavelength division multiplexing (WDM) technology. In WDM systems many (4, 8, 16, or 32) wavelengths carrying data are optically multiplexed and simultaneously transmitted through a single fiber. The received signal with many wavelengths is optically demultiplexed into separate channels, which are then processed electronically in a conventional form. Such a WDM system needs transmitters with many lasers at specific wavelengths. It is desirable to have all of these laser sources on a single chip for compactness and ease of fabrication, like electronic integrated circuits.

Figure 2.40 shows the schematic of a photonic integrated circuit with multiple lasers for a WDM source [113]. This chip has 4 individually addressable DFB lasers, the output of which are combined using a waveguide-based multiplexer. Because the waveguide multiplexer has an optical loss of $\sim$8 dB, the output of the chip is further amplified using a semiconductor amplifier. The laser output in the waveguide is TE polarized and hence an amplifier with a multiquantum well absorption region, which has a high saturation power, is integrated in this chip.

2.7.4.   SPOT SIZE CONVERTER (SSC) INTEGRATED LASER

A typical laser diode has too wide ($\sim$30° $\times$ 40°) an output beam pattern for good mode matching to a single mode fiber. This results in a loss of power coupled to the fiber. Thus a laser whose output spot size is expanded to match an optical fiber is an attractive device for low loss coupling to the fiber without a lens and for wide alignment tolerances. Several researchers have reported such devices [114, 115]. Generally they involve producing a vertically and laterally tapered waveguide near the output facet of the laser. The tapering needs to be done in an adiabatic fashion so as to reduce the scattering losses. The schematic of a SSC laser is shown in Fig. 2.41 [115]. The laser is fabricated using two MOCVD growth steps. The SSC section is

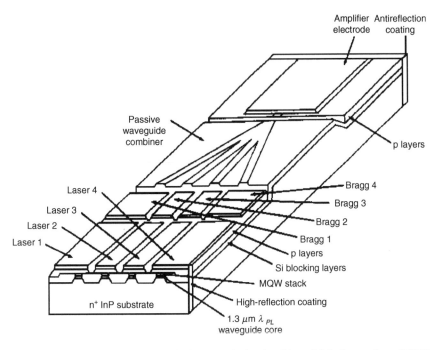

Fig. 2.40 Schematic of a photonic integrated circuit with multiple lasers for a WDM source.

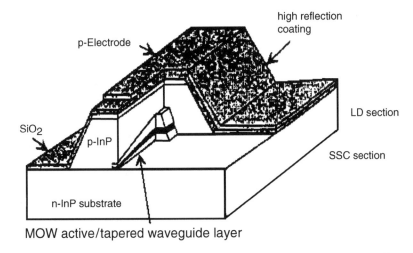

Fig. 2.41 Schematic of a Spot Size Converter (SSC) laser. (Yamazaki *et al.* [115])

about 200 µm long. The waveguide thickness is narrowed along the cavity from 300 nm in the active region to ∼100 nm in the region over the length of the SSC section. The laser emits near 1.3 µm, has a multiquantum well active region, and a laser section length of 300 µm. A beam divergence of 13° was obtained for this device. Beam divergences of 9° and 10° in the lateral and vertical direction have been reported for similar SSC devices [115].

2.8. Summary and Future Challenges

Tremendous advances in semiconductor lasers have occurred over the last decade. The advances in research and many technological innovations have led to the worldwide deployment of fiber optic communication systems that operate near 1.3 µm and 1.55 µm wavelengths and compact storage disks that utilize lasers for read/write purposes. Although most of these systems are based on digital transmission, lasers have also been deployed for carrying high-quality analog cable TV transmission systems. However, many challenges remain.

The need for higher capacity is pushing the deployment of WDM-based transmission, which needs tunable or frequency settable lasers. An important research area would continue to be the development of lasers with very stable and settable frequency. Integration of many such lasers on a single substrate would provide the ideal source for WDM systems.

The laser to fiber coupling is also an important area of research. Recent developments in spot size converter integrated lasers are quite impressive but some more work, perhaps, needs to be done to make them easy to manufacture. This may require more process developments.

Although WDM technology is currently being considered for increasing the transmission capacity, the need for sources with very high modulation capability would remain. Hence research on new mechanisms for very high speed modulation is important.

The surface-emitting laser is very attractive for two-dimensional arrays and for single wavelength operation. Several important advances in this technology have occurred over the last few years. An important challenge is the fabrication of a device with characteristics superior to that of an edge emitter.

Finally, many of the advances in laser development would not have been possible without the advances in materials and processing technology.

The challenges of much of the current laser research are intimately linked with the challenges in materials growth, which include not only the investigation of new material systems but also improvements in existing technologies to make them more reproducible and predictable.

References

1. H. Kressel and J. K. Butler, Semiconductor Lasers and Heterojunction LEDs (Academic Press, NY, 1977).
2. H. C. Casey, Jr. and M. B. Panish, Heterostructure Lasers (Academic Press, NY, 1978).
3. G. H. B. Thompson, Physics of Semiconductor Lasers (John Wiley and Sons, NY, 1980).
4. G. P. Agrawal and N. K. Dutta, Long Wavelength Semiconductor Lasers (van Nostrand Reinhold Co., 1986; second ed. 1993).
5. N. Holonyuk, Jr. and S. F. Bevacqua, Appl. Phys. Lett., 1 (1962) 82.
6. M. I. Nathan, W. P. Dumke, G. Burns, F. H. Dill, Jr., and G. Lasher, Appl. Phys. Lett., 1 (1962) 63.
7. T. M. Quist, R. H. Retiker, R. J. Keyes, W. E. Krag, B. Lax, A. L. McWhorter, and H. J. Ziegler, Appl. Phys. Lett., 1 (1962) 91; R. N. Hall, G. E. Fenner, J. D. Kingsley, T. J. Soltys, and R. O. Carlson, Phys. Rev. Lett., 9 (1962).
8. I. Hayashi, M. B. Panish, P. W. Foy, and S. Sumsky, Appl. Phys. Lett., 47 (1970) 109.
9. J. J. Hsieh, Appl. Phys. Lett., 28 (1976) 283.
10. See Chapter 4, Ref. 4.
11. R. E. Nahory, M. A. Pollack, W. D. Johnston, Jr., and R. L. Barns, Appl. Phys. Lett., 33 (1978) 659.
12. See Chapter 5, Ref. 4.
13. N. Holonyak, Jr., R. M. Kolbas, R. D. Dupuis, and P. D. Dapkus, IEEE J. Quantum Electron., QE-16 (1980) 170.
14. N. Holonyak, Jr., R. M. Kolbas, W. D. Laidig, B. A. Vojak, K. Hess, R. D. Dupuis, and P. D. Dapkus, J. Appl. Phys., 51 (1980) 1328.
15. W. T. Tsang, Appl. Phys. Lett., 39 (1981) 786.
16. W. T. Tsang, IEEE J. Quantum Electron., QE-20 (1986) 1119.
17. S. D. Hersee, B. DeCremoux, and J. P. Duchemin, Appl. Phys. Lett., 44 (1984).
18. F. Capasso and G. Margaritondo, eds., Heterojunction Band Discontinuation: Physics and Applications (Amsterdam: North Holland 1987).
19. N. K. Dutta, S. G. Napholtz, R. Yen, R. L. Brown, T. M. Shen, N. A. Olsson, and D. C. Craft, Appl. Phys. Lett., 46 (1985) 19.

20. N. K. Dutta, S. G. Napholtz, R. Yen, T. Wessel, and N. A. Olsson, Appl. Phys. Lett., 46 (1985) 1036.
21. N. K. Dutta, T. Wessel, N. A. Olsson, R. A. Logan, R. Yen, and P. J. Anthony, Electron. Lett., 21 (1985) 571.
22. N. K. Dutta, T. Wessel, N. A. Olsson, R. A. Logan, and R. Yen, Appl. Phys. Lett., 46 (1985) 525.
23. Y. Arakawa and A. Yariv, IEEE J. Quantum Electron., QE-21 (1985) 1666.
24. Y. Arakawa and A. Yariv, IEEE J. Quantum Electron., QE-22 (1986) 1887.
25. A. Yariv, C. Lindsey, and V. Sivan, J. Appl. Phys., 58 (1985) 3669.
26. A. Sugimura, IEEE J. Quantum Electron., QE-20 (1984) 336.
27. A. Sugimura, Appl. Phys. Lett., 43 (1983) 728.
28. N. K. Dutta and R. J. Nelson, J. Appl. Phys., 53 (1982) 74.
29. L. C. Chiu and A. Yariv, IEEE J. Quantum Electron., QE-18 (1982) 1406.
30. N. K. Dutta, J. Appl. Phys., 54 (1983) 1236.
31. A. Sugimura, IEEE J. Quantum Electron., QE-19 (1983) 923.
32. C. Smith, R. A. Abram, and M. G. Burt, J. Phys., C 16 (1983) L171.
33. H. Temkin, N. K. Dutta, T. Tanbun-Ek, R. A. Logan, and A. M. Sergent, Appl. Phys. Lett., 57 (1990) 1610.
34. C. Kazmierski, A. Ougazzaden, M. Blez, D. Robien, J. Landreau, B. Sermage, J. C. Bouley, and A. Mirca, IEEE J. Quantum Electron., 27 (1991) 1794–1797.
35. P. Morton, R. A. Logan, T. Tanbun-Ek, P. F. Sciortino, Jr., A. M. Sergent, R. K. Montgomery, and B. T. Lee, Electron. Lett., 29 (1993) 1429–1430.
36. P. J. A. Thijs, L. F. Tiemeijer, P. I. Kuindersma, J. J. M. Binsma, and T. van Dongen, IEEE J. Quantum Electron., 27 (1991) 1426.
37. P. J. A. Thijs, L. F. Tiemeijer, J. J. M. Binsma, and T. van Dongen, IEEE J. Quantum Electron., QE-30 (1994) 477–499.
38. H. Temkin, T. Tanbun-Ek, and R. A. Logan, Appl. Phys. Lett., 56 (1990) 1210.
39. W. T. Tsang, L. Yang, M. C. Wu, Y. K. Chen, and A. M. Sergent, Electron. Lett., (1990) 2033.
40. W. D. Laidig, Y. F. Lin, and P. J. Caldwell, J. Appl. Phys., 57 (1985) 33.
41. S. E. Fischer, D. Fekete, G. B. Feak, and J. M. Ballantyne, Appl. Phys. Lett., 50 (1987) 714.
42. N. Yokouchi, N. Yamanaka, N. Iwai, Y. Nakahira, and A. Kasukawa, IEEE J. Quantum Electron., QE-32 (1996) 2148–2155.
43. K. J. Beernik, P. K. York, and J. J. Coleman, Appl. Phys. Lett., 25 (1989) 2582.
44. J. P. Loehr and J. Singh, IEEE J. Quantum Electron., 27 (1991) 708.
45. S. W. Corzine, R. Yan, and L. A. Coldren, "Optical Gain in III-V Bulk and Quantum Well Semiconductors," in Quantum Well Lasers, P. Zory, ed. (Academic Press, NY (to be published)).

46. H. Temkin, T. Tanbun-Ek, R. A. Logan, D. A. Coblentz, and A. M. Sergent, IEEE Photonic. Technol. Lett., 3 (1991) 100.
47. A. R. Adams, Electron. Lett., 22 (1986) 249.
48. E. Yablonovitch and E. O. Kane, J. Lightwave Technol., LT-4 (1986) 50.
49. M. C. Wu, Y. K. Chen, M. Hong, J. P. Mannaerts, M. A. Chin, and A. M. Sergent, Appl. Phys. Lett., 59 (1991) 1046.
50. N. K. Dutta, J. Lopata, P. R. Berger, D. L. Sivco, and A. Y. Cho, Electron. Lett., 27 (1991) 680.
51. J. M. Kuo, M. C. Wu, Y. K. Chen, and M. A. Chin, Appl. Phys. Lett., 59 (1991) 2781.
52. N. Chand, E. E. Becker, J. P. van der Ziel, S. N. G. Chu, and N. K. Dutta, Appl. Phys. Lett., 58 (1991) 1704.
53. H. K. Choi and C. A. Wang, Appl. Phys. Lett., 57 (1990) 321.
54. N. K. Dutta, J. D. Wynn, J. Lopata, D. L. Sivco, and A. Y. Cho, Electron. Lett., 26 (1990) 1816.
55. C. E. Zah, R. Bhat, B. N. Pathak, F. Favire, W. Lin, N. C. Andreadakis, D. M. Hwang, T. P. Lee, Z. Wang, D. Darby, D. Flanders, and J. J. Hsieh, IEEE J. Quantum Electron., QE-30 (1994) 511–523.
56. A. Kusukawa, T. Namegaya, T. Fukushima, N. Iwai, and T. Kikuta, IEEE J. Quantum Electron., QE-27 (1993) 1528–1535.
57. J. M. Kondow, T. Kitatani, S. Nakatsuka, Y. Yazawa, and M. Okai, Proc. OECC 97 Seoul, Korea (Jul. 1997) 168–169.
58. See Chapter 7, Ref. 4.
59. T. L. Koch, U. Koren, R. P. Gnall, C. A. Burrus, and B. I. Miller, Electron. Lett., 24 (1988) 1431.
60. Y. Suematsu, S. Arai, and K. Kishino, J. Lightwave Technol., LT-1 (1983) 161.
61. T. L. Koch and U. Koren, IEEE J. Quantum Electron., QE-27 (1991) 641.
62. N. K. Dutta, A. B. Piccirilli, T. Cella, and R. L. Brown, Appl. Phys. Lett., 48 (1986) 1501.
63. K. Y. Liou, N. K. Dutta, and C. A. Burrus, Appl. Phys. Lett., 50 (1987) 489.
64. T. Tanbun-Ek, R. A. Logan, S. N. G. Chu, and A. M. Sergent, Appl. Phys. Lett., 57 (1990) 2184.
65. H. Soda, K. Iga, C. Kitahara, and Y. Suematsu, Japan J. Appl. Phys., 18 (1979) 2329.
66. K. Iga, F. Koyama, and S. Kinoshita, IEEE J. Quantum Electron., 24 (1988) 1845.
67. J. L. Jewell, J. P. Harbison, A. Scherer, Y. H. Lee, and L. T. Florez, IEEE J. Quantum Electron., 27 (1991) 1332.
68. C. J. Chang-Hasnain, M. W. Maeda, N. G. Stoffel, J. P. Harbison, and L. T. Florez, Electron. Lett., 26 (1990) 940.
69. R. S. Geels, S. W. Corzine, and L. A. Coldren, IEEE J. Quantum Electron., 27 (1991) 1359.

70. R. S. Geels and L. A. Coldren, Appl. Phys. Lett., 5 (1990) 1605.
71. K. Tai, G. Hasnain, J. D. Wynn, R. J. Fischer, Y. H. Wang, B. Weir, J. Gamelin, and A. Y. Cho, Electron. Lett., 26 (1990) 1628.
72. K. Tai, L. Yang, Y. H. Wang, J. D. Wynn, and A. Y. Cho, Appl. Phys. Lett., 56 (1990) 2496.
73. M. Born and E. Wolf, Principles of Optics (Pergamon Press, NY, 1977) Sec. 1.6.5, p. 69.
74. J. L. Jewell, A. Scherer, S. L. McCall, Y. H. Lee, S. J. Walker, J. P. Harbison, and L. T. Florez, Electron. Lett., 25 (1989) 1123.
75. Y. H. Lee, B. Tell, K. F. Brown-Goebeler, J. L. Jewell, R. E. Leibenguth, M. T. Asom, G. Livescu, L. Luther, and V. D. Mattera, Electron. Lett., 26 (1990) 1308.
76. K. Tai, R. J. Fischer, C. W. Seabury, N. A. Olsson, D. T. C. Huo, Y. Ota, and A. Y. Cho, Appl. Phys. Lett., 55 (1989) 2473.
77. A. Ibaraki, K. Kawashima, K. Furusawa, T. Ishikawa, T. Yamayachi, and T. Niina, Japan J. Appl. Phys., 28 (1989) L667.
78. E. F. Schubert, L. W. Tu, R. F. Kopf, G. J. Zydzik, and D. G. Deppe, Appl. Phys. Lett., 57 (1990) 117.
79. R. S. Geels, S. W. Corzine, J. W. Scott, D. B. Young, and L. A. Coldren, IEEE Photonic. Technol. Lett., 2 (1990) 234.
80. D. Vakhshoori, J. D. Wynn, and R. E. Liebenguth, Appl. Phys. Lett., 65 (1994) 144.
81. See for example, Chapter 1 and 2 by D. Deppe and K. Choquette respectively in Vertical Cavity Surface Emitting Lasers, J. Cheng and N. K. Dutta, eds. (Gordon Breach, NY, 2000).
82. N. K. Dutta, L. W. Tu, G. J. Zydzik, G. Hasnain, Y. H. Wang, and A. Y. Cho, Electron. Lett., 27 (1991) 208.
83. H. Soda, K. Iga, C. Kitahara, and Y. Suematsu, Japan J. Appl. Phys., 18 (1979) 2329.
84. K. Iga, F. Koyama, and S. Kinoshita, IEEE J. Quantum Electron., QE-24 (1988) 1845.
85. K. Tai, F. S. Choa, W. T. Tsang, S. N. G. Chu, J. D. Wynn, and A. M. Sergent, Electron. Lett., 27 (1991) 1514.
86. L. Yang, M. C. Wu, K. Tai, T. Tanbun-Ek, and R. A. Logan, Appl. Phys. Lett., 56 (1990) 889.
87. T. Baba, Y. Yogo, K. Suzuki, F. Koyama, and K. Iga, Electron. Lett., 29 (1993) 913.
88. Y. Imajo, A. Kasukawa, S. Kashiwa, and H. Okamoto, Japan J. Appl. Phys. Lett., 29 (1990) L1130–L1132.
89. D. I. Babic, K. Streubel, R. Mirin, N. M. Margalit, J. E. Bowers, E. L. Hu, D. E. Mars, L. Yang, and K. Carey, IEEE Photonic. Technol. Lett., 7 (1995) 1225.
90. Z. L. Liau and D. E. Mull, Appl. Phys. Letts., 56 (1990) 737.

91. F. R. Nash, W. J. Sundburg, R. L. Hartman, J. R. Pawlik, D. A. Ackerman, N. K. Dutta, and R. W. Dixon, AT&T Tech. J., 64 (1985) 809.
92. The reliability requirements of a submarine lightwave transmission system are discussed in a special issue of AT&T Tech. J., 64 (1985) 3.
93. B. C. DeLoach, Jr., B. W. Hakki, R. L. Hartman, and L. A. D'Asaro, Proc. IEEE, 61 (1973) 1042.
94. P. M. Petroff and R. L. Hartman, Appl. Phys. Lett., 2 (1973) 469.
95. W. D. Johnston and B. I. Miller, Appl. Phys. Lett., 23 (1973) 1972.
96. P. M. Petroff, W. D. Johnston, Jr., and R. L. Hartman, Appl. Phys. Lett., 25 (1974) 226.
97. J. Matsui, R. Ishida, and Y. Nannichi, Japan J. Appl. Phys., 14 (1975) 1555.
98. P. M. Petroff and D. V. Lang, Appl. Phys. Lett., 31 (1977) 60.
99. O. Ueda, I. Umebu, S. Yamakoshi, and T. Kotani, J. Appl. Phys., 53 (1982) 2991.
100. O. Ueda, S. Yamakoshi, S. Komiya, K. Akita, and T. Yamaoka, Appl. Phys. Lett., 36 (1980) 300.
101. S. Yamakoshi, M. Abe, O. Wada, S. Komiya, and T. Sakurai, IEEE J. Quantum Electron., QE-17 (1981) 167.
102. K. Mizuishi, M. Sawai, S. Todoroki, S. Tsuji, M. Hirao, and M. Nakamura, IEEE J. Quantum Electron., QE-19 (1983) 1294.
103. E. I. Gordon, F. R. Nash, and R. L. Hartman, IEEE Electron. Device Lett., ELD-4 (1983) 465.
104. R. L. Hartman and R. W. Dixon, Appl. Phys. Lett., 26 (1975) 239.
105. W. B. Joyce, K. Y. Liou, F. R. Nash, P. R. Bossard, and R. L. Hartman, AT&T Tech. J., 64 (1985) 717.
106. S. Yamashita, A. Oka, T. Kawano, T. Tsuchiya, K. Saitoh, K. Uomi, and Y. Ono, IEEE Photonic. Technol. Lett., 4 (1992) 954–957.
107. I. Kotaka, K. Wakita, K. Kawano, H. Asai, and M. Naganuma, Electron. Lett., 27 (1991) 2162.
108. K. Wakita, I. Kotaka, K. Yoshino, S. Kondo, and Y. Noguchi, IEEE Photonic. Technol. Lett., 7 (1995) 1418.
109. F. Koyama and K. Iga, J. Lightwave Technol., 6 (1988) 87–93.
110. J. C. Cartledge, H. Debregeas, and C. Rolland, IEEE Photonic. Technol. Lett., 7 (1995) 224–226.
111. H. Takeuchi, K. Tsuzuki, K. sato, M. Yamamoto, Y. Itaya, A. Sano, M. Yoneyama, and T. Otsuji, IEEE Photonic. Technol. Lett., 9 (1997) 572–574.
112. M. Aoki, M. Takashi, M. Suzuki, H. Sano, K. Uomi, T. Kawano, and A. Takai, IEEE Photonic. Technol. Lett., 4 (1992) 580.
113. T. L. Koch and U. Koren, AT&T Tech. J., 70 (1992) 63–79.
114. R. Y. Fang, D. Bertone, M. Meliga, I. Montrosset, G. Oliveti, and R. Paoletti, IEEE Photonic. Technol. Lett., 9 (1997) 1084–1086.
115. H. Yamazaki, K. Kudo, T. Sasaki, and M. Yamaguchi, Proc. OECC 97 Seoul, Korea, paper 10C1-3, 440–441.

Chapter 3 | High Power Semiconductor Lasers for EDFA Pumping

Akihiko Kasukawa

Yokohama R&D Laboratories,
The Furukawa Electric Co., Ltd.,
2-4-3 Okano, Nishi-ku, Yokohama 220-0073, Japan

ABSTRACT

Optical fiber communication systems using WDM (wavelength-division multiplexing) are being introducing in long-haul networks to manage the explosive increase in transmission capacity. Erbium-doped fiber amplifier (EDFA) is one of the key components to support WDM systems. High power lasers emitting at both 980 nm and 1480 nm are essential for pumping sources for EDFA. In this chapter, state-of-the-art high power pumping lasers are reviewed. An ultra-high output power of over 500 mW has been realized under stable lateral mode operation in both 980 nm and 1480 nm lasers. These high output power laser modules are of great importance for the application of Raman amplifiers as well as EDFA application.

In this chapter, epitaxial growth, design, fabrication, and lasing characteristics will be given.

3.1. Introduction

3.1.1. BACKGROUND

The optical fiber communication systems have realized large transmission capacity. However, bit rate increase utilizing a traditional TDM (time-division multiplexing) cannot manage explosive demands for larger transmission capacity triggered by data communication and the Internet. The urgent demands for larger transmission capacity have driven the system planner to introduce WDM systems instead of the TDM systems to make the best use of installed optical fibers.

WDM systems are thought to be the most cost-effective way to handle the increasing transmission capacity using the established transmission technologies. WDM systems, by using multi-channel single-frequency lasers such as distributed feedback lasers with slightly different wavelengths determined by ITU (International Telecommunication Union), total throughput transmission capacity can be increased by the number of lasers (channels).

59

WDM TECHNOLOGIES: ACTIVE
OPTICAL COMPONENTS
$35.00

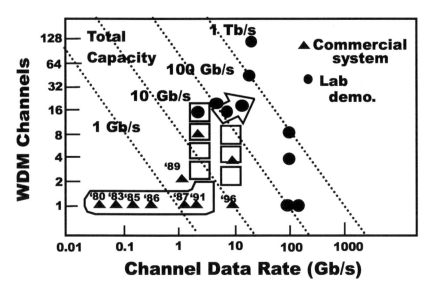

Fig. 3.1 **Transmission capacity trend** Transmission capacity of 1 Tbps can be achieved in wavelength-division multiplexing (WDM) technology by increasing channel counts.

A transmission capacity trend is illustrated in Fig. 3.1 for both commercial and laboratory demonstrations. Total transmission capacity of more than 100 Gbps can easily be realized by using the established 2.5 Gbps and developing 10 Gbps technology. Extremely large transmission capacity of more than 1 Tbps will be realized in the near future utilizing WDM systems.

On the other hand, strict specifications are required for optical devices in terms of wavelength, because WDM systems utilize the wavelength regime. Single-frequency lasers used in the systems, for example, have to control the absolute lasing wavelength to meet the ITU grid as well as wavelength separation. The wavelength separations according to ITU, for example, are set to be 1.6 nm (200 GHz), 0.8 nm (100 GHz) and 0.4 nm (50 GHz), depending on the WDM channels (bit rate). In addition to the strict requirements in wavelength control, the signal light source has to have high output power in order not only to extend the transmission distance but to compensate the insertion loss caused by the many optical components such as wavelength couplers used in WDM systems.

In order to amplify the signal light source in an efficient way, optical amplifiers are the key component to support WDM systems. Erbium-doped fiber amplifiers (EDFAs), which will be described later, have the advantage over the traditional O/E and E/O amplification because of their excellent amplification characteristics such as high speed, simultaneous amplification of many channels, and so forth.

Higher pumping power can make it possible to produce higher signal output power. Therefore, high power semiconductor lasers are the one of the key devices for pumping of EDF. In this chapter, high power semiconductor lasers are described in terms of design, fabrication, characteristics, reliability, and packaging.

3.1.2. *ERBIUM-DOPED OPTICAL FIBER AMPLIFIERS (OFA)*

It is no exaggeration to say that the invention of the optical fiber amplifier is the innovation needed to realize the WDM system. The erbium-doped fiber amplifier especially, is very attractive for practical application because light amplification occurs in the wavelength range of 1500 nm-band, which is the low loss wavelength region of the conventional silica fiber.

Figure 3.2 illustrates the configuration of EDFA. It consists of EDF, pumping laser module, WDM coupler to couple the pumping light into

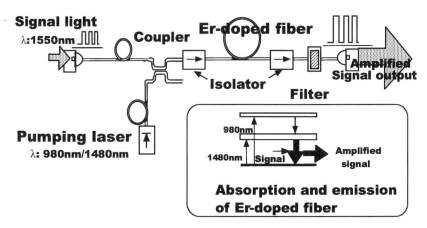

Fig. 3.2   Configuration of Erbium-doped fiber amplifier (EDFA)   EDFA consists of Erbium-doped fiber, pumping laser module, WDM coupler, and optical isolator. In a practical application, 1480 nm pumping is used as a booster amplifier and 980 nm pumping is used as a pre-amplifier. 1550 nm wavelength is effectively amplified by optical pumping.

EDF, optical isolator to prevent the lasing in the EDF, and optical fiber. High power laser modules emitting at wavelengths of 980 nm and 1480 nm are used for pumping sources as an effective pumping of EDF. Efficient power conversion can be realized by a 1480 nm pump, while low noise amplification can be realized by a 980 nm pump. In a practical application, 1480 nm pumping is used as a booster amplifier and 980 nm pumping is used as a pre-amplifier. To increase both channel counts and bit rate, high power operation of pumping power is inevitable. The approaches for high power lasers with single lateral mode operation (narrow stripe geometry) are given, followed by epitaxial growth, laser performance, laser module performance, and future challenges.

3.2.   High Power Semiconductor Lasers

3.2.1.   SEMICONDUCTOR MATERIALS

Figure 3.3 shows the relationship between materials and the wavelength. 1480 nm lasers can be fabricated. We can use the established technology for telecommunication lasers emitting at 1300 nm and 1550 nm—conventional

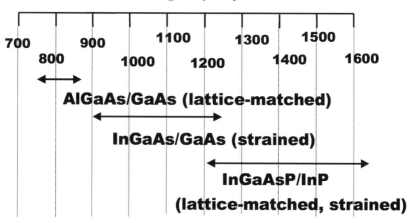

Fig. 3.3   Active layer materials to get a wavelength from 700 nm to 1600 nm GaInAsP quantum well (both lattice-matched and strained-layer) on InP can emit 1480 nm light. InGaAs strained-layer quantum well on GaAs substrate can emit 980 nm light.

$Ga_xIn_{1-x}As_yP_{1-y}$ quaternary compound on InP substrate—to fabricate 1480 nm lasers. On the other hand, 980 nm lasers cannot be fabricated by use of lattice-matched material. The wavelength of around 980 nm emitted from semiconductor lasers has been the forbidden wavelength region that cannot be covered by conventional short wavelength GaAs-based lasers (wavelength shorter than 900 nm) and InP-based long wave-length lasers (wavelength longer than 1200 nm). The concept of strained-layer quantum wells [1,2], however, made it possible to realize 980 nm lasers as well as to make rapid progress of the epitaxial growth tech-nique to grow high-quality very thin strained-layer material. 980 nm lasers can be fabricated by use of intentionally lattice-mismatched In_xGaAs layer on GaAs substrate. In_xGaAs strained-layer has a large lattice con-stant with respect to GaAs. If the layer thickness is controlled within a critical thickness calculated by Mathew's law, the layer can be grown with high crystalline quality on GaAs substrate even if the InGaAs layer has a large amount of strain. This condition is approximately given by $\varepsilon * L_z < 20$ nm% [3], where ε is the amount of strain and L_z is the layer thickness.

Strained-layer GaInAsP instead of lattice-matched system on InP sub-strate is also used for 1480 nm lasers since the concept of strained-layer quantum can improve the lasing characteristics drastically.

Let's explain the quantum wells and strained-layer quantum wells used in high-performance 980 nm and 1480 nm lasers. Quantum wells are made up of two different materials with a layer thickness less than 20 nm to show quantum-confined effect. High material gain can be obtained with fewer carriers because carriers are effectively confined to the quantized state formed by the quantum well. Especially, the use of compressive strain into the quantum wells can change the modification of valence band structure in such a way that the effective mass of heavy-hall becomes light. As a result, the Bernard-Durgaffourg condition (Fermi-level separation larger than energy bandgap) can be satisfied with less carrier density. Thus, lower threshold current density and higher quantum efficiency can be obtained in strained-layer quantum well lasers.

Strained-layer quantum well laser wafer can be grown by metal-organic chemical vapor deposition (MOCVD) for both 980 nm and 1480 nm.

In general, 980 nm lasers are grown by both MOCVD and MBE (molecular beam epitaxy), and 1480 nm lasers are mainly grown by MOCVD because of the presence of phosphorous, which will be explained later.

3.2.2. APPROACH FOR HIGH POWER OPERATION

In a practical application, a laser diode module, the so-called pig-tailed module is used. The laser diode has to be designed in such a way that high-power operation is obtained with a narrow and circular beam for high coupling into a single-mode fiber (SMF). It should be noted that highly reliable operation under high output power has to be realized. For these purposes, buried heterostructure (BH) is widely used for 1480 nm lasers and ridge waveguide (RWG) structure is widely used for 980 nm lasers.

The limiting factors for high-power operation under continuous wave (CW) condition in these narrow stripe lasers are categorized into two phenomena; one is roll-over phenomenon due to the increase of temperature in the active layer or increase of invalid current, and the other is catastrophic phenomenon mainly due to the optical mirror damage. The schematic explanation of these phenomena is shown in Fig. 3.4. The former

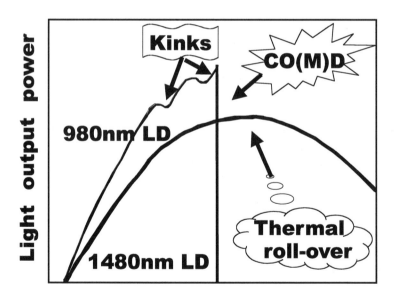

Fig. 3.4   Two mechanisms to limit the maximum light output power of narrow stripe lasers emitting at 980 nm and 1480 nm   "Kink" and catastrophic optical damage are the major factors for 980 nm lasers, while thermal roll-over is most important for 1480 nm lasers. Note that COD is the sudden death failure.

is often observed in 1480 nm lasers, and the latter is the sudden death phenomenon called catastrophic optical mirror damage (COD), observed mainly in 980 nm lasers. Detailed explanation for the COD is reported in [4].

Let's explain how to attain high power operation in solitary lasers. The output power from the front facet, P_f, is given by the following equation if COD has not occurred:

$$P_f = \eta_i \frac{1}{1 + \frac{1-R_f}{1-R_r}\sqrt{\frac{R_r}{R_f}}} \frac{\alpha_m}{\alpha_i + \alpha_m}(I - I_{th})\left(\frac{I_{ac}}{I_{ac} + I_L}\right)\Theta(T) \quad (3.1)$$

where, η_i is the internal quantum efficiency, α_i is internal loss, α_m is the mirror loss, I is the injection current, I_{th} is the threshold current, I_{ac} is the effective current that contributes to the lasing, I_L is the leakage current, R_r is the facet reflectivity of rear facet, R_f is the facet reflectivity of front facet, and $\Theta(T)$ is the parameter related to the decrease of light output power due to the temperature rise of the active layer ($\Theta(T) \leq 1$).

From Eq. (3.1), the following actions are thought to be effective for high power operation.

A. realization of high internal efficiency (η_i)
B. realization of low internal loss (α_i)
C. introduction of asymmetric coating for higher front facet power (α_m)
D. suppression of leakage current (I_L)
E. high thermal dissipation ($\Theta(T)$).

Structural optimizations have to be made because contradictory items are included in the list. The introduction of long cavity, for example, is effective for low thermal resistance, however, increase of the threshold current and decrease of external quantum efficiency are observed in long cavity lasers. The detailed explanations to realize the preceding items are given as follows.

Items A and B are the results expected by use of strained-layer quantum well active layer. The use of compressive strain in the quantum wells can change the modification of valence band structure in such a way that the effective mass of heavy-hall becomes light. As a result, the Bernard-Durgaffourg condition (Fermi-level energy separation larger than energy bandgap) can be satisfied with less carrier density. Thus, fewer carriers are needed for population inversion as compared to lattice-matched QW lasers, resulting in less nonradiative recombination.

A. In order to achieve high internal quantum efficiency, structures of both quantum wells and optical confinement layers such as composition and thickness have to be optimized carefully. Introduction of strained-layer quantum wells into the active layer is indispensable for this purpose. The use of GRIN-SCH (graded-index separate-confinement-structure) is very effective for high internal quantum efficiency, which will be described in detail later.

B. Introduction of quantum well structure, especially strained-layer quantum well, is effective for low internal loss. Strained-layer quantum well laser can provide low threshold current and high differential quantum efficiency operations due to low internal loss even if a long cavity is used.

C. Cleavage of semiconductor material is used to make Fabry-Perot lasers, thus, an equal amount of light is emitted from both sides of the mirror. The output power from the rear facet is not so important because it is used only for back-monitor in a practical application. Therefore, asymmetric facet coatings composed of dielectric mirror are used to increase the output power from the front facet. Low reflective coating (several %) is used for the front facet and high reflective coating (95%) is used for the rear facet.

D. The reduction of leakage current leads to the suppression of temperature rise of the active layer, thus leading to high power operation.

E. It is possible to reduce the temperature rise in the active layer by the use of long cavity (low electric and thermal resistance) and junction down bonding on heat sink material with high thermal conductivity such as diamond and AlN.

The approaches from A to E are the general methods to achieve high light output power for all kinds of semiconductor lasers. 1480 nm lasers are more difficult than 980 nm lasers in terms of high power operation due to poor temperature characteristics of threshold current and quantum efficiency, resulting from poor electron confinement in the wells and non-radiative recombination. We will next examine theoretical consideration for high power operation, i.e., the threshold current and the differential quantum efficiency calculated for 1480 nm lasers, together with output beam profile.

3.2.3.   EFFECTS OF OPTICAL LOSS ON THRESHOLD
CURRENT DENSITY AND QUANTUM EFFICIENCY

Low threshold current density and high quantum efficiency are desirable for high light output power operation with low power consumption. The threshold current density and quantum efficiency are affected by the optical loss. The optical loss in the GaInAsP/InP material ($\lambda = 1300$–1550 nm) is larger than that of the (In)GaAs/AlGaAs system ($\lambda = 800$–1100 nm) due to the intervalence band absorption loss [5]. The gain of the quantum well laser diodes tends to saturate at higher carrier density, therefore, it is very important to investigate the effect of the optical loss on the threshold current density and differential quantum efficiency. The threshold current density, J_{th}, is calculated from Eq. (3.2), and the differential quantum efficiency, η_d, is calculated by (3.3).

$$J_{th} = N_w J_0 \exp\left\{ \frac{1}{N_w \Gamma_{SQW} G_0} \left(\alpha_i + \frac{1}{L} \ln\left(\frac{1}{R_1 R_2} \right) \right) \right\} \qquad (3.2)$$

where N_w is the number of wells, J_0 is the transparent current density, Γ_{SQW} is the optical confinement factor per well (see Appendix), G_0 is the gain coefficient to describe the quantum well gain G as $G = G_0 \ln (J/J_0)$.

$$\eta_d = \eta_i \left(\frac{\alpha_i L}{\alpha_i L + \ln\left(\frac{1}{R}\right)} \right) \qquad (3.3)$$

The threshold current density, calculated using Eq. (3.2), versus cavity length is plotted in Fig. 3.5. The parameter is the internal loss. Small internal loss is effective in the reduction of the threshold current density. For example, the threshold current density of the QW lasers with α_i of 5 cm^{-1} is 20% lower than that of the LD with α_i of 15 cm^{-1}. These internal losses are the actual values for 1480 nm lasers with GRIN-SCH and SCH, as will be described in Section 3.3.6.

From Fig. 3.5, reductions of both internal loss and mirror loss are necessary in order to reduce the threshold current density. It is theoretically verified that cavity loss should be designed to be as low as possible for the low-threshold current density in a quantum well laser. This is due to the logarithmic form of the gain of quantum well lasers resulting from the step-like density of state. Owing to the high material gain and low loss properties in QW lasers, low-threshold current density is obtained for long cavity lasers.

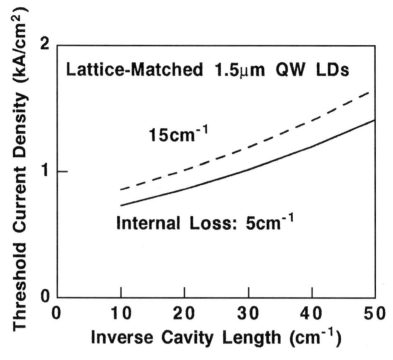

Fig. 3.5 **Threshold current density versus cavity length as a parameter of internal loss**
(1480 nm laser) Threshold current density decreases with cavity length. The threshold
current density of the laser with internal loss of 8 cm^{-1} is about 20% lower than that of the
laser with internal loss of 15 cm^{-1}.

The differential quantum efficiency is plotted in Fig. 3.6 as a function of
cavity length. The parameter is the internal loss. Internal quantum efficiency
of 84% is used in this calculation, which is a reasonable value for 1480 nm
lasers using strained-layer quantum wells. On the other hand, η_i more
than 90% is obtained for 980 nm lasers. Small internal waveguide loss
is effective for improving both threshold current density and differential
quantum efficiency.

Next, let us consider the output beam property of the laser because
laser modules, which include SMF and optics to couple the laser beam
into SMF, instead of solitary lasers are used for EDFA application. It is,
therefore, important to achieve a narrow and circular output beam in order
to get high coupling efficiency. The calculation of far-field pattern (FFP)
will be given in Fig. 3.23.

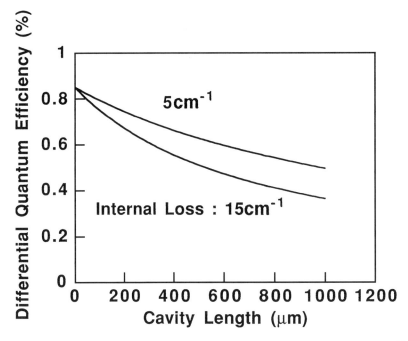

Fig. 3.6 Differential quantum efficiency versus cavity length as a parameter of internal loss (1480 nm laser) Differential quantum efficiency decreases with cavity length. Internal quantum efficiency of 84% is assumed. Internal quantum efficiency more than 90% is obtained for a well-designed 980 nm laser.

From the viewpoint of both electrical and optical properties, the introduction of GRIN-SCH structure as the optical confinement layer in strained-layer quantum well active layer is more suitable for high power operation. By optimization of composition and thickness of the GRIN-SCH layer, high power operation with narrow and circular output beam, which is effective for high coupling efficiency into SMF, can be achieved.

In the following section, fabrication and lasing characteristics of both 1480 nm and 980 nm lasers are separately described.

3.3.   1480 nm Lasers

Fabrication, such as MOCVD growth including buried heterostructure, and lasing characteristics are described.

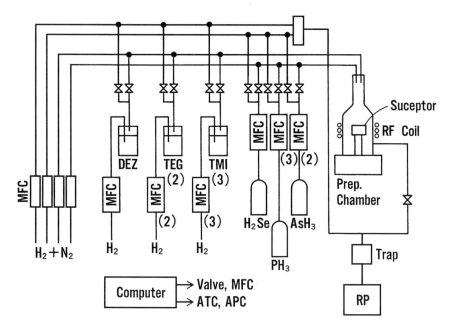

Fig. 3.7 Schematic diagram of MOCVD apparatus Numbers in parentheses indicate the number of source materials and mass flow controllers. The growth sequence is controlled by a computer.

3.3.1. EPITAXIAL GROWTH—MOCVD

The low pressure MOCVD apparatus [6] is shown in Fig. 3.7. MOCVD apparatus has a vertical reactor and a carbon susceptor coated with SiC heated by RF coil. The reactor is designed not to expose apparatus to the air in case of loading the wafer by use of a preparation chamber. Either single or multiple two-inch wafers can be grown in the MOCVD. The MOCVD apparatus is carefully designed to minimize the dead volume, and employs the quick run-vent switching systems. In addition, in order to grow the GaInAs(P) layers ranging from the wide bandgap (bandgap wavelength of 0.95 μm) to the narrow bandgap (bandgap wavelength of 1.65 μm), two or three source materials are equipped for group III and V. Each source material has independent controllable mass flow controllers (MFCs), having different flow rates. Numbers in the parentheses in Fig. 3.7 indicate the number of source materials and MFCs. The procedures of setting growth pressure (Automatic Power Control), growth temperature (Automatic Temperature Control), gas flow rate, and gas switching are controlled by a computer to ensure reproducible growth. MOCVD apparatus designed for mass production is available at present.

Epitaxial growth was carried out at a temperature of around 600°C and a low pressure of 76 Torr. Trimethylindium (TMIn) and triethylgallium (TEGa) for group III, and phosphine (PH3) and arsine (AsH3) for group V are typically used. Hydrogen selenide (H2Se) and diethylzinc (DEZn) are used for n-type and p-type dopants, respectively.

3.3.2. EPITAXIAL GROWTH OF GaInAsP ON InP

Here we describe the results obtained with GaInAs(P) layers with different energy gap wavelengths, grown using the conditions shown in Table 3.1. Figure 3.8 shows the room-temperature photoluminescence (PL) spectra of a GaInAs(P) layer, sandwiched by InP layer, with different composition, that is, 0.95, 1.0, 1.05, 1.1, 1.2, 1.3, 1.50 μm bandgap wavelengths. The full widths at half maximum of the PL spectra are around 45 meV. Lattice mismatching was measured less than 0.1%.

To realize a complicated laser structure with GRIN-SCH, it is necessary to grow GaInAsP multiple-step layer. In the GaInAsP/GaInAsP system, unlike the GaAs/AlGaAs system [7], it is very difficult to grow real graded-index change [8] because the simultaneous control of lattice matching and composition are required for the growth of the quaternary layers that form the GRIN-SCH region. Therefore, either single step or step-like index change [9–11] rather than graded-index change [12] was used. Using the growth condition given in Table 3.1, GRIN-SCH structure consisted of quaternary layers with bandgap wavelengths of 0.95, 1.0, 1.05

Table 3.1 **MOCVD Growth Condition**

Growth temperature	600°C
Growth pressure	76 Torr
Total flow rate	6 l/min.
Group III	TMIn, TEGa
Group V	PH_3, AsH_3
V/III ratio	225
Growth rate	2.3 μm/h (InP)
	1.7 μm/h (GaInAsP)
Dopant	DEZn (p-type)
	H_2Se (n-type)

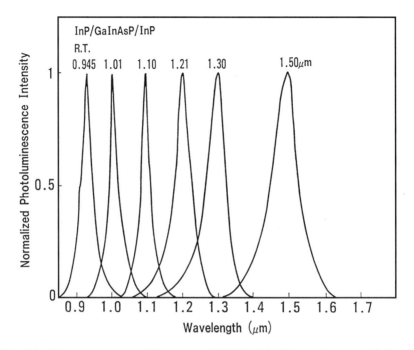

Fig. 3.8   Room-temperature PL spectra   FWHM of the PL spectra are around 45 meV. Lattice mismatching was measured less than 0.1%.

and 1.1 μm, grown on an InP substrate. The SIMS profile and transmission electron microscope (TEM) image of QWs with GRIN-SCH structure using the new gas line system is shown in Fig. 3.9. TEM observation was done using composition analysis by thickness fringes (CAT) method [13]. The step-like changes in the upper and lower GRIN-SCH regions and the periodic change in MQW region were confirmed. In this CAT TEM photograph, composition changes are identified by step-like displacements in the dark/bright thickness fringes that run vertically in Fig. 3.9. Nearly ideal hetero-interfaces between GaInAsP/GaInAsP with different composition could be grown.

3.3.3.   EPITAXIAL GROWTH OF GaInAs(P)/InP QUANTUM WELLS

The GaInAsP/InP quantum wells having different thicknesses are evaluated by 4K photoluminescence. The PL was excited using the 5145 Å line of a Kr laser and detected by a liquid N_2 cooled Ge detector. GaInAsP

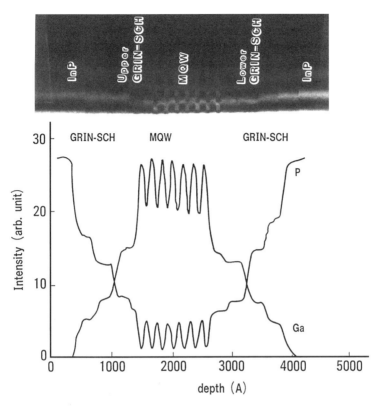

Fig. 3.9   SIMS profile and TEM photograph (CAT) for 1480 nm GRIN-SCH QW region.

single-quantum wells (SQWs), lattice matched to the InP substrate, were prepared on an InP substrate with a 300 nm-thick GaInAs reference layer. The thicknesses of SQWs are 1.2, 2.5, 5 nm, and 10 nm separated by a 5 nm-thick InP layer. The quantum well thickness is determined from transmission electron microscopy (TEM) observation. Figure 3.10 shows the PL energy upshifts of the SQW structures versus the well thickness. The calculated optical transition energy between the first electron level and the first heavy-hole level is also shown. The calculation was made using the envelope function approximation assuming the conduction band offset (ΔE_c) to be 30% and 50% of the bandgap difference (ΔE_g). For GaInAsP, the effective masses used were $0.059\,m_0$ for the electron and $0.50\,m_0$ for the heavy-hole. For InP, the effective masses used were $0.077\,m_0$ for the electron and $0.56\,m_0$ for the heavy-hole. In this experiment, most of the data are in close agreement with the theoretical curves for $\Delta E_c = 0.35\Delta E_g$.

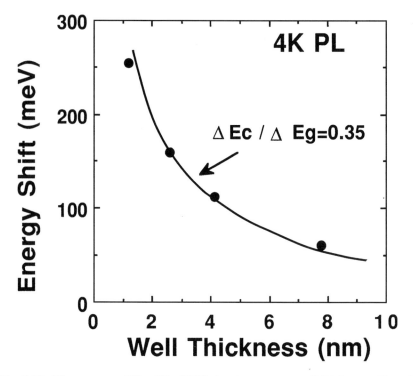

Fig. 3.10   PL energy upshifts of the SQW structures versus well thickness   The calculated optical transition energy between the first electron level and the first heavy-hole
level for $\Delta E_c / \Delta E_g = 0.35$ is shown.

However, it is difficult to determine the actual band offset ratio from the
present data alone.

The PL linewidth versus the well width is shown in Fig. 3.11. The narrow
PL linewidth indicates the sharpness of the QW structure interfaces. The
dotted curve in the figure is the calculated linewidth broadening E due to
a total (both hetero-interfaces) geometric well width fluctuation L_z of one
monolayer ($a_0/2 = 0.293$ nm) using the relationship

$$E = (dE/dL_z)L_z$$

where E is the energy upshift due to quantum size effect in the well with
finite-height barriers. For well width narrower than 2 nm, broadening due to
well width fluctuation becomes very severe and is the dominant contribution
to PL linewidths.

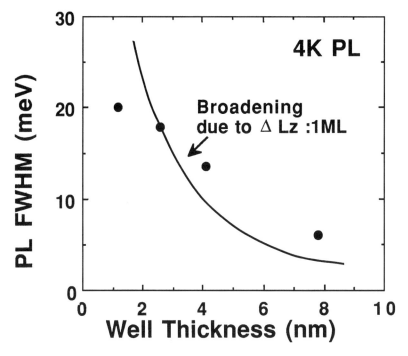

Fig. 3.11   PL line width versus well thickness   Total well width fluctuation of one monolayer obtained.

As described previously, GaInAsP lattice-matched InP with various composition as well as high-quality QWs with sharp hetero-interfaces can be grown in an MOCVD system. Of course, high-quality AlGaAs on GaAs material can be grown in both MOCVD and MBE.

3.3.4.   BURIED HETEROSTRUCTURE LASERS

A buried heterostructure (BH) laser is essential to achieve a low threshold current, fundamental transverse-mode operation. Therefore, BH lasers are widely used for 1480 nm lasers. BH lasers using all MOCVD technique are very attractive in terms of the fabrication of large-scale very uniform characteristics. Here, the fabrication process of the BH laser is given first, then the static characteristics of BH lasers are described.

All epitaxial growths including selective growth for BH structure using MOCVD process ensure the high yield process using a 2-inch wafer. The BH laser is fabricated by three-step MOCVD, as described following.

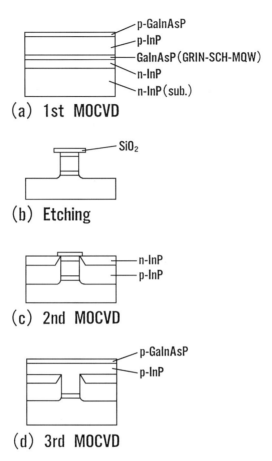

(a) 1st MOCVD

(b) Etching

(c) 2nd MOCVD

(d) 3rd MOCVD

Fig. 3.12   Fabrication procedure of a BH laser by MOCVD   A three-step MOCVD process is required. SiO₂ overhang is important for a flat surface after the 2nd growth.

The fabrication procedure of a BH laser is schematically shown in Fig. 3.12.

(1) DH structure was prepared by first-step MOCVD (Fig. 3.12 (a)). The TEM photograph of a cross-sectional view of the active layer for 1480 nm laser is shown in Fig. 3.13. The active layer is made up of 1% compressively strained quantum wells (4 nm thick) separated by GaInAsP (10 nm thick) with a bandgap wavelength of 1.2 μm. The number of wells is five. GRIN-SCH layer, composed of two different GaInAsP, sandwiches the strained-layer active layer. Very sharp hetero-interfaces are obtained through the MOCVD growth optimization.

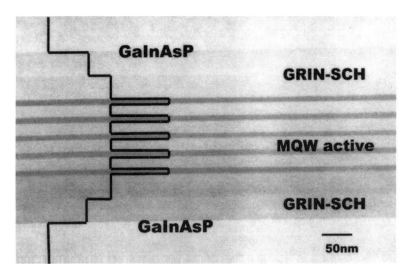

Fig. 3.13   TEM cross-sectional view of a 1480 nm laser   The dark gray area in the center portion shows the GaInAsP compressively strained quantum wells (five QWs) separated by tensile-strained barriers. Step-wise change in the composition layer is used for separate-confinement-heterostructure (SCH).

(2) The narrow stripe mesa with around 2 μm width, having 2 μm height, is prepared by photolithography and wet chemical selective etching using HCl/H_3PO_4 and $H_2SO_4/H_2O/H_2O_2$ solutions for InP and GaInAs(P) layers, respectively, using the SiO_2 as an etching mask. This etching procedure provides the undercutting beneath the SiO_2 mask. It is found that about 1 μm undercutting is essential for achieving planar surface after MOCVD regrowth. The active layer width is set to be around 2.0 μm to achieve fundamental transverse mode operation (Fig. 3.12 (b)).

(3) A current blocking layer consisting of p- and n-InP layers is grown selectively using SiO_2 as a mask in the second MOCVD (Fig. 3.12 (c)). As mentioned in Section 3.2.2, current confinement structure is very important for high power operation. Layer thickness and carrier concentration of current blocking layer have to be designed carefully. This process might also affect the laser long-term reliability.

(4) After removing the SiO_2 mask and p-GaInAsP layer, the p-InP embedding layer and p-GaInAsP contact layer is grown in the third MOCVD growth (Fig. 3.12 (d)).

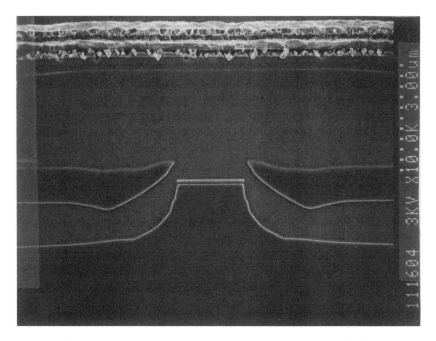

Fig. 3.14   SEM photograph of a BH laser   Almost flat surface is obtained by optimizing the mesa formation and MOCVD regrowth condition.

Figure 3.14 shows a scanning electron microscope photograph of a BH GRIN-SCH-MQW laser. An almost flat surface was realized using all MOCVD growth. Flat surface helps to get a good die attach onto the heat sink.

3.3.5.   DEPENDENCE OF NUMBER OF QUANTUM WELLS ON THRESHOLD CURRENT AND QUANTUM EFFICIENCY

It is important to optimize the number of wells because a theory predicts that threshold current density depends critically on this parameter for GaInAs(P)/InP MQW lasers. In this section, the dependence of the light output characteristics on the number of wells is investigated experimentally for 1480 nm SCH-MQW lasers [14]. The relationship between threshold current, quantum efficiency, and cavity length is shown in Fig. 3.15 as a parameter of the number of wells. The active layer structure investigated is shown in Fig. 3.16. The internal loss of lasers with QW active layer is

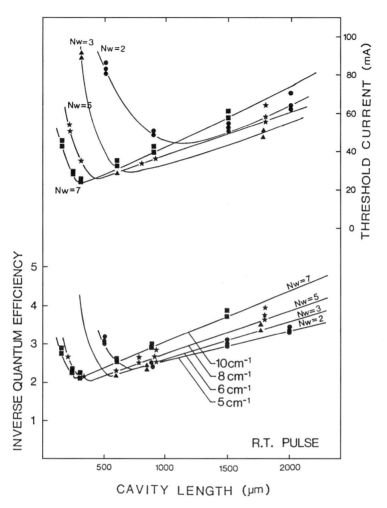

Fig. 3.15 Relationship between threshold current, quantum efficiency, and cavity length
The parameter is number of wells [14].

much smaller than that of lasers with bulk active layer. The internal loss
decreases with small number of wells, however, steep decrease of quantum
efficiency and steep increase of threshold current were observed for a cavity length less than 750 µm. This is due to the increased threshold carrier
density. The carrier overflow into the optical confinement layer induces the
additional loss. The lasers with five quantum wells give the maximum light
output power. A light output power over 250 mW is obtained.

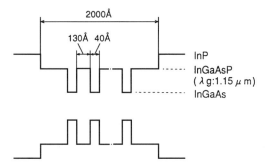

Fig. 3.16   Quantum well active layer structure   Single-step SCH is used for this investigation [14].

3.3.6.   DEPENDENCE OF SCH STRUCTURE ON THRESHOLD CURRENT AND QUANTUM EFFICIENCY

As mentioned in the previous section, internal waveguide loss plays an important role in quantum efficiency. Here, we discuss the effect of SCH structure on internal loss of 1480 nm lasers. As described in Section 3.2, the small internal loss is very important for low-threshold current density and high differential quantum efficiency operations.

In this section, low internal waveguide loss in the GRIN-SCH QW lasers is described. First, the internal waveguide losses of QW lasers with GRIN-SCH structure and QW lasers with SCH structure are compared. Then, the reason for the low internal waveguide loss obtained in GRIN-SCH QW lasers is discussed in detail.

Internal waveguide loss is obtained from the relationship between the inverse differential quantum efficiency and cavity length, because differential quantum efficiency η_d is given by Eq. (3.3). The internal waveguide loss α_i is given by

$$\alpha_i = \alpha_{sc} + \Gamma_{ac}\alpha_{ac} + (1 - \Gamma_{ac})\alpha_{ex}. \qquad (3.4)$$

α_{sc} is the scattering loss resulting from the roughness of the hetero-interfaces and imperfection of the BH mesa. In this case, hetero-interfaces are smooth enough to neglect the scattering loss, and a relatively wide mesa of about 2 μm is used so that the electric field is well confined in the mesa. Therefore, scattering loss is neglected in this consideration. Γ_{ac} is the optical confinement factor in the quantum wells, and α_{ac} and α_{ex} are the absorption losses

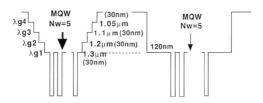

Fig. 3.17   Schematic diagram of quantum well active region   Number of quantum wells is five and SCH thickness is 120 nm for both structures.

of the active layer and cladding layer including the optical confinement layer. First, the internal waveguide loss and internal quantum efficiency are compared between SCH-MQW lasers and GRIN-SCH-MQW lasers, as shown in Fig. 3.17. The MQW structure with GaInAs (6.5 nm each) quantum wells (well number: 3, 5, and 7), separated by GaInAsP ($\lambda_g = 1.3\,\mu$m, 15 nm each) barriers, is prepared. GRIN-SCH structure is made up of the four-step GRIN-SCH ($\lambda_g = 1.3, 1.2, 1.1, 1.05, 1.0\,\mu$m, 30 nm thick each) and SCH structure consists of 120-nm-thick GaInAsP ($\lambda_g = 1.3\,\mu$m) layer. The total thickness of MQW region including the optical confinement layer (SCH and GRIN-SCH) is the same for both structures. The inverse differential quantum efficiency (η_d^{-1}) is plotted in Fig. 3.18 as a function of cavity length for SCH-MQW and GRIN-SCH-MQW lasers with five quantum wells. From this figure, the internal waveguide losses are 13 cm^{-1} and 8 cm^{-1} for SCH-MQW lasers and GRIN-SCH-MQW lasers, respectively. The internal quantum efficiencies are 84% for both structures.

By solving Eq. (3.4) using the different number of wells (different optical confinement factor), we can calculate the absorption coefficients of the active and cladding layers, and obtained values are summarized as follows:

	Internal Waveguide Loss (cm^{-1})	
Number of Wells	**GRIN-SCH**	**SCH**
1	4.3	No lasing
2	4.6	
5	7.0	13.0
7	11.0	15.0

The absorption loss coefficient α_{ac} is calculated to be 120 cm^{-1} for both cases. This value is almost the same as reported for 1.5 μm lasers and it is

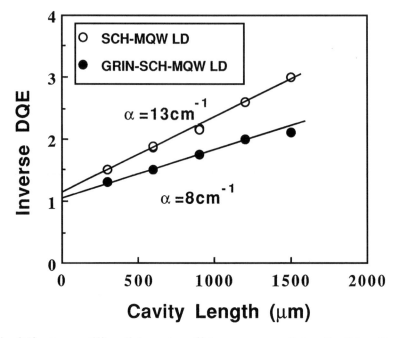

Fig. 3.18 **Inverse differential quantum efficiency versus cavity length**   Internal losses are 13 cm^{-1} and 8 cm^{-1} for SCH and GRIN-SCH QW lasers, respectively.

considered to the intervalence band absorption loss [5]. On the other hand, the absorption loss of the cladding layer including the optical confinement layer is different between SCH and GRIN-SCH structures. The α_{ex} is 2.5 cm^{-1} for the GRIN-SCH structure and 6.5 cm^{-1} for the SCH structure. It should be noted that these values include the free carrier absorption loss in the p-InP cladding layer whose carrier concentration is 1×10^{18} cm^{-3}.

Here, let's discuss the lower α_{ex} observed in GRIN-SCH-MQW lasers. The internal waveguide loss α_i can be rewritten as follows.

$$\alpha_i = \alpha_{sc} + \Gamma_{ac}\alpha_{ac} + \Gamma_{oc}\alpha_{oc} + (1 - \Gamma_{ac} - \Gamma_{oc})\alpha_{ex} \qquad (3.5)$$

The Γ_{oc} and α_{oc} are the optical confinement factor and the absorption loss of the optical confinement layer, respectively. The first term of the righthand side of the equation is almost the same for the same MQW layer because Γ_{ac} is almost the same for both structures. The difference between these structures is the optical confinement factor in the optical confinement layer (Γ_{oc}) with a bandgap wavelength of 1.3 μm (1.3 μm-Q). The optical field is calculated by solving the Maxwell's equations.

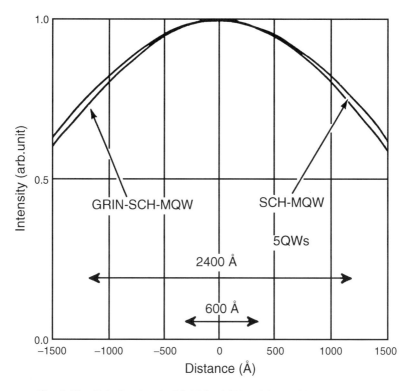

Fig. 3.19 Calculated optical field for SCH and GRIN-SCH QW lasers.

The calculated optical field is shown in Fig. 3.19 for both structures. The optical fields are almost identical for both structures. The optical confinement factors in the MQW (including barriers) layer are 14.7% and 13.2% for SCH-MQW and GRIN-SCH-MQW lasers, respectively. The Γ_{oc} of the SCH-MQW laser is calculated to be 26.7%, which is about 4 times larger than that of a GRIN-SCH-MQW laser (6.8%). Therefore, the absorption loss of the 1.3 μm-Q layer ($\Gamma_{oc}\alpha_{oc}$) in SCH-MQW lasers is larger than that of GRIN-SCH-MQW lasers because the 1.3 μm-Q layer has an optical loss for the lasing wavelength.

From the preceding discussion, the low internal loss in GRIN-SCH-MQW lasers is attributed to the waveguide structure, mainly the difference of the 1.3 μm-Q layer thickness. This smaller α_i gives the larger η_d in GRIN-SCH-MQW lasers.

Next, the detailed explanation was made in order to explain the results obtained to clarify the effect of the loss on the waveguide structure.

The internal loss is rewritten as

$$\alpha_i = \alpha_{sc} + \Gamma_w \alpha_{ac} + (\Gamma_{SCH} + \Gamma_B)\alpha_{SCH} \qquad (3.6)$$

where Γ_w, Γ_{SCH}, and Γ_B are the optical confinement factors in the quantum well, SCH, and barrier layers, respectively. α_{ac} and α_{SCH} are the absorption coefficients in the quantum well and SCH layers.

When the increase of absorption loss due to the injected carrier is taken into account, loss of the quantum well and SCH layers are given by

$$\begin{aligned}
\alpha_{ac} &= \alpha_{ac0} + bn \\
\alpha_{SCH} &= \alpha_{SCH0} + cn
\end{aligned} \qquad (3.7)$$

where α_{ac0}, α_{SCH0} are the intrinsic absorption coefficients of the quantum well and SCH layers, respectively. b and c are the absorption loss coefficients caused by the injected carriers. n is the carrier density. If we assume the linear gain relationship, the gain is expressed by

$$g = a(n - n_t), \qquad (3.8)$$

where a is the gain coefficient and n_t is the value required for transparency.

At threshold, the following condition is satisfied.

$$\Gamma_w g_{th} = \alpha_i + \alpha_m. \qquad (3.9)$$

From Eqs. (3.3), (3.6)–(3.9), by eliminating n,

$$\overline{\eta}_d^{-1} = \overline{\eta}_i^{-1}\left(1 + \frac{\alpha_i L}{\ln\left(\frac{1}{R}\right)}\right)$$

$$\alpha_{iM} = \alpha_{sc} + \Gamma_w(\alpha_{ac} + bn_t) + (\Gamma_{SCH} + \Gamma_B)(\alpha_{SCH0} + cn_t) \qquad (3.10)$$

$$\eta_{iM} = \left(1 - \frac{b}{a} - \frac{c}{a}\frac{\Gamma_{SCH} + \Gamma_B}{F_w}\right)\eta_i \qquad (3.11)$$

From Eq. (3.10), the difference in the internal waveguide loss for the structures would arise from the difference of the product of the optical confinement factor in the optical confinement layer and the absorption loss induced by the carriers in the optical confinement layer, since the optical confinement factor and the absorption loss are almost same for both structures.

First, let us consider the absorption loss induced by the carriers in the optical confinement layer. The carrier distribution of the QW lasers with SCH and GRIN-SCH structure is shown in Fig. 3.20 in the case of carrier

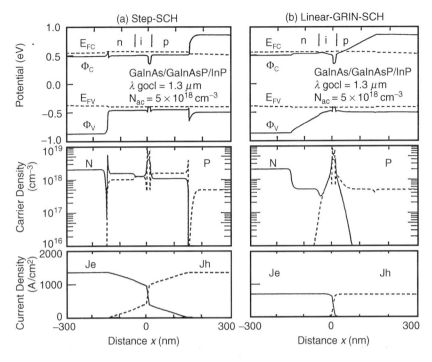

Fig. 3.20 Calculated results of the potential distribution, Fermi-level, and current density distribution for SCH and GRIN-SCH QW lasers [15].

density of 5×10^{18} cm^{-3} in the quantum wells [15]. From this figure, carrier density in the GRIN-SCH layer is much lower than that of SCH layer, because carriers concentrate around the active region in the GRIN-SCH structure due to the potential profile formed by GRIN structure. In this particular case, carrier density in the SCH layer is about as high as 1×10^{18} cm^{-3}, while in the GRIN-SCH is about 5×10^{17} cm^{-3}. It is reported experimentally [16] that the GaInAsP layer with a bandgap wavelength of $1.53\,\mu$m for $1.53\,\mu$m lasing wavelength has an absorption coefficient of $d\alpha/dn = (1-3) \times 10^{-17}$ cm^2. Therefore, the absorption losses induced by the carriers in the optical confinement layer are calculated to be 10–30 cm^{-1} for the SCH structure and 5–15 cm^{-1} for the GRIN-SCH structure. However, the actual value of the absorption is obtained by multiplying the optical confinement factor in the optical confinement layer.

Next, the optical confinement factor in the optical confinement layer is calculated for both structures. The optical confinement factor in the 1.3 μm

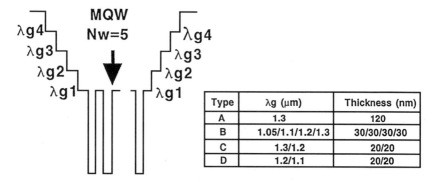

Fig. 3.21 Schematic diagram of 1480 nm GRIN-SCH QW lasers with different waveguide structure.

waveguide layer is 27% for the SCH structure, 6.8% in the GRIN-SCH structure. Therefore, the absorption losses in the optical confinement layer, which is defined as $\alpha_{SCH} = \Gamma_{SCH}(d\alpha/dn)n$, are 2.7–8.1 cm^{-1} for the SCH structure and 0.34–1.0 cm^{-1} for the GRIN-SCH structure. The loss of the GRIN-SCH structure is about eight times smaller than that of SCH structure. The absorption loss of the quaternary layers other than 1.3 μm-Q in the GRIN-SCH layer could be neglected because the optical confinement factor of these layers is less than 5% and the absorption loss coefficient with respect to carriers is small due to the wide bandgap.

In order to verify this assumption experimentally, four types of GRIN-SCH QW BH lasers (referred to as Type A, B, C, and D hereafter), shown in Fig. 3.21, are prepared [17]. The structure of multiple-quantum well is the same in these lasers, that is, five lattice-matched GaInAsP ($\lambda_g = 1.55$ μm; 6.5 nm thick each) quantum wells. The bandgap wavelengths of the barrier layer are 1.3 μm in Type A, B, and C, and 1.2 μm in Type D, respectively. The SCH layer was composed of step GRIN. The compositions of GRIN-SCH layers are as follows;

Type	Bandgap Wavelength (μm)	Thickness (nm)
A	1.3	120
B	1.3/1.2/1.1/1.05	30/30/30/30
C	1.3/1.2	20/20
D	1.2/1.1	20/20

The description "1.3 μm-Q(20 nm)" means the bandgap wavelength (1.3 μm) and thickness (30 nm) of the SCH layer. The threshold current for

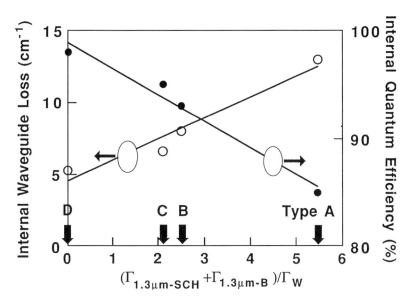

Fig. 3.22   Relationship between $(\Gamma_{1.3Q\text{-}SCH} + \Gamma_{1.3Q\text{-}B})/\Gamma_w$ and η_{iM}, α_{iM}.

these LDs is about 15–20 mA for 900 μm-long cavity without facet coating. However, the LD with a SCH layer consisting of 1.1 μm-Q(20 nm)/1.0 μm-Q(20 nm) showed high threshold current due to a small optical confinement factor. Therefore, this device was not considered in the following discussion.

Figure 3.22 shows the $(\Gamma_{1.3Q\text{-}SCH} + \Gamma_{1.3Q\text{-}B})/\Gamma_w$ versus η_{iM} and α_{iM}. As predicted by Eqs. (3.10) and (3.11), η_{iM} decreases and α_{iM} increases with the increase of $(\Gamma_{1.3Q\text{-}SCH} + \Gamma_{1.3Q\text{-}B})/\Gamma_w$. Thus, it is experimentally demonstrated that the use of a SCH layer with wider bandgap and thinner thickness gives a higher differential quantum efficiency. The enhanced internal quantum efficiency obtained in the GRIN-SCH structure could be attributed to the high current injection efficiency into the quantum wells.

The decrease of optical confinement factor is also effective in obtaining the narrow far field angle. The FWHM of the far field patterns perpendicular to the junction plane $(\theta_\perp)$ is calculated. Figure 3.23 shows the $\theta_\perp$ versus the thickness of each quaternary layer for type A, B, C, and D in Fig. 3.21. In this calculation, each quaternary layer is equal in thickness. The $\theta_\perp$ decreases as the layer thickness decreases. The FWHM of the $\theta_\perp$ is typically in the range of 20–30 when the width of the active layer is 2 μm. The quaternary layer thickness of 20 nm gives the $\theta_\perp$ of 20–30.

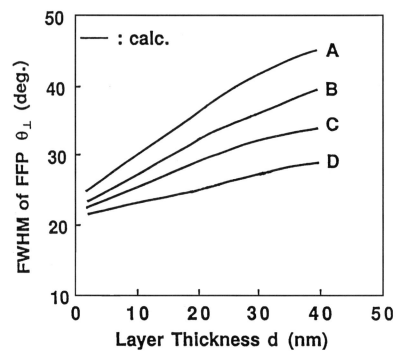

Fig. 3.23 FWHM of FFP perpendicular to the junction plane for Type A, B, C, and D.

Of course, single lateral mode has to be realized. In BH structure, cut-off width for higher lateral mode is given by

$$W_a = \frac{\lambda}{2\sqrt{n_{eff}^2 - n_s^2}}$$

where n_{eff} is the effective refractive index of the active region, n_s is the refractive index of the surrounding layer, and λ is the wavelength. The higher order cut-off width is approximately 2 μm for 1480 nm lasers.

3.3.7. HIGH POWER OPERATION

As a result of high differential quantum efficiency resulting from high internal quantum efficiency and low internal waveguide loss described in the previous section, high power operation can be achieved in BH GRIN-SCH QW lasers.

The dependence of the light output power on the cavity length is shown in Fig. 3.24. The maximum light output power increases with cavity length.

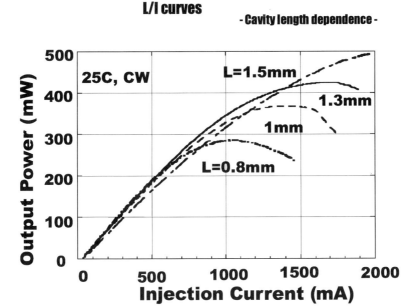

Fig. 3.24 Light output power versus injection current characteristics of 1480 nm laser for various cavity lengths. Maximum light output power increases with increase of cavity length. Light output power of 500 mW is obtained. Both low thermal resistance and low electric resistance help to increase the roll-over current level.

An ultra-high light output power of 500 mW is obtained for a 1.5 mm-long device. The SCH layer consists of two-step GRIN of 1.2 μm-Q(20 nm)/ 1.1 μm-Q(20 nm) (Type D structure). The lasers were coated with AR (8%) and HR (95%) coatings and were bonded with junction down configuration. The FWHM of the FFPs parallel and perpendicular to the junction plane were 20 and 25 degrees, respectively, at a light output power of 100 mW (Fig. 3.25).

Highly reliable operation must be achieved as with conventional FP and DFB lasers because optical fiber amplifiers are used for long-haul trunk line and submarine optical communication systems. Output power of pumping lasers is extremely high as compared to conventional signal lasers. Reliability tests under high temperature and high output power have to be investigated. Figure 3.26 shows the aging test result at high power conditions for 800 μm-long lasers. The aging test has been performed at 35°C, 60°C under automatic power control mode. The output power is set to be 80% of the maximum output power at the given temperature, determined by the thermal rollover, which is 180 mW at 35°C and 120 mW

FFPs of 1480nm BH laser

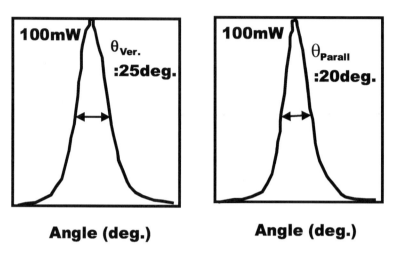

Fig. 3.25   FFPs vertical and parallel to the junction plane for a 1480 nm BH laser at 100 mW   A narrow and circular output beam is obtained for well-designed waveguide structure.

Aging test of 1480nm lasers

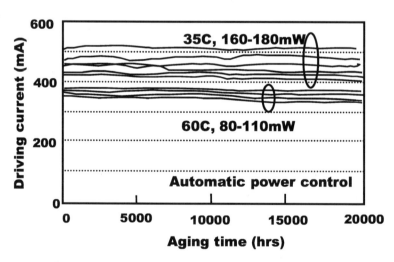

Fig. 3.26   Aging test results for 1480 nm lasers under high temperature and high power   No appreciable change in driving current is observed. Highly reliable operation can be achieved.

at 60°C. Stable operation has been confirmed after 20,000 hours without appreciable increase of driving current. The activation energy of 0.62 eV is derived from various aging test conditions. Extremely high reliability of 100 million hours as a mean time to failure (MTTF) is estimated at an output power of 150 mW, which corresponds to the module output power of 120 mW (reasonable coupling efficiency into an SMF of 80% is assumed). Highly reliable operation under higher light output power is obtained using long cavity lasers more than 1 mm.

3.4. 980 nm Lasers

3.4.1. LASER STRUCTURE

The noise figure of a fiber amplifier pumped by 980 nm wavelength is lower than that of fiber amplifiers pumped by 1480 nm wavelength, making it very attractive for practical applications.

Material candidates for 980 nm lasers are compressively strained InGaAs quantum wells for the active layer, while for the cladding layer there is a choice of conventional AlGaAs [18–22] layer or novel InGaP [23] layers. The use of InGaP cladding layer material might have the advantage over AlGaAs in terms of less oxidized property and slow surface recombination velocity, which are thought to be effective methods for suppressing COD. Either MOCVD or MBE is used or epitaxial growth.

As mentioned previously, transverse-mode stabilized laser structure is used for efficient coupling to SMF. Ridge waveguide structure is widely used as shown in Fig. 3.27. This laser structure is grown by MOCVD. Recently, other laser structures such as buried ridge waveguide structure and self-aligned-structure (SAS) have been investigated [24].

First, 980 nm ridge waveguide laser is explained. The active layer is made up of double InGaAs compressive-strained quantum wells separated by a GaAsP tensile-strained barrier layer to realize strain compensation. The ridge width is around 4 μm. The ridge is formed by wet chemical etching. The ridge width and off-ridge height have to be controlled precisely for stable lateral mode operation. As with 1480 nm lasers, highly reliable operation is required. One of the key items for high reliability operation is to overcome the COD. Detailed explanation for COD is found in Ref. [4]. Key technology for this is facet passivation [25]. Several approaches have been reported; cleavage laser bars in the ultra-high-vacuum atmosphere and in-situ facet passivation, facet passivation using high thermal

Schematic diagram of 980nm RWG laser

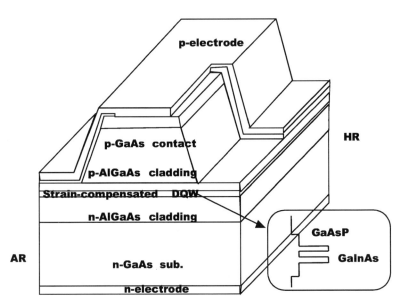

Fig. 3.27   Schematic diagram of a 980 nm ridge waveguide laser   Ridge width around 4 μm is used for stable lateral mode operation. InGaAs/GaAsP Al-free strain-compensated QWs are used for highly reliable operation at high output power. AR/HR asymmetric facet coatings are used for high output power from front facet.

conductivity material, and formation of so-called "window structure" near the facet.

The self-aligned structure (SAS), shown in Fig. 3.28, is the other candidate for high power operation. In the first growth, DH structure, which includes the first cladding layer and current blocking layer, is prepared. After etching the current blocking layer, the second cladding layer and contact layer are grown in the second growth. Because the etching has to be stopped above the active layer, an etch-stopping layer is introduced for better controllability. In general, overgrowth on Al-containing layer is very difficult; either GaAs or InGaP, lattice matched to GaAs, is used as the surface layer to avoid the oxidation. In addition, regrowth of the second AlGaAs cladding layer on the AlGaAs first cladding and current blocking layers has to be investigated carefully for high reliability. The lateral optical confinement can be controlled by a refractive index step between center channel and outer regions. The thickness of the first cladding layer

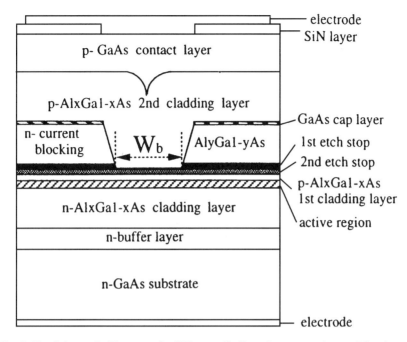

Fig. 3.28 Schematic diagram of a 980 nm self-aligned structure laser The channel width (bottom width) is 2.2 µm.

and composition of first cladding, second cladding, and current blocking layers are the parameters to control the refractive index step. The refractive index difference is set to be around 3.5×10^{-3} under a channel width of around 2 µm.

3.4.2. LASER CHARACTERISTICS

Light output power versus injection current characteristics is shown in Fig. 3.29 for a 980 nm ridge waveguide laser with 1500 µm-long cavity. Light output power more than 600 mW is achieved with a slope efficiency as high as 0.94 W/A. FFP parallel to the junction plane is also shown in Fig. 3.29 at various output powers. Stable lateral mode operation up to 500 mW is confirmed, however, "beam steering" is observed at higher output power. Although output beam is elliptical, high coupling efficiency into an SMF is achieved by optimizing a coupling scheme. Using wedge-shaped fiber that is specially designed for beam shape, coupling efficiency as high as 80% was achieved.

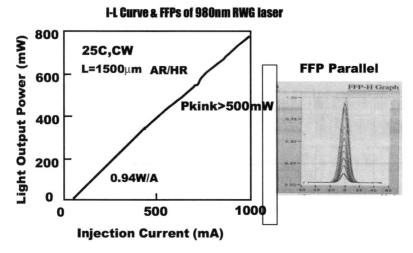

Fig. 3.29 Light output power versus injection current characteristics for a 1500 mm-long 980 nm laser. Light output power over 500 mW is obtained. Kink current is 700 mA in this case. FFPs parallel to the junction plane are also shown at various output powers. Stable single-lobe FFP is obtained up to 500 mW. "Beam steering" phenomenon is observed at 550 mW for this device.

Note that highly stable lateral mode operation has to be realized. FFP perpendicular to the junction plane is almost determined by the layer structure, and it is very stable for QW laser structure. However, FFP parallel to the junction plane depends on a lateral confinement scheme, i.e., buried heterostructure, ridge waveguide structure, and so on. A BH laser provides strong optical confinement (strong index guiding; index step in the order of several %), therefore, refractive index change (decrease of refractive index of the active layer induced by plasma effect) does not affect the FFP. However, ridge waveguide structure provides weak optical confinement (weak index guiding; index step in the order of 10^{-3}), so lateral FFP could change due to the temperature rise in the active layer and carrier distribution, which can cause the so-called "beam steering." The beam steering phenomenon is shown in Fig. 3.29. FFP parallel to the junction plane remains single-lobe, however, peak position shifts slightly. A slight change in FFP degrades the coupling efficiency dramatically and a large "kink" in the L-I curve is observed.

Long-term aging results are shown in Fig. 3.30 under condition of 60C, 250 mW. No appreciable change in driving current has been observed after 5000 hours. SAS lasers also can provide high power operation with highly reliable operation.

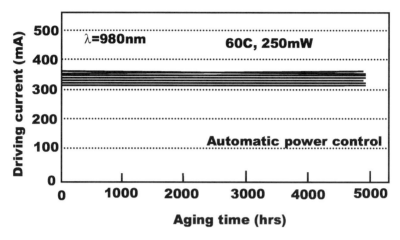

Fig. 3.30   Aging test results for 980 nm lasers under high temperature and high power
No appreciable change in driving current is observed. Highly reliable operation can be achieved by use of proper facet passivation.

3.5.   Laser Modules

Figure 3.31 shows a photograph of the 1480 nm package. It consists of a laser, a back monitoring photodiode, a thermistor, a thermoelectric cooler, and lens. Two lenses, collimating and focusing lenses, are used for the coupling into the SMF. A very high coupling efficiency of 80% is obtained by optimizing the coupling scheme. When a ridge waveguide structure laser, which has an elliptical beam profile, is used, lensed fiber is used for high coupling efficiency.

The characteristics of 980 nm and 1480 nm laser modules are described. Light output power versus injection current characteristics for 980 nm and 1480 nm laser modules are shown in Fig. 3.32 and Fig. 3.33, respectively. Very high coupled powers of 300 mW are used for both 980 nm laser module and 1480 nm laser module.

Several types of laser modules have been fabricated for WDM application. Wavelength stabilized laser module is one example. Lasing wavelength can be stabilized by use of fiber Bragg grating. In general, lasing wavelength of FP lasers shifts to longer wavelength with a temperature coefficient of 0.5 nm/degree in the 1480 nm wavelength. By using a fiber Bragg grating with narrow pass band, stable lasing wavelength for both

Photograph of laser diode module

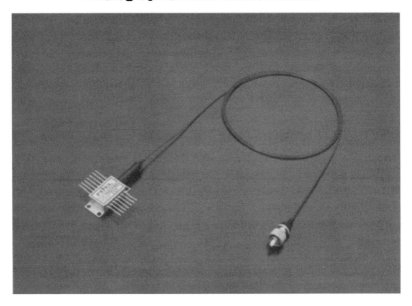

Fig. 3.31 Photograph of a 1480 nm pig-tailed laser module Compact size (30 mm (L) × 13 mm (W) × 8 mm (H)) packaging is widely used. Over 300 mW coupled power is obtained.

980nm Module L-I

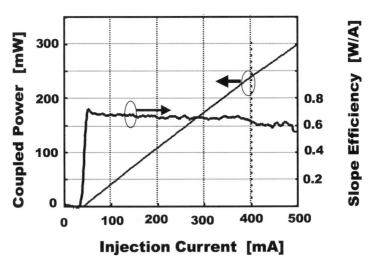

Fig. 3.32 Light output power versus injection current characteristics for a 980 nm pig-tailed laser module. A coupling efficiency of approximately 75% is obtained using a lensed fiber. Over 300 mW coupled power is obtained.

1480nm pig-tailed module performance

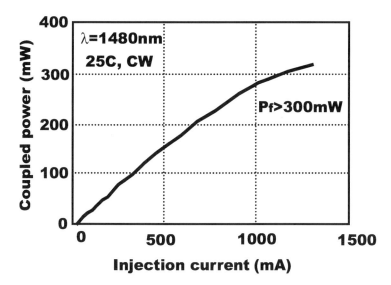

Fig. 3.33   Light output power versus injection current characteristics for a 1480 nm pig-tailed laser module. A coupling efficiency of approximately 85% is obtained.

current and temperature can be achieved. Especially, high power, wavelength stabilized laser modules in the 1420–1520 nm wavelength range are attractive as pumping sources for Raman amplification.

3.6.   Future Prospects

The history for power improvement in narrow stripe (single lateral mode operation) 1480 nm lasers is shown in Fig. 3.34. Drastic power improvement has been achieved by technological innovations such as the introduction of quantum well active and strained-layer quantum well active layers. By using a strained-layer quantum well active layer, maximum output power of about 500 mW was achieved.

The importance of the temperature dependence of the output characteristics of 1480 nm is shown in Fig. 3.35. Here, two kinds of active layer materials, conventional GaInAsP and AlGaInAs, are used. Due to the difference of the so-called characteristic temperature, T_0, lasers with AlGaInAs active layer can provide higher output. Similarly, lower thermal resistance

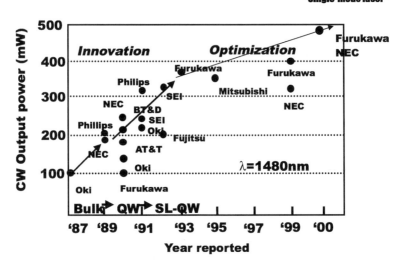

Fig. 3.34 **Light output power improvement for 1480 nm narrow stripe lasers** Technological innovation such as quantum well and strained-layer QW can improve the light output power.

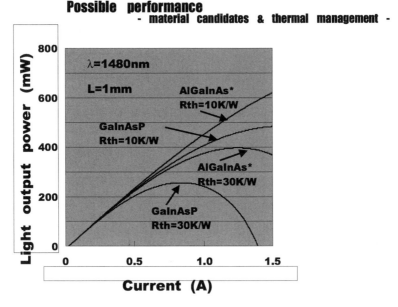

Fig. 3.35 **Calculated light output power versus injection current characteristics for 1480 nm lasers** AlGaInAs active layer on InP substrate is the possible candidate to get higher power operation. This is due to the temperature insensitive threshold current and quantum efficiency. The reduction of thermal resistance can improve the light output power.

can improve the output drastically. This is because longer wavelength laser is influenced by the poor temperature characteristics due to the Auger recombination and other nonradiative recombination. New technological innovations such as novel active layer structure and some consideration for active layer material can lead to further output power improvement in the future. Other material candidates, for example, GaInNAs quantum wells on GaAs substrate, are promising due to excellent temperature character-istics, even though much attention has to be paid for the realization of high crystalline quality for highly strained material over 2%. Lower dimensional quantum well structure such as quantum dot is also promising.

ACKNOWLEDGMENTS

The author would like to thank Drs. K. Ohkubo, Y. Suzuki, and M. Shibata for encouragement and also thanks to all the members of the pump laser team at Furukawa Electric Co., Ltd.

Appendix

OPTICAL CONFINEMENT FACTOR

Optical confinement factor Γ is the very important parameter that describes the laser action as given in the content. It is derived from Maxwell's equation for the multi-layer waveguide. The calculation model of the multiplayer waveguide structure is shown in Fig. 3.36. x- and y-axes are perpendicular and parallel to multilayers, respectively. Optical field propagates along the z-axis.

In lattice-matched and compressive-strained MQW lasers, lasing mode is fixed to be transverse electric field (TE) mode, where electric field E has only y-component and magnetic field H for y-direction is zero, i.e., $E = (0, E_y, 0)$ and $H = (H_x, 0, H_z)$. Thus the Maxwell's equation for TE mode can be given as

$$H_x = \frac{1}{i\omega\mu_0} \frac{\partial E_y}{\partial z}, \tag{A.1}$$

$$H_z = -\frac{1}{i\omega\mu_0} \frac{\partial E_y}{\partial x}, \tag{A.2}$$

$$\frac{\partial H_z}{\partial z} - \frac{\partial H_z}{\partial x} = i\omega E_y, \tag{A.3}$$

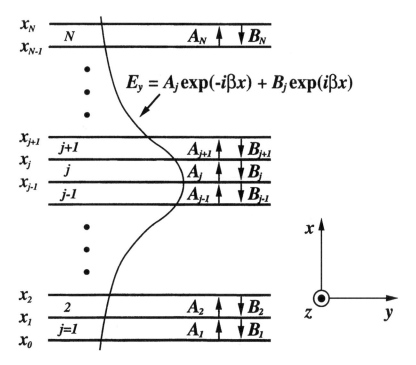

Fig. 3.36 Multi-layer waveguide structure for the calculation of the optical confine-ment factor A_j and B_j corresponding to the amplitude of electric field propagating upward and downward, as appeared in the equation, are computed with the boundary conditions of the continuity of electric and magnetic fields.

where ω, ε, μ_0 are angular frequency of propagated light, permittivity, and permeability in vacuum, respectively. Assuming z-dependence of E_y as $\exp(-i\beta z)$ with propagation constant of β, we can obtain

$$\frac{\partial^2 E_y}{\partial x^2} + \left(k_0^2 n^2 - \beta^2\right)E_y = 0 \qquad (A.4)$$

where n is the refractive index. k_0 is defined by using the relationship of $\varepsilon = \varepsilon_0 n^2$ as

$$k_0^2 = \omega^2 \mu_0 \varepsilon_0 = \left(\frac{2\pi}{\lambda}\right)^2. \qquad (A.5)$$

From Eq. (A.4), electric field E_y can be given by using constants A and B as

$$E_y = A \exp(-i\gamma x) + B \exp(i\gamma x), \qquad (A.6)$$

where γ is defined as

$$\gamma = \sqrt{k_0^2 n^2 - \beta^2}. \tag{A.7}$$

Substituting Eq. (A.6) into Eq. (A.2), H_z is expressed as

$$H_z = -\frac{1}{i\omega\mu_0}\{-i\gamma A \exp(-i\gamma x) + i\gamma B(i\gamma x)\}. \tag{A.8}$$

Equations (A.6) and (A.8) are consistent in each layer. Boundary conditions for the jth layer are the continuity of both E_y and H_z at the boundaries with the $(j-1)$th layer and $(j+1)$th layer, i.e. $x = x_{j-1}$ and $x = x_j$. We must consider one interface because another one can be the boundary condition for the adjacent layer. Here, we consider the interface at $x = x_{j-1}$. The boundary conditions are

$$
\begin{aligned}
A_j &\exp(-i\gamma_j x_{j-1}) + B_j \exp(i\gamma_j x_{j-1}) \\
&= A_{j-1} \exp(-i\gamma_{j-1} x_{j-1}) + B_{j-1} \exp(i\gamma_{j-1} x_{j-1}), \\
\gamma_j\{&-A_j \exp(-i\gamma_j x_{j-1}) + B_j \exp(i\gamma_j x_{j-1})\} \\
&= \gamma_{j-1}\{-A_{j-1} \exp(-i\gamma_{j-1} x_{j-1}) + B_{j-1} \exp(i\gamma_{j-1} x_{j-1})\},
\end{aligned}
\tag{A.9}
$$

where j means that the parameter is for material of the jth layer. This equation can be simplified by using the matrix formula as

$$M1_j \begin{pmatrix} A_j \\ B_j \end{pmatrix} = M2_{j-1} \begin{pmatrix} A_{j-1} \\ B_{j-1} \end{pmatrix}, \tag{A.10}$$

where $M1_j$ and $M2_j$ are defined as

$$M1_j = \begin{pmatrix} \exp(-i\gamma_j x_{j-1}) & \exp(i\gamma_j x_{j-1}) \\ -\gamma_j \exp(-i\gamma_j x_{j-1}) & \gamma_j \exp(i\gamma_j x_{j-1}) \end{pmatrix},$$

$$M2_j = \begin{pmatrix} \exp(-i\gamma_j x_j) & \exp(i\gamma_j x_j) \\ -\gamma_j \exp(-i\gamma_j x_j) & \gamma_j \exp(i\gamma_j x_j) \end{pmatrix}. \tag{A.11}$$

Then, we can obtain

$$\begin{pmatrix} A_j \\ B_j \end{pmatrix} = M_{j-1} \begin{pmatrix} A_{j-1} \\ B_{j-1} \end{pmatrix}, \tag{A.12}$$

and the matrix M_j is defined as

$$M_j = M_{j+1}^{-1} M2_j. \tag{A.13}$$

Consequently, (A_j, B_j) for electric field with any j can be obtained by using

$$\begin{pmatrix} A_j \\ B_j \end{pmatrix} = M_{j-1} M_{j-2} \dots M_0 \begin{pmatrix} A_0 \\ B_0 \end{pmatrix}. \tag{A.14}$$

Assuming the total number of layer is N, A_0 and B_{N+1} must be zero. Then we can obtain

$$A_{N+1} = M_{12} B_0, \tag{A.15}$$
$$0 = M_{22} B_0, \tag{A.16}$$

where

$$\begin{pmatrix} M_{11} & M_{12} \\ M_{21} & M_{22} \end{pmatrix} = M_N M_{N-1} \dots M_0. \tag{A.17}$$

From Eq. (A.15), $B_0 \neq 0$ is required. Therefore, it is satisfies Eq. (A.16) that M_{22} must be zero. The propagation constant b that results in $M_{22} = 0$ is the solution of this waveguide. Here, if we assume that the 0th layer is the infinite cladding layer, M_0 can be approximated to be unity. After β is computed, we will know any electric field E_y by substituting the obtained β into Eq. (A.14). The optical confinement layer $\Gamma(N_w \Gamma_{SQW})$ will be obtained by

$$\Gamma = \frac{\int_{QW} |E_y|^2 \, dx}{\int_{-\infty}^{\infty} |E_y|^2 \, dx}, \tag{A.18}$$

where the numerator means summation of the integration for all QWs and the denominator is the normalization factor.

References

1. A. R. Adams, "Band-structure engineering for low-threshold high-efficiency semiconductor lasers," Electron. Lett., 22:5 (Feb. 1986) 249–250.
2. E. Yablonovitch and E. O. Kane, "Reduction of lasing threshold current density by the lowering of valence band effective mass," J. Lightwave Technol., LT-4:5 (May 1986) 504–506.

3. T. J. Andersson *et al.*, Appl. Phys. Lett., 51 (1987) 752.
4. M. Fukuda, M. Okayasu, J. Temmyo, and J. Nakano, "Degradation behavior of 0.98μm strained quantum well InGaAs/AlGaAs lasers under high-power operation," IEEE J. Quantum Electron., 30 (1994) 471–476.
5. M. Asada, A. R. Adams, K. E. Stubkjaer, Y. Suematsu, Y. Itaya, and S. Arai, "The temperature dependence of the threshold current of GaInAsP/InP DH lasers," IEEE J. Quantum Electron., QE-17:5 (May 1981) 611–619.
6. A. Kasukawa, Y. Imajo, I. J. Murgatroyd, and H. Okamoto, "Effect of well number in 1.3μm GaInAsP graded-index separate-confinement-heterostructure multiple quantum well (GRIN-SCH-MQW) laser diodes," Trans. IEICE E 73 (1990) 59–62.
7. W. T. Tsang, "Extremely low threshold (AlGa)As modified multiquantum well heterostructure lasers grown by molecular-beam epitaxy," Appl. Phys. Lett., 39:10 (Nov. 1981) 786–788.
8. W. T. Tsang, "Extremely low threshold (AlGa)As graded-index waveguide separate-confinement heterostructure lasers grown by molecular-beam epitaxy," Appl. Phys. Lett., 40:3 (Feb. 1982) 217–219.
9. U. Koren, B. I. Miller, Y. K. Su, T. L. Koch, and J. E. Bowers, "Low internal loss separate confinement heterostructure InGaAs/InGaAsP quantum well laser," Appl. Phys. Lett., 51:21 (1987) 1744–1746.
10. P. J. A. Thjis, L. F. Tiemeijer, J. J. M. Binsma, and T. van Dogen, "High-performance 1.5μm wavelength InGaAs-InGaAsP strained quantum well lasers and amplifiers," IEEE J. Quantum Electron., 27:6 (Jun. 1991) 1426–1439.
11. A. Kasukawa, I. J. Murgatroyd, Y. Imajo, T. Namegaya, H. Okamoto, and S. Kashiwa, "1.5μm GaInAs/GaInAsP graded index separate confinement heterostructure multiple quantum well (GRIN-SCH-MQW) laser diodes grown by metalorganic chemical vapor deposition (MOCVD)," Electron. Lett., 25 (1989) 659–660.
12. T. Tanbun-Ek, R. A. Logan, N. A. Olsson, H. Temkin, A. M. Sergent, and K. W. Wecht, "High power output 1.48-1.51μm continuously graded index separate confinement strained quantum well lasers," Appl. Phys. Lett., 57:16 (Jul. 1990) 224–226.
13. H. Kakibayashi and F. Nagata, "Composition analysis of GaxAl1-xAs superstructure by equal thickness fringes," Proc. XI Int. Cong. on Electron Microscopy, Kyoto, Japan (1986) 1495–1496.
14. H. Asano, S. Takano, M. Kawaradani, M. Kitamura, and I. Mito, "1.48μm high-power InGaAs/InGaAsP MQW LD's for Er-doped fiber amplifiers," IEEE Photonic. Technol. Lett., 3:5 (1991) 415–417.
15. H. Hirayama, Y. Miyake, and M. Asada, "Analysis of current injection efficiency of separate-confinement-heterostructure quantum-film lasers," IEEE J. Quantum Electron., 28:1 (Jan. 1992) 68–74.

16. P. Brosson, C. Rabourie, L. Le Gouezigou, J. L. Lievin, J. Jacqest, F. Leblond, A. Olivier, and D. Leclerc, "Experimental determination of carrier-induced differential loss in 2-section GaInAsP/InP laser-waveguide," Electron. Lett., 25:24 (Nov. 1989) 1623–1624.

17. A. Kasukawa, T. Namegaya, N. Iwai, N. Yamanaka, Y. Ikegami, and N. Tsukiji, "Extremely high power 1.48μm GaInAsP/InP GRIN-SCH strained MQW lasers," IEEE Photonic. Technol. Lett., 6:1 (1994) 4–6.

18. D. Fekete, K. Chan, J. M. Ballantyne, and L. F. Eastman, "Graded-index separate-confinement InGaAs/GaAs strained-layer quantum well laser grown by metalorganic chemical vapor deposition," Appl. Phys. Lett., 49:24 (1986) 1659–1660.

19. M. Okayasu, M. Fukuda, T. Takeshita, and S. Uehara, "Stable operation (over 5000 h) of high-power 0.98μm InGaAs-GaAs strained quantum well ridge waveguide lasers for pumping Er^{3+}-doped fiber amplifiers," IEEE Photonic Technol. Lett., 2:10 (1990) 689–691.

20. S. Ishikawa, K. Fukagai, H. Chida, T. Miyazaki, H. Fujii, and K. Endo, "0.98-1.02μm strained InGaAs/AlGaAs double quantum-well high-power lasers with GaInP buried waveguide," IEEE J. Quantum Electron., 29:6 (1993) 1936–1942.

21. H. Meier, Technical Digest of 20th European Conference on Optical Communications, Firenze (Sep. 1994) 974.

22. J. S. Major, W. E. Plano, D. F. Welch, and D. Scifres, "Single-mode InGaAs-GaAs laser diodes operating at 980nm," Electron. Lett., 27:6 (1991) 539–541.

23. M. Ohkubo, T. Ijichi, A. Iketani, and T. Kikuta, "980nm aluminum-free InGaAs/InGaAsP/InGaP GRIN-SCH SL-SQW lasers," IEEE J. Quantum Electron., 30 (1994) 408–414.

24. H. Horie, N. Arai, Y. Mitsuishi, N. Komuro, H. Kaneda, H. Gotoh, M. Usami, and Y. Matsushima, "Greater than 500-mW CW kink-free single transverse-mode operation of weakly index guided buried-stripe type 980nm laser diodes," IEEE Photonic Technol. Lett., 12:10 (2000) 1304–1306.

25. H. Horie, H. Ohta, and T. Fujimori, "Reliability improvement of 980nm laser diodes with a new facet passivation process," IEEE J. Select. Topics Quantum Electron., 5 (1999) 832–838.

Chapter 4 | Tunable Laser Diodes

Gert Sarlet

Orkanvägen 35, 17771 Järfälla, Sweden

Jens Buus

*Gayton Photonics Ltd., 6 Baker St., Gayton, Nothants,
NN7 3EZ, UK*

Pierre-Jean Rigole

*ADC–Sweden, Bruttov. 7, SE-175 43 Järfälla-Stockholm,
Järfälla, Sweden*

This chapter gives an overview of tunable laser diodes that are at this writing (autumn 2001) considered suitable for applications in wavelength-division multiplexed (WDM) fiber optic communications. For a more comprehensive review of tunable laser diodes, the reader is referred to the book by Amann and Buus [1].

Most of these tunable lasers consist of a longitudinal integration of sections with different functionality. Typically, one has an active section providing the optical gain for the laser oscillator, one or more filter sections with a (tunable) frequency selective transmission or reflection characteristic, and a phase shifter section for fine-tuning of the cavity resonance frequencies.

Lasers in which the frequency selection and tuning functions are external to the semiconductor structure are discussed separately. Typically, the frequency selectivity in an external cavity tunable laser is provided by an external diffraction grating, which reflects only a small fraction of the optical spectrum back into the semiconductor optical amplifier (SOA) supplying the gain (cf. Chapter 8 in [1]). The center frequency of the light that is reflected back into the SOA can be tuned by rotating the diffraction grating.

In the first section of the chapter, the relevant physical mechanisms enabling electronic control of the emission frequency of a monolithic semiconductor laser are presented. Next, we run through the main requirements posed on tunable lasers for WDM applications. The remaining sections

WDM TECHNOLOGIES: ACTIVE
OPTICAL COMPONENTS
$35.00

present various types of tunable lasers, beginning with the basic integrated tunable laser, the distributed Bragg reflector (DBR) laser, followed by more advanced DBR-type lasers with wider tuning range, such as the sampled or super-structure grating DBR laser or the grating coupler with sampled reflector laser. Another section deals with mechanically tuned lasers, which include external-cavity lasers and vertical-cavity surface-emitting lasers. We also look at selectable sources and arrays, which provide an alternative to tunable lasers. Finally, we address the integration of additional functions such as modulation, amplification, and wavelength control, and we conclude the chapter with a comparison of the various technologies.

4.1. Electronic Frequency Control

Figure 4.1(a) shows a simplified equivalent circuit of a laser oscillator.

For laser operation at a frequency v two requirements need to be fulfilled simultaneously: the roundtrip cavity gain $G(v)$ should be unity and the roundtrip phase $\phi(v)$ should be an integer multiple of 2π.

$$G(v) = 1$$

$$\phi(v) = \frac{2\pi v}{c} 2 \sum_l n_l L_l = 2\pi k \tag{4.1}$$

Here c is the speed of light in vacuum. The summation over l represents the optical length of the laser cavity, i.e., the sum of the optical lengths of the different concatenated sections, where the optical length of a section l is defined as the product of the physical length L_l with the effective refractive index n_l.

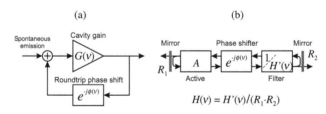

Fig. 4.1 Simplified representations of a (tunable) laser oscillator: equivalent circuit (a) and block diagram showing the concatenation of active (gain), phase shifter, and filter sections (b).

We assume for simplicity that the roundtrip cavity gain G can be written as the division of a frequency-independent gain factor A by a frequency dependent loss factor $H(v)$ (Fig. 4.1(b)).

$$G(v) = \frac{A}{H(v)} \qquad (4.2)$$

The phase condition defines a set of discrete frequencies v_k, the so-called cavity modes. The laser will oscillate at the frequency v_n among this set of frequencies, which requires minimal pumping of the laser to fulfil the gain condition $G(v_n) = 1$, i.e., the cavity mode frequency v_n for which the loss $H(v_n)$ is minimal. The emission frequency can be tuned either by changing the frequency of minimal cavity loss H or by shifting the set of frequencies v_k, which requires altering the roundtrip phase condition. Both mechanisms are illustrated in Fig. 4.2.

If the frequency of minimal loss is adjusted while maintaining a pre-set roundtrip phase, the lasing frequency at first remains fixed at v_n

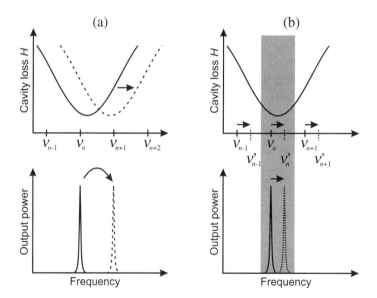

Fig. 4.2 Two basic mechanisms to tune the emission frequency of a laser: changing the frequency of minimal cavity loss (a) and/or changing the roundtrip phase (b). In (a) the lasing mode hops from one cavity mode frequency v_k to the next as the loss minimum is tuned. In (b) the cavity mode frequencies v_k change continuously. The lasing mode is the longitudinal mode with lowest roundtrip loss, i.e. the cavity mode that falls within the shaded region.

(Fig. 4.2(a)). The gain level A adapts automatically to compensate for the varying losses $H(\nu_n)$, such that the gain condition continues to be fulfilled. At a certain point, though, an adjacent cavity mode (mode $n + 1$) will experience a lower loss and the oscillation frequency will jump to that mode.

Alternatively, the optical length of the cavity can be adjusted, which shifts the position of the cavity modes in a continuous manner (Fig. 4.2(b)). Accordingly, the oscillation frequency will also change continuously, albeit within a limited range. Because of the gain condition, the tuning range is limited to a range centered on the frequency of minimal loss with a width equal to the mode spacing $\Delta\nu = \nu_n - \nu_{n-1}$ (indicated by the shaded area in Fig. 4.2(b)). When the emission frequency reaches the edge of this range, it jumps discontinuously to the cavity mode at the other edge.

In a tunable laser the three principal functions are providing optical gain, tunable frequency selective filtering, and changing the optical length of the cavity to achieve phase resonance at the appropriate frequency. This is illustrated by the block diagram in Fig. 4.1(b). Note that in practice the different functions cannot always be entirely separated. Tuning the loss minimum will, for example, often also affect the roundtrip phase. In the following paragraphs, we briefly describe the physical mechanisms that can be used to electronically tune the filter transfer function or the phase resonance frequencies.

In optical filters, two types of parameters govern the filter transfer function: the physical dimensions and the refractive indices of the different elements that make up the filter. The same applies to the roundtrip phase condition, as can be seen from Eq. (4.1). For monolithically integrated lasers no change of any mechanical parameter can be used for tuning. This only leaves the effective refractive index of the optical waveguide as a tuning parameter.

Figure 4.3 shows a cross-section of a typical InGaAsP/InP waveguide, as is used in various laser structures. For application in the tuning sections, the bandgap energy $E_g (= E_C - E_V)$ of the quaternary InGaAsP material is chosen sufficiently larger than the energy $h\nu$ of the photons generated by the laser, such that the material is transparent for these photons. The effective refractive index can be seen as a weighed average of the refractive indices of the different layers that build the waveguide. The weight of a certain layer's refractive index is related to the fraction of the optical power of the waveguide mode that is confined within this layer. Thus, the effective index is a function of the refractive indices of the different layers that build the

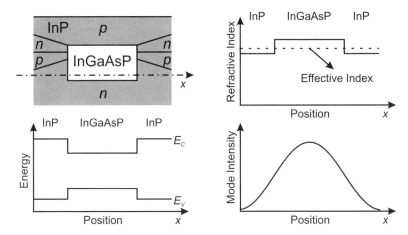

Fig. 4.3 Typical InGaAsP/InP waveguide structure with energy band diagram, refractive index profile, and intensity distribution of the waveguide mode. The core InGaAsP layer has both a higher refractive index, yielding optical confinement, as well as a lower band-gap energy ($E_g = E_C - E_V$), yielding confinement of injected carriers.

optical waveguide and their physical dimensions. Electronic variation of the refractive index of any of these layers, e.g. the InGaAsP core layer, yields a change of the effective refractive index of the waveguide. If this waveguide is part of an optical filter, the frequency of maximal transmission/reflection of the filter will vary accordingly. If, on the other hand, the waveguide is part of a phase shifter section, then the optical length of this section will change and the phase resonance frequencies will move.

In practice, three physical mechanisms can be used to change the refractive index of a semiconductor: injecting free carriers, applying an electrical field, and changing the temperature.

4.1.1. CARRIER-INDUCED INDEX CHANGE

In the InGaAsP/InP double heterostructure waveguide as depicted in Fig. 4.3, the undoped InGaAsP core layer is sandwiched between a p-doped and an n-doped InP cladding layer. If the resulting p-i-n diode is forward-biased, majority carriers flow from the doped cladding layers to the undoped core layer (holes from the p-doped layer, electrons from the n-doped layer). Because of the higher bandgap of the InP cladding layers, the injected carriers are confined to the InGaAsP layer and thus high hole densities P and electron densities N can be attained ($N = P$ because of charge neutrality).

The injected electron-hole plasma is the source of a number of effects influencing the refractive index [2–3].

Band-Filling

The injected electrons occupy the lowest energy states in the conduction band, just as the injected holes fill the states in the valence band closest to the band edge (which is the same as saying that electrons are removed from the upper energy states in the valence band). Consequently, on average higher photon energies $h\nu$ are required to excite electrons from occupied energy states in the valence band to empty energy states in the conduction band. This causes a reduction of the absorption coefficient $\alpha(\nu)$ (i.e., the material absorption per unit length) for photon energies $h\nu$ slightly above the nominal bandgap energy.

Bandgap Shrinkage

Electron–electron interactions at the densely populated states at the bottom of the conduction band reduce the energy of the conduction band edge E_C. A similar correlation effect for holes increases the energy of the valence band edge E_V. The sum of these effects causes a bandgap shrinkage, which lowers the minimum photon energy for which significant absorption occurs.

Free-Carrier Absorption (Intra-Band Absorption)

A free carrier can absorb a photon and move to a higher energy state within a band. The excess energy is released in the form of lattice vibrations as the carrier relaxes toward its equilibrium state.

Inter-Valence Band Absorption

For holes, another important absorption mechanism exists: inter-valence band absorption (IVBA). In this case, transitions occur between the heavy-hole valence band and the spin-orbit split-off valence band [4–5].

Through the combination of these effects, the absorption as a function of the photon frequency $\alpha(\nu)$ changes significantly with carrier density. The absorption coefficient is proportional to the imaginary part of the refractive index n''.

$$\alpha(\nu) = -2\frac{2\pi\nu}{c}n''(\nu) \tag{4.3}$$

Because changes in the imaginary part ($\Delta n''$) and the real part ($\Delta n'$) of the refractive index are linked through the Kramers-Kronig relations,

$$\Delta n'(v) = \frac{2}{\pi} P \int_0^\infty \frac{v' \Delta n''(v')}{v'^2 - v^2} dv' = -\frac{c}{2\pi^2} P \int_0^\infty \frac{\Delta \alpha(v')}{v'^2 - v^2} dv' \quad (4.4)$$

these effects also yield a change in refractive index, even at photon energies below the bandgap. It appears that for typical carrier densities ($\sim 10^{18}$ cm^{-3}), and for photon energies sufficiently (~ 0.1 eV) below the bandgap energy, the refractive index decreases linearly with the carrier density N [3].

$$n'(N) = n_0' + \frac{\partial n'}{\partial N} N \quad \text{with} \quad \frac{\partial n'}{\partial N} < 0 \quad (4.5)$$

Normal values for $\partial n'/\partial N$ at frequencies around 193 THz (wavelengths around 1.55 μm) are on the order of -10^{-20} cm^3. The effective refractive index n_{eff} of the waveguide obviously also decreases linearly with carrier density.

$$n_{eff}(N) = n_{eff,0} + \Gamma \frac{\partial n'}{\partial N} N \quad (4.6)$$

Here Γ is the optical confinement factor of the mode propagating along the waveguide, i.e., the ratio of the mode power in the core layer to the total mode power (see e.g. [1]). Through careful design of the core material composition, the core dimensions, and the cladding doping, effective index variations up to -0.04 can be achieved [3].

Unfortunately, at the same time the absorption losses in the core layer increase linearly with carrier density, mainly due to free-carrier and inter-valence band absorption (for the same photon energies of about 0.1 eV below the bandgap). The coupling between changes in the real and the imaginary part of the refractive index (at a particular photon frequency v) is commonly described by means of the linewidth-enhancement factor (also called chirp parameter, or alpha factor).

$$\alpha_H = -\frac{\partial n'/\partial N}{\partial n''/\partial N} = 2 \frac{2\pi v}{c} \frac{\Delta n'}{\Delta \alpha} \quad (4.7)$$

A typical value for InGaAsP material with a bandgap wavelength of 1.3 μm in the 193 THz frequency range is -20. Because the loss and (effective) index changes of the waveguide both scale with the confinement factor, the alpha factor of the waveguide equals the alpha factor of the core material (except for certain extreme cases).

The carrier density N is determined by the current I through the forward-biased hetero-junction, as described by the carrier density rate equation.

$$\frac{dN}{dt} = \frac{I}{qLwd} - (AN + BN^2 + CN^3) \tag{4.8}$$

Here q is the electron charge and L, w, and d are the length, width, and thickness of the core respectively. The terms between brackets describe the nonradiative, radiative, and Auger recombination processes respectively. Because the injected electron-hole pairs recombine, a sustained current must be applied to the tuning section in order to maintain a certain carrier density. The injection-recombination process has a time constant in the nanosecond range, which limits the tuning speed.

Another disadvantage is the parasitic heating of the waveguide due to the nonzero series resistance of the tuning diode (Joule heating) and the non-radiative recombination processes. The resultant thermal tuning partially counteracts the carrier-induced tuning (see Section 4.1.3). If very accurate tuning is required, the thermal effects will reduce the tuning speed. Indeed, because of the large time-constants of thermal processes (ranging from microseconds to milliseconds), it will take a long time before the refractive index has completely stabilized.

4.1.2. ELECTRIC-FIELD-INDUCED INDEX CHANGE

In bulk III/V-semiconductors, the electric-field dependence of the absorption and the refractive index is rather weak. Two effects change the refractive index when an electric field is applied: the linear electrooptic (or Pockels) effect and the quadratic electrooptic (or Franz-Keldysh) effect [6–7]. Typical strengths for both effects in InGaAsP material, at photon energies below the bandgap, are [7]

$$\begin{aligned}
\Delta n_{\text{Pockels}} &\approx -3 \cdot 10^{-11} \text{m/V} \cdot \xi \\
\Delta n_{FK} &\approx +1 \cdot 10^{-18} \text{m}^2/\text{V}^2 \cdot \xi^2
\end{aligned} \tag{4.9}$$

Hence, these effects counteract each other and even with a strong applied field ($\xi \approx 10^7$ V/m) the refractive index change is only on the order of 10^{-4}.

However, in multiquantum well (MQW) structures these changes are greatly enhanced by the quantum confined Stark (QCSE) effect [8–9].

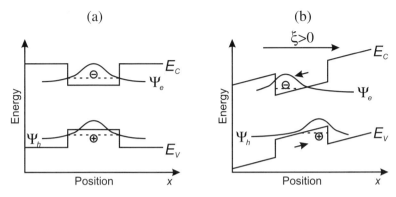

Fig. 4.4 The quantum confined Stark effect. By applying an electric field ξ perpendicular to a quantum well structure (a), the band edges are tilted (b). Consequently, the electron and hole wavefunctions are shifted with respect to each other and their energy difference (i.e. the effective bandgap energy) is reduced [1, 8].

In this case, the core of the tuning waveguide consists of a number of thin "well" layers (a few nanometers thick) with low bandgap, separated by thicker "barrier" layers with higher bandgap (Fig. 4.4). In the narrow potential wells quantization effects occur: the conduction and valence band are split up into a number of sub-bands. The effective bandgap energy, i.e., the energy difference between the first order sub-bands in conduction and valence band, is larger than the bandgap of the corresponding bulk material.

If an electric field is applied perpendicular to the quantum wells, the band edges are tilted. Consequently, the electron and hole wavefunctions are shifted with respect to each other and their energy difference (i.e., the effective bandgap energy) is reduced. This also shifts the absorption edge to lower frequencies and thus modifies the refractive index, even at frequencies below the absorption edge, as can be seen from the Kramers-Kronig relation (Eq. 4.4). Of course, significant changes of the refractive index only occur for photon energies near the bandgap energy, where significant absorption also occurs. Hence, a careful adjustment of the laser frequency and quantum well structure is required. The refractive index changes are normally on the order of 10^{-3} to 10^{-2}, depending on how close the laser wavelength is to the bandgap wavelength of the quantum well. The chirp parameter α_H is usually only about 10 [10]. Using more complicated quantum well structures, consisting of two asymmetric coupled quantum wells, somewhat larger refractive index changes are achievable with a higher alpha factor and therefore lower losses [11–12].

What matters for applications in tunable lasers is the change in effective refractive index of the waveguide (Eq. 4.6). Because quantum wells are very thin, the optical confinement Γ in a single quantum well is low (only a few percent). By stacking a number of quantum wells separated by barrier layers into a multiquantum well (MQW) structure, the confinement factor is roughly multiplied by the number of wells. Nevertheless, even then the confinement is still much lower than in the bulk waveguide core that is used for the carrier-induced tuning (0.1–0.2 versus 0.4–0.7), because a significant part of the mode field is located within the barrier layers. Thus, the maximum effective index change that can be achieved with electrooptic effects is only about $2 \cdot 10^{-3}$.

On the other hand, electric-field-induced tuning also has some advantages. Just as the bulk waveguide core used for carrier-induced tuning, the (undoped) MQW structure is placed between a p-doped and an n-doped cladding layer. In order to apply a strong electric field to the MQW, the p-i-n diode is now reverse-biased, which means that almost no current flows and no extra heat is generated. Furthermore, no carrier concentrations have to be built up, so the tuning can be almost instantaneous. The tuning speed is only limited by parasitic capacitances and inductances and the rise time can be on the order of a few tens of picoseconds.

4.1.3.   THERMALLY-INDUCED INDEX CHANGE

The refractive index of III/V-semiconductors also exhibits considerable temperature dependence. A well-known rule of thumb is that the emission wavelength of a single-mode InGaAsP/InP laser emitting in the 1550 nm (193 THz) region increases with temperature at a rate of approximately 0.1 nm/K [13]. Accordingly, the temperature coefficient of the refractive index $\partial n'/\partial T$ is about $2 \cdot 10^{-4}\,\text{K}^{-1}$. Note that with thermal tuning the confinement factor is always 1, because both core and cladding of the waveguide are heated. Heating the entire laser, however, has the disadvantage that the threshold current increases and the differential efficiency (change in output power per unit change in drive current) decreases. Moreover, driving a laser at high temperatures for longer periods reduces the lifetime of the device. Hence, the temperature variation usually has to be limited to a few tens of degrees.

If the heating is only applied to the tuning section(s) of the laser, and the thermal isolation is sufficient to avoid excessive heating of the active section, higher temperatures can be accepted. Usually resistive heating is used. In practice this is done by reverse-biasing the tuning diode as described in Section 4.1.1 [14], by integrating a resistor in the top InP cladding layer [15], or by placing thin-film resistive heaters on top of the waveguide [16]. A major advantage of this method is that the heating only has a limited influence on the absorption losses in the tuning waveguide, yielding a high alpha parameter. Probably the largest drawback of thermal tuning is the slow response speed, which can range from microseconds to milliseconds.

Some heat is also generated when carrier-induced tuning is used, due to the nonzero resistance of the tuning diode and the nonradiative recombination processes (see Section 4.1.1). It should be noted that the carrier effects decrease the refractive index, whereas the thermal effect increases the refractive index. This ultimately limits the tuning range achievable using carrier injection. At low tuning currents, the carrier-induced refractive index change is dominant and the refractive index decreases. At some point, though, the thermal tuning efficiency becomes larger than the carrier-induced tuning efficiency, since carrier density increases sub-linearly with current (see Eq. (4.8)), whereas the temperature increases super-linearly with current. If the current is raised beyond that point, the refractive index starts to increase again.

4.1.4. COMPARISON OF TUNING MECHANISMS

The tuning mechanisms are compared in Table 4.1, which summarizes the typical parameter values mentioned in the preceding paragraphs. The electric-field-induced tuning has the advantages of low power consumption

Table 4.1   Comparison of the Physical Mechanisms for Electronic Refractive Index Variation, Quoting Typical Parameter Values [1]

Parameter	Carriers	Electric Field	Temperature
$\Delta n'$	−0.05	−0.01	0.01
Γ	0.5	0.2	1
Δn_{eff}	−0.025	−0.002	0.01
α_H	−20	−10	Large
3-dB bandwidth	100 MHz	>10 GHz	<1 MHz
Power consumption	Large	Negligible	Very large

and very high tuning speed, but on the other hand, only small effective index changes are achievable, with considerable absorption losses. Thermal tuning yields larger index changes and is the easiest to implement, but it requires a very high electrical input power and has a low response speed. At present, the preferred mechanism seems to be carrier-induced tuning, which has so far yielded the largest tuning ranges, at reasonable tuning speeds (if the parasitic thermal effects can be neglected), yet at the cost of considerable power consumption.

4.2. Characteristics of Tunable Lasers

4.2.1. TUNING RANGE—TUNING ACCURACY

The frequency tuning range is naturally the first property by which a tunable laser is evaluated. When tuning ranges of different lasers are compared, care has to be taken though that comparisons are made on the same basis. Normally three different types of tuning are distinguished: continuous, discontinuous, and quasi-continuous tuning. Figure 4.5 illustrates the basic frequency versus control current (or voltage) characteristics for these tuning schemes.

Continuous Tuning

Continuous tuning is the ideal scheme from a practical point of view (Fig. 4.5(a)). The laser frequency can be tuned smoothly, in arbitrarily small steps, by adjusting a single control parameter (or multiple control parameters, provided there is a 1-to-1 relation between any two of these parameters). Continuous tuning over a small range can, for example, be achieved by

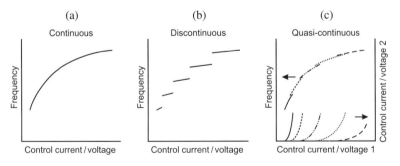

Fig. 4.5   Emission frequency versus control current(s) or voltage(s) for continuous (a), discontinuous (b) or quasi-continuous tuning (c).

merely adjusting the optical length of the cavity, without changing the frequency of minimal roundtrip loss (Fig. 4.2(b)). In that case, the continuous tuning range is limited to somewhat less than the cavity mode spacing Δv.

If the cavity modes and the frequency of minimal loss are tuned simultaneously, larger continuous tuning ranges are possible. In most tunable lasers, however, this requires synchronized adjustment of at least two control parameters. Owing to the stringent requirement that the same cavity mode has to remain the lasing mode across the entire tuning range, the tuning range is smallest in the continuous tuning scheme. The present record value is about 13 nm or 1.6 THz for lasers emitting in the 193 THz frequency region [17]. On the other hand, for the longitudinally integrated tunable lasers that will be described further on, the continuous tuning range is usually limited to a few 100 GHz [18].

Discontinuous Tuning

Larger tuning ranges can be achieved if sudden, discontinuous frequency changes are allowed for. An example was already given in Fig. 4.2(a). When the frequency of minimal roundtrip loss is tuned without adjusting the roundtrip phase, the emission frequency initially remains constant. Only when a cavity mode adjacent to the lasing mode experiences a lower loss does the laser frequency jump to this adjacent mode. In practice the tuning of the loss minimum is always accompanied by some tuning of the cavity modes, as is illustrated in Fig. 4.5(b).

Here the tuning range is not limited by the tunability of a single longitudinal mode, but rather by the tunability of the roundtrip loss minimum. An upper limit is naturally also imposed by the bandwidth of the optical gain in the active section of the laser. Nonetheless, if the active section consists of a multiquantum well (MQW) structure, this bandwidth can be more than 100 nm (12.5 THz around 193 THz). Discontinuous tuning ranges of more than 100 nm have indeed already been demonstrated [19–20].

Quasi-Continuous Tuning

Quasi-continuous tuning is achieved by joining overlapping continuous tuning ranges in order to get full frequency coverage over a wider range. Quasi-continuous tuning thus requires the adjustment of at least two control parameters. The principle is illustrated in Fig. 4.5(c) for two controls. By setting an appropriate combination of the two parameter values, the laser can be tuned to any frequency within a wide range. Still, there is no 1-to-1 relation between the two controls across the entire range, so there is no possibility to tune smoothly from one frequency to any other frequency. Continuous tuning is only achievable over the ranges corresponding to individual control 2 versus control 1 curves. Referring to Fig. 4.2, quasi-continuous tuning is for example accomplished by tuning a cavity mode and the loss minimum synchronous over a range equal to the longitudinal mode spacing, then resetting the cavity modes to their original locations, and subsequently tuning the next cavity mode simultaneously with the loss minimum, etc.

For dense wavelength-division multiplexing (DWDM) applications, a typical requirement for the tuning range is complete frequency coverage across the entire C-band (i.e., roughly from 192 to 196 THz) or L-band (i.e., from 187 to 191 THz). With monolithically integrated tunable laser diodes, this has so far only been achieved in the quasi-continuous tuning regime (see e.g. [21–22]). With mechanically tuned lasers, described in Section 4.5, truly continuous tuning is possible over such a wide range. The number of frequency channels that can effectively be used in these bands is mainly limited by the accuracy with which the laser can be tuned to a particular channel. The present version of ITU-T Recommendation G.692, "Optical interfaces for multi-channel systems with optical amplifiers," which contains specifications for WDM systems, proposes a channel grid with 50 or 100 GHz channel spacing, anchored at 193.1 THz [23]. At 50 GHz spacing, about 80 channels would be available in both C- and L-band. For these multi-channel systems, a frequency accuracy of $\pm 10\%$ of the channel separation is commonly required. Therefore, if one wants to reduce the channel spacing by a factor N, the frequency accuracy has to be improved by the same amount.

In the case of quasi-continuous tuning, the control of a tunable laser can be quite complicated, because two or more parameters have to be adjusted simultaneously to change the laser frequency. Therefore, the laser is usually built into a module containing a microprocessor and drive electronics that allow easy, command-based control of the laser frequency and output power. The set-points for the different frequency channels are stored in a look-up table in an EPROM (erasable, programmable read-only memory). When the laser has to be tuned to a particular channel, the microprocessor controller interprets the incoming command, reads the appropriate values from the look-up table, and adjusts the control currents/voltages accordingly. The initial frequency error is hence mainly limited by the accuracy of the procedure that was used to generate this look-up table (assuming the current/voltage sources have ample resolution and are sufficiently accurate). Some form of feedback control can also be applied to improve the accuracy and stability of the emission frequency.

4.2.2. OTHER CHARACTERISTICS

Side Mode Suppression Ratio

Previously, we implicitly assumed that the laser was always emitting in a single longitudinal mode. Still, this is only true if all cavity modes other than the lasing mode experience a roundtrip loss that is sufficiently higher than the roundtrip loss of the lasing mode. For applications in optical communication systems, this is an essential requirement because fiber dispersion is proportional to the spectral width of the carrier wave, which means that multi-mode operation would seriously limit the achievable transmission distance. The spectral purity of a laser is quantified by the side mode suppression ratio (SMSR), which is defined as the ratio of the power in the dominant mode to the power in the strongest side mode. The SMSR is usually expressed in decibels (Fig. 4.6). For telecom applications, a SMSR of at least 30 dB, preferably even 40 dB, is required.

It was already mentioned that semiconductor lasers have a very wide gain bandwidth (several THz). Because typical mode spacings are below 100 GHz (corresponding to a cavity length of 400 μm or more), this means that without any filtering the laser may oscillate in more than one longitudinal mode simultaneously. How this filtering can be implemented will be illustrated in the following sections. Apart from designing a sufficiently

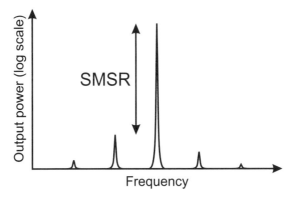

Fig. 4.6 Definition of the side mode suppression ratio (SMSR): ratio of the power in the main (lasing) mode to the power in the strongest side mode (usually expressed in dB).

narrow intra-cavity filter, care also has to be taken that a cavity mode is more or less aligned with the loss minimum of the filter (see Fig. 4.2). Even with a filter bandwidth comparable to the mode spacing, one can still get two-mode operation, namely in the case when the loss minimum lies halfway between two cavity modes.

Output Power

For communication applications, a fiber-coupled power of a few mW is needed. If tunable lasers are used as spares, or replacements, for fixed wavelength single frequency lasers, they will have to satisfy the same power requirements as these lasers. In addition, all of the tuning mechanisms described in Section 4.1 not only change the refractive index, but also to some extent the losses, so one has to consider the variation of output power across the tuning range. These variations have to be kept as low as possible, either by cleverly designing the laser or by adjusting the active section drive current (e.g., by using some form of feedback control).

Linewidth

For many applications, the spectral linewidth is an important characteristic of single-mode laser diodes, e.g., for coherent communication systems using optical heterodyne detection. For coherent systems, a linewidth of no more than a few MHz is required. However, for systems using direct detection, the linewidth can be at least an order of magnitude larger. Of course, a prerequisite for narrow linewidths is that the current and/or voltage sources that drive the laser exhibit low noise levels.

In lasers that use the quasi-continuous or discontinuous tuning schemes, the linewidth can vary significantly. The linewidth is relatively low as long as the lasing mode and the loss minimum are more or less aligned, but singularities in the linewidth arise at the mode boundaries, where frequency jumps occur [24].

With respect to linewidth, the three tuning mechanisms behave quite differently. Thermal and field-induced tuning have negligible effects on the linewidth, provided the lasing mode coincides with the loss minimum. However, when carrier-induced tuning is used, considerable linewidth broadening is observed [25], more than can be expected from the classic Schawlow-Townes-Henry theory [26] (even when taking into account the broadening due to increased losses). This excess broadening is attributed to injection-recombination shot noise in the tuning section(s) [27]. The shot noise of the carrier injection and recombination processes causes carrier density fluctuations, which lead to refractive index and loss variations. These in turn produce fluctuations of the instantaneous laser frequency that finally lead to a broadened spectral line. This broadening can be avoided to a large degree by using a voltage source (low internal resistance) instead of a current source (high internal resistance) to bias the tuning section(s) [28].

4.3. Distributed Bragg Reflector Laser

After the distributed feedback (DFB) laser, the distributed Bragg reflector (DBR) laser is the most common design for a single-mode laser diode [29]. The basic DBR laser consists of two longitudinally integrated sections: an active section and a reflector section (Fig. 4.7). The waveguide core of the active section has a bandgap matching the desired emission frequency and hence provides optical gain if sufficient carriers are injected. The core material of the reflector in contrast has a higher bandgap, such that the material is transparent (passive) for the laser light. Along the reflector section, a diffraction grating is embedded in the waveguide, yielding a periodic modulation of the effective refractive index of the waveguide.

This grating can, for example, be obtained by periodically varying the thickness of the waveguide core. The grating pattern is commonly defined by a holographic process, in which a photo-resist layer is exposed by two interfering beams of ultraviolet light. The angle of the beams is chosen such that an interference pattern with period Λ is generated. Subsequently, the resist is developed and the grating pattern is etched into the semiconductor material. Finally, the etched grating structure is overgrown with the top cladding layer, which has a higher bandgap and a lower refractive index than the core layer. Because the effective index of the waveguide increases with the thickness of the high-index core layer, it exhibits the same periodic variation as the thickness of the core layer.

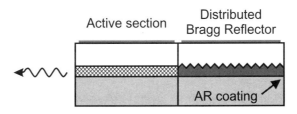

Fig. 4.7 Longitudinal cross-section of a 2-section distributed Bragg reflector (DBR) laser.

Because of the grating, the passive section reflects light back in a narrow frequency band. This can be understood intuitively as follows. Every tooth of the grating reflects a small amount of light. Reflections from consecutive teeth of the grating have a phase difference that depends on the ratio of the wavelength in the material λ/n_d to the grating period Λ. Here $\lambda = c/v$ is the vacuum wavelength, c is the speed of light in vacuum, and n_d is the average effective index. If the wavelength in the material equals twice the grating period, successive reflections interfere constructively and a strong overall reflection is obtained. This condition defines the Bragg frequency v_B:

$$v_B = \frac{c}{2n_d(v_B)\Lambda} \tag{4.10}$$

If there is a mismatch between the frequency of the incident light and the Bragg frequency, the reflection is much lower. To avoid interference between reflections from the grating and reflections from the end facet, an anti-reflection (AR) coating is usually applied to the facet.

A more quantitative analysis of the reflectivity of a Bragg grating is usually performed using the coupled-mode theory [30]. The modulated effective refractive index is written as

$$n'(z) = n_d + \mathrm{Re}\left[\Delta n_1 e^{j(2\beta_0 z + \phi)}\right] + \cdots \tag{4.11}$$

where $\beta_0 = \pi/\Lambda$, and Δn_1 is first-order Fourier component of the refractive index modulation, which is assumed to be much smaller than n_d. The model is based on the scalar wave equation for the electric field

$$\frac{\partial^2 E}{\partial z^2} + [(n' + jn'')k_0]^2 E = 0 \tag{4.12}$$

where E is the complex amplitude of a field with frequency v, which is assumed to be independent of the x and y coordinates, and $k_0 = 2\pi v/c$ is the free-space propagation constant. Assuming that $n'' \ll n'$ and $\Delta n_1 \ll n_d$, we have

$$[(n' + jn'')k_0]^2 \approx \beta^2 + 2j\beta\zeta + 4\beta\,\mathrm{Re}\left[\kappa e^{j(2\beta_0 z + \phi)}\right], \tag{4.13}$$

with $\beta = n_d k_0$ the mode propagation constant, $\zeta = n''k_0 = -\alpha/2$ the mode *field* gain coefficient, and κ the so-called coupling coefficient (usually expressed in cm^{-1}):

$$\kappa = \frac{\pi v \Delta n_1}{c} \tag{4.14}$$

The coupling coefficient is a measure for the strength of the backward scattering by the grating structure. In principle, the periodic index modulation generates an infinite set of diffraction orders, but in the vicinity of the Bragg frequency, i.e., $\Delta\beta = \beta - \beta_0 \ll \beta_0$, only two modes are more or less in phase synchronism. These are the two counter-propagating waves, which are coupled due to the Bragg scattering. We can therefore expand the electric field in the forward and backward propagating modes.

$$E(z) = R(z)e^{-j\beta_0 z} + S(z)e^{j\beta_0 z} \qquad (4.15)$$

where the functions $R(z)$ and $S(z)$ vary slowly as a function of z, so that their second derivatives in Eq. (4.12) can be neglected. If we insert Eq. (4.15) into the wave equation Eq. (4.12), take into account all of the above assumptions, and collect terms with identical phase factors ($\exp(-j\beta_0 z)$ and $\exp(j\beta_0 z)$, respectively), we obtain the coupled-mode equations

$$
\begin{aligned}
-\frac{dR}{dz} + (\zeta - j\Delta\beta)R &= j\kappa^* e^{-j\phi} S \\
\frac{dS}{dz} + (\zeta - j\Delta\beta)S &= j\kappa e^{j\phi} R
\end{aligned}
\qquad (4.16)
$$

By solving these coupled first-order differential equations, it can be shown that the field reflectivity of a distributed Bragg reflector of length L is given by

$$r(\nu) = \frac{-j\kappa^* e^{-j\phi} \sinh(\gamma L)}{\gamma \cosh(\gamma L) - (\zeta - j\Delta\beta) \sinh(\gamma L)} \qquad (4.17)$$

where

$$\gamma^2 = |\kappa|^2 + (\zeta - j\Delta\beta)^2 \qquad (4.18)$$

Figure 4.8 shows a characteristic power reflectivity spectrum $R(\nu) = |r(\nu)|^2$ of a distributed Bragg reflector with $\kappa L = 1.5$.

In 1977, Okuda and Onaka proposed the first integrated tunable laser diode, which was essentially a tunable 2-section DBR laser [31]. Indeed, if the effective refractive index of the DBR section can be changed electronically, the frequency of minimal roundtrip loss (i.e., the Bragg frequency ν_B) can be shifted, thus enabling discontinuous tuning of the laser frequency [32] (see Fig. 4.2(a)). In order to make quasi-continuous tuning possible, an additional phase shifter section is needed [33] (Fig. 4.9). This section has the same structure as the DBR, except for the grating. The roundtrip

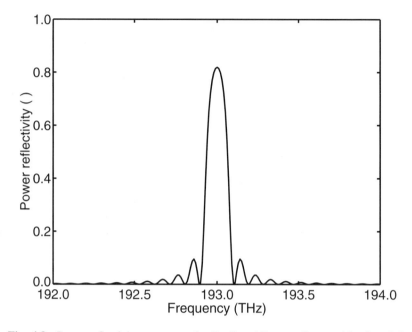

Fig. 4.8 Power reflectivity spectrum of a distributed Bragg reflector with $\kappa L = 1.5$.

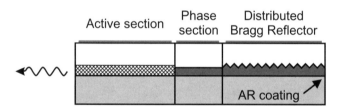

Fig. 4.9 Longitudinal cross-section of a 3-section tunable distributed Bragg reflector (DBR) laser.

phase condition in this case reads (see Eq. (4.1)):

$$\frac{2\pi \nu_k}{c}(2n_a L_a + 2n_p L_p) + \phi_d(\nu_k) = 2\pi k \qquad (4.19)$$

Here n_x and L_x are the effective index and length of the active and phase section; $\phi_d = -\arg(r)$ is the phase of the reflection from the grating.

If the laser is biased above threshold, the carrier density in the active section is clamped. Changing the active section current therefore has no effect on the index of the active section (if thermal effects are neglected). The index of the phase section, on the other hand, can be adjusted electronically,

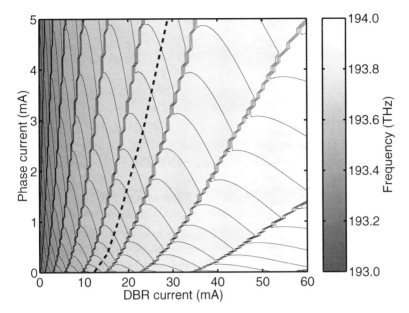

Fig. 4.10 Tuning characteristic of a 3-section DBR laser. Contour plot of frequency as a function of DBR and phase current (20 GHz increments). The dashed line indicates a possible trajectory for continuous tuning.

e.g., through current injection. Using this a cavity mode can be tuned to the Bragg frequency.

Figure 4.10 displays a tuning characteristic of such a 3-section DBR laser. The cavity mode hops are easily discernible by the fact that at the hops multiple frequency contours coincide. Continuous tuning is possible along curves that lie approximately halfway between two mode hop contours, as the one indicated by the dashed line. For these points, the SMSR is normally higher than 30 dB. This particular laser has a quasi-continuous tuning range of approximately 1 THz (8 nm), which is a typical value for DBR lasers. The tuning range is limited by the maximum index change Δn_d that can be reached in the DBR section (see Section 4.1.2).

$$\frac{\Delta \nu_{B,\max}}{\nu_B} = -\frac{\Delta n_{d,\max}}{n_{d,g}(\nu_B)} \tag{4.20}$$

where

$$n_{d,g}(\nu) = n_d(\nu) + \nu \frac{\partial n_d}{\partial \nu} \tag{4.21}$$

represents the group effective index, which includes the dispersion of $n_d(\nu)$ around ν_B. Through careful optimization of the waveguide structure [3], the tuning range for carrier-induced tuning can be increased to about 2 THz [34–35].

Note again that absorption losses in the DBR and phase sections increase with the applied current. If the active section current is kept constant during tuning, the output power from the front facet generally varies by 1 to 2 dB.

4.4.   Increasing the Tuning Range of DBR-Type Lasers

The tuning ranges of conventional DBR lasers, 1 to 2 THz, are significantly smaller than the available gain bandwidth of MQW semiconductor material (more than 10 THz) and Erbium-doped fiber amplifiers (about 4 THz in the C- or L-band). Consequently, a lot of research effort has been devoted to the development of integrated lasers with extended tuning ranges, beyond the $\Delta\nu/\nu = -\Delta n/n_g$ limit. The basic principle behind all schemes that have been developed for wide tuning is that a refractive index *difference* is changed rather than the index itself. Therefore, the relative frequency change is equal to a relative change in index difference, which can be significantly larger for similar absolute refractive index variations. In the following paragraphs, we will describe the two most common schemes: one using a Vernier effect between two comb reflectors and the other using the broad tunability of a grating assisted co-directional coupler.

4.4.1.   USING A VERNIER EFFECT BETWEEN TWO COMB REFLECTORS

The Vernier caliper, invented by the French scientist Pierre Vernier (1580–1637), is a well-known tool for high-resolution length measurements. The principle is illustrated in Fig. 4.11. The caliper consists of two graduated scales, a main scale like a ruler and a second scale, the Vernier, which slides parallel to the main scale. The two scales have a small relative pitch difference, e.g. $1/20$, such that a shift of the slider by an amount δx leads to a shift of the point where tick marks on both scales coincide by an amount $\Delta x = 20 \cdot \delta x$. In other words, any change in position is enhanced by a factor equal to the inverse of the relative pitch difference.

The same principle can be applied to a tunable laser, if the laser has two mirrors with comb-shaped reflectivity spectra [36] (Fig. 4.12). The mirrors are designed such that the peak spacing of the front mirror (δ_f) and the rear mirror (δ_r) differ by a small amount $\Delta\delta$. Lasing can then only occur in the

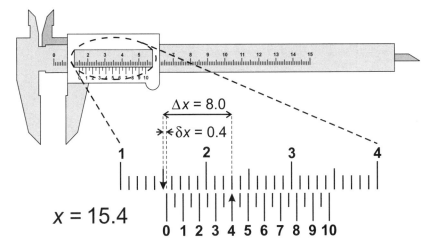

Fig. 4.11 A Vernier caliper, using two scales with a pitch difference of 1/20. A shift of the lower scale by an amount δx leads to a shift of the point where tick marks on both scales coincide by an amount $\Delta x = 20 \cdot \delta x$.

frequency range where two peaks coincide, since the cavity roundtrip loss is inversely proportional to the product of both mirror reflectivities. The phase section can again be used to adjust the longitudinal modes, such that a mode can be aligned with the loss minimum.

If one of the mirrors is tuned by $\pm\Delta\delta$, two adjacent peaks coincide and a large change in frequency is obtained. If the phase section is simultaneously adjusted such that a cavity resonance is aligned with the coincident peaks, the frequency changes by an amount $\delta_f(-\delta_r)$ if the rear (front) mirror is tuned by $\Delta\delta$. Thus, the tuning enhancement is either $F+1$ or $-F$, where F is defined by:

$$\Delta\delta = \frac{\delta_r}{F} = \frac{\delta_f}{F+1} \qquad (4.22)$$

The coincidence of two particular peaks is often called a "super-mode" and the large frequency changes observed when applying the Vernier tuning mechanism are consequently called "super-mode" jumps.

A few simple design criteria for the mirrors can easily be derived:

- The reflection peaks should be sufficiently narrow (relative to the mode spacing $\Delta\nu$) to suppress all cavity modes but one.
- The difference in peak spacing $\Delta\delta$ should be comparable with the width of the reflection peaks. If $\Delta\delta$ is too large, there is a region

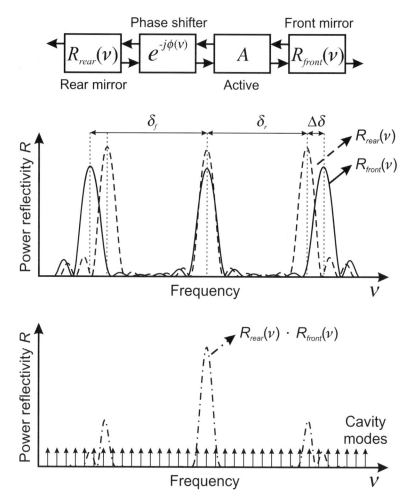

Fig. 4.12 Vernier principle applied to a tunable laser, which has two mirrors with comb-shaped reflectivity spectra. Lasing can only occur in the frequency range where reflection peaks coincide.

where the laser frequency is unpredictable during the tuning of a single reflector over an amount $\Delta\delta$. If $\Delta\delta$ is too small, the overlap of adjacent reflector peaks is too large and cavity modes at these frequencies are insufficiently suppressed.

• Care has to be taken that only one pair of peaks coincides at the same time within the gain bandwidth of the laser (especially if F is an integer). This can be done either by designing the mirrors such that strong reflection peaks only occur within a sufficiently limited bandwidth, or by making sure that all other possible coincidences fall outside of the gain bandwidth. Ideal reflectors have a limited number of uniform reflection peaks.

Intermediate tuning, from one longitudinal mode to the adjacent one, is obtained by tuning both reflectors simultaneously. True continuous tuning (fine tuning), requires synchronized adjustment of the two reflectors and the phase section. For full frequency coverage over a wide range (i.e., quasi-continuous tuning), a number of requirements have to be met. The front and rear reflectors must be tunable over at least δ_f and δ_r, respectively. At the same time, the phase section should allow tuning of the cavity modes by more than the mode spacing $\Delta\nu$.

Practical implementations of this Vernier tuning scheme look much like the 3-section DBR laser in Fig. 4.9, but with distributed Bragg reflectors on both sides. The gratings in the DBR sections are modified to obtain multiple reflection peaks. Two examples are discussed following.

4.4.2. SAMPLED GRATING DBR

The sampled grating (SG) is technologically the simplest way to obtain a reflectivity spectrum that has periodic maxima [37]. The sampled grating is nothing more than a conventional uniform grating with an appropriate grating pitch Λ, multiplied by a sampling function with period Λ_s, as shown in Fig. 4.13. It can be fabricated essentially in the same way as an ordinary Bragg grating. The grating pattern is again defined by exposing a photo-resist layer with the interference pattern of two ultraviolet beams. By periodically masking off part of the photo-resist layer (period Λ_s), the resist is only exposed in the regions with width Λ_g. In the following etch step, the sampled interference pattern is then transferred into the semiconductor.

A qualitative idea of the shape of the reflectivity spectrum can easily be derived from the coupled-mode theory, which says that every spatial Fourier component of the refractive index modulation contributes a peak to the reflection spectrum (see Section 4.3). The Fourier components of the sampled grating are of course obtained by convolution of the Fourier transforms of the uniform grating and of the sampling function. The uniform grating has a single Fourier component with a coupling strength κ_u given by Eq. (4.14), at a spatial frequency $1/\Lambda$, which corresponds to the Bragg frequency ν_B according to Eq. (4.10).

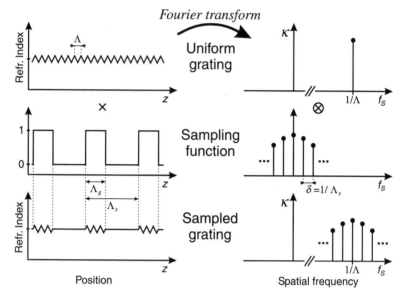

Fig. 4.13 Principle of the sampled grating. In real space, the uniform grating is multiplied by a sampling function. In Fourier space, this corresponds to a convolution of the respective spectra. According to the coupled-mode theory, every spatial Fourier component of the refractive index modulation contributes a peak to the reflection spectrum [37].

The Fourier transform of the sampling function, on the other hand, consists of a comb of peaks with a spatial frequency spacing of $1/\Lambda_s$.

The modulation function is given by (l is an integer)

$$F(z) = \begin{cases} 1 & l\Lambda_s < z < l\Lambda_s + \Lambda_g \\ 0 & l\Lambda_s + \Lambda_g < z < (l+1)\Lambda_s \end{cases} \tag{4.23}$$

from which the amplitudes of the Fourier components are easily obtained as

$$F_k = \frac{1}{\Lambda_s} \int_0^{\Lambda_s} F(z) e^{-j\frac{2\pi k}{\Lambda_s}z} dz = \frac{\Lambda_g}{\Lambda_s} \frac{\sin(\pi k \Lambda_g / \Lambda_s)}{\pi k \Lambda_g / \Lambda_s} e^{-j\pi k \Lambda_g / \Lambda_s} \tag{4.24}$$

The convolution of these Fourier transforms exhibits peaks centered at $1/\Lambda$, with spacing $1/\Lambda_s$. This leads to strong reflections at frequencies ν_k (see Eq. (4.10)):

$$\nu_k = \frac{c}{2n(\nu_k)} \left(\frac{1}{\Lambda} + \frac{k}{\Lambda_s} \right) \tag{4.25}$$

The reflection peak spacing is determined by the sampling period Λ_s.

$$\delta = \nu_{k+1} - \nu_k = \frac{c}{2n_g \Lambda_s} \tag{4.26}$$

where n_g is the group refractive index.

$$n_g(\nu) = n(\nu) + \nu \frac{\partial n}{\partial \nu} \tag{4.27}$$

The coupling coefficients at the frequencies ν_k are equal to the product of the coupling coefficient of the unsampled grating κ_u with the Fourier components F_k of the sampling function

$$\kappa_k = \kappa_u F_k \tag{4.28}$$

If we assume that only one diffracted order is phase matched at any frequency, then the overall reflectivity can be written as the sum of the contributions of the separate Fourier components. In other words, the overall reflectivity is the sum of the reflectivities of individual gratings with Bragg frequencies ν_k and coupling coefficients κ_k (see Eq. (4.17)):

$$r(\nu) = \sum_k \frac{-j\kappa_k^* \sinh(\gamma_k L)}{\gamma_k \cosh(\gamma_k L) - (\zeta - j\Delta\beta_k)\sinh(\gamma_k L)} \tag{4.29}$$

with

$$\gamma_k^2 = |\kappa_k|^2 + (\zeta - j\Delta\beta_k)^2$$
$$\Delta\beta_k = \frac{2\pi\nu}{c}n(\nu) - \pi\left(\frac{1}{\Lambda} + \frac{k}{\Lambda_s}\right) \tag{4.30}$$

Here L is the length of the sampled grating. Fig. 4.14 shows the result of a more detailed calculation for a sampled grating DBR consisting of 8 periods with length $\Lambda_s = 65\,\mu m$, which leads to a peak spacing $\delta = 0.6$ THz ($n_g = 3.85$).

If waveguide losses are neglected ($\zeta = 0$), then the peak power reflectivity at the frequency ν_k is simply given by

$$R(\nu_k) = |r(\nu_k)|^2 = \tanh^2(|\kappa_k|L) = \tanh^2(|F_k|\kappa_u L) \tag{4.31}$$

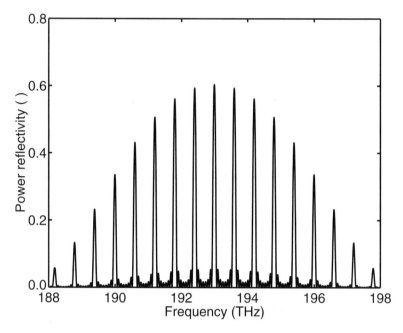

Fig. 4.14 Reflection spectrum of a sampled grating DBR, consisting of 8 periods with length $\Lambda_s = 65\,\mu\text{m}$. Parameters: sampling duty cycle $\Lambda_g/\Lambda_s = 10\%$, unsampled coupling coefficient $\kappa_u = 200\,\text{cm}^{-1}$, peak spacing $\delta = 0.6$ THz ($n_g = 3.85$).

For order zero, and with $L_g = L\Lambda_g/\Lambda_s$ the total length of grating the peak power reflectivity becomes

$$R(\nu_0) = \tanh^2(|F_0|\kappa_u L) = \tanh^2(\kappa_u L\Lambda_g/\Lambda_s) = \tanh^2(\kappa_u L_g) \quad (4.32)$$

This is nothing less than the peak reflectivity of a uniform Bragg reflector with coupling coefficient κ_u and length L_g. If $\kappa_u L_g < 0.5$, we can approximate the reflectivity of peak k by $(|\kappa_k|L)^2$. From Eq. (4.24) it is clear that the envelope of the reflectivity peaks becomes broader as the sampling duty cycle Λ_g/Λ_s is reduced. For small duty cycles, the number of peaks within the 3 dB bandwidth of the envelope is approximately equal to the inverse of the duty cycle [37]. Reducing the duty cycle requires increasing the unsampled coupling coefficient κ_u to maintain the same peak reflectivity (at constant sampled grating length L). Unfortunately, the technological limit for gratings in InGaAsP/InP waveguides is about 300 cm^{-1}. The example in Fig. 4.14 assumes a coupling coefficient $\kappa_u = 200\,\text{cm}^{-1}$ and a sampling duty cycle $\Lambda_g/\Lambda_s = 10\%$, resulting in about 11 peaks within the 3 dB bandwidth.

4.4.3. SUPER-STRUCTURE GRATING DBR

Although it is technologically the simplest way to obtain a comb reflector, the sampled grating does not exhibit the optimum reflection spectrum. Ideally one should have a number of equally spaced, equally strong reflectivity peaks within a limited bandwidth, with close to zero reflectivity outside that bandwidth. The above approach of periodically sampling a uniform Bragg grating in order to obtain multiple reflection peaks around the Bragg frequency can be extended to other types of periodic modulation [38]. Any modulation function that has a comb-shaped Fourier spectrum can be applied. These more general periodically modulated gratings are commonly called super-structure gratings (SSGs). The sampled grating consists of a digital on/off variation of the coupling coefficient κ, so the first option one could think of is a smoother periodic modulation of κ. However, with existing etching technology this is very difficult to implement. A second option is a variation of the grating frequency (or phase). The drawback is that this requires direct writing of the grating pattern on the semiconductor substrate with an electron beam, a slow and hence costly procedure.

The first demonstration of such a super-structure grating consisted of a periodic linear chirp of the grating frequency $f = 1/\Lambda$ [39] (Fig. 4.15). Within the super-period of length Λ_s the grating frequency f is varied from $f_a = 1/\Lambda_a$ to $f_b = 1/\Lambda_b$.

$$f(z) = \frac{1}{2}(f_a + f_b) + \frac{z - l\Lambda_s}{\Lambda_s}(f_b - f_a) = f_0 + \frac{z - l\Lambda_s}{\Lambda_s}\Delta f \quad (4.33)$$

for

$$\left(1 - \frac{1}{2}\right)\Lambda_s < z < \left(l + \frac{1}{2}\right)\Lambda_s$$

The grating modulation function has a magnitude of one and a quadratic phase term:

$$F(z) = \exp[j2\pi(f(z) - f_0)z] = \exp\left[j2\pi\Delta f \frac{z - l\Lambda_s}{\Lambda_s}z\right] \quad (4.34)$$

As a result, the Fourier coefficients are given in terms of Fresnel integrals [40].

$$F_k = \frac{1}{\Lambda_s}\int_{-\Lambda_s/2}^{\Lambda_s/2} F(z)e^{-j\frac{2\pi k}{\Lambda_s}z}dz = A\int_a^b e^{ju^2}du \quad (4.35)$$

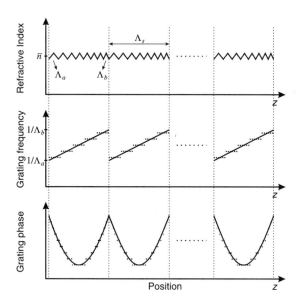

Fig. 4.15 Schematic of a linearly chirped super-structure grating. Solid lines indicate the ideal linear frequency chirp, equivalent to a quadratic relative phase variation. Dotted lines indicate practical implementations: discrete frequency changes or discrete phase changes (with uniform grating pitch).

where

$$A = \frac{1}{\sqrt{2\pi \, \Delta f \, \Lambda_s}} \exp\left[-j \frac{(k\pi)^2}{2\pi \, \Delta f \, \Lambda_s}\right]$$

$$a = \frac{1}{2}\sqrt{2\pi \, \Delta f \, \Lambda_s}\left(-1 - \frac{2\pi k}{2\pi \, \Delta f \, \Lambda_s}\right) \qquad (4.36)$$

$$b = \frac{1}{2}\sqrt{2\pi \, \Delta f \, \Lambda_s}\left(+1 - \frac{2\pi k}{2\pi \, \Delta f \, \Lambda_s}\right)$$

Strong reflection peaks are obtained in the frequency interval determined by the minimal and maximal grating frequency (see the Bragg condition, Eq. (4.10)):

$$f_a < 2n(v)v/c < f_b \qquad (4.37)$$

Due to the limited resolution of the e-beam lithography, a truly continuous chirp cannot be accomplished. For this reason, the first super-structure gratings were made by varying the grating frequency in discrete steps [39–40],

as indicated by the dotted lines in Fig. 4.15. For these first implementations, reflection characteristics were rather poor due to the limitations of
the e-beam lithography [20]. Fabrication becomes simpler and more reliable if the same grating pitch can be used throughout the SSG. This can
be achieved if the discrete frequency steps are replaced by discrete phase-
shifts. Indeed, any frequency modulation can also be regarded as a phase
modulation.

We already noticed previously that a linear frequency chirp is equivalent
to a quadratic change in the phase ϕ of the super-structure grating (measured
relative to the phase of a uniform grating with frequency f_0), as follows:

$$\phi(z) = \int_{-\Lambda_s/2}^{z} 2\pi (f(z') - f_0)dz' = \frac{\pi}{4}\Delta f \Lambda_s[(2z/\Lambda_s - 2l)^2 - 1] \quad (4.38)$$

In practice, this phase variation is approximated by discrete phase-shifts
of e.g. $\pi/10$ with a quadratically varying density [20]. Figure 4.16 shows

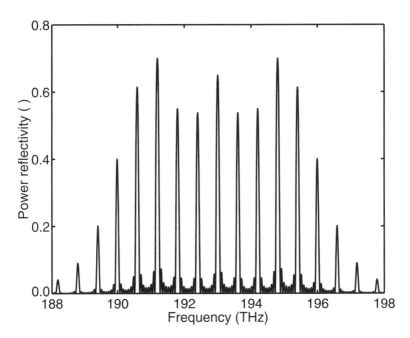

Fig. 4.16   Reflection spectrum of a super-structure grating with quadratic phase variation,
consisting of 8 periods with length $\Lambda_s = 65\,\mu$m. Parameters: peak-to-peak phase variation
of 3.05π, coupling coefficient $\kappa_u = 67$ cm^{-1}, peak spacing $\delta = 0.6$ THz ($n_g = 3.85$).

the reflection spectrum of an 8-period grating with a peak-to-peak phase variation of 3.05 π. Note that although the coupling coefficient κ_u is only one-third of the value used for the sampled grating in Fig. 4.14 and both gratings have the same length L, the 11 reflection peaks within the 3 dB bandwidth are on average stronger for the SSG.

For a sampled grating, the number of peaks N is inversely proportional to the sampling duty cycle Λ_g/Λ_s, while the peak reflectivities are directly proportional to the duty cycle (for constant length L and coupling coefficient κ_u). In other words, the peak reflectivities of the SG decrease as $1/N$ when the number of peaks N increases. In a frequency- or phase-modulated SSG on the other hand, reflectivities only decrease approximately as $1/\sqrt{N}$, because in this case, according to Parseval's theorem,

$$\sum_k |F_k|^2 = 1 \tag{4.39}$$

In order to get similar reflectivities for the SG as for the SSG (for a given number of peaks N), either the grating length L or the coupling coefficient κ_u has to be increased by roughly $\sqrt{N}$ (i.e., ≈ 3.3 for 11 peaks).

Still, with a quadratic phase variation, the envelope of the reflection peaks is not yet rectangular. To obtain this, the phase variation can be optimized numerically [41–42]. The target reflectivity spectrum consists of N uniform peaks, with zero reflectivity outside the band. Because of Eq. (4.31) and Eq. (4.39), this can be expressed by following trial function (for given length L and coupling coefficient κ_u).

$$C(\phi(z)) = \sqrt{\frac{1}{N}\sum_k^N \left(1 - \frac{R(\nu_k)}{R_T}\right)^2} \tag{4.40}$$

where

$$R_T = \tanh^2(\kappa_u L/\sqrt{N}) \tag{4.41}$$

The target reflectivity R_T is the maximum reflectivity that can be achieved for N uniform peaks, and if all N peaks have reflectivity R_T, out-of-band reflectivity will automatically be zero.

Figure 4.17 shows the optimized phase variation for $N = 11$. Reflectivity spectra were calculated using a transfer matrix method. The super-period was divided into 51 sections of equal length, with a discrete phase-shift at

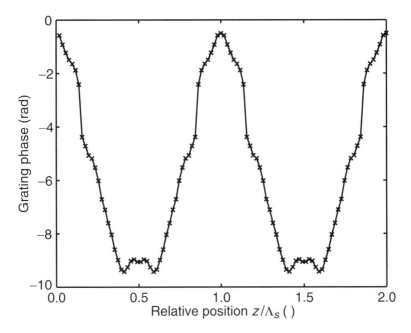

Fig. 4.17 Optimized grating phase variation for a multi-phase-shift (MPS) SSG with 11 reflectivity peaks. Discrete phase-shifts were equally spread over the super-period (51 per period), and their values were optimized numerically in order to obtain a rectangular reflectivity envelope.

each interface. The phase variation was initialized to the parabolic curve used to calculate the spectrum in Fig. 4.16. Subsequently, the phase-shifts were optimized numerically using a simulated annealing algorithm, combined with the downhill simplex method of Nelder and Mead [43–44]. The optimization procedure was repeated a few times, and the best result was retained. At first sight, the result in Fig. 4.17 looks quite different from the curve in [41]. On closer inspection though, an excellent match is found provided the phase curve is inverted and is shifted by half the super-period. Changing the sign of the phase of course merely corresponds to calculating the reflectivity from the other end of the SSG-DBR (replace z by $-z$ in Eq. (4.11)), and should therefore yield the same result.

The corresponding reflection spectrum is plotted in Fig. 4.18. The peak reflectivities are clearly highly uniform, and very close to the target value of 0.61. Consequently, the reflectivity is also negligible outside the band of 11 peaks.

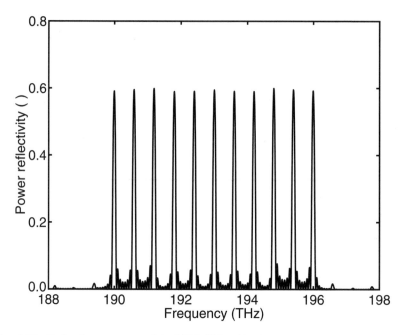

Fig. 4.18   Reflection spectrum of the MPS-SSG with phase variations as shown in Fig. 4.17, consisting of 8 periods with length $\Lambda_s = 65$ μm. Parameters: coupling coefficient $\kappa_u = 67$ cm^{-1}, peak spacing $\delta = 0.6$ THz ($n_g = 3.85$). The target reflectivity R_T is 0.61.

4.4.4.   SG-DBR AND SSG-DBR LASERS

The longitudinal cross-section of a SG-DBR laser is sketched in Fig. 4.19. Both cavity mirrors consist of sampled grating Bragg reflectors. The difference in reflection peak spacing δ, which is required for the Vernier-tuning, is obtained by applying different sampling periods Λ_s to front and rear reflector (see Eq. (4.26)). Apart from the specifics of the grating design, the cross-section of a super-structure grating (SSG)-DBR laser is evidently identical.

Shown in Fig. 4.20 is the Vernier-tuning characteristic of a SSG-DBR laser, in which both reflectors have 7 uniform reflectivity peaks. It is clearly visible that if either the front reflector or the rear reflector is tuned, large frequency jumps (super-mode jumps) of about 0.7 THz are obtained. At one particular super-mode jump the frequency jumps from one end of the spectrum to the other. If both reflectors are tuned simultaneously, such that a particular pair of reflector peaks is kept aligned, smaller frequency hops of approximately 50 GHz are observed. These correspond to longitudinal

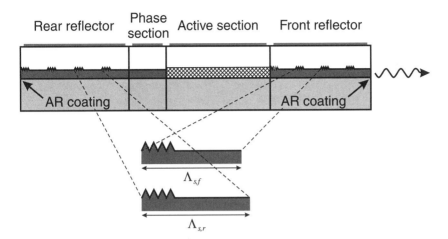

Fig. 4.19 Longitudinal cross-section of a tunable sampled grating (SG) DBR laser. Apart from the specifics of the grating design, the cross-section of a super-structure grating (SSG) DBR laser is identical.

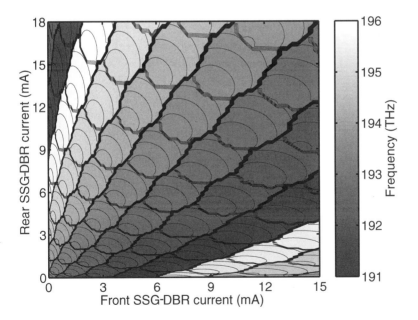

Fig. 4.20 Tuning characteristic of a SSG-DBR laser. Contour plot of frequency as a function of front and rear reflector currents (10 GHz increments). Both SSG reflectors have 7 uniform reflection peaks.

mode hops. If one runs through the diagram from top-left to bottom-right, one finds more or less the same frequency for points that are 7 super-mode jumps apart, consistent with the number of reflectivity peaks.

In order to accomplish quasi-continuous tuning over the depicted range of about 5 THz, one has to add the third tuning dimension: the phase section control. Tuning a SG-DBR or SSG-DBR laser to a particular frequency is hence achieved in three steps. First, either the front reflector or the rear reflector is tuned to align the appropriate pair of reflector peaks (coarse tuning). Then, both reflectors are tuned simultaneously to bring the coincident peaks to the correct frequency (medium tuning). Finally, the phase section current is adjusted to align a cavity mode with the coincident reflector peaks (fine tuning).

4.4.5. COMBINING A CO-DIRECTIONAL GRATING COUPLER WITH A COMB REFLECTOR

Instead of using a Bragg reflector as intra-cavity tunable filter, one could look for alternative filters that are more widely tunable. One option is the grating-assisted co-directional coupler (GACC) filter [45–46] (Fig. 4.21). This filter consists of two parallel, asymmetric waveguides. The dual-waveguide structure supports two guided modes R and S. Because of the asymmetry, one mode (R) is mainly confined in the lower waveguide, while the other (S) has most of its power in the upper waveguide. Parallel to both waveguides, there is a grating layer with a periodically varying refractive index. At the input, mainly mode R is excited. Without the grating, there would only be a weak coupling between the two modes, and only little power would be transferred from the lower to the upper waveguide. With the grating, efficient coupling is obtained in a limited frequency band.

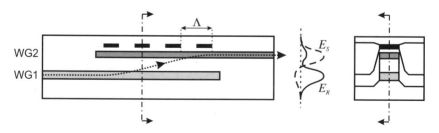

Fig. 4.21 Longitudinal and lateral cross-section of a grating-assisted co-directional coupler (GACC) filter. Also shown are the field distributions of the two modes R and S of the asymmetric dual-waveguide structure.

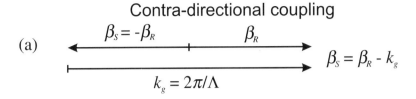

Fig. 4.22 Vector diagram for contra-directional (a) and co-directional (b) coupling between two waveguide modes with propagation constants β_R and β_S, by a periodic structure with period Λ (propagation constant $k_g = 2\pi/\Lambda$).

In Section 4.3 we described the coupling in a single waveguide between a right-propagating mode R and a left-propagating mode S, by a periodic refractive index modulation with period Λ. Both modes have the same transverse field distribution, but opposite propagation constants $\beta_R = -\beta_S = 2\pi\nu n/c$. Efficient coupling between the modes is obtained if the periodic structure provides phase matching between the two modes, i.e. $\beta_S = \beta_R - k_g$, where $k_g = 2\pi/\Lambda$ is the propagation constant of the grating (Fig. 4.22(a)). This condition immediately translates to the Bragg condition (4.10).

In the co-directional coupler, the periodic structure should provide coupling between the two modes R and S propagating in the same direction, with different propagation constants $\beta_R = 2\pi\nu n_R/c$ and $\beta_S = 2\pi\nu n_S/c$ (Fig. 4.22(b)). Here n_R and n_S are the effective indices of modes R and S. Phase matching again occurs when $\beta_S = \beta_R - k_g$.

The coupling frequency ν_c is hence given by:

$$\nu_c = \frac{c}{\Lambda[n_R(\nu_c) - n_S(\nu_c)]} \tag{4.42}$$

Because $(n_R - n_S) \ll n_R$, it is clear that the period Λ required for co-directional coupling is much longer than the period needed for contra-directional coupling. As the power transfer from lower to upper waveguide is only efficient near the coupling frequency, this structure can be used as frequency selective filter. In a laser cavity one could have the light in the upper waveguide reflect back at the facet, while making sure that the light that remains in the lower waveguide is either absorbed or lost through diffraction. In that way, only light with frequencies close to the coupling frequency is efficiently coupled back to the gain section, which in the configuration of Fig. 4.21 would be connected to the lower waveguide on the lefthand side.

Because the coupling frequency depends on an index *difference*, a small change of either n_R or n_S can yield a large tuning [46–48].

$$\frac{\Delta v_{c,\max}}{v_c} = -\frac{\Delta(n_R - n_S)_{\max}}{n_{R,g} - n_{S,g}} \qquad (4.43)$$

Here $n_{R,g}$ and $n_{S,g}$ are the group effective indices of modes R and S (see Eq. (4.27)).

Let us assume that $\Delta(n_R - n_S) = \Delta n_R$ is equal to the effective index change Δn_d in the DBR structure. Then the tuning enhancement factor F, i.e. the ratio of the tuning range of the GACC to the tuning range of a DBR (Eq. 4.20), is found as

$$F = \frac{n_{d,g}}{n_{R,g} - n_{S,g}} > 1 \qquad (4.44)$$

By means of a coupled-mode analysis, it can be shown that the power transfer through the coupler (neglecting absorption and scattering losses) is described by [49]

$$T(v) = \frac{\kappa^2}{\kappa^2 + (\Delta\beta)^2} \sin^2\left[L\sqrt{\kappa^2 + (\Delta\beta)^2}\right] \qquad (4.45)$$

where

$$\Delta\beta = \frac{\beta_R - \beta_S}{2} - \frac{\pi}{\Lambda} \qquad (4.46)$$

In Eq. (4.45) κ represents the grating coupling coefficient. Complete power transfer is possible at zero detuning ($\Delta\beta = 0$), when the grating length is equal to the coupling length L_c.

$$L_c = \frac{\pi}{2\kappa} \qquad (4.47)$$

Figure 4.23 shows a typical power transfer characteristic of a GACC. With the parameters used in the figure, and assuming $n_{d,g} = n_{R,g}$, we obtain a tuning enhancement factor F of about 9.6. Unfortunately, the GACC filter bandwidth is also much larger than that of a DBR (compare with Fig. 4.8). From (Eq. (4.45)), the filter full width at half maximum (FWHM)

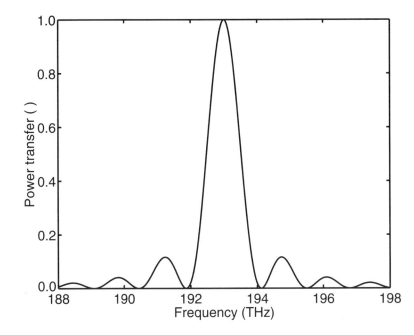

Fig. 4.23 Power transfer of a grating-assisted co-directional coupler (GACC) filter. Parameters: $n_R = 3.307$, $n_{R,g} = 3.95$, $n_S = 3.205$, $n_{S,g} = 3.54$, $\nu_c = 193$ THz and $\kappa = 28$ cm^{-1}.

bandwidth is calculated as [49–50]

$$\Delta \nu_{FWHM} \approx 0.8 \frac{c}{L_c(n_{R,g} - n_{S,g})} \qquad (4.48)$$

With the parameters from Fig. 4.23 this gives a bandwidth of approximately 1 THz. On closer inspection, both the tuning range (Eq. (4.43)) and the 3 dB bandwidth (Eq. (4.48)), are inversely proportional to the group index difference, and the ratio is

$$\frac{\Delta \nu_{FWHM}}{\Delta \nu_{c,\max}} \approx -0.8 \frac{c}{\nu_c L_c \Delta (n_R - n_S)_{\max}} \qquad (4.49)$$

The only parameter that allows reducing the bandwidth without affecting the tuning range is the coupler length L_c. In a laser cavity, the filter bandwidth has to be measured relative to the longitudinal mode spacing. Because increasing the coupler length at the same time reduces the mode spacing, it is difficult to obtain both a large tunability and sufficient mode selectivity in a tunable laser with a GACC filter [49–50]. Typically, the side mode suppression ratio is only about 20 dB.

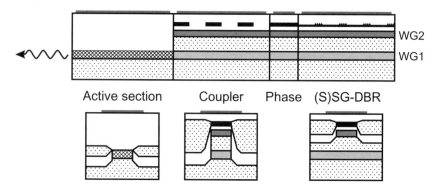

Fig. 4.24 Longitudinal and lateral cross-sections of a grating-assisted coupler with rear sampled or super-structure grating reflector (GCSR) laser.

In order to take advantage of the broad tunability of the GACC filter, without having to trade-off tunability for selectivity, it was proposed to combine the GACC filter with a sampled or super-structure grating DBR [46, 51, 52]. Due to its broad tunability, the GACC can be used to filter out one of the reflectivity peaks of the SG- or SSG-DBR. The narrow reflectivity peaks on the other hand supply the required selectivity.

When this dual filter structure is integrated into a laser cavity, together with a phase shifter section for fine tuning of the cavity modes, one obtains the so-called GCSR laser depicted in Fig. 4.24 [53]. Note that in the phase and reflector sections, the lower waveguide is planar, such that the light that is not coupled to the upper waveguide in the GACC diffracts and cannot couple back to the active section. Alternatively, one could replace the lower waveguide core material with absorptive material in those sections. Figure 4.25 plots typical reflection and transmission characteristics of the intra-cavity filters. The photograph in Fig. 4.26 shows the laser (with a length of approximately 2 mm) mounted on a ceramic carrier, with bonding pads for the different laser contacts. On the carrier, one also finds a thermistor, which is used in the control loop stabilizing the laser temperature. The coplanar line can be used to apply a high-frequency modulation signal to the active section.

As for a SG- or SSG-DBR laser, tuning the GCSR laser to a particular frequency is done in three steps. First, the coupler is tuned to filter out the appropriate reflector peak (coarse tuning). Then, coupler and reflector are tuned simultaneously to bring the reflector peak to the correct frequency (medium tuning). Finally, the phase section current is adjusted to align a cavity mode with the selected reflector peak (fine tuning).

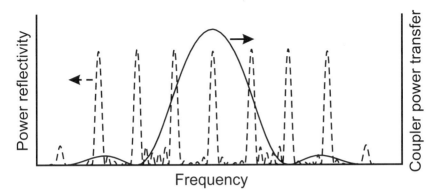

Fig. 4.25 Power reflectivity of the super-structure grating DBR and power transfer of the grating-assisted co-directional coupler of a GCSR laser.

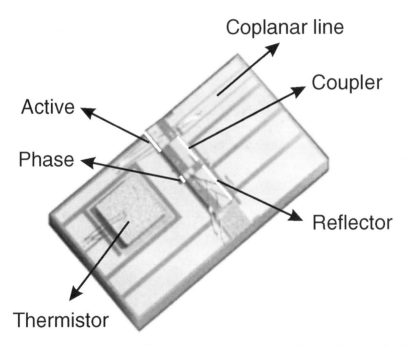

Fig. 4.26 Photograph of a GCSR laser on a ceramic carrier with bonding pads for the different laser contacts and a thermistor (for temperature control). The coplanar line can be used to apply a high-frequency modulation signal to the active section. The actual laser chip is about 2 mm long.

One of the main advantages of the GCSR laser with respect to the SG- and SSG-DBR lasers is the lower variation of the output power across the tuning range. From Section 4.1.1 we know that when a section is tuned, the absorption losses in that section increase. Because the light generated in a SG- or SSG-DBR laser has to propagate through the front reflector, the output power will vary more with tuning in these lasers. On the other hand, the combination of two narrowband filters in the SG- and SSG-DBR devices provides better suppression of neighboring cavity modes than the combination of a narrowband and a broadband filter in the GCSR. This effect is enhanced by the fact that the effective cavity length is typically shorter in these lasers (and hence the cavity mode spacing is larger). Another disadvantage of the GCSR laser is that the laser structure is more difficult to fabricate (compare Fig. 4.19 and Fig. 4.24).

4.5.   External Cavity Tunable Lasers

Instead of using completely monolithic structures, tunable lasers can be based on hybrid structures where the frequency selective element is placed externally to the laser. These external cavity lasers (ECLs) will be described in this section. A detailed discussion of cavity design and coupling optics, as well as an extensive list of references can be found in [54].

In addition to the "bulk optics" external cavity lasers we will also consider lasers using MEM (micro electromechanical) technology, as well as tunable vertical cavity surface emitting lasers (VCSELs).

4.5.1.   CONVENTIONAL EXTERNAL CAVITY LASERS

Traditional external cavity lasers consist of a laser chip and an external reflector. By using a grating as the external reflector, turning of the grating will lead to a tuning of the lasing wavelength. In the past, tuning ranges in excess of 240 nm have been reported [55].

Because the cavity length is much larger than for a solitary semiconductor laser, the photon lifetime is much longer, resulting in a very narrow spectral linewidth. Values in the kHz range can be obtained, as opposed to MHz for a solitary laser. In order to suppress multi-cavity effects, the laser facet facing the external cavity is usually anti-reflection (AR) coated. The laser output is usually taken from the facet at the other end of the laser, the reflectivity of this facet may also be modified by a coating in order to increase the available power.

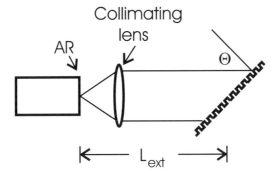

Fig. 4.27 Schematic of an external cavity laser.

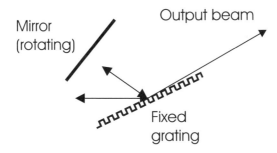

Fig. 4.28 Littman-Metcalf cavity.

If the external feedback is provided by a simple grating with a grating period Λ, and the angle of incidence on the grating is θ (Fig. 4.27), then the lasing wavelength λ is determined by the Bragg condition

$$\lambda = 2\Lambda \sin(\theta) \tag{4.50}$$

Turning the grating changes the angle of incidence and hence tunes the wavelength. However, when the wavelength changes, the ratio between wavelength and cavity length changes, leading to hops between cavity modes. In order to achieve phase continuous tuning (i.e., tuning with the laser remaining in the same longitudinal mode), it is necessary to change the cavity length by exactly the same relative amount as the wavelength. Simultaneous change of the cavity length and the grating angle can be achieved with a special mechanical mounting of the grating, e.g. [56], or by rotating the grating around an optimized pivot point, e.g. [57].

An alternative cavity design, the Littman-Metcalf cavity [58], uses a fixed grating and a rotating mirror (Fig. 4.28). In this configuration the part of the beam directly reflected from the grating (0th order) forms the output beam, and the 1st order diffracted beam is reflected by the mirror. Again tuning without longitudinal mode hops can be achieved by selecting the pivot point for the mirror. The reflectivity of the rear facet of the laser can be increased by applying a high reflectivity (HR) coating, which will increase the power efficiency of the laser.

When a semiconductor laser is tuned away from the maximum gain, the threshold current will increase. Consequently, in the case of wide tuning, the output power will vary during tuning, unless the laser current is varied to compensate. Constant power operation can be achieved by using a monitor diode and a relatively simple control circuit.

A particular advantage of ECLs is that they can use semiconductor lasers, which are specifically designed for high output power. In addition, there is a degree of freedom in the selection of facet reflectivities; this makes it possible to have a structure with a high power efficiency. However, the traditional ECLs involve delicate mechanics, they tend to be quite bulky, and in order to ensure spectral stability the demand on mechanical stability is very high. Consequently, they have remained a specialist, low-volume product with a relatively high unit price.

4.5.2. *MEM EXTERNAL CAVITIES*

A relatively new development is the use of a micro electromechanical (MEM) structure to form a micro-ECL. The device described in [59], and shown in Fig. 4.29, has a footprint of only about 2 mm by 3 mm. The small size means that the device is mechanically robust. Although this MEM-ECL

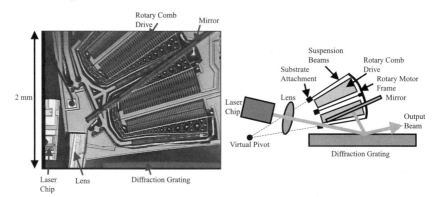

Fig. 4.29   MEM-ECL with Littman-Metcalf cavity [59].

is clearly aimed at the telecom transmitter market, its performance (40 nm continuous tuning, +7 dBm fiber coupled power over the whole range) certainly makes it a candidate for test and measurement applications as well. Switching from one WDM channel to another is relatively slow (15 ms), but wavelength stabilization, using a wavelength locker, is simple. Truly continuous tuning is possible, and probably a good deal faster than for a standard ECL.

4.5.3. TUNABLE VCSELs

Since the late 1980s there has been a rapid development of vertical-cavity surface-emitting lasers (VCSELs). In these lasers, the light propagates perpendicular to the plane defined by the active layer. The optical feedback is provided by Bragg reflectors, consisting of layers with alternating high and low refractive indices, instead of the cleaved facets of edge-emitting lasers. Because of the very short cavity length, very high (>99%) reflectivities are required, and the reflectors typically have 20 to 30 layer pairs. An advantage of the short cavity length is that the mode spacing is large compared with the width of the gain curve, such that, if the resonant wavelength is close to the gain peak, single-longitudinal-mode operation occurs. As an example, a cavity length of about 10 μm will give a mode spacing of about 30 nm. It should be noted, however, that if the diameter of the active region is large, multi transverse-mode operation might occur.

One of the particular advantages of VCSELs is that the spot size can be made compatible with that of a single-mode optical fiber, making the coupling from laser to fiber easier and more efficient. The VCSEL structure also makes it possible to fabricate very high-density two-dimensional laser arrays. Most VCSELs are fabricated using the AlGaAs material system, with one or more strained InGaAs quantum wells as the active material; for these lasers, the wavelength is usually close to 1 μm. However, VCSELs are now also being fabricated for the longer wavelengths of interest for fiber optics.

A tunable VCSEL can be made by having an electrostatically deflectable mirror suspended over the active region. A wide continuous tuning range (limited by the longitudinal mode spacing), with a single voltage control is then possible. An example of a tunable VCSEL is shown in Fig. 4.30.

A tuning range of 40 nm with 7 mW fiber coupled power has been achieved with this laser. One of the special features of the device is the use of optical pumping using a 980 nm pump laser incorporated into the tunable

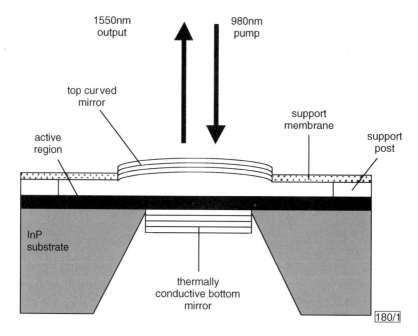

Fig. 4.30   Tunable VCSEL [60].

laser module. Wavelength control is obviously simple because it depends on a single tuning voltage only, but tuning speed may be an issue.

4.6.   Selectable Sources and Arrays

Laser arrays, where each laser in the array operates at a particular wavelength (or in a limited wavelength range), are an alternative to tunable lasers. In their simplest form, these arrays have separate outputs for each array element. More sophisticated structures incorporate a combiner element, which makes it possible to couple the output to a single optical fiber without the use of complicated external coupling optics. If each laser in the array can be tuned by an amount exceeding the wavelength difference between the array elements, a very wide total wavelength range can be achieved.

This section reviews various diode laser array structures. Some array designs can, at least in principle, work at several wavelengths simultaneously. However, this is likely to give rise to cross-talk problems, and most array structures are therefore designed to work at a single wavelength at a time.

It is an advantage of a laser array that each element operates at a particular wavelength. This makes the control easier than that of a monolithic tunable laser (e.g., SG-DBRs or GCSRs). However, many array designs require a larger or more complicated chip.

4.6.1. DFB ARRAYS

There have been several reports on arrays of DFB lasers where all the array elements operate at different optical frequencies. The different frequencies can be obtained either by varying a structural parameter (e.g., stripe width) from laser to laser, thereby changing the effective refractive index of the structure, or by changing the grating period (this requires e-beam writing). In order to form a practical device the lasers must be integrated with a combiner in order to have a common output waveguide.

Standard DFB lasers usually have one AR-coated and one HR-coated facet. For integrated lasers in an array the facets will have to be non-reflecting. This is necessary to avoid the yield problem that occurs in AR/HR devices because the relative position of a facet relative to the grating cannot be controlled. In order to have a single, well-defined lasing mode, a DFB laser with two nonreflecting facets must have a quarter wavelength phase shift in the center of the grating. Note that an equivalent phase shift may be introduced by other means, for example by varying the stripe width.

The structure described in [61] has six DFB lasers integrated with a combiner, an amplifier, and a modulator, as well as monitor detectors (Fig. 4.31).

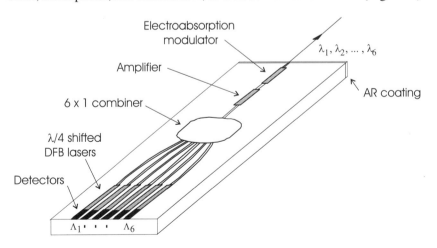

Fig. 4.31 DFB (selectable) array [61].

The amplifier is included in order to compensate for the splitting loss caused by the combiner, and the insertion loss due to the modulator. The emission frequency of a given laser can be aligned to the ITU channel plan by a moderate degree of temperature tuning.

Fabrication of a DFB laser to a specified wavelength is very difficult, but in an array, the accuracy of the wavelength spacing can be very high. This means that if one array element is fine-tuned to its design frequency (e.g., thermally), then all the other array elements will automatically be at, or very close to, their respective design frequencies. Use of an array also makes it possible to have redundancy in order to improve the reliability, by having two lasers for each wavelength.

4.6.2.   CASCADED DFB LASERS

An alternative to a parallel array of DFB lasers is the cascading of lasers. Several grating sections with different periods and separate electrodes can be formed on a single active waveguide. Only one section is biased high above threshold at any time, with the sections in front of it being biased close to threshold and working as amplifiers. In an extension of this concept, two sets of three cascaded DFB lasers were integrated with a combiner to form a single output. Using a 50°C temperature change, a total tuning range of 30 nm (i.e., 5 nm tuning per laser, consistent with 0.1 nm tuning per °C) was demonstrated [62]. This structure is shown in Fig. 4.32.

Arrays or cascaded DFB lasers are obviously not practical for addressing a large number of channels unless a high degree of temperature tuning

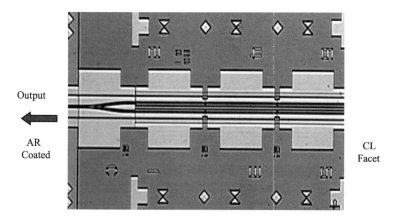

Fig. 4.32   Structure with two sets of three cascaded DFB lasers [62].

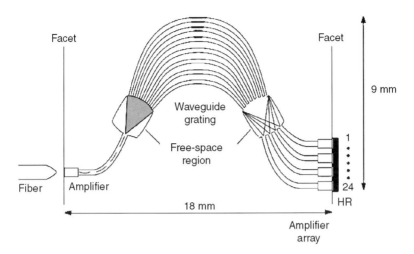

Fig. 4.33   AWG laser structure [63].

is used. This will in turn reduce the tuning speed, and some of the control simplicity advantage will be lost.

4.6.3.   AWG BASED STRUCTURES

The arrayed waveguide grating (also known as phased array) multi-wavelength laser has an array of semiconductor optical amplifiers (SOAs) on one side of a waveguide grating (Fig. 4.33).

The SOAs are coupled to the waveguide grating via a star coupler. On the other side of the waveguide grating another coupler provides output through a single waveguide, which may contain a common amplifier. Wavelength selectivity is provided by the AWG, and the lasing frequency is selected by turning on the appropriate element in the SOA array.

The operation of the AWG can be explained as follows. Because the array elements have different lengths, light (with a given wavelength) will be subject to different delays; consequently the phase front of the combined light at the output of the array will be tilted. The amount of tilt is wavelength dependent, and light at different wavelengths will be focused on different output waveguides.

In [64] a slightly more elaborate structure is described. This structure has 5 SOAs on one side of the AWG and 8 on the other. The structure is designed in such a way that 40 (= 5 × 8) optical channels with a 100 GHz frequency spacing are available.

Simultaneous operation of an AWG laser at several optical frequencies has been demonstrated, but cross-talk is likely to prevent this from being a practical proposition. Other functions, such as modulation, may be integrated on the chip as well.

AWG laser chips are usually quite large, with a side length ranging from some millimeters to more than a centimeter. In spite of the long cavity length, and corresponding small mode spacing, it has been found experimentally that the spectral properties are surprisingly good, with clean longitudinal single-mode operation. It is thought that this is due to a nonlinear wave-mixing phenomenon, which is actually helped by the small mode spacing.

The AWG laser becomes increasingly attractive over a traditional array as the element number N increases because it does not suffer from the inherent $1/N$ combiner loss present in a conventional combiner.

4.7. Integration Technology

Sections 4.3 and 4.4 of this chapter describe tunable lasers consisting of a longitudinal integration of sections with different functionality. These have an active section providing the optical gain and one or more filter sections with a tunable frequency selective characteristic. For optical transmitter applications, more functions can be added, such as modulation for encoding of data, power amplification for power management, or wavelength locking to ensure frequency stability during the lifetime of the device. Integrating these features on chip is an attractive cost-effective solution compared to hybrid integration.

Monolithic integration requires that the sections added to the device are optically and electrically decoupled from the laser. Indeed, any external feedback to the laser cavity might lead to unacceptable frequency or output power variations. Consequently, any air-semiconductor facets external to the laser cavity have to be antireflection coated to reduce the power reflectivity to the order of 10^{-5}. For DBR and GCSR lasers, one of the cavity mirrors is a cleaved facet. To enable integration on that side of the cavity, the facet mirror has to be replaced by an "on-chip" mirror, e.g., by using a deeply etched Bragg grating [65] (i.e., a grating with a high coupling coefficient and hence wide reflectivity bandwidth) or an etched mirror [66]. On the Bragg reflector side, the integration is of course straightforward. Consequently, SG-DBR lasers are well suited for integration. Figure 4.34 illustrates the integration of a DBR with a semiconductor optical amplifier and a modulator on the front and a detector on the back.

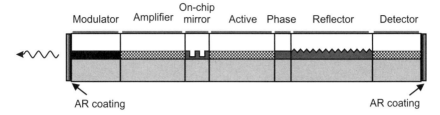

Fig. 4.34 DBR laser with integrated amplifier, modulator and detector.

Technologically, integration challenges the state-of-the-art in fabrication, involving several epitaxial growths and regrowths as well as techniques such as selective area growth [67].

Modulation

The modulation functionality can be added either by using direct modulation of the gain current or by integrating a modulator. High-speed direct modulation in DBR lasers has been demonstrated [68]. However, it causes significant frequency excursions during the rising and falling edges of the optical pulses, which limits the transmission distance due to the dispersion in standard single-mode fiber and can even degrade the side mode suppression ratio to an unacceptable level. In GCSR and SG-DBR lasers, direct modulation is limited to bit rates of about 2.5 Gbit/s. This is a consequence of the lower intrinsic bandwidth of these lasers compared to ordinary DBR lasers, because of the longer cavity [69–70].

There are several types of modulators that are suitable for integration. Most common is the electroabsorption modulator, which is often integrated with DFB lasers but has also successfully been integrated with DBR [71] and SG-DBR lasers [72]. In order to achieve a sufficient extinction ratio across a wide wavelength range, an electroabsorption modulator can require quite large bias voltages. Consequently, modulators based on refractive index changes, such as Mach-Zehnder and guiding/antiguiding modulators, might be preferable for widely tunable lasers. For more details on modulators the reader is referred to Chapter 8 in [73].

Amplification

Amplification is required to boost the output power and/or equalize the output power of widely tunable lasers over the tuning range (without equalization the power can vary by 3 to 6 dB). Integration of a semiconductor optical amplifier (SOA) has been demonstrated both for DBR [71] and SG-DBR lasers [74]. The integration of an SOA is simplified by the fact that the same material can be used for both the active section and the SOA. Using an extra SOA for power control has the advantage of decoupling the power control and frequency control. Indeed, adjustment of the power through control of the gain current of the active section of the laser also affects the frequency of the laser and therefore requires a frequency control loop to be included.

It should be noted that the inclusion of an SOA adds amplified spontaneous emission (ASE) to the emission spectrum over a wide frequency range, which might have an impact on system design.

Wavelength Locker – Power Monitor

Wavelength, mode, and power stability are important issues for tunable lasers. Typically, it is required to maintain stable single-mode emission (i.e., no mode hops) and a constant output power, with less than 3 GHz frequency drift, over a 20-year lifetime. This in turn sets high requirements on the material quality (i.e., a low density of defects) in multi-section tunable lasers. Such high material quality is unfortunately not readily achieved with today's fabrication technology. Hence, control loops for power, frequency, and mode stabilization are necessary.

Wavelength stabilization requires a wavelength dispersive element (a wavelength filter). An example of such a filter is shown in Fig. 4.35. By taking the ratio of the signals from the two detectors, a wavelength-dependent but power-independent signal is obtained.

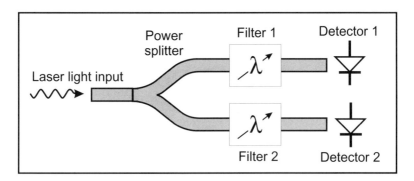

Fig. 4.35   Schematic of a wavelength locker.

A good wavelength locker should give a frequency resolution of about 0.1 GHz and have a sufficient locking range around each frequency at which the laser is aimed to operate (e.g., all multiples of 50 GHz within the C-band). A wavelength locker, enabling wavelength measurements within a 30 nm band, has been integrated with a SG-DBR laser [75]. The dispersive element consisted of a two-mode interference waveguide and a Y-branch splitter. The frequency resolution was limited to 55 GHz due to a parasitic reflection at the Y-branch. In order to reach the appropriate frequency resolution, both the filter design and the fabrication process have to be improved.

The aging of the integrated wavelength locker itself might be an issue, but because only passive waveguides and simple detectors are used, the degradation should be much less than that of the laser sections with current injection.

4.8.   Comparison of State-of-the-Art Tunable Lasers

Table 4.2 summarizes state-of-the-art characteristics of some of the laser types that were introduced previously. All lasers can achieve a side mode suppression ratio (SMSR) of more than 30 dB across their entire tuning range. A very noticeable difference between the laser types is the tuning speed. This difference is due to the different tuning mechanisms— electronic (changes in the carrier density), thermal, or mechanical—used in the different laser types.

OPTO+ in France have demonstrated a DBR laser with an output power of more than 13 dBm across a 2 THz tuning range [76]. To obtain this high output power, the phase section was omitted and thermal tuning was used instead to align a cavity mode with the Bragg reflectivity peak. This of course has the disadvantage that the tuning becomes very slow. The same applies to the temperature-tuned DFB cascade [82].

External cavity lasers [59,79] tend to have higher output power and narrower linewidth than the DBR-type lasers, but the fact that they are tuned mechanically has several disadvantages. The laser cavity is built up from discrete components that have to be precisely aligned, which increases assembly and packaging costs. The mechanical tuning also makes the devices sensitive for shock and vibration. Furthermore, the tuning speed is still quite slow. Finally, it still remains to be proven that these lasers can live up to the stringent reliability requirements imposed on lasers for telecom applications. Similar comments apply to the optically pumped MEMS-VCSELs described in [80–81].

Table 4.2 Comparison of State-of-the-Art Tunable Laser Characteristics (*Single Cascade of 3 DFB Lasers)

Recent References	DBR [76]	SG-DBR [77]	GCSR [22, 78]	ECL [59, 79]	MEMS-VCSEL [80, 81]	DFB Cascade [82]*
Tuning mechanism	Thermal + electronic	Electronic	Electronic	Mechanical	Mechanical	Thermal
Tuning range (THz)	<2	>4	>4	>4	>4	<2
# channels (50 GHz channel spacing)	<40	>80	>80	>80	>80	<40
Freq. stability with locker (GHz)	±3	±3	±3	±3	±3	±3
Freq. stability without locker (GHz)	Good	Good	Good	Poor	Poor	?
Output power (dBm)	>13	>3	>3	>10	>6	>3
Power uniformity without control (dB)	2–3	4–5	2–3	?	2–3	?
SMSR (dB)	>35	>35	>35	>40	>40	>40
Linewidth (MHz)	<25	<25	<25	<5	<10	<10
Tuning speed	>1 s	<20 ns	<20 ns	>1 ms	>10 µs	>1 s
Reliability	Good	Good	Good	?	?	?
Power consumption	Low	Low	Low	High	High	High

Electronically tuned DBR-type widely tunable lasers—like the GCSR, SG-DBR, and SSG-DBR lasers—have demonstrated quasi-continuous tuning ranges exceeding the bandwidth of (either C- or L-band) Erbium-doped fiber amplifiers. (In the case of 3-section DBR lasers, typically 3 different lasers are required to cover this bandwidth.) These widely tunable lasers generally have somewhat lower output power than the 3S-DBR and the ECL. Additionally, the SG-DBR and SSG-DBR have the disadvantage of relatively large output power variation across the tuning range. As was explained previously, this is because in these lasers light generated in the active section has to traverse a reflector section in which absorption losses increase with tuning (see Section 4.1). The fact that these lasers are all monolithic keeps assembly and packaging costs low. Furthermore, they allow the integration of additional components like an optical amplifier (to boost the output power) or an electroabsorption modulator. Because these lasers are the only ones switching at nanosecond speeds, they are also key-enablers for future optical packet switching systems.

References

1. M.-C. Amann and J. Buus, Tunable laser diodes (Artech House, Norwood, MA, 1998).
2. B. R. Bennett, R. A. Soref, and J. A. Del Alamo, "Carrier-induced change in refractive index of InP, GaAs and InGaAsP," IEEE J. Quantum Electron., 26:1 (Jan. 1990) 113–122.
3. J.-P. Weber, "Optimization of the carrier-induced effective index change in InGaAsP waveguides—Application to tunable Bragg filters," IEEE J. Quantum Electron., 30:8 (Aug. 1994) 1801–1816.
4. M. Asada, A. Kameyama, and Y. Suematsu, "Gain and intervalence band absorption in quantum-well lasers," IEEE J. Quantum Electron., 20:7 (Jul. 1984) 745–753.
5. I. Joindot and J. L. Beylat, "Intervalence band absorption coefficient measurements in bulk layer, strained and unstrained multiquantum well 1.55 μm semiconductor lasers," Electron. Lett., 29:7 (Apr. 1993) 604–606.
6. A. Sneh and C. R. Doerr, "Indium Phosphide-based photonic circuits and components," Chapter 7 in Integrated optical circuits and components, ed. E. J. Murphy (Marcel Dekker, New York, 1999).
7. J. E. Zucker, "Linear and quadratic electro-optic coefficients in InGaAsP," in Properties of lattice-matched and strained Indium Gallium Arsenide, ed. P. Bhattacharya (INSPEC, the Institution of Electrical Engineers, London, UK, 1993).

8. D. A. B. Miller, D. S. Chemla, T. C. Damen, A. C. Gossard, W. Wiegmann, T. H. Wood, and C. A. Burrus, "Electric field dependence of optical absorption near the band gap of quantum-well structures," Phys. Rev., B 32:2 (Jul. 1985) 1043–1060.

9. H. Yamamoto, M. Asada, and Y. Suematsu, "Electric-field-induced refractive index variation in quantum-well structure," Electron. Lett., 21:13 (Jun. 1985) 579–580.

10. J. E. Zucker, I. Bar-Joseph, B. I. Miller, U. Koren, and D. S. Chemla, "Quaternary quantum wells for electro-optic intensity and phase modulation at 1.3 and 1.55μm," Appl. Phys. Lett., 54:1 (Jan. 1989) 10–12.

11. N. Susa and T. Nakahara, "Enhancement of change in refractive index in asymmetric quantum well," Appl. Phys. Lett., 60:20 (May 1992) 2457–2459.

12. N. Susa, "Electric-field-induced refractive index changes in InGaAs-InAlAs asymmetric coupled quantum wells," IEEE J. Quantum Electron., 31:1 (Jan. 1995) 92–100.

13. J. Hong, M. Cyr, H. Kim, S. Jatar, C. Rogers, D. Goodchild, and S. Clements, "Cascaded strongly gain-coupled (SGC) DFB lasers with 15-nm continuous wavelength tuning," IEEE Photonic. Technol. Lett., 11:10 (Oct. 1999) 1214–1216.

14. M. Öberg, S. Nilsson, T. Klinga, and P. Ojala, "A three-electrode distributed Bragg reflector laser with 22 nm wavelength tuning range," IEEE Photonic. Technol. Lett., 3:4 (Apr. 1991) 299–301.

15. S. L. Woodward, U. Koren, B. I. Miller, M. G. Young, M. A. Newkirk, and C. A. Burrus, "A DBR laser tunable by resistive heating," IEEE Photonic. Technol. Lett., 4:12 (Dec. 1992) 1330–1332.

16. T. Kameda, H. Mori, S. Onuki, T. Kikugawa, Y. Takahashi, F. Tsuchiya, and H. Nagai, "A DBR laser employing passive-section heaters, with 10.8 nm tuning range and 1.6 MHz linewidth," IEEE Photonic. Technol. Lett., 5:6 (Jun. 1993) 608–610.

17. M.-C. Amann, "Broad-band wavelength-tunable twin-guide lasers," Optoelectronics—Devices and Technologies 10:1 (Mar. 1995) 27–38.

18. M.-C. Amann and W. Thulke, "Continuously tunable laser diodes: longitudinal versus transverse tuning scheme," IEEE J. Selected Areas in Commun., 8:6 (Aug. 1990) 1169–1177.

19. P.-J. Rigole, S. Nilsson, L. Bäckbom, T. Klinga, J. Wallin, B. Stålnacke, E. Berglind, and B. Stoltz, "Access to 20 evenly distributed wavelengths over 100 nm using a single current tuning in a four-electrode monolithic semiconductor laser," IEEE Photonic. Technol. Lett., 7:11 (Nov. 1995) 1249–1251.

20. H. Ishii, Y. Tohmori, Y. Yoshikuni, T. Tamamura, and Y. Kondo, "Multiple-phase-shift super structure grating DBR lasers for broad wavelength tuning," IEEE Photonic. Technol. Lett., 5:6 (Jun. 1993) 613–615.

21. H. Ishii, F. Kano, Y. Tohmori, Y. Kondo, T. Tamamura, and Y. Yoshikuni, "Broad range (34 nm) quasi-continuous wavelength tuning in super-structure grating DBR lasers," Electron. Lett., 30:14 (Jul. 1994) 1134–1135.
22. P.-J. Rigole, S. Nilsson, L. Bäckbom, B. Stålnacke, E. Berglind, J.-P. Weber, and B. Stoltz, "Quasi-continuous tuning range from 1560 to 1520 nm in a GCSR laser, with high power and low tuning currents," Electron. Lett., 32:25 (Dec. 1996) 2352–2354.
23. International Telecommunication Union—Telecommunication Standardization Sector (ITU-T), Recommendation G.692: "Optical interfaces for multichannel systems with optical amplifiers," *Series G: Transmission systems and media, digital systems and networks (Transmission media characteristics—Characteristics of optical components and sub-systems)*, October 1998.
24. T. L. Koch, U. Koren, R. P. Gnall, C. A. Burrus, and B. I. Miller, "Continuously tunable 1.5 μm multiple-quantum-well GaInAs/GaInAsP distributed-Bragg-reflector lasers," Electron. Lett., 24:23 (Nov. 1988) 1431–1433.
25. Y. Kotaki and H. Ishikawa, "Spectral characteristics of a three-section wavelength-tunable DBR laser," IEEE J. Quantum Electron., 25:6 (Jun. 1989) 1340–1345.
26. C. H. Henry, "Theory of the linewidth of semiconductor lasers," IEEE J. Quantum Electron., 18:2 (Feb. 1982) 259–264.
27. M.-C. Amann and R. Schimpe, "Excess linewidth broadening in wavelength-tunable laser diodes," Electron. Lett., 26:5 (Mar. 1990) 279–280.
28. M.-C. Amann, S. Illek, and H. Lang, "Linewidth reduction in wavelength tunable laser diodes by voltage control," Electron. Lett., 27:6 (Mar. 1991) 531–532.
29. K. Kobayashi and I. Mito, "Single frequency and tunable laser diodes," J. Lightwave Technol., 6:11 (Nov. 1988) 1623–1633.
30. H. Kogelnik and C. V. Shank, "Coupled-wave theory of distributed feedback lasers," J. Appl. Phys., 43:5 (Mar. 1972) 2327–2335.
31. M. Okuda and K. Onaka, "Tunability of distributed Bragg-reflector laser by modulating refractive index in corrugated waveguide," Japan J. Appl. Phys., 16 (1977) 1501–1502.
32. M. Yamaguchi, M. Kitamura, S. Murata, I. Mito, and K. Kobayashi, "Wide range wavelength tuning in 1.3 μm DBR-DC-PBH-LDs by current injection into the DBR region," Electron. Lett., 21:2 (Jan. 1985) 63–65.
33. S. Murata, I. Mito, and K. Kobayashi, "Over 720 GHz (5.8 nm) frequency tuning by a 1.5 μm DBR laser with phase and Bragg wavelength control regions," Electron. Lett., 23:8 (Apr. 1987) 403–405.
34. F. Delorme, S. Slempkes, G. Alibert, B. Rose, and J. Brandon, "Butt-jointed DBR laser with 15 nm tunability grown in three MOVPE steps," Electron. Lett., 31:15 (Jul. 1995) 1244–1245.

35. F. Delorme, S. Grosmaire, A. Gloukhian, and A. Ougazzaden, "High power operation of widely tunable 1.55 μm distributed Bragg reflector lasers," Electron. Lett., 33:3 (Jan. 1997) 210–211.

36. L. A. Coldren, "Multi-section tunable laser with differing multi-element mirrors," US Patent, no. 4896325, filed August 23, 1988, published January 23, 1990.

37. V. Jayaraman, Z.-M. Chuang, and L. A. Coldren, "Theory, design and performance of extended tuning range semiconductor lasers with sampled gratings," IEEE J. Quantum Electron., 29:6 (Jun. 1993) 1824–1834.

38. Y. Tohmori, Y. Yoshikuni, H. Ishii, F. Kano, and T. Tamamura, "Distributed reflector and wavelength-tunable semiconductor laser," US Patent, no. 5325392, filed March 3, 1993, published June 28, 1994.

39. Y. Tohmori, Y. Yoshikuni, T. Tamamura, H. Ishii, Y. Kondo, and M. Yamamoto, "Broad-range wavelength tuning in DBR lasers with super structure grating (SSG)," IEEE Photonic. Technol. Lett., 5:2 (Feb. 1993) 126–129.

40. H. Ishii, Y. Tohmori, T. Tamamura, and Y. Yoshikuni, "Super structure grating (SSG) for broadly tunable DBR lasers," IEEE Photonic. Technol. Lett., 4:4 (Apr. 1993) 393–395.

41. H. Ishii, F. Kano, Y. Tohmori, Y. Kondo, T. Tamamura, and Y. Yoshikuni, "Narrow spectral linewidth under wavelength tuning in thermally tunable super-structure-grating (SSG) DBR lasers," IEEE J. Selected Topics Quantum Electron., 1:2 (Jun. 1995) 401–407.

42. H. Ishii, H. Tanobe, F. Kano, Y. Tohmori, Y. Kondo, and Y. Yoshikuni, "Quasi-continuous wavelength tuning in super-structure-grating (SSG) DBR lasers," IEEE J. Quantum Electron., 32:3 (Mar. 1996) 433–440.

43. W. H. Press, S. A. Teukolsky, W. T. Vetterling, and B. P. Flannery, Numerical recipes in C: the art of scientific computing (Cambridge University Press, Cambridge, UK, 1992).

44. J. A. Nelder and R. Mead, "A Simplex Method for Function Minimization," Computer Journal, 7 (1964) 308–313.

45. R. C. Alferness, T. L. Koch, L. L. Buhl, F. Storz, F. Heismann, and M. J. R. Martyak, "Grating-assisted InGaAsP/InP vertical codirectional coupler filter," Appl. Phys. Lett., 55:19 (Nov. 1989) 2011–2013.

46. J. Willems, Modeling and design of integrated optical filters and lasers with a broad wavelength tuning range, Ph.D. Thesis (University of Gent, January 1995).

47. R. C. Alferness, L. L. Buhl, U. Koren, B. I. Miller, M. G. Young, T. L. Koch, C. A. Burrus, and G. Raybon, "Broadly tunable InGaAsP/InP buried rib waveguide vertical coupler filter," Appl. Phys. Lett., 60:8 (Feb. 1992) 980–982.

48. R. C. Alferness, L. L. Buhl, T. L. Koch, and U. Koren, "Tunable optical waveguide coupler," US Patent, no. 5253314, filed January 31, 1992, published October 12, 1993.

49. Z.-M. Chuang and L. A. Coldren, "Design of widely tunable semiconductor lasers using grating-assisted codirectional-coupler filters," IEEE J. Quantum Electron., 29:4 (Apr. 1993) 1071–1080.

50. Z.-M. Chuang and L. A. Coldren, "On the spectral properties of widely tunable lasers using grating-assisted codirectional coupling," IEEE Photonic. Technol. Lett., 5:1 (Jan. 1993) 7–9.

51. J. Willems, G. Morthier, and R. Baets, "Novel widely tunable integrated optical filter with high spectral selectivity," Proc. ECOC '92, paper WeB9.2 (Berlin, Germany, September 1992) pp. 413–416.

52. J. Willems and R. Baets, "Integrated tunable optical filter," European Patent, no. EP 0613571 B1, filed September 17, 1993, published July 29, 1998.

53. M. Öberg, S. Nilsson, K. Streubel, L. Bäckbom, and T. Klinga, "74 nm wavelength tuning range of an InGaAsP/InP vertical grating assisted codirectional coupler laser with rear sampled grating reflector," IEEE Photonic. Technol. Lett., 5:7 (Jul. 1993) 735–738.

54. P. Zorabedian, "Tunable external-cavity lasers," in Tunable lasers handbook (Academic Press, San Diego, CA, 1995).

55. M. Bagley, R. Wyatt, D. J. Elton, H. J. Wickes, P. C. Spurdens, D. M. Coper, and W. J. Devlin, "242 nm continuous tuning from a GRIN-SC-MQW-BH InGaAsP laser in an extended cavity," Electron. Lett., 26:4 (Feb. 1990) 267–269.

56. F. Favre and D. Le Guen, "82 nm of continuous tunability for an external cavity semiconductor laser," Electron. Lett., 27:2 (Jan. 1991) 183–184.

57. W. R. Trutna, Jr. and L. F. Stokes, "Continuously tuned external cavity semiconductor laser," J. Lightwave Technol., 11:8 (Aug. 1993) 1279–1286.

58. M. Littman and H. Metcalf, "Spectrally narrow pulsed dye laser without beam expander," Appl. Opt., 17 (1978) 2224–2227.

59. J. D. Berger, Y. Zhang, J. D. Grade, H. Lee, S. Hrinya, and H. Jerman, "Widely tunable external cavity diode laser based on a MEMS electrostatic rotary actuator," Proc. OFC'2001, paper TuJ2 (Anaheim, CA, USA, March 2001).

60. D. Vakhshoori, P. Tayebati, C.-C. Lu, M. Azimi, P. Wang, J.-H. Zhou, and E. Canoglu, "2 mW CW singlemode operation of a tunable 1550 nm vertical cavity surface emitting laser with 50 nm tuning range," Electron. Lett., 35:11 (May 1999) 900–901.

61. M. G. Young, U. Koren, B. I. Miller, M. Chien, T. L. Koch, D. M. Tennant, K. Feder, K. Dreyer, and G. Raybon, "Six wavelength laser array with integrated amplifier and modulator," Electron. Lett., 31:21 (Oct. 1995) 1835–1836.

62. J. Hong, H. Kim, F. Shepherd, C. Rogers, B. Baulcomb, and S. Clemens, "Matrix-grating strongly gain-coupled (MG-SGC) DFB lasers with 34 nm continuous wavelength tuning range," IEEE Photonic. Technol. Lett., 11:5 (May 1999) 515–517.

63. M. Zirngibl, C. H. Joyner, C. R. Doerr, L. W. Stulz, and H. M. Presby, "An 18-channel multifrequency laser," IEEE Photonic. Technol. Lett., 8:7 (Jul. 1996) 870–872.

64. C. R. Doerr, C. H. Joyner, and L. W. Stulz, "40-wavelength rapidly digitally tunable laser," IEEE Photonic. Technol. Lett., 11:11 (Nov. 1999) 1348–1350.

65. H. Bissessur, C. Graver, O. Le Gouezigou, A. Vuong, A. Bodéré, A. Pinquier, and F. Brillouet, "WDM operation of a hybrid emitter integrating a wide-bandwidth on-chip mirror," IEEE J. Selected Topics Quantum Electron., 5:3 (May/Jun. 1999) 476–479.

66. T. Shiba, S. Funaba, T. Nakayama, E. Ishimura, K. Takagi, T. Sakaino, T. Aoyagi, K. Kagahama, and M. Aiga, "High performance laser diode integrated with a monitoring photodiode fabricated with precisely controlled dry etching technique," Proc. ECOC'96, paper ThD.3.6 (Oslo, Norway, September 1996) pp. 5.75–5.78.

67. M. Aoki, M. Suzuki, H. Sano, T. Kawano, T. Ido, T. Taniwatari, K. Uomi, and A. Takai, "InGaAs/InGaAsP MQW electroabsorption modulator integrated with a DFB laser fabricated by band-gap energy control selective area MOCVD," IEEE J. Quantum Electron., 29:6 (Jun. 1993) 2088–2096.

68. O. Kjebon, R. Schatz, S. Lourdudoss, S. Nilsson, B. Stålnacke, and L. Bäckbom, "30 GHz direct modulation bandwidth in detuned loaded InGaAsP DBR lasers at 1.55µm wavelength," Electron. Lett., 33:6 (Mar. 1997) 488–489.

69. G. Morthier, G. Sarlet, R. Baets, R. O'Dowd, H. Ishii, and Y. Yoshikuni, "The direct modulation bandwidth of widely tunable DBR laser diodes," Proc. XVII International Semiconductor Laser Conference, paper P12 (Monterey, CA, USA, September 2000).

70. A. A. Saavedra, P.-J. Rigole, E. Goobar, R. Schatz, and S. Nilsson, "Amplitude and frequency modulation characteristics of widely tunable GCSR lasers," IEEE Photonic. Technol. Lett., 10:10 (Oct. 1998) 1383–1385.

71. J. Johnson, L. Ketelsen, J. M. Geary, F. Walters, J. Freund, M. Hybertsen, K. Glogovsky, C. Lentz, W. Asous, P. Parayanthal, T. Koch, and R. Hartman, "10 Gb/s transmission using an electro-aborption-modulated distributed Bragg reflector laser with integrated semiconductor optical amplifier," Proc. OFC'2001, paper TuB3 (Anaheim, CA, USA, March 2001).

72. B. Mason, G. Fish, S. DenBaars, and L. Coldren, "Widely tunable sampled grating DBR laser with integrated electroaborption modulator," IEEE Photonic. Technol. Lett., 11:6 (Jun. 1999) 638–640.

73. L. Coldren and S. Corzine, Diode lasers and photonic integrated circuits (Wiley, New York, 1995).

74. S.-L. Lee, M. E. Heimbuch, D. A. Cohen, L. A. Coldren, and S. P. DenBaars, "Integration of semiconductor laser amplifiers with sampled grating tunable

lasers for WDM applications," IEEE J. Selected Topics Quantum Electron., 3:2 (Apr. 1997) 615–627.

75. B. Mason, S. Denbaars, and L. Coldren, "Tunable sampled-grating DBR lasers with integrated wavelength monitors," IEEE Photonic. Technol. Lett., 10:8 (Aug. 1998) 1085–1087.

76. H. Debrégeas-Sillard, A. Vuong, F. Delorme, J. David, V. Allard, A. Bodéré, O. LeGouezigou, F. Gaborit, J. Rotte, M. Goix, V. Voiriot, and J. Jacquet, "DBR module with 20-mW constant coupled output power, over 16 nm (40 × 50-GHz spaced channels)," IEEE Photonic. Technol. Lett., 13:1 (Jan. 2001) 4–6.

77. G. A. Fish, "Monolithic, widely-tunable, DBR lasers," Proc. OFC'2001, paper TuB1 (Anaheim, CA, USA, March 2001).

78. P.-J. Rigole, S. Nilsson, L. Bäckbom, B. Stålnacke, T. Klinga, E. Berglind, B. Stoltz, D. J. Blumenthal, and M. Shell, "Wavelength coverage over 67 nm with a GCSR laser. Tuning characteristics and switching speed," Proc. XV IEEE Int. Semiconductor Laser Conference, pp. 125–126, paper W1.1 (Haifa, Israel, October 1996).

79. T. Day, "External-cavity tunable diode lasers for network deployment," Proc. OFC'2001, paper TuJ4 (Anaheim, CA, USA, March 2001).

80. D. Vakhshoori, J.-H. Zhou, M. Jiang, M. Azimi, K. McCallion, C.-C. Lu, K. J. Knopp, J. Cai, P. D. Wang, P. Tayebati, H. Zhu, and P. Chen, "C-band tunable 6 mW vertical-cavity surface-emitting lasers," Proc. OFC'2000— Postdeadline papers, paper PD13, Baltimore, MD, USA, March 2000.

81. D. Vakhshoori, P. D. Wang, M. Azimi, K. J. Knopp, and M. Jiang, "MEMS-tunable vertical-cavity surface-emitting lasers," Proc. OFC'2001, paper TuJ1 (Anaheim, CA, USA, March 2001).

82. J. Hong, M. Cyr, H. Kim, S. Jatar, C. Rogers, D. Goodchild, and S. Clements, "Cascaded strongly gain-coupled (SGC) DFB lasers with 15-nm continuous wavelength tuning," IEEE Photonic. Technol. Lett., 11:10 (Oct. 1999) 1214–1216.

Chapter 5 | Vertical-Cavity Surface-Emitting Laser Diodes

Kenichi Iga

The Japan Society for the Promotion of Science
6 Ichibancho, Chiyodaku, 102-8471, Japan

Fumio Koyama

Precision & Intelligence Lab.,
Tokyo Institute of Technology,
4259 Nagatsuta, Midoriku, Yokohama, 226-8503, Japan

ABSTRACT

The vertical-cavity surface-emitting laser (VCSEL) becomes a key laser device in optical high-speed LANs by taking the advantage of low power consumption and high speed modulation capability. This device also enables ultra-parallel data transfer in digital equipment and computer systems. Another important feature is its wide range of continuous wavelength tunability, which is utilized in single-mode silica fiber systems for metropolitan area networks (MANs). In this chapter, we will review its history, structures, and design concept. Then, we introduce the progress of VCSELs, covering the spectral band for optical communication by looking at their fabrication technology, and performance issues such as threshold, output power, polarization, modulation, reliability, and so on. Lastly, we will touch on some applied systems.

Key Words: Surface-emitting laser, Vertical-cavity surface-emitting laser, VCSEL, Laser array, Distributed Bragg reflector, DBR, Gigabit Ethernet, LAN, Interconnect, Microlens.

5.1.   Introduction

The structure of a surface-emitting (SE) laser or vertical-cavity surface-emitting laser (VCSEL) is substantially different from that of conventional stripe lasers. For example, the vertical cavity is formed with the surfaces of epitaxial layers, and light output is taken from one of the mirror surfaces as shown in Fig. 5.1.

As seen from Table 5.1, the vertical-cavity surface-emitting laser (VCSEL) [1, 2] is meeting the 3rd generation of development as we enter a new information technology era in the 3rd millennium. The VCSEL is

167

WDM TECHNOLOGIES: ACTIVE
OPTICAL COMPONENTS
$35.00

Table 5.1   **History of VCSEL Research**

I	1977	First idea and initial demonstrations
II	1988	CW and device feasibility study
III	1999	Production and extension of applications

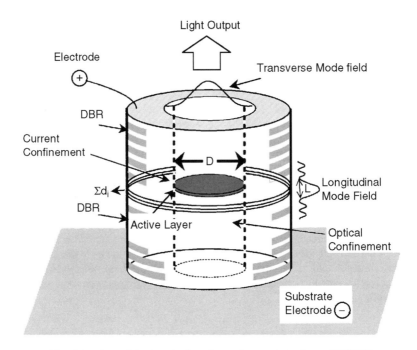

Fig. 5.1   A model of vertical-cavity surface-emitting laser (VCSEL).

being applied in various optical systems such as optical fiber networks, parallel optical interconnects, laser printers, high-density optical disks, and so on. We first review its history and the progress of VCSELs in wide spectral ranges covered by various III-V compound semiconductors.

It is recognized that one of the present authors (K. Iga) suggested a VCSEL device in 1977, and the first device came out in 1979, where we used GaInAsP/InP for the active region, emitting 1300 nm-wavelength light [3]. In 1986, we made a 6 mA threshold GaAs device [4]. Then we employed the metal-organic chemical vapor deposition (MOCVD) for its crystal growth, and the first room-temperature continuous wave (CW) laser using GaAs material was demonstrated in 1988 [5]. After that, in 1989, Jack Jewell of AT&T demonstrated a GaInAs VCSEL exhibiting a few mA threshold [6]. These two experiments encouraged researchers to get into the technical field of vertical-cavity surface-emitting lasers. Sub-milli-ampere threshold devices were demonstrated by improving the quality of the active region and laser cavity.

Since 1992, VCSELs based on GaAs have been extensively studied [7–9] and some 980, 850, and 780 nm devices have been commercialized and utilized in various optical systems. In 1993, the author's group demonstrated a 1300 nm room-temperature CW device [10]. A wafer fusion technique enabled us to operate 1550 nm VCSELs at higher temperatures [11]. In 1993, a room-temperature high-performance CW red color AlGaInAs device was demonstrated [12]. Since 1996, green-blue-ultraviolet device research has been ongoing [13, 14]. Since 1999, VCSEL-based optical tranceivers have been introduced into Giga-bit/sec Ethernet and high speed local area networks.

The initial motivation of surface-emitting laser invention was fully mono-lithic fabrication of the laser cavity. The current issues include, based on this concept, high speed modulation capability at very low power consumption level, reproducible array production, inexpensive moduling, and so on. The VCSEL structure may provide a number of advantages as follows:

1. Laser devices can be fabricated by a fully monolithic process, yielding very low-cost production.
2. Laser cavity can be completed before separation into individual chips.
3. Ultra-low threshold operation is expected from its small cavity volume reaching micro-Ampere levels.
4. Dynamic single-mode operation is possible.
5. High-speed modulation beyond 10 GBits/s is possible even at low driving ranges.
6. Wide and continuous wavelength tuning is possible.
7. Temperature independent operation is allowable, which yields no power controller operation.
8. Power conversion efficiency is greater than 50%.
9. High power and low power devices are subject to design.
10. Device has high reliability due to completely embedded active region and passivated surfaces.
11. Vertical and circular beam is inherently provided.
12. Coupling to optical fibers is easy due to good mode matching from single mode through thick multi-mode fibers.

13. Bonding and mounting are easy.
14. Modules and packages costs are cheap.
15. Densely packed and precisely arranged two-dimensional laser arrays can be formed.
16. Vertical stack integration of multi-thin-film functional optical devices can be made intact to VCSEL resonator, taking the advantage of micro-machining technology(MEMS) providing polarization independent characteristics.
17. Integration is compatible together with LSIs.

In this chapter, we will review the progress of VCSELs in a wide range of optical spectra based on GaInAsP, AlGaInAs, GaInNAs, GaInAs, AlGaAsSb, GaAlAs, AlGaInP, ZnSe, GaInN, and some other materials.

5.2. Scaling Laws

5.2.1. THRESHOLD CURRENT

The physical difference of vertical-cavity surface-emitting lasers (VCSELs) and conventional stripe geometry lasers is summarized in Table 5.2. The major point is the cavity length. That of VCSELs is on the

Table 5.2   **Comparison of Parameters Between Stripe Laser and VCSEL**

Parameter	Symbol	Stripe Laser	Surface Emitting Laser
Active layer thickness	d	100 Å–0.1 μm	80 Å–0.5 μm
Active layer area	S	$3 \times 300\,\mu m^2$	$5 \times 5\,\mu m^2$
Active volume	V	$60\,\mu m^3$	$0.07\,\mu m^3$
Cavity length	L	300 μm	$=1\,\mu m$
Reflectivity	R_m	0.3	0.99–0.999
Optical confinement	ξ	=3%	=4%
Optical confinement (transverse)	ξ_t	3–5%	50–80%
Optical confinement (longitudinal)	ξ_l	50%	$2 \times 1\% \times 3$ (3QW's)
Photon lifetime	τ_p	=1 ps	=1 ps
Relaxation frequency (Low current levels)	f_r	<5 GHz	>10 GHz

Table 5.3 **Applications of VCSELs**

Technical Fields	Systems
1. Optical communications	LANs, Optical links, Mobile links, etc.
2. Computer optics	Computer links, Optical interconnects, High speed/Parallel data transfer, etc.
3. Optical memory	CD, DVD, Near field, Multi-beam, Initializer, etc.
4. Optoelectronic equipments	Printer, Laser pointer, Mobile tools, Home appliances, etc.
5. Optical information processing	Optical processors, Parallel processing, etc.
6. Optical sensing	Optical fiber sensing, Bar code readers, Encoders, etc.
7. Displays	Array light sources, Multi-beam search-lights,
8. Illuminations	High efficiency sources, Micro illuminators, etc. Adjustable illuminations, etc.

order of the wavelength, whereas that of stripe lasers is about 300 μm. This provides us with substantial differences in laser performance.

The threshold current I_{th} of vertical-cavity surface-emitting lasers can be expressed by the equation with threshold current density J_{th} as

$$I_{th} \cong \frac{e V B_{eff}}{\eta_i \eta_{spon}} N_{th}^2 \qquad (5.1)$$

where e is electron charge, V is the volume of active region, N_{th} is the threshold carrier density, B_{eff} is the effective recombination coefficient, η_i is injection efficiency (sometimes referred to as internal efficiency), and η_{spon} is spontaneous emission efficiency.

As seen from Eq. (5.1), we recognize that it is essential to reduce the volume of the active region in order to decrease the threshold current. Assume that the threshold carrier density does not change much, if we reduce the active volume, we can decrease the threshold as we make a small active region. We compare the dimensions of surface-emitting lasers and conventional stripe geometry lasers as already shown in Table 5.2. It is noticeable that the volume of VCSELs could be $V = 0.06 \, \mu m^3$, whereas that for stripe lasers it remains as $V = 60 \, \mu m^3$. This directly reflects that the typical threshold of stripe lasers is ranging mA or higher, but that for

VCSELs is able to be less than mA by a simple carrier confinement scheme such as proton bombardment. It could even be as low as micro-Ampere by implementing sophisticated carrier and optical confinement structures as will be introduced later.

An early stage estimation of threshold showed that the threshold current can be reduced proportional to the square of the active region diameter. However, there should be a minimum value originating from the decrease of optical confinement factor that is defined by the overlap of optical mode field and gain region when the diameter is becoming small. In addition to this, the extreme minimization of volume, in particular in the lateral direction, is limited by the optical and carrier losses due to optical scattering, diffraction of lightwave, and nonradiative carrier recombination, and other technical imperfections.

5.2.2.   OUTPUT POWER AND QUANTUM EFFICIENCY

If we use a non-absorbing mirror for the front reflector of the VCSEL, the differential quantum efficiency η_d from the front mirror is expressed as

$$\eta_d = \eta_i \frac{(1/L)\ln(1/R_f)}{\alpha + (1/L)\ln\left(1/\sqrt{R_f R_r}\right)} \tag{5.2}$$

where α is the total internal loss, and R_f and R_r are front and rear mirror reflectivity.

The optical power output is expressed by

$$P_o = \eta_d \eta_{spon} C E_g I \qquad (I \leq I_{th})$$

$$= \eta_d \eta_{spon} C E_g I_{th} + \eta_d E_g (I - I_{th}) \quad (I \geq I_{th}) \tag{5.3}$$

where E_g is bandgap energy, C is spontaneous emission factor, and I is driving current. On the other hand, the power conversion efficiency η_P far above the threshold is given by

$$\eta_P = \frac{P_o}{V_b I} = \eta_d \frac{E_g}{V_b}\left(1 - \frac{I_{th}}{I}\right) \cong \eta_d \frac{E_g}{V_b} \cong \eta_d \tag{5.4}$$

where V_b is a bias voltage and the spontaneous component has been neglected. In the case of a surface-emitting laser, the threshold current could be very small, and therefore, the power conversion efficiency can be relatively large, i.e., higher than 50%. (The power conversion efficiency is sometimes called wall-plug efficiency.)

The modulation bandwidth is given by

$$f_{3dB} = 1.55 f_r \tag{5.5}$$

where f_r denotes the relaxation frequency, which is expressed by the equation,

$$f_r = \frac{1}{2\pi\tau_s}\sqrt{\frac{\tau_s}{\tau_p}\left(\frac{I}{I_{th}} - 1\right)} \tag{5.6}$$

The photon lifetime τ_p is given by

$$\tau_p = \frac{n_{eff}/c}{\alpha + \alpha_m} \tag{5.7}$$

When the threshold current I_{th} is negligible to the driving current I, f_r can be expressed as

$$f_r \cong \frac{1}{2\pi\tau_s}\sqrt{\frac{\tau_s}{\tau_p}\frac{I}{I_{th}}}$$

$$= \frac{1}{2\pi\tau_s}\sqrt{\frac{\tau_s}{\tau_p}\frac{\eta_i\eta_{spon}I}{e B_{eff}N_{th}^2 V}} \tag{5.8}$$

The relaxation frequency is inversely proportional to the square root of the active volume and it can be larger, if we can reduce the volume as small as possible.

The photon lifetime is normally on the order of pico-sec, which can be made slightly smaller than stripe lasers. Because the threshold current can be very small in VCSELs, the relaxation frequency could be relatively higher than stripe lasers even in low driving ranges. The threshold carrier density N_{th} can be expressed in terms of photon lifetime, which represents the cavity loss and is given by using Eqs. (5.3) and (5.7);

$$N_{th} = N_t + \frac{1}{(c/n_{eff})}\frac{1}{(dg/dN)}\frac{1}{\xi}\frac{1}{\tau_p} \tag{5.9}$$

It is noted that the threshold carrier density can be small when we make the differential gain dg/dN, confinement factor ξ, and photon lifetime τ_p large.

The tuning wavelength bandwidth $\delta\lambda$ of semiconductor laser is determined by its free spectrum range, as follows:

$$\delta\lambda = \frac{\lambda^2}{2n_{eff}L_{eff}} \tag{5.10}$$

This is inversely proportional to an effective cavity length L_{eff} and an effective refractive index n_{eff}. This means that the effective cavity length in VCSELs can be as large as almost one wavelength and very wide continuous tuning range is available.

5.2.3. CRITERIA FOR CONFIRMATION OF LASING

When we face a new or ultra-low threshold device, the existence of a break from the linear increase of light output versus injection current (I-L) characteristic is an easy observation for checking the lasing operation. Sometimes, a nonlinearity is observed in the I-L characteristic, but this does not necessarily confirm laser oscillation. Even with a non-lasing sample this can be seen owing to "filtering effect" and electron-hole plasma emission, nonradiating floor, and so on. The methods to definitely confirm the laser operation of the vertical cavity, for example, are as follows;

1. Break or kink in current vs. light output (I-L) characteristic
2. Narrow spectral linewidth (<1 Å).
3. Difference of near-field pattern (NFP) and far-field pattern (FFP) between the emissions below and above the threshold.
4. Linearly polarized light of the emission above the threshold.

5.3. Device Structures and Design

5.3.1. DEVICE CONFIGURATION

As already shown in Fig. 5.1, the structure common to most VCSELs consists of two parallel reflectors that sandwich a thin active layer. The reflectivity necessary to reach the lasing threshold should normally be higher than 99.9%. Together with the optical cavity formation, the scheme for injecting electrons and holes effectively into small volume of active region is necessary for current injection device. The ultimate threshold current depends on how to make the active volume small as introduced in the previous section and how well the optical field can be confined in the cavity to maximize the overlap with the active region. These confinement structures will be presented in the later sections.

5.3.2. MATERIALS

We show some choices of the materials for VCSELs in Fig. 5.2. Here are some of the problems that should be considered for making vertical-cavity VCSELs, as discussed in the previous section.

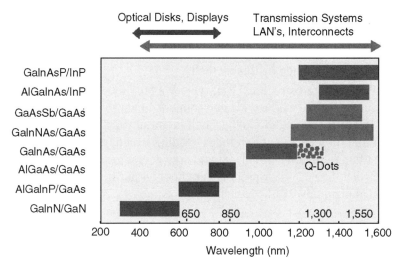

Fig. 5.2   Materials for VCSELs in wide spectral bands.

1. design of resonant cavity and mode-gain matching
2. multi-layered distributed Bragg reflectors (DBRs) to realize high reflective mirrors
3. optical losses such as Auger recombination, intervalence band absorption, scattering loss, and diffraction loss
4. p-type doping to reduce the resistivity in p-type materials for CW and high efficiency operation. If we wish to form multi-layer DBRs, this will become much more severe.
5. heat sinking for high temperature and high power operation
6. COD (Catastrophic Optical Damage) level, very important for high power operation
7. Crystal growth at reasonably high temperatures (e.g., higher than half of melting temperatures)

5.3.3.   CURRENT INJECTION SCHEME

Let us consider the current confinement for VCSELs. Some typical models of current confinement schemes reported so far are as follows:

1. Ring electrode type: This structure can limit the current flow in the vicinity of the ring electrode. The light output can be taken out from the center window. This is easy to fabricate, but the current cannot completely be confined in a small area due to diffusion.

2. Proton bombardment type: We make an insulating layer by proton (H+) irradiation to limit the current spreading toward the surrounding area. The process is rather simple and most commercialized devices are made by this method.

3. Buried-Heterostructure (BH) type: We bury the mesa, including the active region, with a wide gap semiconductor to limit the current. The refractive index can be small in the surrounding region, resulting in formation of an index-guiding structure. This is an ideal structure in terms of current and optical confinement. The problem is that the necessary process is rather complicated, in particular, in making a tiny 3D device.

4. Air-post type: The circular or rectangular air-post is used to make a current confinement. It is the simplest means of device fabrication, but nonradiative recombination at the outer wall may deteriorate the performance.

5. Selective AlAs oxidation type: We oxidize AlAs layer to make a transparent insulator.

6. Oxidized DBR type: The same method is applied to oxidize DBR consisting of AlAs and GaAs. This is one of the volume confinement methods and can reduce the nonradiative recombination.

By developing fine process technology, we could reach laser performance expected from the theoretical limit.

5.3.4. OPTICAL GUIDING

Some optical confinement schemes were developed for VCSELs. The fundamental concept is to increase the overlap of the optical field with the gain region.

1. Fabry-Perot type: The optical resonant field is determined by two reflectors, which form a plane parallel to the Fabry-Perot cavity. The diffraction loss increases if the mirror diameter gets too small.

2. Gain-guide type: We simply limit the field at the region where the gain exists. The mode may be changed at higher injection levels due to spatial hole burning.

3. Buried Heterostructure (BH): As introduced in the previous section, an ideal index-guiding structure can be formed.

4. Selective AlAs oxidation type: Due to the index difference between AlAs and the oxidized region, we can confine the optical field as well by a kind of lens-effect.

5. Anti-guiding type: The index is designed to be lower in the surrounding region to make a so-called anti-guiding scheme. The threshold is rather high, but this structure is good for stable mode in high driving levels.

5.3.5. TRANSVERSE AND LONGITUDINAL MODE

The resonant mode in most surface-emitting lasers can be expressed by the well-known Fabry-Perot TEM mode. The near-field pattern (NFP) of fundamental mode can be given by the Gassiann function

$$E = E_0 \exp\left[-\frac{1}{2}(r/s)^2\right] \tag{5.11}$$

where E is optical field, r is lateral distance, and s denotes the spotless.

The spot size of normal surface-emitting lasers is several microns and relatively large compared with stripe lasers, say, 2–3 μm. In the case of multi-mode operation, the mode behaves like the combination of multiple TEM_{pq}. The associated spectrum is broadened due to different resonant wavelengths.

The far-field pattern (FFP) associated with Gaussian near field can be expressed by Gaussian function and spreading angle $\Delta\theta$ as given by the equation

$$2\Delta\theta = 0.64(\lambda/2s) \tag{5.12}$$

Here, if $s = 3\,\mu\text{m}$ and $\lambda = 1\,\mu\text{m}$, $\Delta\theta = 0.05\,(\text{rad}) = \cong 3°$. This kind of angle is narrower than conventional stripe lasers.

5.3.6. POLARIZATION MODE

The VCSEL generally has linear polarization without exception. This is due to a small amount of asymmetric loss coming from the shape of the device or material. The device grown on (100)-oriented substrate polarizes in (110) or equivalent orientations. The direction cannot be identified definitely and sometimes switches over due to spatial hole burning or temperature variation. In order to stabilize the polarization mode, special care should be taken. This issue will be discussed later.

5.4.   Surface-Emitting Laser in Long Wavelength Band

5.4.1.   GaInAsP/InP VCSEL

The first surface-emitting laser device was demonstrated by using GaInAsP/InP system in 1978 and published in 1979 [3]. The importance of 1300 or 1550 nm devices is currently increasing, because parallel lightwave systems are really needed to meet the rapid increase of information transmission capacity in local area networks (LANs). However, the GaInAsP/InP system conventionally used in trunk communication systems with the help of a temperature controller has some substantial difficulties for making VCSELs due to the following reasons;

1. The Auger recombination and inter-valence band absorption (IVBA) are noticeable.
2. The index difference between GaInAsP and InP is relatively small to make DBR mirrors.
3. Conduction band offset is small.

A hybrid mirror technology is being developed. One technique is to use a semiconductor/dielectric reflector [15]. Thermal problems for CW operation are extensively studied. A MgO/Si mirror with good thermal conductivity was demonstrated to achieve the first room temperature CW operation at 1300 nm surface-emitting lasers [10]. Better results have been obtained by using Al_2O_3/Si mirrors [16].

The other technique is epitaxial bonding of GaInAsP/InP active region and GaAs/AlAs mirrors, where 144°C pulsed operation was achieved by optical pumping. The CW threshold of 0.8 mA [17] and the maximum operating temperature up to 69°C [11] have been reported for 1550 nm VCSELs with double-bonded mirrors [17]. More recently, the maximum operation was achieved at 71°C [18]. In 1998, a tandem structure of 1300 nm VCSEL optically pumped by 850 nm VCSEL was demonstrated to achieve 1.5 mW of output power [19]. However, the cost of wafer consumption in wafer fusion devices may become the final bottleneck of low-cost commercialization.

For the purpose of improving the performance of 1300 nm and 1550 nm wavelength VCSELs, a good thermal conductive mirror consisting of MgO/Si was employed and the first room-temperature CW operation was achieved at 1300 nm [4]. The AlGaInAs/InP system can provide a good temperature characteristic due to large conduction band offset, and together with an AlAs/AlInAs superlattice MQB for oxide aperture and a good temperature characteristic was demonstrated in edge emitters [20].

The AlAs/GaAs mirror has advantages in both electrical and thermal conductivity. A wafer fusion has been adopted to combine the GaInAsP/InP active layer and AlAs/GaAs DBRs. A 1550 nm VCSEL exhibiting 66°C CW operation and 0.8 mA threshold using selective AlAs oxidation was reported [11]. The direct growth of AlAs/GaAs DBR on InP based active layer was also demonstrated [21]. A photo-pumped 1300 nm VCSEL with an integrated 850 nm pump VCSEL was demonstrated exhibiting a few mW and operation up to 80°C [19].

Recently, a GaAsSb QW on GaAs substrate has been demonstrated for the purpose of 1300 nm VCSELs [22]. An AlGaAsSb/GaAs system has been found to form a good DBR [23]. A tunnel junction and AlAs oxide confinement structures may be very helpful for long wavelength VCSEL innovation [24, 25].

5.4.2. AlGaInAs/AlGaInAs VCSEL

The AlGaInAs lattice matched to InP is also considered. This system may exhibit a larger conduction band offset than the conventional GaInAsP system. Moreover, we can grow a thin AlAs layer to make the native oxide for current confining aperture like the GaAs/AlAs system. The preliminary study has been made to demonstrate a stripe laser in the author's group, where a large T_0 was demonstrated [20]. By using this system the first monolithic VCSEL was fabricated demonstrating room-temperature CW operation [26].

5.4.3. LONG WAVELENGTH VCSELs ON GaAs SUBSTRATE

The long-wavelength VCSEL formable on GaAs as shown in Fig. 5.3 will have a great impact on the realization of high-performance devices [27]. Every GaAs based structure can be applied and a large conduction band offset is expected. GaInNAs system has been pioneered by Kondow et al. [28] by a gas source molecular beam epitaxy (GSMBE) and 1190 nm stripe lasers were fabricated, where the nitrogen content is 0.4%. Room-temperature CW operation of horizontal cavity lasers has recently been obtained exhibiting the threshold current density of 1.5 kA/cm^2. Also, stripe geometry lasers were demonstrated having the threshold of 24 mA at room temperature [29]. It is reported that the characteristic temperature

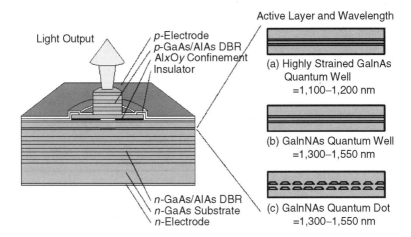

Fig. 5.3 Long-wavelength VCSELs on GaAs substrate (after T. Miyamoto, unpublished).

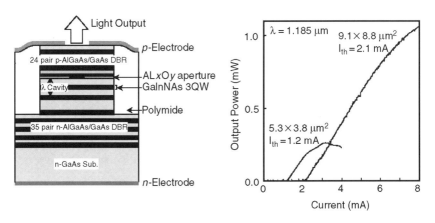

Fig. 5.4 Structure and performances of GaInNAs/GaAs VCSEL (after Kageyama *et al.* [33]).

is 120 K at around room temperature [29]. Some 1300 nm edge-emitting lasers and a 1186 nm VCSELs were demonstrated [30].

If we can increase the nitrogen content up to 5%, the wavelength band of 1300–1550 nm may be covered. In particular, GaAs/AlAs Bragg reflectors can be incorporated on the same substrate, and AlAs oxidation is utilized. Some consideration of device design was presented [31]. In any case, this system will substantially change the surface-emitting laser performances in the long wavelength range. We achieved a lasing operation in GaInNAs edge emitters grown by chemical beam epitaxy (CBE) demonstrating $T_0 >$ 270 K and a VCSEL as shown in Fig. 5.4 [32, 33].

During the research of GaInNAs lasers we found that a highly strained GaInAs/GaAs system containing large In-content ($\cong$40%) can provide an excellent temperature characteristic [34], i.e., operating with $T_0 > 200$ K [35]. This system should be viable for $\lambda > 1200$ nm VCSELs for silica-fiber-based high speed LANs [36].

A quantum dot structure is considered as a long-wavelength active layer on GaAs substrate. Room temperature continuous wave operation of 1300 nm GaInAs-dot VCSEL was reported with a threshold current of 0.5 mA [37].

5.5. Surface-Emitting Laser in Mid-Wavelength Band

5.5.1. 980–1200 nm GaInAs/GaAs VCSEL

The GaInAs/GaAs strained pseudomorphic system grown on a GaAs substrate emitting at 980 nm of wavelength exhibits a high laser gain and has been introduced into surface-emitting lasers together with using GaAs/AlAs multi-layer reflectors [38]. A low threshold (1 mA at CW) has been demonstrated by Jewell *et al.* [6] The threshold current of vertical-cavity surface-emitting lasers has been reduced down to sub-milliampere orders in various institutions in the world. Very low thresholds reported before 1995 were 0.7 mA [7], 0.65 mA [8], 0.2 mA [39]. Moreover, a threshold of 91 µA at room-temperature CW operation was reported by introducing the oxide current and optical confinement [40]. The theoretical expectation was 10 µA or less, if some good current and optical confinement structure could be introduced.

It has been made clear that the oxide aperture can function as a focusing lens, since the central window has a higher index and the oxide region exhibits a lower index. This provides us some phase shift to focus the light toward the center axis to reduce the diffraction loss. The Al-oxide is effective both for current and optical confinements and solves the problems on surface recombination of carriers and optical scattering. The author's group demonstrated 70 µA [41, 42] of threshold by using oxide DBR structure, the university of Texas achieved 40 µA [43], and USC reported 8.5 µA [44].

In 1995, we developed a novel laser structure employing a selective oxidizing process applied to AlAs, which is one of the members of the multi-layer Bragg reflector [41, 42, 45]. The active region is three quantum wells consisting of 80 Å GaInAs strained layers. The Bragg reflector consists of GaAs/AlAs quarter wavelength stacks of 24.5 pairs. After etching the epitaxial layers, including the active layer and two Bragg reflectors, the sample was treated in the high-temperature oven with water vapor, which is bubbled by nitrogen gas. The AlAs layers are oxidized preferentially with this process and native oxide of aluminum is formed at the periphery of the etched mesas. It is recognized from the SEM picture that only AlAs layers in DBR have been oxidized [41]. The typical size is a 20 μm core starting from a 30 μm mesa diameter. We achieved about 1 mW of power output and submicro-ampere threshold. The nominal lasing wavelength is 980 nm. We have made a smaller diameter device having a 5 μm core started from a 20 μm mesa. The minimum threshold achieved is 70 μA at room-temperature CW operation [41].

A relatively high power, higher than 50 mW, is becoming possible [46]. Power conversion efficiency of 50% is reported [47]. Also, high efficiency operation at relatively low driving levels, i.e., a few mA, became possible, which has been hard to achieve in stripe lasers. This is due to the availability of low resistive DBRs incorporating an Al-oxide aperture. Actually, in devices of about 1 μm in diameter, higher than 20% of power conversion efficiencies was reported [48, 49].

Regarding the power capability, near 200 mW has been demonstrated by a large size device at the University of Ulm [50]. In a two-dimensional array involving 1000 VCSELs with active cooling, more than 2 W of CW output was achieved [51].

In these low power consumption devices, high-speed modulation is possible in low driving currents around 1 mA as well. This is especially important in low power interconnect applications enabling > 10 Gbits/s transmission or 1 Gb/s zero-bias operation [52]. Actually, transmission experiments over 10 Gbits/s and zero-bias transmission have been reported. We measured an eye diagram for 10 Gbits/s transmission experiment through a 100 m multimode fiber [53].

Finally, VCSELs in this wavelength may find a market in 10 Gigabit LANs together with high speed detectors and silica fibers. In many ways, GaInAs VCSELs show the best performance and research to challenge the extreme characteristics will be continued.

5.5.2.   980–1200 nm GaInAs/GaAs VCSEL ON GaAs (311) SUBSTRATE

Most VCSELs grown on GaAs (100) substrates show unstable polarization states due to isotropic material gain and symmetric cavity structures. VCSELs grown by MBE on GaAs (311)A substrates, however, show a very stable polarization state [54]. Also, trials of growth on (GaAs)B substrates

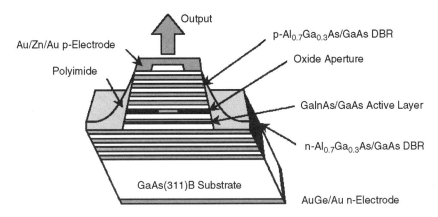

Fig. 5.5   Schematic structure of polarization controlled VCSEL on (311) GaAs substrate (after Nishiyama *et al.* [60]).

by using MOCVD have been performed [55–57]. Single transverse mode and polarization mode controlled VCSELs had not been realized at the same time.

In this section, we introduce a single transverse mode and polarization controlled VCSEL grown on a GaAs (311)B substrate. Both higher-order transverse modes and a non-lasing orthogonal polarization mode are well suppressed with a suppression ratio of over 25 dB [58].

The schematic structure of a fabricated top-emitting VCSEL grown on GaAs (311)B by low pressure MOCVD is shown in Fig. 5.5 [59, 60]. The bottom n-type distributed Bragg reflector (DBR) consists of 36 pairs of $Al_{0.7}Ga_{0.3}As$/GaAs doped with Se. The top p-type DBR consists of 21 pairs of Zn-doped $Al_{0.7}Ga_{0.3}As$/GaAs and a 70 Å thick AlAs carbon high-doping layer inserted at the upper AlGaAs interface by the carbon auto-doping technique proposed by us.[53] The active layer consists of three 8 nm-thick $In_{0.2}Ga_{0.8}As$ quantum wells and 10 nm GaAs barriers surrounded by $Al_{0.2}Ga_{0.8}As$ to form a cavity. An 80 nm-thick AlAs was introduced on the upper cavity spacer layer to form an oxide confinement. We oxidized the AlAs layer of etched 50 μm × mx 50 μm mesa at 480°C for 5 minutes in an N_2/H_2O atmosphere by bubbling in 80°C water and formed an oxide aperture of 2.5 μm × 3.0 μm.

Figure 5.6 shows a typical current-light (I-L) of a 1150 nm highly strained GaInAs/GaAs VCSEL under uncooled operation [61]. The threshold current is below 1 mA at room temperature, which is comparable to the value reported for non-(100) substrate VCSELs. The threshold voltage

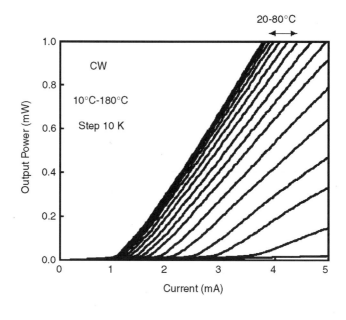

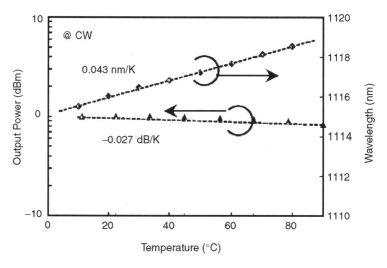

Fig. 5.6 (a) *L/I* characteristics and (b) temperature dependences of output power and wavelength (after Nishiyama *et al.* [61]).

is 1.5 V and the maximum output power is 1 mW at 4 mA. We changed the ambient temperature by maintaining drive current so as to obtain 1mW output at room temperature as shown in Fig. 5.6(a). The change of output power was not so large and the result indicates no necessity of a thermo cooler and power controller in system applications.

In other devices, a large side mode suppression ratio (SMSR) of over 35 dB and an orthogonal polarization suppression ratio (OPSR) of over 25 dB were achieved at the same time against the entire tested driving range ($I < 16I_{th}$). The single polarization operation was maintained at 5 GHz modulation condition [57, 60].

The selective oxidation of AlAs is becoming a standard current and optical confinement scheme for mA threshold devices. Technology of mode-stable lasers using (311)B substrate is demonstrated for polarization control [56]. We have obtained completely single-mode VCSEL by employing most of the available advanced techniques. We performed a transmission experiment using 1200 nm VCSEL and single-mode silica fiber as shown in Fig. 5.7 [62]. It should be noted that we could use a high-performance

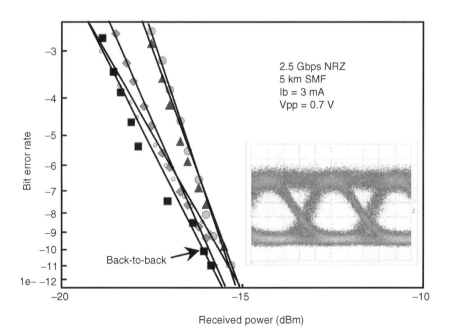

Fig. 5.7 Data transmission experiment of 1.2 μm GaInAs/GaAs multi-wavelength VCSEL (after Arai *et al.* [62]).

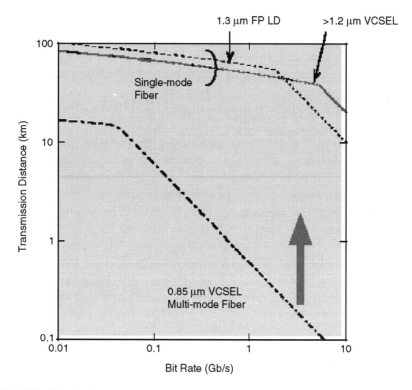

Fig. 5.8   Calculated transmission bandwidth for various light sources (after Koyama *et al.* [36]).

VCSEL and silica fiber for single-mode transmission. We show its potentiality in Fig. 5.8.

5.6.   Surface-Emitting Lasers in Near Infrared-Red Band

5.6.1.   *850 nm GaAlAs/GaAs VCSEL*

A GaAlAs/GaAs laser can employ almost the same circular buried heterostructure (CBH) as the GaInAsP/InP laser. In order to decrease the threshold, the active region is also constricted by the selective meltback method. In 1986, the threshold of 6 mA was demonstrated for the active region 6 μm diameter under pulsed operation [4]. It is noted that a micro-cavity of 7 μm long and 6 μm in diameter was realized.

The MOCVD grown CBH VCSEL was demonstrated by a two-step MOCVD growth and fully monolithic technology [63]. The first room-temperature CW operation was achieved [5]. The lowest CW threshold current was 20 mA. The differential quantum efficiency is typically 10%. The maximum CW output power is about 2 mW. The saturation of the output power is due to a temperature increase of the device. Stable single-mode operation is observed with neither sub-transverse modes nor other longitudinal modes. The spectral linewidth above the threshold is less than 1 Å, which is limited by the resolution of the spectrometer. The mode spacing of this device was 170 Å. The side mode suppression rate (SMSR) of 35 dB is obtained at $I/I_{th} = 1.25$. This is comparable to that of well-designed DBR or DFB dynamic single-mode lasers.

Sub-mA thresholds and 10 mW outputs have been achieved. A power conversion efficiency of 57% has been demonstrated [64]. Some commercial optical links have already been to market. The price of low-skew multimode fiber ribbons may be a key issue for inexpensive multimode-fiber-based data links.

As for the reliability of VCSELs, 10^7 hours of room-temperature operation is estimated based on the acceleration test at high temperature using proton-defined devices [65]. In 1998, some preliminary test results began to be reported on oxide-defined devices exhibiting no substantial negative failures.

5.6.2. 780 nm GaAlAs/GaAs VCSEL

The VCSEL in this wavelength was demonstrated in 1987 by optical pumping, and the first current injection device was developed by Y. H. Lee of AT&T Bell [9]. He moved to KAIST and continues to study VCSELs in this wavelength. If we choose the Al content x to be 0.14 for $Ga_{1-x}Al_xAs$, the wavelength can be as short as 780 nm. This is common for compact disc lasers. When the quantum well is used for the active layer, blue shift should be taken into account. The active layer $Ga_{0.86}Al_{0.14}As$ is formed by a superlattice consisting of GaAs (33.9 Å), and AlAs (5.7 Å), with 14 periods. The DBR is made of $AlAs$-$Al_{0.35}Ga_{0.65}As$-$Al_{0.3}Ga_{0.7}As$-$Al_{0.35}Ga_{0.65}As$ as 1 period. The n-DBR has 28.5 pairs and p-DBR consists of 22 pairs. The threshold in 1991 was 4–5 mA and the output was 0.7–0.8 mW. Later on, the MQW made of Al content ($x = 0.1/0.3$) was introduced and a threshold of 200 μA and output of 1.1 mW were demonstrated [66].

5.6.3. AlGaInP RED VCSEL

Generally, the light-emitting device may have more severe operating problems in short wavelength regions than in longer ones, because the photon energy is large, and p-type doping is technically harder to perform. If aluminum (Al) is included in the system, the degradation due to Al-oxidation is appreciable. The AlGaInP/GaAs system emitting red color in the 630–670 nm range is considered as a laser for the first-generation digital video disc system. GaInAlP/GaAs VCSELs were developed and room-temperature operation exhibiting a submilliampere threshold and 8 mW output and 11% of conversion efficiency have been obtained [67]. The wavelength is 6720 nm with oxide aperture of $2 \, \mu\text{m} \times 3 \, \mu\text{m}$. The threshold is 0.38 mA, the output is 0.6 mW, and the maximum operation temperature is 85°C [12]. The red color VCSEL emitting 650 nm can match to the low loss band of plastic fibers. Short distance data links using 1 mm diameter plastic fibers having a graded index have been developed. This system provides very easy optical coupling. VCSELs can match nicely to this application.

5.7. Surface-Emitting Lasers in Green-Blue-UV Band

Visible surface-emitting lasers are extremely important for disk, printers, and display applications, in particular, red, green, and blue surface emitters may provide much wider technical applications, if realized. The ZnSe system is the material to provide CW operation of green-blue semiconductor lasers operating over 1000 hours. It is supposed to be good for green lasers and the metal-organic chemical vapor deposition (MOCVD) may be a key to getting reliable devices into mass production. We have developed a simple technique to get a high p-doping by an ample diffusion of LiN to ZnSe. Also, a dielectric mirror deposition was investigated and relatively high reflectivity was obtained to provide an optical pumped vertical cavity. Some trials regarding optical pumped and current injection surface-emitting lasers have been made [14].

The GaN and related materials can cover wide spectral ranges green to UV. The reported reliability of GaN-based LEDs and LDs [68, 69] appears to indicate a good material potentiality for surface-emitting lasers as well. The optical gain is one of the important parameters to estimate the threshold current density of GaN-based VCSELs. The estimation of linear gain for $GaN/Al_{0.1}Ga_{0.9}N$ quantum well is carried out using the density-matrix theory with intraband broadening. The transparent carrier density of GaN is higher than other III–V materials such as GaAs, presumably originating from its heavy electron and hole masses. Generally, the effective masses of electrons and holes depend on the bandgap energy. Thus it seems that the wide-bandgap semiconductors require higher transparent carrier densities than do narrow-bandgap materials. The introduction of quantum wells for wide-bandgap lasers is really effective. This result indicates that the $GaN/Al_{0.1}Ga_{0.9}N$ QW is useful for low-threshold operation of VCSELs.

The trial for realizing green to UV VCSELs has just started. Some optical pumping experiments have been reported [14, 70]. It is necessary to establish some process technologies for device fabrication such as etching, surface passivation, substrate preparation, metalization, current confinement formation, and so on. We have made a preliminary study to search for dry etching of a GaN system by a chlorine-based reactive ion beam etch, and it was found to be possible.

The GaN system has large potentialities for short wavelength lasers. AlN/GaN DBR and ZrO/SiO_2 DBR are formed for VCSELs [71], and some selective growth techniques were attempted [72]. A photo-pumped GaInN VCSEL was reported [14]. Also we are trying to grow GaInN/GaN on silica glass for large-area light emitters [73].

5.8. Innovating Technologies

5.8.1. ULTIMATE CHARACTERISTICS

By overcoming any technical problems, such as making tiny structures, ohmic resistance of electrodes, and improving heat sinking, we believe that we can obtain a $1\,\mu A$ device. Many efforts toward improving the characteristics of surface-emitting lasers have been made, including surface passivation in the regrowth process for buried heterostructure, microfabrication, and fine epitaxies.

As has been previously introduced, very low thresholds of around $70\,\mu A$, $40\,\mu A$, and $10\,\mu A$ were reported by employing the aforementioned oxidation techniques. Therefore, by optimizing the device structures, we can expect a threshold lower than micro-amperes in the future [74, 75].

The efficiency of devices is another important issue for various applications. By introducing the oxide confinement scheme the power conversion efficiency has been drastically improved due to the effective current confinement and the reduction of optical losses. Also, the reduction of driving voltage by innovating the contacting technology helped a lot. As already mentioned, higher than 57% of power conversion efficiency (sometimes called wall-plug efficiency) has been realized. The noticeable difference from the conventional stripe laser is that high efficiency can be obtained at relatively low driving ranges in the case of VCSELs. Further improvement may enable us to achieve very high efficiency arrayed devices not attained in any other types of lasers.

The high-speed modulation capability is very essential for communication applications. In VCSELs, 10 Gbits/s or higher modulation experiments have been reported. It is a big advantage for VCSELs systems that over 10 Gbits/s modulation is possible at around 1 mA driving levels. This characteristic is very preferable for low power consumption optical interconnect applications.

The reliability of devices is a final screening of applicability of any components and systems. A high-temperature acceleration life test of proton-implanted VCSELs showed an expected room-temperature lifetime of over 10^7 hours. There is no reason why we cannot have very long life devices with VCSELs, because the active region is completely embedded in wide gap semiconductor materials and the mirror is already passivated.

The lasing performance of VCSELs will be improved by optimizing and solving the following issues; a) improvement of crystal quality, b) quantum structures (strain, wire/dot, modulation doping), c) polarization control, d) wavelength control, e) high power and low operation voltage.

Micro-etching technology is inevitably required to make reproducible arrayed VCSELs. We have prepared ICP (Inductively Coupled Plasma) etching for well controlled and low damage etching of GaAs and InP systems [76].

In order to further achieve substantial innovations in surface-emitting laser performances, the following technical issues remain unsolved or not yet optimized;

1. AlAs oxidation and its application to current confinement and optical beam focusing
2. Modulation doping, p-type and n-type modulation doping to quantum wells/barriers
3. Quantum wires and dots for active engines

4. Strained quantum wells and strain compensation
5. Angled substrates such as (311A), (311B), (411), etc.
6. New material combinations such as GaInNAs/GaAs for long wavelength emission, etc.
7. Wafer fusion technique to achieve optimum combination of active region and mirrors
8. Transparent mirrors to increase quantum efficiency and output power
9. Multi-quantum barriers (MQBs) to prevent carrier leakage to p-cladding layer
10. Tunnel junction.

Among them, the AlAs oxidation technology looks to be the most important technology to confine the current to reduce the threshold. Moreover, the oxidized layer works to give some amount of phase shift to focus the beam, providing an index-guiding cavity.

A tunnel junction was introduced in surface-emitting lasers [77]. Recently, the reverse tunnel junction began to be utilized for effective carrier injection and a novel self-aligned current aperture was proposed [24].

5.8.2. *POLARIZATION STEERING*

A wide variety of functions, such as polarization control, amplification, detecting, and so on can be integrated along with surface-emitting lasers by stacking. The polarization control will become very important for VCSELs [78]. One of the methods is to incorporate a grating terminator to a DBR. The other method includes the utilization of quantum wires and off-angled substrate, where we can differentiate the optical gain between one lateral direction and the perpendicular direction [54]. As already introduced, reasonably low threshold and well-controlled polarization behavior has been demonstrated by a (311)A and (311)B substrate. The device formation on (311)B GaAs substrates employing MOCVD methods has been attempted by solving the difficulties of crystal growth and p-type doping. We have achieved $260\,\mu A$ of CW threshold and single transverse and single polarization operation. The orthogonal Polarization Suppression Ratio (OPSR) of about $30\,dB$ was obtained. We have shown single-mode operation in a (311)B-based InGaAs/GaAs VCSEL under DC and 5 Gbits/s modulation condition [60]. At high-speed modulation conditions, some deterioration of OPSR was observed. Later, we achieved more stable operation by optimizing the device structure. The use of angled substrates, which provides us differential gain in two orthogonal polarizations, will be very effective to control the polarization independent of structures and the size of devices.

5.9. VCSEL-Based Integration Technology

A wide variety of functions, such as frequency tuning, amplification, and filtering, can be integrated along with surface-emitting lasers by stacking. Another possible way of moduling is to use the micro-optical bench (MOB) concept [79] to ease the assembling of components, as shown in Fig. 5.9.

A tunable VCSEL [80] is attractive as a widely tunable laser because of its short cavity structure. The first wide continuous wavelength tuning was demonstrated using a micromechanical external mirror [81]. Following that, tunable Fabry-Perot filters and VCSELs with a micromachined distributed Bragg reflector (DBR) were demonstrated [82, 83]. Micromachined filters/VCSELs have various advantages, such as wide wavelength

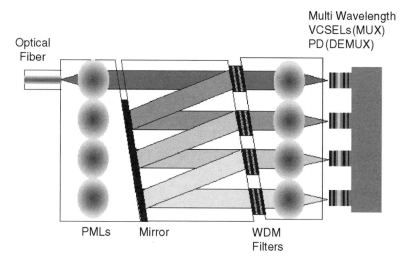

Fig. 5.9   WDM module based on stacked planar optics (after Aoki *et al.* [79]).

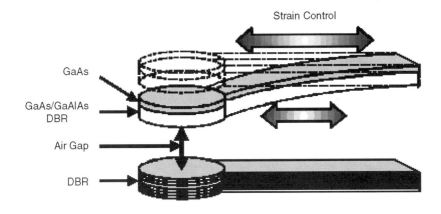

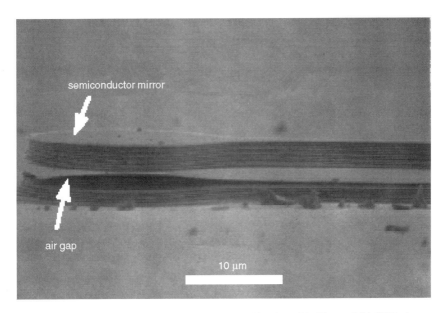

Fig. 5.10 (a) Schematic structure of (a) micromachined tunable filter and (b) SEM picture of MEMS GaAlAs/GaAs vertical cavity filter (after Amano *et al.* [85]).

tuning ranges and two-dimensional array integration. We previously proposed a novel technique for wavelength stabilization and wavelength trimming in a VCSEL using a micromachined DBR mirror tuned by differential thermal expansion [84]. We demonstrated an optical filter using vertical cavity configuration [85]. In Fig. 5.10, its concept is shown, which

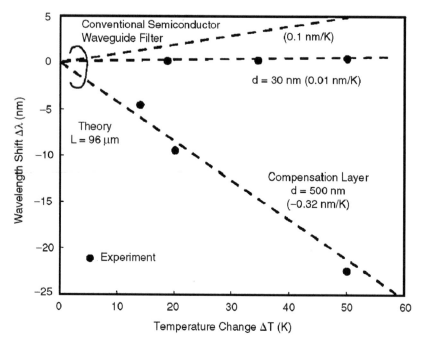

Fig. 5.11 Temperature dependence of MEMS GaAlAs/GaAs vertical cavity filter (after Amano *et al.* [85]).

demonstrates a Fabry-Perot filter fabricated by micro electromechnical system (MEMS) technology. By employing an additional compensation layer, a temperature independent filtering characteristic was obtained as shown in Fig. 5.11 [85].

5.10. VCSEL Application to WDM Networks

Lastly, we consider some possible applications including optical interconnects, parallel fiber-optic subsystems, WDM networks, and so on. We performed an experiment of >10 Gbits/s modulation of VCSELs and transmission via 100 m multi-mode fibers. The bit-error-rate (BER) is shown in Fig. 5.12 [53].

Multi-wavelength lasers are very important in massive transmission of optical signals. By using a selective crystal growth in metal-organic chemical vapor deposition (MOCVD), we achieved a multi-wavelength VCSEL array as shown in Fig. 5.13(a). The cavity length of each device can be tuned

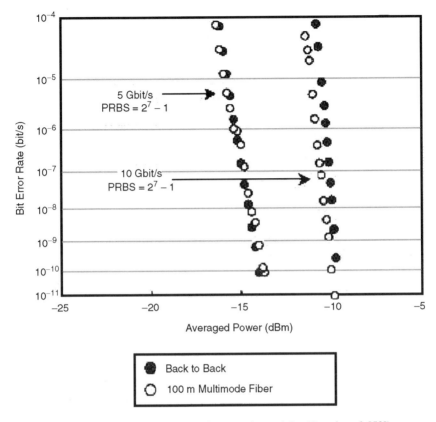

Fig. 5.12   10 Gb/s data transmission experiment (after Hatori *et al.* [53]).

during the crystal growth. The experimental result is shown in Fig. 5.13(b). Multi-wavelength transmission was also demonstrated.

Long wavelength VCSELs should be useful for silica-based fiber links providing ultimate transmission capability by taking the advantage of single wavelength operation and massively parallel integration. The development of 1200–1550 nm VCSELs may be one of the most important issues in surface-emitting laser research for metropolitan area networks (MAN). 1550 nm VCSELs with MEMS tunable functions have been attracting much interest for use in high end MAN systems. Electrical pumped tunable VCSELs with a tuning range of 20 nm [86] and photo-pumped tunable VCSELs with a tuning range of over 50 nm were demonstrated [87].

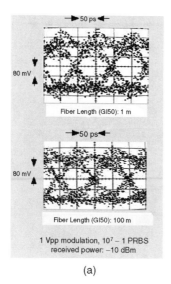

(a)

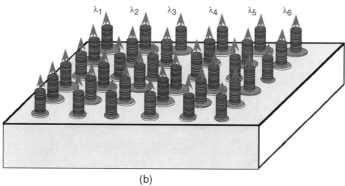

(b)

Fig. 5.13   (a) Schematic and (b) lasing spectra of multiwavelength VCSEL array on patterned substrate (after Arai *et al.* [62]).

5.11. Summary

The technology for high-performance VCSELs has matured. In practical 850 nm devices, sub-mA thresholds and 10 mW outputs have been achieved. A power conversion efficiency of >50% has been demonstrated. As for the reliability of VCSELs, 10^7 hours of room-temperature operation are estimated. Life test results on oxide-defined devices exhibited higher reliability. The Gigabit Ethernet has already been in the market by the use of multimode-fiber-based optical links. This system is being extended to 10 Gigabits/s Ethernets.

The importance of 1300 or 1550 nm devices is currently increasing for metropolitan area networks (MAN). A 1550 nm VCSEL with MEMS tunable functions began to be introduced into a high-end MAN system. One of the viable materials for long wavelength emitters is a GaInNAs system that can be formed on GaAs substrate.

In order to control the polarization of VCSEL output, (311)B substrate has been introduced and >30dB of orthogonal polarization suppression ratio (OPSR) was obtained even in high-speed modulation conditions.

The VCSEL itself is basically an exploratory device which has generated a Gigabit Ethernet and fiber channel applications. It is emerging into a higher class of data communication system such as 10 Gigabit Ethernet, high-speed LANs, optical interconnects, optical links, and so on. Moreover, long-wavelength VCSELs have been developed toward long-distance metropolitan area networks (MANs). It is noted that a continuous and wide-range wavelength tunability is a unique solution among many other candidates for this purpose. It may be a disruptive technology to replace distributed feedback (DFB) lasers.

It is found that temperature dependence upon threshold and quantum efficiency could be removed by properly designing the device structure and material. The highly strained GaInAs/GaAs emitting 1200 nm band is one of the candidates. The GaInNAs system may follow.

ACKNOWLEDGMENTS

The authors would like to thank to Prof. T. Miyamoto and other laboratory members for collaboration and assistance for preparing the original drawings. The related works have been supported by Grant-in Aid for COE Program #07CE2003 from MEXT.

References

1. K. Iga, F. Koyama, and S. Kinoshita, "Surface emitting semiconductor laser," IEEE J. Quantum Electron., QE-24:9 (Sep. 1988) 1845–1855.
2. K. Iga, "Surface emitting laser," Trans. of IEICE, C-I, JBl-C-1:9 (Sep. 1998) 483–493.
3. H. Soda, K. Iga, C. Kitahara, and Y. Suematsu, "GaInAsP/InP surface emitting injection lasers," Japan J. Appl. Phys., 18 (1979) 2329–2330.

4. K. Iga, S. Kinoshita, and F. Koyama, "Microcavity GaAlAs/GaAs surface-emitting laser with Ith=6 mA," Electron. Lett., 23:3 (Jan. 1987) 134–136.
5. F. Koyama, S. Kinoshita, and K. Iga, "Room-temperature continuous wave lasing characteristics of GaAs vertical cavity surface-emitting laser," Appl. Phys. Lett., 55:3 (1989) 221–222.
6. J. L. Jewell, S. L. McCall, A. Scherer, H. H. Houh, N. A. Whitaker, A. C. Gossard, and J. H. English, "Transverse modes, waveguide dispersion and 30 ps recovery in submicron GaAs/AlAs microresonators," Appl. Phys. Lett., 55 (1989) 22–24.
7. R. Geels and L. A. Coldren, "Narrow-linewidth, low threshold vertical-cavity surface-emitting lasers," 12th IEEE Intl. Semi-conductor Laser Conference, B-1 (1990) 16–17.
8. T. Wipiejewski, K. Panzlaf, E. Zeeb, and K. J. Ebeling, "Submilliamp vertical cavity laser diode structure with 2.2 nm continuous tuning," 18th European Conf. on Opt. Comm., ECOC'92, PDII-4 (1992).
9. Y. H. Lee, B. Tell, K. F. Brown-Goebeler, R. E. Leibenguth, and V. D. Mattera, "Deep-red continuous wave top-surface-emitting vertical-cavity AlGaAs superlattice lasers," IEEE Photonic. Technol. Lett., 3:2 (1991) 108–109.
10. T. Baba, Y. Yogo, K. Suzuki, F. Koyama, and K. Iga, "Near room temperature continuous wave lasing characteristics of GaInAsP/InP surface emitting laser," Electron. Lett., 29:10 (May 1993) 913–914.
11. D. I. Babic, K. Streubel, R. P. Mirin, J. Pirek, N. M. Margalit, J. E. Bowers, E. L. Hu, D. E. Mars, L. Yang, and K. Carey, "Room temperature performance of double-fused 1.54μm vertical-cavity lasers," IPRM '96, ThA1–2 (April 1996).
12. K. D. Choquette, R. P. Schneider, M. H. Crawford, K. M. Geib, and J. J. Figiel, "Continous wave operation of 640-660 nm selectively oxidised AlGaInP vertical cavity lasers," Electron. Lett., 31 (1995) 1145.
13. K. Iga, "Possiblility of Green/Blue/UV surface emitting lasers," Int'l. Symp. Blue Laser and Light Emitting Diodes (Mar. 1996).
14. T. Someya, K. Tachibana, Y. Arakawa, J. Lee, and T. Kamiya, "Lasing oscillation in InGaN vertical cavity surface emitting lasers," 16th International Semiconductor Laser Conference, PD-1 (1998) 1–2.
15. T. Miyamoto, T. Uchida, N. Yokouchi, Y. Inaba, F. Koyama, and K. Iga, "A study on gain-resonance matching of CBE grown l=1.5μm surface emitting lasers," IEEE/LEOS Annual DLTA13.2 (Nov. 1992).
16. S. Uchiyama, N. Yokouchi, and T. Ninomiya, "Room-temperature CW operation of 1.3μm GaInAsP SL-MQW surface emitting laser," The 43th Spring Meeting of Jpn. Soc. Appl. Phys., 26p-C-7 (1996).
17. N. M. Margalit, D. I. Babic, K. Streubel, R. P. Mirin, R. L. Naone, J. E. Bowers, and E. L. Hu, "Submilliamp long wavelength vertical cavity lasers," Electron. Lett., 32 (1996) 1675.

18. K. A. Black, N. M. Margalit, E. R. Hegblom, P. Abraham, Y.-J. Chiu, J. Piprek, J. E. Bowers, and E. L. Hu, "Double-fused 1.5μm vertical cavity lasers operating continuous wave upto 71°C," 16th International Semiconductor Laser Conference, ThA8 (1998) 247–248.

19. V. Jayaraman, J. C. Geske, M. H. MacDougal, F. H. Peters, T. D. Lowes, and T. T. Char, "Uniform threshold current, continuous-wave, singlemode 1300nm vertical cavity lasers from 0 to 70°C," Electron. Lett., 34:14 (1998) 1405–1406.

20. N. Ohnoki, G. Okazaki, F. Koyama, and K. Iga, "Record high characteristic temperature (T_0=122 K) of 1.55μm strain-compensated AlGaInAs/ AlGaInAs MQW lasers with AlAs/AlInAs multiquantum barrier," Electron. Lett., 35:1 (Jan. 1999) 51–52.

21. J. Boucart, C. Starck, A. Plais, E. Derouin, C. Fortin, F. Gaborit, A. Pinquier, L. Goldstein, D. Carpentier, and J. Jaquert, "RT pulsed operation of metamorphic VCSEL at 1.55μm," Electron. Lett., 34 (1998) 1508–1509.

22. M. Yamada, T. Anan, K. Kurihara, K. Nishi, K. Tokutome, A. Kamei, and S. Sugou, "Room temperature low threshold cw operation of 1.23μm GaAsSb VCSELs on GaAs substrate," Electron. Lett., 36 (2000) 483–484.

23. E. Hall, G. Almineau, J. K. Kim, O. Sjolund, H. Kromer, and L. A. Coldren, Electrically-pumped single-epitaxial VCSELs at 1.55μm with Sb-based mirrors," Electron. Lett., 35 (1999) 1337–1338.

24. S. Sekiguchi, T. Miyamoto, T. Kimura, F. Koyama, and K. Iga, "Self-arraigned current confinement structure using AlAs/InP tunnel junction in GaInAsP/ InP semiconductor lasers," Appl. Phys. Lett., 75:11 (Sep. 1999) 1512–1514.

25. N. Ohnoki, T. Mukaihara, N. Hatori, A. Mizutani, F. Koyama, and K. Iga, "GaInAs/AlGaInAs semiconductor lasers on InP substrate with AlAs oxide current confinement," Int. Conf. on Solid State Devices and Materials, Yokohama, 304 (Aug. 1996).

26. C. Kazmierski, J. P. Debray, R. Madani, N. Bouadma, J. Etrillard, I. Sagnes, F. Alexandre, and M. Quillec, "First all-monolithic VCSELs on InP:+55°C pulse lasing at 1.56μm with GaInAlAs/InP system," 16th International Semiconductor Laser Conference, PD-3 (1998) 5–6.

27. K. Iga, "Vertical cavity surface emitting lasers based on InP and related compounds—bottleneck and corkscrew," Conference on Indium Phosphide and Related materials, Schwabisch Gmund, Germany (Apr. 1996).

28. M. Kondow, K. Uomi, A. Niwa, T. Kitatani, S. Watahiki, and Y. Yazawa, "GaInNAs: a novel material for lomg wavelength range laser diodes with excellent high-temperature performance," Japan J. Appl. Phys., 35 (1996) 1273–1275.

29. K. Nakahara, M. Kondow, T. Kitatani, M. C. Larson, and K. Uomi, "1.3μm continuous-wave lasing operation in GaInNAs quantum-well lasers," IEEE Photonic. Technol. Lett., 10:4 (1998) 487–488.

30. M. C. Larson, M. Kondow, T. Kitatani, K. Nakahara, K. Tamura, H. Inoue, and K. Uomi, "Room temperature pulsed operation of GaInNAs/GaAs long-wavelength vertical cavity lasers," IEEE/LEOS'97, PD1.3 (1997).

31. T. Miyamoto, K. Takeuchi, F. Koyama, and K. Iga, "Novel GaInNAs/GaAs quantum well structure for long wavelength semiconductor lasers," IEEE Photonic. Technol. Lett., 9:11 (Nov. 1997) 1448–1450.

32. T. Kageyama, T. Miyamoto, S. Makino, N. Nishiyama, F. Koyama, and K. Iga, "High-temperature operation up to 170°C of GaInNAs-GaAs quantum-well lasers grown by chemical beam epitaxy," IEEE Photonic. Technol. Lett., 12:1 (2000) 10–12.

33. T. Kageyama, T. Miyamoto, S. Makino, Y. Ikenaga, N. Nishiyama, A. Matsutani, F. Koyama, and K. Iga, "Room temperature continuous-wave operation of GaInNAs/GaAs VCSELs grown by chemical beam epitaxy with output power exceeding 1mW," Electron. Lett., 37 (2001) 225–226.

34. D. Schlenker, T. Miyamoto, Z. Chen, F. Koyama, and K. Iga, "1.17µm highly strained GaInAs-GaAs quantum-well laser," IEEE Photonic. Technol. Lett., 11:8 (Aug. 1999) 946–948.

35. Z. Chen, D. Schlenker, T. Miyamoto, T. Kondo, M. Kawaguchi, F. Koyama, and K. Iga, "High temperature characteristics of near 1.2µm InGaAs/AlGaAs lasers," Japan J. Appl. Phys., 38:10B (Oct. 1999) L1178–L1179.

36. F. Koyama, S. Schlenker, T. Miyamoto, Z. Chen, A. Matsutani, T. Sakaguchi, and K. Iga, "1.2µm highly strained GaInAs/GaAs quantum well lasers for singlemode fibre datalink," Electron. Lett., 35:13 (Jun. 1999) 1079–1081.

37. D. G. Deppe, D. L. Huffaker, Q. Deng, T.-H. Oh, and L. A. Graham, "Oxide-confined VCSEL's with quantum well quantum dot active region," IEEE/LEOS'97, ThA 1 (1997).

38. T. Sakaguchi, F. Koyama, and K. Iga, "Vertical cavity surface emitting laser with and AlGaAs/AlAs bragg reflector," Electron. Lett., 24 (1988) 928–929.

39. T. Numai, T. Kawakami, T. Yoshikawa, M. Sugimoto, Y. Sugimoto, H. Yokoyama, K. Kasahara, and K. Asakawa, "Record low threshold current in microcavity surface-emitting laser," Japan J. Appl. Phys., 32:10B (Oct. 1993) L1533–L1534.

40. D. L. Huffaker, D. G. Deppe, C. Lei, and L. A. Hodge, "Sealing AlAs against oxidative decomposition and its use in device fabrication," JTuH5 (1996).

41. Y. Hayashi, T. Mukaihara, N. Hatori, N. Ohnoki, A. Matsutani, F. Koyama, and K. Iga, "Lasing characteristics of low-threshold oxide confinement InGaAs-GaAlAs vertical-cavity surface-emitting lasers," IEEE Photonic. Technol. Lett., 7 (Nov. 1995) 1234–1236.

42. Y. Hayashi, T. Mukaihara, N. Hatori, N. Ohnoki, A. Matsutani, F. Koyama, and K. Iga, "Record low-threshold index-guided InGaAs/GaAlAs vertical-cavity surface-emitting laser with a native oxide confinement structure," Electron. Lett., 31:7 (Mar. 1995) 560–561.

43. D. L. Huffaker, J. Shin, and D. G. Deppe, "Low threshold half-wave vertical-cavity lasers," Electron. Lett., 31 (1994) 1946.

44. G. M. Yang, M. MacDougal, and P. D. Dupkus, "Ultralow threshold current vertical cavity surface emitting lasers obtained with selective oxidation," Electron. Lett., 31 (1995) 886–888.

45. N. Hatori, A. Mizutani, N. Nishiyama, F. Motomura, F. Koyama, and K. Iga, "P-type delta doped InGaAs/GaAs quantum well vertical-cavity surface-emitting lasers," IEICE C-I, J81-C-I:7 (1998) 410–416.

46. F. H. Peters, M. G. Peters, D. B. Young, J. W. Scott, B. J. Tibeault, S. W. Corzine, and L. A. Coldren, "High power vertical cavity surface emitting lasers," 13th IEEE Semiconductor Laser Conf., PD-1 (1992) 1–2.

47. K. L. Lear, R. P. Schneider, Jr., K. D. Choquette, S. P. Kilcoyne, and K. M. Geib, "Selectively oxidised vertical cavity surface emitting lasers with 50% power conversion efficiency," Electron. Lett., 31 (1995) 208.

48. K. D. Choquette, A. A. Allerman, H. Q. Hou, G. R. Hadley, K. M. Geib, and B. E. Hammons, "Improved efficiency of small area selectively oxidized VCSELs," 16th International Semiconductor Laser Conference, ThA3 (1998) 237–238.

49. L. A. Coldren, E. R. Hegblom, and N. M. Margalit, "Vertical cavity lasers with record efficiency at small sizes using tapered apertures," 16th International Semiconductor Laser Conference, PD-2 (1998) 3–4.

50. B. Weigl, G. Reiner, M. Grabherr, and K. J. Ebeling, "High-power selectively oxidized vertical-cavity surface-emitting lasers," CLEO'96, Anaheim, JTuH2 (1996).

51. D. Francis, H.-I. Chen, W. Yuen, G. Li, and C. Chang-Hasnain, "Monolithic 2D-VCSEL array with .2W CW output power," 16th International Semiconductor Laser Conference, TuE3 (1998) 99–100.

52. B. J. Thibeault, K. Bertilsson, E. R. Hegblom, P. D. Floyd, and L. A. Coldren, "High-speed modulation characteristics of oxide-apertured vertical-cavity lasers," 15th IEEE International Semicon. Laser Conf., M3.2 (1996) 17–18.

53. N. Hatori, A. Mizutani, N. Nishiyama, A. Matsutani, T. Sakaguchi, F. Motomura, F. Koyama, and K. Iga, "An over 10 Gbits/s transmission experiment using p-type d-doped InGaAs/GaAs quantum-well vertical cavity surface-emitting laser," Photonic. Technol. Lett., 10:2 (Feb. 1998) 194–196.

54. M. Takahashi, N. Egami, T. Mukaihara, F. Koyama, and K. Iga, "Lasing characteristics of GaAs (311)A substrate based InGaAs/GaAs vertical cavity surface emitting lasers," IEEE J. Select. Top. Quantum Electrron., 3:2 (1997) 372–378.

55. K. Tateno, Y. Ohiso, C. Amano, A. Wakatsuki, and T. Kurokawa, "Growth of vertical-cavity surface-emitting laser structures on GaAs (311)B substrates

by metalorganic chemical vapor deposition," Appl. Phys. Lett., 70:25 (Jun. 1997) 3395–3396.

56. A. Mizutani, N. Hatori, N. Ohnoki, N. Nishiyama, N. Ohtake, F. Koyama, and K. Iga, "P-type doped AlAs growth on GaAs (311)B substrate using carbon auto-doping for low resistance GaAs/AlAs distributed bragg reflectors," Japan J. Appl. Phys., 36:11 (Nov. 1997) 6728–6729.

57. A. Mizutani, N. Hatori, N. Nishiyama, F. Koyama, and K. Iga, "A low threshold polarization-controlled vertical cavity laser grown on GaAs (311) substrate," Photonic. Technol. Lett., 10:5 (May 1998) 633–635.

58. N. Nishiyama, A. Mizutani, N. Hatori, F. Koyama, and K. Iga, "Single mode and stable polarization InGaAs/GaAs surface emitting laser grown on GaAs (311)B substrate," 16th International Semiconductor Laser Conference, ThA1 (1998) 233–234.

59. N. Nishiyama, A. Mizutani, N. Hatori, F. Koyama, and K. Iga, "Single transversemode and stable polarization operation under high-speed modulation of InGaAs/GaAs vertical cavity surface emitting laser grown on GaAs (311)B substrate," Photonic. Technol. Lett., 10:12 (Dec. 1998) 1676–1678.

60. N. Nishiyama, A. Mizutani, N. Hatori, M. Arai, F. Koyama, and K. Iga, "Lasing characteristics of InGaAs/GaAs polarization controlled vertical-cavity surface emitting laser grown on GaAs (311) B Substrate," Selected Topics of Quantum Electron., 5:3 (1999) 530–536.

61. N. Nobuhiko, M. Arai, S. Shinada, M. Azuchi, T. Miyamoto, F. Koyama, and K. Iga, "Highly strained GaInAs/GaAs quantum well vertical-cavity surface-emitting laser on GaAs (311)B substrate for stable polarization operation," IEEE J. Select. Top. Quantum Electron. (2001), in press.

62. M. Arai, T. Kondo, N. Nishiyama, A. Matsutani, T. Miyamoto, F. Koyama, and K. Iga, "1.1–1.2μm multiple-wavelength vertical cavity surface emitting laser array with highly strained GaInAs/GaAs QWs on patterned substrate," IEEE LEOS Annual Meeting, 2001 WY4 (2001).

63. F. Koyama, K. Tomomatsu, and K. Iga, "GaAs surface emitting lasers with circular buried heterostructure grown by metalorganic chemical vapor deposition and two-dimensional laser array," Appl. Phys. Lett., 52:7 (1988) 528–529.

64. R. Jager, M. Grabherr, C. Jung, R. Michalzik, G. Reiner, B. Weigl, and K. J. Ebeling, "57% wallplug efficiency oxide-confined 850nm wavelength GaAs VCSELs," Electron. Lett., 33:4 (Feb. 1997) 330–331.

65. J. K. Guenter, R. A. Hawthorne, D. N. Granville, M. K. Hibbs-Brenner, and R. A. Morgan, "Reliability of proton-implanted VCSELs for data communications," Proc. of SPIE, 2683:1 (1996).

66. H.-E. Shin, Y.-G. Ju, H.-H. Shin, J.-H. Ser, T. Kim, E.-K. Lee, I. Kim, and Y.-H. Lee, "780nm oxidised vertical-cavity surface-emitting lasers with Al0.11Ga0.89As quantum wells," Electron. Lett., 32 (1996) 1287–1288.

67. M. H. Crawford, K. D. Choquette, R. J. Hickman, and K. M. Geib, "Performance of selectively oxidized AlGaInP-based visible VCSEL's," OSA Trends in Optics and Photonics Series, 15 (1998) 104–105.

68. S. Nakamura, M. Senoh, S. Nagahama, N. Iwasa, T. Yamada, T. Matsushita, H. Kiyoku, and Y. Sugimoto, "InGaN-based multi-quantum-well-structure laser diodes," Japan J. Appl. Phys., 35(2): 1B (1996) L74–L76.

69. S. Nakamura, M. Senoh, S. Nagahama, N. Iwasa, T. Yamada, T. Matsushita, H. Kiyoku, Y. Sugimoto, T. Kozaki, H. Umemoto, M. Sano, and K. Chocho, "High-power, long-lifetime InGaN/GaN/AlGAN-based laser diodes grown on pure GaN substrates," Japan J. Appl. Phys., 37 (1998) L309–L312.

70. T. Sakaguchi, T. Shirasawa, N. Mochida, A. Inoue, M. Iwata, T. Honda, F. Koyama, and K. Iga, "Highly reflective AlN/GaN and ZrO2/SiO2 multilayer distributed Bragg reflectors for InGaN/GaN surface emitting lasers," LEOS 1998 11th Annual Meeting Conference Proceedings, TuC4 (Dec. 1998).

71. T. Shirasawa, N. Mochida, A. Inoue, T. Honda, T. Sakaguchi, F. Koyama, and K. Iga, "Interface control of GaN/AlGaN quantum well structures in MOVPE growth," Journal of Crystal Growth, 180/190 (1998) 124–127.

72. M. Iwata, T. Sakaguchi, Y. Moriguchi, Y. Uchida, T. Miyamoto, F. Koyama, and K. Iga, The 60th Autumn Meeting, Japan Soc. of Appl. Phys., 4a-V-2 (1999).

73. Y. Moriguchi, T. Miyamoto, T. Sakaguchi, M. Iwata, Y. Uchida, F. Koyama, and K. Iga, "GaN polycrystal growth on silica substrate by metalorganic vapor phase epitaxy (MOVPE)," 3rd Int. Symp. on Blue Laser and Light Emitting Diodes (ISBLLED2000), WeP-19, Berlin (2000).

74. D. G. Deppe, D. L. Huffaker, J. Shin, and Q. Deng, "Very-low-threshold index-confined planar microcavity lasers," IEEE Photonic. Technol. Lett., 7:9 (Sep. 1995) 965–967.

75. H. Bissessur, F. Koyama, and K. Iga, "Modeling of oxide-confined vertical cavity surface emitting lasers," IEEE J. Quantum Electron., 3:2 (Apr. 1997) 344–352.

76. A. Matsutani, H. Ohtsuki, F. Koyama, and K. Iga, "C60 resist mask of electron beam lithography for chlorine-based reactive ion beam etching," Jpn. J. Appl. Phys., 37:7 (1998) 4211–4212.

77. Y. Kotaki, S. Uchiyama, and K. Iga, "GaInAsP/InP surface emitting laser with two active layers," Extended Abstract, 1984 Internat'l Conference on Solid State Devices and Materials, C-2-3 (Aug. 1984).

78. T. Mukaihara, N. Ohnoki, Y. Hayashi, N. Hatori, F. Koyama, and K. Iga, "Polarization control of vertical-cavity surface-emitting lasers using a bire-fringent metal/dielectric polarizer loaded on top distributed bragg reflector," J. Selected Topics in Quantum Electronics, 1:2 (Jun. 1995) 667–673.

79. Y. Aoki, R. J. Mizuno, Y. Shimada, and K. Iga, "Parallel and bi-directional optical interconnect module using vertical cavity surface emitting lasers

(VCSELs) and 3-D micro optical bench (MOB)," 1999 IEEE/LEOS Summer Topical Meetings, WA2.4 (Jul. 1999).

80. C. J. Chang-Hasnain, J. P. Harbison, C. E. Zah, L. T. Florez, and N. C. Andreadakis, "Continuous wavelength tuning of two-electrode vertical cavity surface emitting lasers," Electron. Lett., 27 (1991) 1002.

81. N. Yokouchi, T. Miyamoto, T. Uchida, Y. Inaba, F. Koyama, and K. Iga, "40Å continuous tuning of a GaInAsP/InP vertical-cavity surface-emitting laser using an external mirror," IEEE Photonic. Technol. Lett., 4:7 (Jul. 1992) 701–703.

82. M. S. Wu, E. C. Vail, G. S. Li, W. Yuen, and C. J. Chang-Hasnain, "GaAs micromachined widely tunable Fabry-Perot filter," Electron. Lett., 31 (1995) 1671–1672.

83. M. S. Wu, E. C. Vail, G. S. Li, W. Yuen, and C. J. Chang-Hasnain, "Tunable micromechanical vertical cavity surface emitting laser," Electron. Lett., 31 (1995) 1671–1672.

84. F. Koyama and K. Iga, "Wavelength stabilization and trimming technologies for vertical-cavity surface emitting laser arrays," OSA Technical Digest Series of Quantum Optoelectronics, 9:QTh-14 (1997) 90–92.

85. T. Amano, F. Koyama, N. Nishiyama, and K. Iga, "Temperature-insensitive micromachined AlGaAs-GaAs vertical cavity filter," IEEE Photon. Technol. Lett., 12:5 (May 2000) 510–512.

86. W. Yuen, G. S. Li, R. F. Nabiev, M. Jansen, D. Davis, and C. J. Chang-Hasnain, "Electrically-pumped directly-modulated tunable VCSEL for metro DWDM applications," IEEE LEOS Annual Meeting, MI3 (2001).

87. D. Vakhshoori, P. Tayebati, C.-C. Lu, M. Azimi, P. Wang, J.-H. Zhou, and E. Canoglu, "2mW CW singlemode operation of a tunable 1550nm vertical cavity surface emitting laser with 50nm tuning range," Electron. Lett., 35:11 (1999) 661–662.

Part 2 | Optical Modulators

Chapter 6 | Lithium Niobate Optical Modulators

Rangaraj Madabhushi

Optoelectronics Center, Room 31-153, Agere Systems Inc.,
9999 Hamilton Blvd., Breinigsville, PA, 18031, USA

6.1. Introduction and Scope

With the advent of the laser, a great interest in communication at the optical frequencies was created. A new era of optical communication was launched in 1970, when an optical fiber, having 20 dB/km attenuation, was fabricated at the Corning Glassworks. Dr. Kaminow and his team from Bell Labs reported the concept of electrooptic light modulators [1]. At the same time, Miller [2], coined the term "Integrated Optics" and heralded the beginning of efforts in development of a number of optical components, including light sources, waveguide devices, and detectors. The demand for fiber-optic telecommunication systems and larger bandwidth requirements has increased tremendously in the past 10 years. External modulators are extensively developed and used for these systems, with the bandwidth ranging from 2.5, to 10, and presently 40 GHz. High-speed LiNbO$_3$ (Lithium Niobate, LN) optical waveguide external modulators have the advantages of superior chirping characteristics, wider bandwidths, and low insertion losses over the direct modulation of the lasers.

The system performance of high-speed digital communication systems is limited by fiber dispersion. The optical communication system degradations caused by the fiber dispersion problems can be reduced by the zero-chirp or negative-chirp capabilities of the LiNbO$_3$ optical modulators. The LiNbO$_3$ modulator technology, which started in the late 1960s at

207

WDM TECHNOLOGIES: ACTIVE
OPTICAL COMPONENTS
$35.00

Bell Labs (AT&T/Lucent Technologies), advanced in terms of the material properties, fabrication process, and various modulation schemes over the years [3–15]. Until the middle of the 1980s, a number of researchers in universities, research laboratories, and corporations all over the world made tremendous contributions in improving the optical and electrooptical characteristics [16–55].

Although the electrooptic characteristics were greatly improved over the years, practical realization of the Lithium Niobate modulators for use in actual systems was not possible until the early 1990s. The stable operation of these devices was highly limited by the bias-voltage-induced drift (DC drift) and the temperature-induced drift, which made these devices unsuitable for practical applications in optical communication systems. This caused many corporations/companies to slowly reduce the development activities, with a few exceptions. Limited development activities were being carried in the United States, Europe, and Japan, with a view to solving these reliability problems in addition to improving the bandwidth. The breakthrough came in the late 1980s and early 1990s. Bell Labs-Lucent Technologies in the United States and Fujitsu/NTT/NEC in Japan were successful in developing low-drift modulator technologies. In the last few years, the business opportunities have increased, and there is a strong demand and supply for these $LiNbO_3$ modulators, as per the optical communication industry's requirements.

The increase in use of the WDM (wavelength-division multiplexing) in high-speed and long-haul fiber systems necessitates the use of high-speed modulators. Significant advances have been made in recent years in the design and performance of these modulators. The aim of this work is to review the basic results and progress of the technology. A total review of the field is out of the scope of this work, and several books and special papers [3–15] can be referred to for more details. In this chapter, the emphasis is more on the design, fabrication, and various characteristics that are required by the system applications and the state-of-the-art in achieving these characteristics.

The chapter is organized to give the progress, general methods, design aspects, fabrication, and reliability of the lithium niobate optical modulators. Section 6.2 treats the types of optical modulation and various physical effects that control the modulation. Section 6.3 gives the general structures, principle of operation, and design methods. The emphasis is on velocity matching and microwave attenuation reduction. Section 6.4 deals with the fabrication methods and the reliability aspects.

6.2. Optical Modulation

6.2.1. INTRODUCTION

It is possible to realize various optical devices by externally controlling the lightwave propagating in the optical waveguide. Optical modulators are the devices, made of optical waveguides on some material with special properties, where the information is placed on the lightwave externally, imposing time-varying change on the lightwave. The information content is then related to the bandwidth of the imposed variation. Similarly, the switches are the devices that change the spatial location of the lightwave with respect to the switching signal. These modulators and switches are the important components in almost all the optical communication systems. The materials have physical properties such as electrooptic effect, acousto-optic effect, magnetooptic effect, and thermooptic effect. In this section the basic modulation types and the electrooptic effect are discussed. The details for other physical effects, such as, acoustooptic and magnetooptic, can be obtained from reference [5].

6.2.2. TYPES OF MODULATIONS

The optical waveguide structures are formed in a material with a physical effect such as electrooptic effect, and it is then possible to achieve modulation by external application of the signal. This is an operation where the information content is placed on a coherent lightwave. A modulator alters the detectable properties of a lightwave, in response to the applied external signal. The desired modulation characteristics depend on the system characteristics and system applications.

The modulation types include intensity or amplitude modulation, phase modulation, frequency modulation, and polarization modulation.

The intensity modulators are those in which the intensity or amplitude of the coherent lightwave varies with a time-varying signal. In this case, for the plane wave of $E(t)\exp\{j(\omega t - \varphi)\}$, the intensity will be given by $E(t) \times E^*(t)$ or E^2, which will vary as a function of the applied signal.

In phase modulation, the phase of the lightwave responds to the applied signal. If the electric field of the lightwave when no signal is applied is $E\exp\{j(\omega t - \varphi)\}$, when the signal is applied the field is shifted in phase by an amount $\Delta\varphi$. The field will then become $E(t)\exp\{j(\omega t - \varphi + \Delta\varphi)\}$. If the signal is time varying, the $\Delta\varphi$ also varies with time. The amplitude of the first side band and carrier amplitude are related to the Bessel functions.

In polarization modulation, using the electrooptic effect, the polarization states of the lightwave respond to the signal applied. In general, when there is no signal applied, the lightwave emerges as a linearly polarized light. The vector amplitude of such a plane-polarized wave can be represented as $E = xEx \exp\{j(\omega t - \varphi)\} + yEy \exp\{j(\omega t - \varphi)\}$. For $Ex = Ey$, the polarization is $45°$ to the x axis. The amplitude when the signal is applied can be represented as $E = xEx \exp\{j(\omega t - \varphi + \Phi x)\} + yEy \exp\{j(\omega t - \varphi + \Phi y)\}$, with Φx and Φy functions of the applied signal and time. This expression is that of an elliptical polarized light. The phase shift is $\Delta \Phi = \Phi x - \Phi y$. If $\Delta \Phi$ is zero the light is plane polarized, and if $\Delta \Phi = \pi$ then the light is also plane polarized, but rotated through $90°$ from the previous plane-polarized light of $\Delta \Phi = 0$. These changes, from linear to elliptical polarization, are the characteristics of polarization modulators that use the electrooptic effect. In the case of magnetooptic polarization modulators, the light remains linearly polarized, but rotated in direction as a function of the applied signal. These polarization modulators are usually used as switches.

The last method is frequency modulation, in which the frequency or the wavelength is changed with the applied signal. The detection of such frequency shifts gives rise to more complicated heterodyne system applications.

6.2.3. ELECTROOPTIC EFFECT

The electrooptic effect is in general defined as a change of refractive index inside an optical waveguide in optical aniisotropic crystals, when an external electric field is applied. If the refractive index changes linearly with the amplitude of the applied field, it is known as the linear electrooptic effect or the Pockels effect. This effect is the most widely used physical effect for the waveguide modulators. The details can be learned from the existing literature (for example, Ref. [5]); here some of the basic fundamentals are given.

The dielectric tensor $[\varepsilon]$ of these anisotropic crystals such as $LiNbO_3$ can be represented as follows, when the asymmetric diagonal components are made zero:

$$[\varepsilon] = \begin{bmatrix} \varepsilon_{11} & & 0 \\ & \varepsilon_{22} & \\ 0 & & \varepsilon_{33} \end{bmatrix} \tag{6.1}$$

Assuming the relation between the dielectric constant and the refractive index n_j, as $\varepsilon_{jj} = n_j^2$ $(j = 1, 2, 3)$, the index of the ellipsoid can be written as

$$\frac{x^2}{n_x^2} + \frac{y^2}{n_y^2} + \frac{z^2}{n_z^2} \qquad (6.2)$$

When the x, y, z are chosen to be parallel to the principal dielectric axis of the crystal, the linear change in the refractive index coefficients due to the applied electric field E is given by

$$\Delta\left[\frac{1}{n^2}\right] = \sum_{j=1}^{3} r_{ij} E_j \qquad (6.3)$$

where $i = 1, 2, 3, 4, 5, 6$ and $j = 1, 2, 3$ are associated with the x, y, z axes respectively and r_{ij} is known as the electrooptic constant. Equation 6.3, when written in a matrix form (the 6×3 $[r_{ij}]$ matrix),

$$\begin{bmatrix} r_{11} & r_{12} & r_{13} \\ r_{21} & r_{22} & r_{23} \\ r_{31} & r_{32} & r_{33} \\ r_{41} & r_{42} & r_{43} \\ r_{51} & r_{52} & r_{53} \\ r_{61} & r_{62} & r_{63} \end{bmatrix} \qquad (6.4)$$

is called the electrooptic tensor.

In the case of an anisotropic crystal such as LiNbO$_3$, $n_x = n_y = n_o$ represents the ordinary refractive index and $n_z = n_e$ is the extraordinary refractive index. Then the electrooptic tensor becomes

$$\begin{bmatrix} 0 & -r_{22} & r_{13} \\ 0 & r_{22} & r_{13} \\ 0 & 0 & r_{33} \\ 0 & r_{51} & 0 \\ r_{51} & 0 & 0 \\ -r_{22} & 0 & 0 \end{bmatrix} \qquad (6.5)$$

Assuming $Ex = Ey = 0$ and $Ez \neq 0$, and the light is propagating in the x direction, it can be written, from Eqs. (6.2) to (6.5),

$$\frac{y^2}{n_o^2\left(1 - 0.5\, r_{13}\, n_o^2\, Ez\right)} + \frac{z^2}{n_e^2\left(1 - 0.5\, r_{33}\, n_e^2\, Ez\right)} \qquad (6.6)$$

The change in refractive index can be

$$\Delta n_o = 0.5\, r_{13}\, n_o^3\, Ez, \qquad \Delta n_e = 0.5\, r_{33}\, n_e^3\, Ez \qquad (6.7)$$

For LiNbO$_3$, $r_{33} = 30.8 \times 10^{-12}$ m/V, $r_{13} = 8.6 \times 10^{-12}$ m/V, $r_{22} = 3.4 \times 10^{-12}$ m/V, $r_{33} = 28.0 \times 10^{-12}$ m/V, $n_o = 2.2$, and $n_e = 2.15$ at $\lambda = 1.5\,\mu$m.

6.3. Basic Principles of the Modulator Design and Operation

6.3.1. INTRODUCTION

In this section, the basic principles of operation and design considerations are discussed, vis-à-vis the characteristics of optical modulators based on the dielectric crystals, such as Lithium Niobate (LiNbO$_3$) [15].

The block diagram of a single-channel time-division multiplexing (TDM) communication system and the multi-channel wavelength-division multiplexing systems (WDM) are given in Fig. 6.1. TDM systems basically consist of the transmitter and receivers, connected through the fibers. The transmitter part consists of a laser, which provides the coherent optical (light) wave, and the modulator (either external or the direct modulation of lasers), where the desired signal is modulated and is placed on the coherent lightwave. The light is then propagated/transmitted through the fiber, using amplifiers/boosters for transporting to the destination. At the receiver end the light is demodulated and the signal is separated from the coherent lightwave for the final processing, depending on the application.

In the case of wavelength-division multiplexing (WDM), a number of channels are used in a way similar to a TDM system, except that each channel is propagated at a single wavelength/frequency. At the transmitter

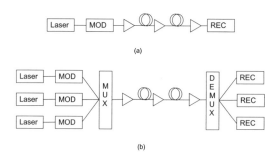

(a)

(b)

Fig. 6.1 The high-speed long-haul optical communication system components, for (a) time-division multiplexing (TDM) and (b) wavelength-division multiplexing (WDM) applications.

side the individual channels are multiplexed into a single path and transmitted through the fiber. At the receiver end, the optical wave is demultiplexed into various wavelength channels and detected, as in TDM.

The direct modulation of lasers is limited by the achievable bandwidth, chirp, or dispersion, and the ability to be transmitted to longer distances. The advantages for short-distance applications include small device size and cost effectiveness. On the other hand, external modulators are bulky and costly and increase the system requirements. But the advantages, such as large bandwidths and capability to propagate long distances, make these external modulators the winners in optical communication systems. The external modulators include devices made of the dielectric crystals such as lithium niobate and lithium tantalite, semiconductors including GaAs, InP, InGaAs etc., and polymers such as PMMA. The lithium-niobate-based modulators have the advantages of large bandwidth capabilities, low chirp characteristics, low insertion loss, better reliability, and mature manufacturing capabilities. The disadvantages include higher driving voltages, large size of the device, and cost. The semiconductor modulators have the advantages of smaller size, low driving voltages, relatively low cost (for large volumes), and compatibility of future integration with other semiconductor devices. The disadvantages include large insertion loss, smaller transmission distances, chirp, and manufacturing yields. The polymers are just emerging and although they can achieve large bandwidths and low driving voltages, their long-term reliability is still being investigated. In today's marketplace, lithium niobate external modulators are widely used, especially for applications of more than a few Gb/s.

Figure 6.2 shows a generic trend of the speed of the systems that are in practical use, in terms of the date of deployment. Starting with 0.4/1.6 Gb/s systems in the early/late 1980s, through 2.4/10 Gb/s in the early/late 1990s, systems in the early 2000s need 40 Gb/s optical components. Due to the advantages of the lithium niobate external modulators, it is expected that a large market share will be held by these devices at 40 Gb/s.

6.3.2. BASIC STRUCTURE AND CHARACTERISTICS OF THE MODULATORS

In general, the Mach-Sender interferometer type structure is used in the lithium-niobate-based intensity modulators [55–60]. The basic operational structure is shown in Fig. 6.3. The modulator basically consists of an input straight waveguide, an input Y branch waveguide, which divides the incoming light into two parts, then an interferometer consisting of

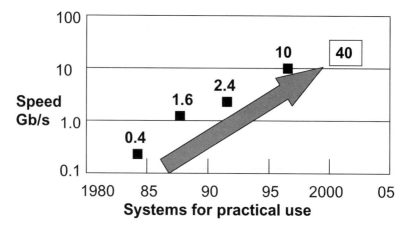

Fig. 6.2 The progress of the systems in practical use as commercial systems, as a function of time since 1980, vis-a-vis the speed or bandwidth of these systems.

two arms, to which the signal can be applied in the form of voltage, then another output Y branch waveguide, which combines the two waves from the interferometer arms, and finally an output straight waveguide.

When there is no signal/voltage applied ($V = 0$), the input wave (field) will be divided in two equal parts, E_A and E_B. At the interference arms these waves propagate with the same amplitude and phase and recombine at the output Y branch and propagate in the output waveguide without any change in the intensity (Fig. 6.3 (a)). When voltage is applied, it changes the phase of the two waves at the interferometer arms, and when the applied voltage, V, equals the voltage required to achieve a π phase shift, V_π, the output waves from the interferometer have the same amplitude, but a phase difference of π. The output light will become zero by destructive interference (Fig. 6.3 (b)). For the values of the voltage between V and V_π, output power varies as

$$P_{\text{out}} = 0.5\{(IE_A I - IE_B I)^2 + 2IE_A I \cdot IE_B I \cos^2 \Delta\varphi\} \qquad (6.8)$$

$$= 0.5 P_{\text{in}} \cdot (K1 + K2 \cos^2(\pi V/2V_\pi) \qquad (6.9)$$

where the phase shift

$$2\Delta\varphi = \pi V / V_\pi \qquad (6.10)$$

Figure 6.3.(c) shows the output intensity as the function of switching/driving voltage, which is represented by Eq. (6.9).

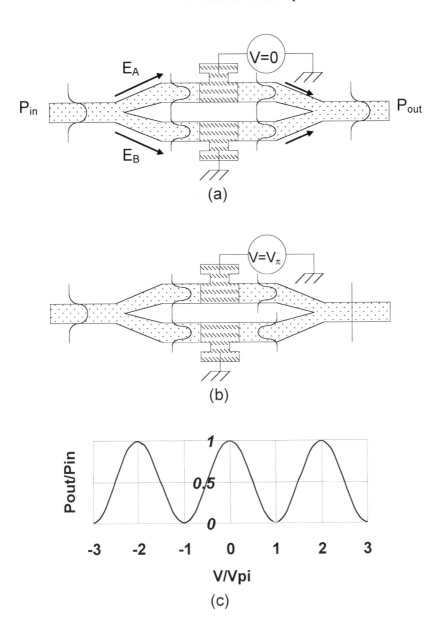

Fig. 6.3 The principle of operation of the Mach-Zehnder intensity modulators (a) when no voltage is applied across the two arms, on-state, (b) when a voltage equivalent to a π phase shift, between the arms is applied, off-state, (c) with the output power of the modulator as a function of the applied voltage.

Driving Voltage

The change in the index as a function of voltage is

$$\Delta n(V) = n_e^3 3 r_{33} V \Gamma / 2G \qquad (6.11)$$

The phase difference in each arm of the interferometer will be φ and as the voltage is applied on both arms, the push–pull effect can be used and the total phase difference will be 2φ, and

$$2\varphi = \pi V / V_\pi$$

The driving/switching voltage $\quad V_\pi = \lambda G \big/ 2 n_e^3 r_{33} \Gamma L \quad$ (6.12)

The voltage length product $\quad V_\pi L = \lambda G \big/ 2 n_e^3 r_{33} \Gamma \quad$ (6.13)

where λ is the wavelength of operation (say, 1.5), n_e is the extraordinary refractive index of the $LiNbO_3$ waveguide (say, 2.15 at $\lambda 1.5$ μm), r_{33} is the electrooptic coefficient, 30.8×10^{-12} m/V, V is the voltage applied, Γ is the overlap integral between optical and electric (RF) fields (usually a value of 0.3 to 0.5), G is the gap between the electrodes, and L is the electrode length.

Depending on the crystal orientation (z-cut, x-cut, or y-cut), the electrode configuration, whether they are placed on the waveguides or on the sides of the waveguide, will result in the use of vertical or horizontal fields (Fig. 6.4). The overlap integral, Γ, is better for the z-cut compared to an x-cut. The driving voltage will be less in the case of the z-cut crystal orientation/vertical field, due to the large overlap factor. But, there is a need to place a dielectric layer between the electrode and the waveguides,

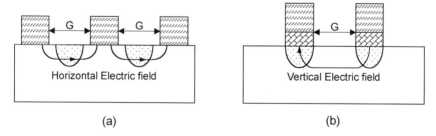

(a)   (b)

Fig. 6.4 The normally used electrode configurations and the respective field conditions, (a) horizontal field used for the x- or y-cut crystal orientations, (b) vertical field used for the z-cut crystal orientation, with the electrodes placed above the waveguides.

to minimize the waveguide insertion loss, for a TM mode propagation. This will increase the driving voltage. The parameters of the dielectric layer, usually a SiO_2 layer, can be used as a design parameter to achieve larger bandwidths.

Extinction Ratio and Insertion Loss

If I_0 is the intensity at the output of the modulator, when no voltage is applied, I_{max} is the maximum intensity, and I_{min} is the minimum intensity when the voltage is applied, then

$$\text{the insertion loss is defined as } 10 \log(I_{max}/I_0), \tag{6.14}$$

and

$$\text{the extinction ratio, ER, is } 10 \log(I_{min}/I_{max}). \tag{6.15}$$

Chirp

In the case of small-signal applications, the dynamic chirp $\acute{\alpha}(t)$ is the instantaneous ratio of the phase modulation to amplitude modulation of the transmitted signal and expressed as [53]

$$\acute{\alpha}(t) = \frac{d\Psi/dt}{[(1/2I)(dI/dt)]} \tag{6.16}$$

where Ψ and I are the phase and intensity of the optical field and t denotes the time. In case of the intensity modulator using a Mach-Zehnder type (Fig. 6.3), the α can be represented in a simplified form,

$$\acute{\alpha} \sim \frac{\Delta\beta2 + \Delta\beta1}{\Delta\beta2 - \Delta\beta1}$$

$$= \frac{\Delta V2 + \Delta V1}{\Delta V2 - \Delta V1} \tag{6.17}$$

$\Delta\beta1$, $\Delta\beta2$ are the electrooptically induced phase shifts, and $\Delta V1$, $\Delta V2$ are the peak-to-peak applied voltages of the two arms of the interferometer.

Although this expression can be applied in general to the small-signal region, it can also be applicable, to a large extent, for the large signal region, due to the shape of the switching curve. Also, the value of α can take $-$ or $+$ values and the chirp can be used to advantage, depending on the optical transmission system. For systems that operate, away from the zero-dispersion wavelength region, and depending on the fiber used for transmission, a negative chirp can be advantageous to achieve low dispersion penalties [53, 64]. In general, for a z-cut lithium niobate intensity modulator, with a traveling wave type electrode, the value can be -0.7. Depending on the crystal orientation and type of electrode structure, the value can be zero (x-cut modulator).

Common Electrode Structures

The simple electrode structure, consisting of two symmetric electrodes on both interferometer waveguides, otherwise known as lumped electrode structure, is shown in Fig. 6.5 (a). Figure 6.5 (b) shows the equivalent circuit of such a lumped electrode structure. The source part is represented by a voltage source, V_{source}, and load resistance, R_{load}. The modulator part is represented as a capacitance, C_{mod}. The bandwidth in this case is limited by the RC (load resistance and modulator capacitance) and can be

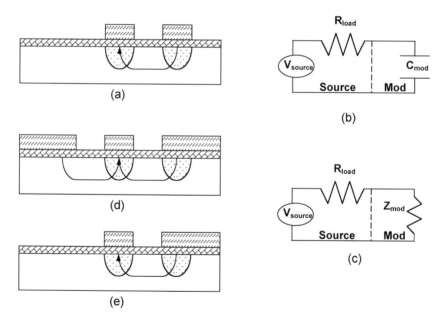

Fig. 6.5   The commonly used electrode structures, for a z-cut oriented lithium niobate optical modulator, (a) a symmetrical lumped electrode structure, (b) the RC equivalent circuit of the lumped electrode structure, (c) the equivalent circuit of a traveling wave electrode structure. The commonly used traveling wave electrode structures (d) a CPW (Coplanar Waveguide) electrode structure and (e) the ACPS or Asymmetric Coplanar Stripline structure.

given by $1/2V_\pi R_{load}C_{mod}$, hence, it is difficult to achieve large bandwidths. The widely used electrode structure for larger bandwidths is the traveling wave electrode structure, where the modulator electrode structure is designed to be an extension of load resistance. Figure 6.5 (c) shows the equivalent structure of this design. The modulator is designed to have the characteristic impedance, Z_{mod}. The widely used traveling wave electrode structures are shown in Fig. 6.5 (d) and (e). Figure 6.5. (d) shows the structure of a CPW Coplanar electrode structure [56, 57], which consists of a central signal electrode and two ground electrodes on both sides of the signal electrode, whose widths are assumed to be sufficiently larger than the signal (or central) electrode structure. Figure 6.5 (e) shows the ACPS, Asymmetric Coplanar Stripline or ASL, Asymmetric stripline electrode structure [58, 62], which consists of a central signal and one ground electrode, where the ground electrode width is assumed to be sufficiently larger than that of the signal electrode. In both of these cases the bandwidth is no more limited by the capacitance of the modulator, but is dependent on the velocity matching and microwave attenuation of the electrode structures.

The other important characteristics include

Optical: Wavelength of operation, optical return loss, maximum power, polarization dependency, etc.

Electrooptic and microwave characteristics: Bandwidth (frequency response), microwave attenuation, characteristic impedance, etc.

Mechanical and long-term stability: size, temperature, and DC drift stability, humidity and shock and vibration stability, fiber pull strength, etc.

These are the characteristics that need to be addressed by the modulator designer, from the initial stage. The waveguide technology is mature enough to satisfy most of the characteristics. The most important characteristics that need special attention from the design point of view are larger bandwidths and lower driving voltages, which will be discussed in detail in the next section.

6.3.3. DESIGN CONSIDERATIONS OF MODULATORS

The usual system requirements are larger bandwidths with lower driving voltages, due to the limitations of available low driving voltage drivers. Bandwidth and driving voltage of lithium niobate modulators are in a trade-off relationship. One has to be sacrificed for the other. Modulator design concentrates on optimizing various parameters and finding ways to achieve both larger bandwidths and lower driving voltages [65–97].

6.3.4. BANDWIDTH

The bandwidth of a modulator is dependent on the velocity mismatch between the optical and microwave (RF) and the microwave attenuation of the electrode structure. The velocity mismatch can be controlled by the electrode and buffer layer parameters. But once the electrode/buffer layer parameters get fixed, the microwave attenuation (α) gets fixed. In other words, the microwave attenuation, which gets fixed by the electrode/buffer layer parameters, limits the achievable bandwidth, even though perfect velocity matching is achieved. The driving voltage or V_π is also dependent on the electrode/buffer layer parameters.

Velocity Matching

As the effective refractive indices of the optical wave (2.15, for TM mode, at 1.55 μm) and the microwave (4.2) are different, a velocity mismatch exists between the two fields, which are propagating simultaneously. This mismatch limits the achievable optical bandwidth value. It is possible to reduce the microwave refractive index to that of the optical refractive index, by optimizing the electrode and buffer layer parameters. Figure 6.6 shows the cross-sectional view of the Ti-diffused LiNbO₃ Mach-Zehnder modulator with CPW electrode structure. The parameters that are controlled and

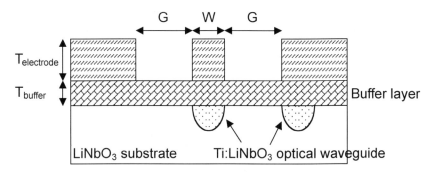

Fig. 6.6   The cross-sectional view of the Ti-diffused LiNbO₃ Mach-Zehnder modulator with a CPW electrode structure.

optimized are W, the width of the signal electrode, G, the gap between the signal and ground electrodes, $T_{electrode}$, the thickness of the electrode, T_{buffer}, the thickness of the buffer layer, and ε, the dielectric constant of the lithium niobate crystal. Highly accurate optical waveguide simulation methods and tools are needed to design these modulators. The WKB method, finite element method, etc., are used to solve the two-dimensional analysis. There is some commercial software available for three-dimensional analysis also. But most of the optical and microwave field analysis tools give a general trend as the calculation, and the actually measured characteristics usually do not agree completely. There is a need to improve the simulation techniques, incorporating the measured experimental values of various parameters, in order to achieve a better simulation tool. Here the general design criteria are discussed, where the various parameters used here are optimized using the experimental results. Care should be taken in modifying these parameters with the measured and fabricated conditions. In the following design, two-dimensional finite element analysis is used for the microwave analysis, to calculate the capacitance, effective microwave index, and the characteristic impedance. The BPM (beam propagation method) or PBM (propagation beam method) is used for optical field analysis. The details can be obtained from various references, including, [58–60, 62, 66–69]. Here the results are given.

The parameters used include the refractive index (TM modes) at 1.55 μm of wavelength, n_e at 2.15, the dielectric constants of the z-cut $LiNbO_3$ 28 for z direction and 43 in other directions. The buffer layer is assumed to be SiO_2 with the dielectric constant of 3.9. Figure 6.7 (a) and (b) show the microwave refractive index n_m and the characteristic impedance Z as functions of the electrode width to gap ratio, W/G, buffer layer thickness, and electrode thickness. It can be observed that n_m decreases by increasing the buffer layer and the electrode thickness. Similar results can be observed for the characteristic impedance. Figure 6.8 shows the results as functions of the buffer layer and electrode thickness for a fixed W/G of $7/28$. A set of optimized design parameters to achieve the velocity matching and characteristic impedance of nearly equal to 50 Ω are $W = 7$ μm, $G = 28$ μm, $T_{buffer} = 1.1$ μm, $T_{electrode} = 1.1$ μm. For these values the $n_m = 2.15$ ($n_m - n_o = 0$) and $Z = 48.5$ Ω. These values depend on various experimental factors and fabrication conditions. Hence, care should be taken that the necessary optimization is achieved in incorporating the experimental values with the modulator design parameters.

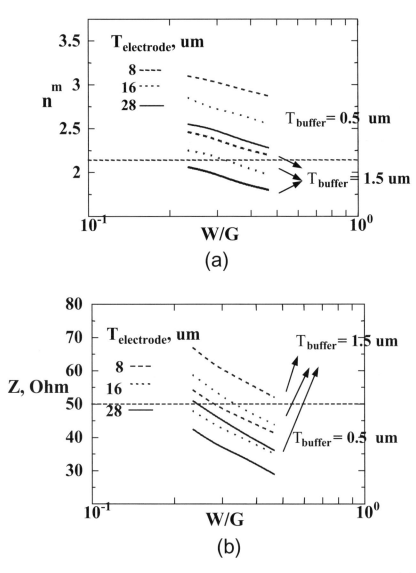

Fig. 6.7 The calculated values of (a) microwave refractive index n_m, and (b) characteristic impedance Z as functions of the electrode width-to-gap ratio, W/G, buffer layer thickness, and electrode thickness.

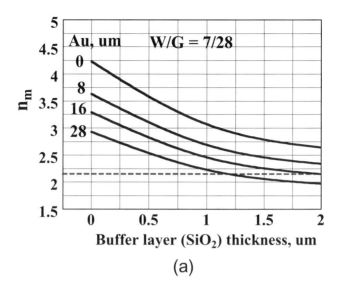

(a)

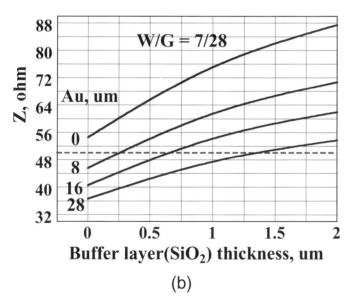

(b)

Fig. 6.8   The calculated values of (a) microwave refractive index n_m, and (b) characteristic impedance Z as functions of the buffer layer thickness and the electrode thickness for a fixed value of the electrode width-to-gap, W/G, ratio.

Optical Response Function

The bandwidth of a modulator can be obtained from the optical response function, which can be defined as

$$H(f) = \frac{[1 - 2e^{-\alpha L}\cos 2u + e^{-2\alpha L}]^{1/2}}{[(\alpha L)^2 + (2u)^2]^{1/2}} \qquad (6.18)$$

where

$$u = \pi f L(n_m - n_o)/C \qquad (6.19)$$

$$\alpha = \alpha_0 f^{1/2}/(20\log e) \qquad (6.20)$$

α_0 = microwave attenuation constant, f = frequency, n_m = effective microwave index, n_o = effective optical index, $(n_m - n_o)$ = velocity mis−match, L = length of electrode, C = velocity of light.

The units of optical response in dB, are given as

$$\text{Optical response in dB, electrical is } 20\log\{H(f)\} \qquad (6.21)$$

and

$$\text{Optical response in dB, optical is } 10\log\{H(f)\} \qquad (6.22)$$

S_{21} response is the electrical or microwave response, which can be approximated by Eq. (6.20), under the assumption that the microwave attenuation is mainly due to the stripline conductor loss. The approximate relation between the microwave response S_{21} and the optical response can be represented as follows:

6 dB value of S_{21} corresponds to approximately to the

3 dB value of the optical response in dB, electrical or

1.5 dB value of the optical response in dB, optical,

taking into consideration the exponential factors of Eq. (6.18).

Care should be taken in understanding the difference between the bandwidths (optical response) represented in dB, electrical, and dB, optical, as they are different in system use. The 3 dB bandwidth, when represented in dB, optical, shows very large bandwidths, but, in reality, the bandwidth values will be much smaller when they are seen in 3 dB, electrical, values. For example, referring to Fig. 6.9 (a) and (b), for $\alpha_0 = 0.4$ dB/{cm (GHz)$^{1/2}$}, the 6 dB, S_{21} bandwidth is approximately 16 GHz. The corresponding optical bandwidth, when represented, under dB, electrical, is

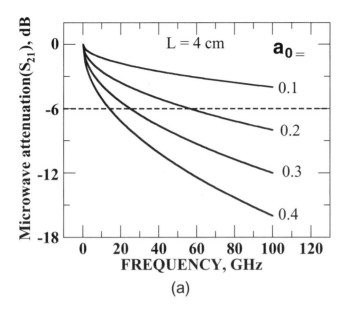

(a)

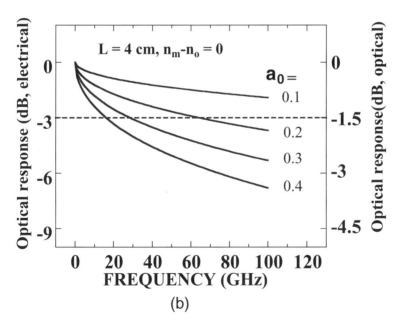

(b)

Fig. 6.9 The relation between (a) the electrical response and (b) the optical response, as functions of the frequency of operation and the microwave attenuation constant, for a fixed electrode length of 4 cm.

18 GHz, whereas the optical bandwidth, when represented in dB, optical, is 72 GHz. Under these values, the modulator will be useful for system applications at 20 Gb/s, and not at 80 Gb/s. It will be misleading to say that the modulator can be used for an 80 Gb/s system, just because the optical bandwidth in dB, optical, is 72 GHZ. Figure 6.9 (a) shows the calculated values of S_{21}, the microwave attenuation, as functions of frequency, and various assumed values of the microwave attenuation, α_0, for a fixed electrode length of 4 cm. Figure 6.9 (b) is the corresponding optical response, the left axis in dB, electrical, and the right axis in dB, optical. In this case, the perfect velocity matching condition ($n_m - n_o = 0$), is assumed. It is seen that the bandwidth increases tremendously when the microwave attenuation is decreased. From both these figures, it is very evident that, even when a perfect velocity matching is achieved, the bandwidth is limited by the microwave attenuation. Thus, reduction of microwave attenuation is the key in achieving very large bandwidths.

The velocity matching using the thick electrodes and thick buffer layer is in ACPS electrode structure (Fig. 6.5) reported by Ref. [58]. For an electrode length of 2 cm, a driving voltage of 5.4 V, a bandwidth of 20 GHz, and a microwave attenuation of 0.67 dB/{cm (GHz)$^{1/2}$} was achieved. One drawback of the ACPS structure is the resonance problem at higher frequencies, and there is a need to reduce the chip thickness and width. In the case of CPW electrode structure, thick electrodes and buffer layer are utilized in refs. [68, 69]. For an electrode length of 2.5 cm, a driving voltage of 5 V, a bandwidth of 20 GHz with a microwave attenuation of 0.54 dB/{cm (GHz)$^{1/2}$} was achieved. The issue with a CPW electrode was higher microwave loss due to the higher order mode propagation. Reduction of chip thickness is needed.

Another structure where velocity matching is achieved uses a shielded plane above the electrode structure [57]. The idea is to put a metal above the electrode structure, with an air gap between of the order of 5 microns. For an electrode length of 2 cm, a driving voltage of 5.2 V, bandwidth of 20 GHz and a microwave attenuation of .67 dB/{cm (GHz)$^{1/2}$} was achieved. The issues include difficulty of manufacturing and higher microwave loss.

Microwave Attenuation Components and Reduction Techniques

Reduction of microwave attenuation is the main factor to achieve very large bandwidths. Figure 6.10 shows the general structure of the CPW electrode. As mentioned earlier, to achieve velocity matching and required

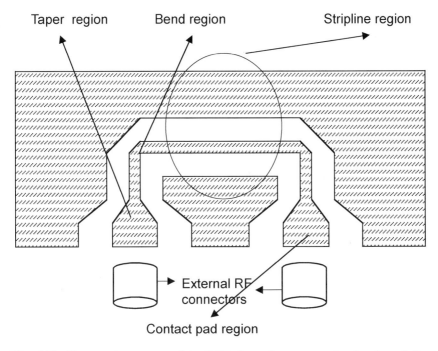

Fig. 6.10 The generic structure of the CPW electrode structure, showing the stripline region, the bends, and the tapered regions.

characteristic impedance, the design deals mainly with the stripline region. In this region the signal electrode width and gap are of the order of a few microns to a few tens of microns. When this structure is to be connected to outside RF connectors, having dimensions of a few hundreds of microns order (say, the K-connector is 280 μm, the V-connector 220 μm), there is a need to include a bend region, tapered region, and the connector matching contact pad region. The external RF connector can be directly placed, or wire/ribbon bonded, with or without ceramic CPW structures in between. All these components are sources of microwave attenuation and need to be designed with minimum loss.

The total microwave attenuation of the electrode structure can be subdivided broadly into the following factors [77, 81, 83, 89–91, 96]:

1. Stripline conductor loss: This is the main component of the loss, and is dependent on the electrode and buffer layer parameters. These parameters are already optimized when the velocity matching design is completed. This loss has a root frequency relationship with the frequency (as given by Eq. (6.20)). In general, the surface area/volume of the electrode structure limits this loss. Increase of the electrode thickness and width decreases this loss.

Simply, increasing the electrode width, W, may decrease the microwave attenuation, but it will also make the gap, G, become large, in order to maintain the same W/G ratio for velocity matching. This in turn increases the driving voltage. A novel electrode structure, with a two-layer structure, with different widths in the upper and lower portions, is reported [77]. This structure has two layers of electrodes, the lower layer with the standard W, G values say, 7 and 28 μm, and another upper portion, with W', G' as say, 25 and 100 μm. This structure keeps the driving voltage small, as the lower portion has a smaller gap, but, at the same time, has minimum microwave attenuation, due to increase in the signal electrode dimensions. One issue of this structure is the degradation of S_{11}, the RF return loss, due to proximity of the upper signal electrode of width W' to the lower ground electrode gap, G. The reason for this is the degradation of the characteristic impedance from the designed 50 ohm value, as the W/G changes, in effect to W'/G. Improved structures are shown in Ref. [15]. A bandwidth of 18 GHz and DC driving voltage of 3.3 V, with a microwave attenuation constant of 0.36 dB/{cm (GHz)$^{1/2}$}, was the lowest reported achievable value at that time [77].

2. Dielectric loss: This loss is an inherent loss of the crystal, and depends on the dielectric constant and tan δ of lithium niobate and cannot be controlled by design. This loss is directly proportional to frequency. This becomes a real issue, usually after 30 GHz. Reduction of the above-mentioned stripline loss will ease the impact of this loss.

3. Higher order mode propagation loss: This loss is more prominent for the traveling wave electrode structure, including the CPW structure. It is always desirable to have a single-mode propagation structure, but, as explained earlier, the tapers and the contact pad regions, due to their large dimensions, always contribute to the multimode propagation, which is lossy. This loss can be reduced by reducing the substrate thickness [27, 69], but this reduction in substrate thickness necessitates the need of extra packaging techniques for fiber attachment, etc.

4. Losses due to bends, tapers: The bends and tapers increase the microwave attenuation. These can be reduced to some extent by proper design considerations. The bends can be designed as straight 90 degree bends, with a 45 degree cut at the edge, or curved bends with smaller bend radius. The tapers are designed with smooth and unabrupt transition regions.

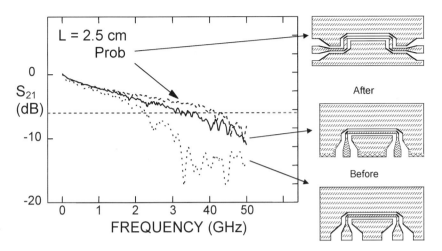

Fig. 6.11 The electrical response characteristics of the CPW electrode structure, as the function of frequency, for the cases of the structure that is fully connectorized, before/after the modifications, and that with a probe measurement.

5. The connector to contact pad loss and other package related loss: These losses are very critical for bandwidths above 25 GHz, and proper design considerations are needed. These problems are more severe for the modulators at 40 GHz applications. Reduction in size of the contact pad, taper widths, taper lengths, and the bends can be optimized [88] to achieve low loss. Figure 6.11 shows the S_{21} microwave attenuation measurements. The experiments were carried out with a probe measurement, the packaged device without the above-mentioned improvements for the losses, and that with the improvements. The probe measurements give the chip's S_{21} characteristics, without the losses associated with the connector, the connector to contact pad and other package losses. It can be observed that the improvements on the package-related losses increase the achievable bandwidths to a large extent, especially above 25 GHz. Figure 6.12 shows the results of Ref. [88], with all the above-mentioned microwave attenuation improvements. For an electrode length of 4 cm, the driving voltage is 3.3 V, and the bandwidth is 26 GHz with the microwave attenuation of 0.3 dB/{cm (GHz)$^{1/2}$}.

It is also possible to reduce the microwave attenuation by increasing the buffer layer thickness [89]. The big problem will be the increase of driving voltage.

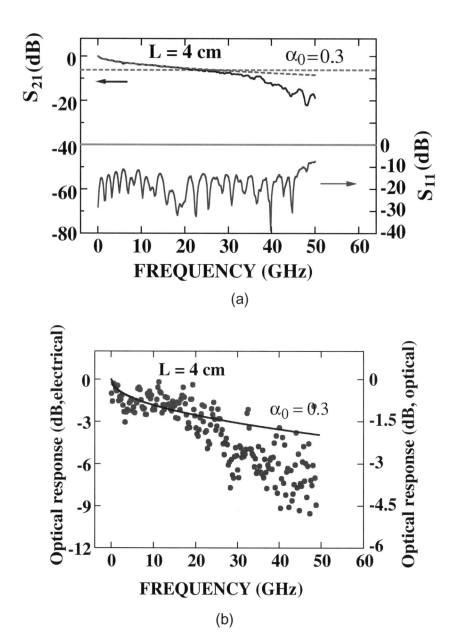

Fig. 6.12 (a) The electrical and (b) optical characteristics of a modulator with the improvements (ref. ECOC 97), for an electrode length of 4 cm.

Further reduction of the microwave attenuation is needed to achieve much larger bandwidths of 80 GHz/160 GHz.

6.3.5. *DRIVING VOLTAGE REDUCTION*

The driving voltage is given by Eq. 6.12. The driving voltage reduction can be realized, mainly by increasing the electrode length or Γ, the overlap integral, between the optical and RF waves, or decreasing G, the gap between the two arms of the interferometer. There is a limit to decreasing G. If the arms are too close, there is a problem of mode coupling between these two arms. This will cause a degradation of extinction ratio. Also G is the parameter that became fixed in earlier velocity matching design. Increasing the electrode length poses problems on the achievable bandwidth due to microwave attenuation problems.

The overlap integral, Γ, can be represented as

$$\frac{G \iint E^2(x, y)\varepsilon(x, y)\, dx\, dy}{V \iint E^2(x, y)\, dx\, dy} \tag{6.23}$$

where, $E(x, y)$ and $\varepsilon(x, y)$ are the optical and microwave/electrical fields at a point $P(x, y)$ in the crystal. V is the applied voltage across the electrodes, G is the electrode gap. The optical field can be calculated using BPM calculations and the electric field, using the finite element analysis [55–60, 68]. The overlap integral needs to be as large as possible and it depends on the waveguide fabrication parameters and diffusion parameters. The waveguide parameters include the titanium thickness, titanium concentration, and the gap between the electrodes. The diffusion parameters include the diffusion time and temperature. All these parameters are to be optimized, in order to achieve strong mode confinement. Also, the position of electrodes vis-à-vis the waveguide position dictates the overlap integral value. The other important parameter is the buffer layer thickness. As the buffer layer thickness is increased, the driving voltage increases as the overlap integral decreases. Thicker buffer layers are needed to achieve the velocity matching, as explained previously. Figure 6.13 shows the driving voltage as a function of the buffer layer thickness. Once the velocity matching condition is obtained, the buffer layer thickness and the achievable driving voltage get fixed. The optimization of the waveguide/electrode parameters to achieve a strong confinement remains to achieve the lower driving voltages, in the usual cases.

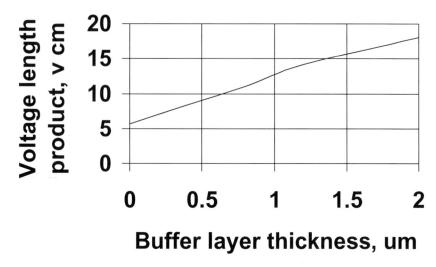

Buffer layer thickness, um

Fig. 6.13   The driving voltage as the function of the buffer layer thickness.

Another method to reduce the driving voltage is the use of a dual electrode structure [65, 86]. In this structure (Fig. 6.14 (a)), where the two
arms of the interferometer are driven by two independent signal electrode structures, the driving voltage can be reduced by approximately half.
This structure has the advantage of controlling chirp value. As can be observed from Eq. (6.17), by individually controlling the voltages applied to
the two arms, it is possible to obtain a zero chirp or a negative/positive
chirp.

Another method (Fig. 6.14. (b)) of reducing the driving voltage is use of
a ridge waveguide structure, at the two arms of the interferometer [73, 78,
79, 84]. By etching rides in the region, the overlap integral can be made
larger. At the same time, it possible to design a modulator to achieve both
velocity matching and required characteristic impedance. A bandwidth of
30 GHz, driving voltage of 3.3 V for an electrode length of 3 cm is achieved.

Another method (Fig. 6.14. (c)) of controlling the thickness of the buffer
layer, across the waveguides, to achieve both large bandwidth and low
driving voltage is reported in [83, 88, 90, 95]. The thickness is varied so
that both the velocity matching condition and the low driving voltage are
achieved at the same time. For the electrode lengths of 4 cm and 3 cm,
driving voltages of 2.5 V and 3.3 V and bandwidths of 25 GHz and 32 GHz
were achieved respectively.

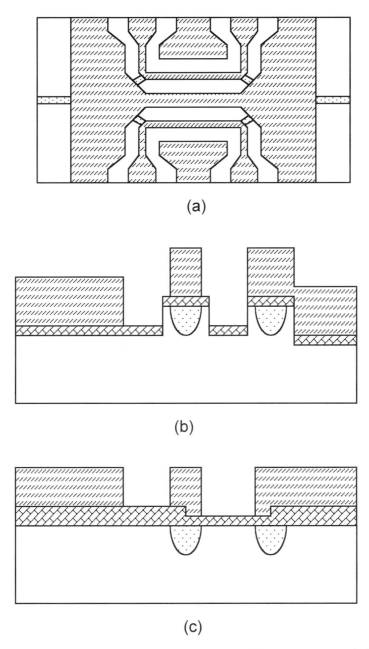

Fig. 6.14 The reported structures, with voltage reduction improvements: (a) a dual electrode structure, (b) a ridge waveguide structure, and (c) a step buffer layer structure.

Thus, the various design parameters can be optimized to achieve large bandwidth, low driving voltage modulators. Further innovations and improvements will be needed to achieve further increase in bandwidths.

6.4.   Modulator Fabrication Methods and Reliability

6.4.1.   FABRICATION METHODS

Lithium niobate wafers, of optical grade, with high surface homogeneity and flatness are used to make these devices. These wafers are obtained from the crystals grown using the Czochralski method, under controlled conditions. These wafers are available commercially, in sizes of 3 or 4 inches (75 or 100 mm), in various crystal orientations, x- or y- or z-cut, in different thicknesses.

There are two methods to make waveguides using the Lithium Niobate wafers for these modulators: the thermal in-diffusion and the proton exchange.

Proton Exchange waveguides: A method of making the waveguides in $LiNbO_3$ is the Proton Exchange (PE) [33–36]. Using acid baths (proton rich) such as C_6H_5COOH, at lower temperatures of 120–250°C, it is possible to make waveguides. The Li^- ions in $LiNbO_3$ are exchanged with protons, H^+, from the acids, resulting in the higher refractive index part on the surface of the wafer, in the form of $H_xLi_{1-x}NbO_3$, where x usually >0.5. The proton exchange waveguides need to be annealed at higher annealing temperatures, after proton exchange. The initial issues of the proton exchange waveguides were non-uniformity stability problems of the refractive index and the degradation of electrooptic characteristics. Annealing operation is very critical in solving these problems. In proton exchanged layers, there is an increase in the extraordinary refractive index and no change, or in some cases a decrease, in the ordinary refractive index. Also, the proton exchange is possible in z-cut and x-cut orientations, as the acid etches, chemically, the y-cut wafers.

Thermal in-diffusion: The waveguides are usually made by in-diffusion of titanium metal strips, at temperatures ranging from 950 to 1050°C, for 5 to 10 hours [17, 32, 37]. The fabrication of the modulators involves the fabrication of the optical waveguides, buffer layer formation, and electrode layer formation. The general steps are shown in Figs. 6.15 and 6.16. The optical grade $LiNbO_3$ wafers (substrates) are inspected for surface flatness, in order to achieve uniformity of modulator characteristic across the wafer

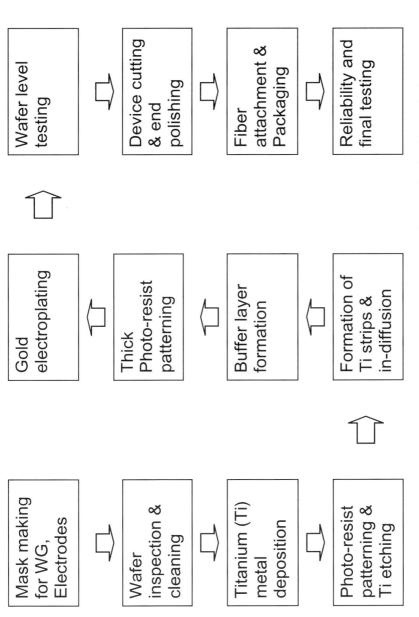

Fig. 6.15 The general fabrication steps of the typical z-cut modulator, by the Ti etching method.

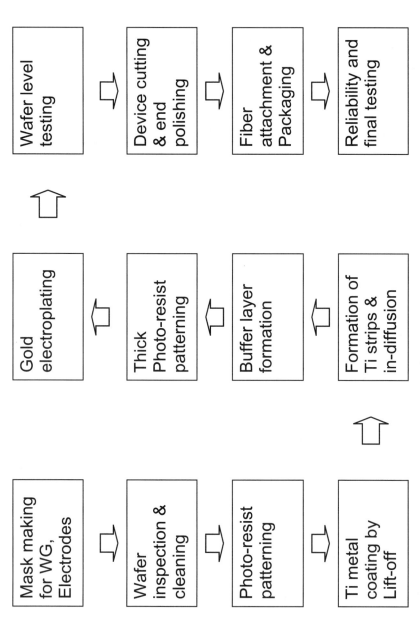

Fig. 6.16 The general fabrication steps of the typical z-cut modulator, by the Ti liftoff method.

and inside each modulator element chip. High yield and uniform characteristics are the key factors for volume manufacturing. The various photo masks for the waveguide structure and electrode structures are made. The titanium metal strips fabrication can be done in two ways. One is the lift-off method, in which the photo resist pattern is first made and the titanium metal is coated, resulting in the required titanium strip structure. The other method is first coating the substrate with titanium metal, then making the photo resist pattern, and finally etching the extra Ti metal. Figures 6.15 and 6.16 show the difference in these two methods. The latter process is explained in detail.

Titanium metal of thickness a few hundred microns (the thickness is dependent on the single mode requirements, Ti concentration etc.), is coated on the substrate, either using electron beam evaporation or sputtering method. Then the pattern of the modulator waveguide structure is transferred on to the titanium metal using photo resist patterning technique. Then the titanium is etched, except the areas where the modulator pattern is required. After removing the photo resist and cleaning, the substrate is placed in the diffusion chamber and the titanium in-diffusion is carried out. During this process, care is taken in suppressing the out-diffusion of lithium ions from the surface. The out-diffusion of lithium results in unwanted surface planar waveguides, which degrades the waveguide characteristics. The methods include the use of covered platinum enclosures, water vapor atmosphere during the diffusion, etc. The in-diffused titanium forms the waveguide structure with increased refractive index, vis-à-vis the non-diffused substrate region.

The next step will be deposition of the buffer layer, which is usually SiO_2 or doped SiO_2. It can be deposited by electron beam evaporation or sputtering. Then a thick photo resist is coated and the patterning of the electrode structure (either CPW or ACPS) is carried out. Usually, the thickness is of the order of 15 to 35 μm and care is taken in making the walls of the pattern as straight as possible. Then, the gold electrodes, of thickness 10 to 30 μm, are electroplated. The edge straightness and the surface grain size of the electroplated gold electrode play an important role in the performance of the modulator in terms of microwave attenuation and characteristic impedance. Once the wafer processing is done, the wafer is inspected and tested for performance, depending on the requirements of the modulator. Then, various modulator chips are cut and edge polished. The chips are packaged in the hermetically sealed packages, after the input/output fibers are attached and the necessary connections to the external connectors are made. The presence of OH ions on the surface will degrade the reliability of the device. Hermetical packaging is needed to avoid contamination of the surface by OH ions.

6.4.2. RELIABILITY

The long-term reliability was the main performance parameter that is vital for using these devices for commercial and practical systems. The DC drift and temperature stability (and humidity drift) are the main long-term reliability issues [31, 45, 70, 78, 80, 85].

DC Drift

DC drift is the optical output power variation under the constant DC bias voltage application. Figure 6.17 (a) shows the output power of the modulator as a function of the applied voltage. The broken line shows the output power as a function of applied voltage when only AC voltage is applied (and no DC is applied, at $t = 0$) and the solid line shows the same, after $t = t_1$, when DC voltage is also applied in addition to the previous AC signal voltage. The shift between these two curves, ΔV, is the measure of the DC drift. When these types of modulators are used in practical systems, the signal is usually applied at the center of the switching curve (i.e., middle of the maximum/minimum), which is known as the driving point. Once the shift due to DC drift occurs, there is a need to bring back the driving point voltage to the previous operating point, using an automatic bias control circuit (ABC circuit) or feedback control circuits (FBC circuit). It is desirable to minimize this shift, and in most cases, a negative shift is more desirable, as it facilitates smaller voltage application, through the ABC circuit. The actual mechanism of the DC drift and its causes are not yet well understood. But the cause can be attributed to the movement of ions, including OH ions, inside the lithium niobate substrate and the buffer layer. It is influenced by the balance of the RC time constants, in both horizontal and vertical directions in the equivalent circuit model, as shown in Fig. 6.17 (b). It was also found that the DC drift is more affected by the buffer layer. In the circuit model of Fig. 6.17 (b), all layers, including the LiNbO$_3$ substrate, the Ti:LiNbO$_3$ optical waveguide and the buffer layer, are represented in terms of resistances R and capacitances C, in both vertical and horizontal directions of the crystal.

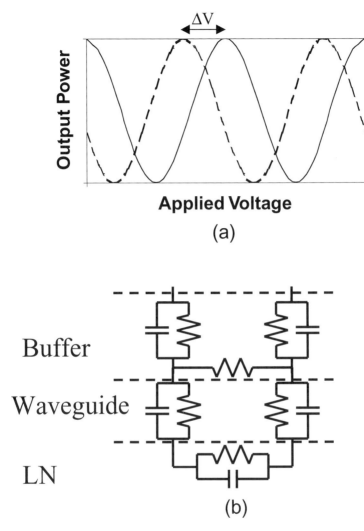

Fig. 6.17 The DC drift, (a) the output power of the modulator as the function of the driving voltage, with and without the DC applied voltage, and (b) the equivalent RC circuit model of the structure, with vertical and horizontal components.

It is experimentally proved that the DC drift can be reduced by decreasing the vertical resistivity of the buffer layer or by increasing the horizontal resistivity of the buffer layer (or that of the surface layer). The surface layer is the boundary layer between the buffer layer and the substrate.

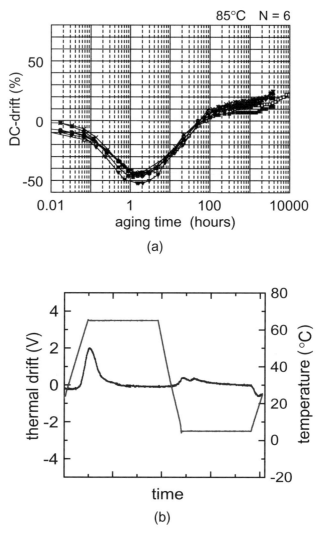

Fig. 6.18   The long-term reliability of the z-cut lithium niobate modulator, (a) the DC drift characteristics as a function of the aging time at 85°C, and (b) the thermal (temperature) drift as a function of time.

The reduction of the vertical resistivity is obtained by doping the SiO_2 buffer layer using TiO_2 and In_2O_3 [80]. The increase of the horizontal surface resistivity can be obtained by making a slit in Si/SiO_2 [87]. In both cases, the movement of ions, especially between the two interferometer arms (waveguides) is arrested. Fig. 6.18 (a) [87] shows the DC drift characteristics of the z-cut modulators at 85°C, for a few thousand hours for an applied DC voltage of 10 V. The accelerated DC drift measurement was done at higher temperatures to assess the modulator's operation at room temperature. It can be observed that, between 0 and 50 hours, the DC drift stays negative and tends to saturate after 1000 hours. The DC drift variation is less than 30% and the saturation shows that the modulator can be operated for longer periods. It can be estimated, from the accelerated DC drift measurements and the activation energy of approximately 1.0 eV, that the long-term stability can be proved, in the form of DC drift variations of less than 30% for more than 25 years at room-temperature operation. These values are more than sufficient for using the feedback control circuits.

Thermal Drift

Thermal drift is the optical output power variations under temperature changes. Once the temperature changes, the Piezoelectric charges are induced on the surface of the LN substrate. This causes a surface charge distribution across the two arms of the interferometer, affecting the electric field. This results as a shift in the switching curve and the driving point/operation point shifts, similar to the DC drift. Hence, in order to reduce this thermal drift, there is a need to distribute or dissipate the charges that are accumulated on the LN surface and between the electrodes.

A method in which the charges are dissipated using the semiconductor layers like Si [48] and other materials was proved to reduce the thermal drift. Another method, using a Si double slit and reduction of the resistivity by one order of magnitude, was reported [92]. Figure 6.18 (b) shows the thermal drift, the horizontal axis denotes the time in minutes, the right vertical axis shows the thermal drift in V, and the left vertical axis shows the temperature. The temperature of the modulator is increased from room temperature (25°C) to 65°C in around 5 minutes. The modulator is then kept at that temperature for around 15 minutes and then the temperature is reduced to 5°C. The modulature is then kept at that temperature for around 15 minutes and brought back to room temperature. During this cycle, the thermal drift is changed from +3.5 V to −3 V. Without the Si double slit stricture and without reducing the Si resistivity, the modulators had much higher drift values (3 to 4 times higher than that given above), and were previously unusable for practical system applications.

6.5. Summary and Conclusion

In this chapter, the progress of the lithium niobate optical modulator technology is reviewed. The research and development in this technology, all over the world, has produced tremendous improvement in device characteristics, especially long-term reliability. This in turn, established the lithium niobate modulators as the most promising external modulators, increasing the system functionality. The deployment of $LiNbO_3$ external modulators for high-speed long-distance fiber optic communication systems [91–98] is proof of the technological evolutions in all these years. The versatility, and development of various other devices, such as polarization controllers, switches, and wavelength filters, mean the lithium niobate device technology will greatly influence the long-term usage and marketplace superiority in future years.

ACKNOWLEDGMENTS

The present chapter, on the progress of lithium niobate devices technology, represents the work and achievements of a number of researchers and developers all over the world for many years. Due to limited space available, it is not possible to refer to and acknowledge all of their valuable contributions or to thank them individually. The author is grateful to all of them.

Most of this work was done when the author was working with NEC Corporation, Japan. The author is thankful to all the colleagues, including M. Kondo, T. Hosoi, T. Miyakawa, T. Kambe, Y. Uematsu, and other researchers and the management at NEC Corporation, Central Research labs, Kawasaki, Japan. The author is also thankful to Prof. S. Kawakami and Prof. Minakata of Tohoku University, Sendai, Japan, who introduced the author to the lithium niobate technology, during his Doctorate period. The author is also thankful to Agere Systems (formerly Lucent Technologies Microelectronics group) for their kind support since the author joined them in 1999.

Finally, I am very grateful for all their support, in all these years, to my wife Hitomi and my parents, Krishnamachari and Ranganayaki.

References

1. I. P. Kaminow, T. J. Bridges, and E. H. Turner, "Electrooptic light modulators," Appl. Opt., 5 (1966) 1612–1614.
2. S. E. Miller, "Integrated Optics: an introduction," Bell Sys. Tech. Journal, 48 (1969) 2059–2069.
3. I. P. Kaminov, ed., An Introduction of Electrooptic Devices (Academic Press, NY, 1974).
4. H. F. Taylor and Y. Yariv, "Guided wave optics," Proc. IEEE, 62 (1974) 1044–1060.

5. T. Tamir, ed., "Integrated Optics," Topics in Applied Physics, 2nd ed., (Springer-Verlag, NY, 1979).
6. R. G. Hunsperger, Integrated Optics: Therory and Technology (Springer-Verlag, NY, 1982).
7. R. C. Alferness, "Waveguide electrooptic modulators," IEEE Trans. Microwave Theory Tech., MT-30 (1982) 1121–1137.
8. R. C. Alferness and J. N. Walpol, eds., "Special issue on Integrated Optics," IEEE J. Quantum Electron., QE-22 (1986) 803–988.
9. L. D. Hutcheson, Integrated Optical Circuits and Componets (Marcel Decker, NY, 1987).
10. J. T. Boyd, ed., "Special issue on integrated optics," J. Lightwave Technol., 6 (1988) 741–1159.
11. S. K. Korotky, J. C. Campbell, and H. Nakajima, "Special issue on photonic devices and integrated optics," IEEE J. Quantum Electron., QE-27 (1991) 516–849.
12. K. Kawano, "High-speed Ti:LiNbO$_3$ and semiconductor optical modulators," IEICE Trans. Electron., E76-C (1993) 183–190.
13. K. Komatsu and R. Madabhushi, "Gb/s range Semiconductor and Ti:LiNbO$_3$ guided-wave optical modulators," IEICE Trans. Electron., E79-C (1996) 3–13.
14. F. Heismann, S. K. Korotky, and J. J. Veslka, "Lithium niobate integrated optics: selected contemporary devices and system applications," Optical Fiber Telecommunications IIIB (Academic Press, NY, 1997).
15. R. Madabhushi, "High speed modulators for coding and encoding," Short course, SPIE Photonics west, international conference (Jan. 2001).
16. I. P. Kaminov, J. R. Caruthers, E. H. Turner, and L. W. Stulz, "Thin-Film LiNbO$_3$ electro-optic light modulator," Appl. Phys. Lett., 22:10 (1973) 540–542.
17. R. V. Schmidt and I. P. Kaminov, "Metal-diffused optical waveguides in LiNbO$_3$," Appl. Phys. Lett., 25:18 (1974) 458–460.
18. W. E. Martin, "A new waveguide switch/modulator for integrated optics," Appl. Phys. Lett., 26 (1975) 562–564.
19. M. Papuchon, Y. Combemale, X. Mathieu, D. B. Ostrawski, I. Reober, A. M. Roy, B. Sejourne, and M. Werner, "Electrically switched optical directional couplers," Appl. Phys. Lett., 27 (1975) 289–291.
20. H. Kogelnik and R. V. Schmidt, "Switched directional couplers with alternating $\Delta\beta$," IEEE J. Quantum Electron., QE-12 (1976) 396–401.
21. W. K. Burns, A. B. Lee, and A. F. Milton, "Active branching waveguide modulator," Appl. Phys. Lett., 29 (1976) 790–792.
22. T. R. Ranganath and S. Wang, "Ti-diffused LiNbO$_3$ branched-waveguide modulators: performance and design," IEEE J. Quantum Electron., QE-13 (1977) 290–295.

23. M. Izutsu, Y. Yamane, and T. Sueta, "Broad-band traveling wave optical modulator using LiNbO$_3$ optical waveguide," IEEE J. Quantum Electron., QE-13 (1977) 287–290.
24. V. Ramaswamy, M. D. Divino, and R. D. Standley, "Balanced bridge modulator switch," Appl. Phys. Lett., 32 (1978) 644–645.
25. M. Minakata, T. Yamada, and S. Uehara, "Optical intensity modulator using a pair of optical gate couplersand conventional phase shifters," Trans. IEICE Japan, E-61 (1978) 148–151.
26. K. C. Gupta, R. Garg, and J. Bahl, Microstrp Lines and Slotlines (Artech House, Dedham, MA, 1979).
27. M. Minakata, S. Uehara, K. Kibota, and S. Saito, "Temperature stabilized optical waveguide modulation," Rev. ECL 26 (1978) 1139–1151.
28. M. Minakata, "Efficient LiNbO$_3$ balanced bridge modulator/switch with an ion-etched slot," Appl. Phys. Lett., 35 (1979) 40–42.
29. O. Mikami and S. Zenbutsu, "Modified balanced bridge switch with two straight waveguides," Appl. Phys. Lett., 35 (1979) 145–147.
30. K. Kubota, J. Noda, and O. Mokami, "Traveling wave optical modulator using a directional coupler LiNbO$_3$ waveguide," IEEE J. Quantum Electron., QE-16 (1980) 754–762.
31. S. Yamada and M. Minakata, "DC drift Phenomenon in LiNbO$_3$ optical waveguide devices," Japan J. Appl. Phys., 20 (1981) 733–737.
32. J. L. Jackal, V. Ramaswamy, and S. P. Lyman, "Elimination of out-diffused surface guiding in titanium-diffused LiNbO$_3$," Appl. Phys. Lett., 38:7 (1981) 509–511.
33. J. L. Jackal, C. E. Rice, and J. J. Veselka, "Proton-exchange for high index waveguides in LiNbO$_3$," Appl. Phys. Lett., 41:7 (1982) 607–608.
34. A. C. G. Nutt, K. K. Wong, D. F. Clarke, P. J. R. Laybourn, and R. M. De La Rue, "Proton-exchange lithium niobate slab and stripwaveguides; characterization and comparisions," 2nd ECIO conference, Tech. Dig., (1983) 53–56.
35. M. D. Michell, J. Botineau, S. Neveu, S. Sibillot, D. B. Ostrowski, and M. Papucheon, "Independent control of refractive index and profiles in proron exchange Lithium Niobate guides," Opt. Lett., 8:2 (1983) 114–115.
36. A. Yi-Yan, "Index instabilities in proton exchanged LiNbO$_3$ waveguides," Appl. Phys. Lett., 42:7 (1983) 633–635.
37. R. A. Becker, "Comparision of guided-wave interferometric modulators fabricated on LiNbO$_3$ via Ti indiffucion and proton exchange," Appl. Phys. Lett., 41:9 (1983) 131–133.
38. C. M. Gee, G. D. Thurmond, and H. W. Yen, "17-GHz bandwidth electrooptic modulator," Appl. Phys. Lett., 43 (1983) 998–1000.
39. R. C. Alferness, C. H. Joyner, L. L. Buhl, and S. K. Korotky, "High-speed traveling-wave directional coupler switch/modulator for $\lambda = 1.32\,\mu$m," IEEE J. Quantum Electron., QE-19 (1983) 1339–1343.

40. S. K. Korotky and R. C. Alferness, "Time and frequency domain response of directional coupler traveling wave optical modulators," J. Lightwave Technol., LT-1 (1983) 244–251.

41. R. A. Becker, "Traveling wave electrooptic modulator with maximum bandwidth-length product," Appl. Phys. Lett., 45 (1984) 1168–1170.

42. R. C. Alferness, S. K. Korotky, and E. A. Marcatili, "Velocity-matching techniques for integrated optical traveling wave switch/modulators," IEEE J. Quantum Electron., QE-20 (1984) 301–309.

43. C. H. Bulmer and W. K. Burns, "Linear interferometric modulators in Ti:LiNbO$_3$," J. Lightwave Technol., LT-2 (1984) 512–523.

44. S. K. Korotky, G. Eisenstein, R. C. Alferness, J. J. Veselka, L. L. Buhl, G. T. Harvey, and R. H. Read, "Fully connectorized high-speed Ti:LiNbO$_3$ switch/modulator for time division multiplexing and data encoding," IEEE J. Lightwave Technol., LT-3 (1985) 1–6.

45. R. A. Becker, "Circuit effect in LiNbO$_3$ channel waveguide modulators," Opt. Lett., 10 (1985) 417–419.

46. H. Haga, M. Izutsu, and T. Sueta, "LiNbO$_3$ traveling wave light modulator/ switch with an etched groove," IEEE J. Quantum Electron., QE-22 (1986) 902–906.

47. I. Sawaki, H. Nakajima, M. Seino, and K. Asama, "Thermally stabilized z-cut Ti:LiNbO$_3$ waveguide switch," Proc. CLEO'86, Paper MF2 (1986) 46–47.

48. L. McCaughan and S. K. Korotky, "Three electrode Ti:LiNbO$_3$ optical switch," J. Lightwave Technol., LT-4 (1986) 324–329.

49. J. P. Donnelly and A. Gopinath, "A comparision of power requirements of traveling-wave LiNbO$_3$ optical couplers and interferometric modulators," IEEE J. Quantum Electron., QE-23 (1987) 30–41.

50. S. K. Korotky, G. Eisenstein, R. S. Tucker, J. J. Veselka, and G. Raybon, "Optical intensity modulation to 40 GHz using a wave-guide electrooptic switch," Appl. Phys. Lett., 50 (1987) 1631–1633.

51. K. Miura, M. Minakata, and S. Kawakami, "A broad band traveling optical modulator with high efficiency," IEICE Technical report, OQE87-26 (1987) (in Japanese).

52. C. R. Giles and S. K. Korotky, "Stability of Ti:LiNbO$_3$ waveguide modulators in an optical transmission system," Technical digest topical meeting on Integrated and Guided-wave opticsm, Santa Fe, Washington, Paper ME5 (1988).

53. K. Koyama and K. Iga, "Frequency chirping in external modulators," J. Lightwave Technol., 6 (1988) 87–95.

54. D. W. Dolfi, M. Nazarathy, and R. L. Jungerman, "40 GHz electrooptic modulator with 7.5 V drive voltage," Electron. Lett., 24 (1988) 528–529.

55. R. Madabhushi, "A study of an antenna coupled optical Y branch and its applications," Doctorate thesis for Doctor of Electronics Engineering, Tohoku University, Sendai, Japan, 1989.

56. K. Kawano, T. Nozawa, M. Yanagibashi, and H. Jumonji, "Broad band and low driving power LiNbO$_3$ external optical modulators," NTT Review, 1 (1989) 103–113.

57. K. Kawano, T. Kitoh, H. Junmoji, T. Nozawa, and M. Yanagibashi, "New traveling wave electrode Mach-Zehnder optical modulator with 20GHz bandwidth and 4.7V driving voltage at 1.52μm wavelength," Electron. Lett., 25 (1989) 1382–1383.

58. M. Seino, N. Mekada, T. Namiki, and H. Nakajima, "33-GHz-cm broadband Ti:LiNbO$_3$ Mach-Zehnder modulator," Proc. ECOC, Paper ThB22-5 (1989) 433–435.

59. M. Rangaraj and M. Minakata, "A new type of Ti:LiNbO$_3$ integrated optical Y branch," IEEE Photonic. Technol. Lett., 1 (1989) 230–231.

60. M. Rangaraj, M. Minakata, and S. Kawakami, "Low loss integrated optical Y branch," J. Lightwave Technol., 7 (1989) 753–758.

61. N. Kuwata, H. Nishimoto, T. Harimatsu, and T. Touge, "Automatic bias control circuit for Mach-Zehnder modulators," (In Japanese), IEICE Spring National Convention, paper B-976 (1990).

62. M. Seino, N. Mekada, T. Yamane, Y. Kubota, M. Doi, and T. Nakazawa, "20-GHz 3dB-bandwidth Ti:LiNbO$_3$ Mach-Zehnder modulator," Proc. ECOC'90 (Amsterdam, Post deadline session, 1990) 999–1002.

63. R. L. Junferman, C. Johnson, D. J. McQuate, K. Saloma, M. P. Zurakowski, R. C. Bray, G. Conrad, D. Cropper, and P. Hernday, "High speed optical modulator for application in instrumentation," J. Lightwave Technol., 8 (1990) 1363–1370.

64. A. H. Gnauck, S. K. Korotky, J. Veselka, J. Nagel, C. T. Kemmerer, W. J. Minford, and D. T. Moser, "Dispersion penalty reduction using an optical modulator with adjustable chirp," IEEE Photonic. Technol. Lett., 3 (1991) 916–928.

65. S. K. Korotky, J. J. Veselka, C. T. Kemmerer, W. J. Minford, D. T. Moser, J. E. Watson, C. A. Mattoe, and P. L. Stoddard, "High-speed low-power optical modulator with adjustable chirp parameter," Proceedings of Top. Meet. Integrated Photon. Res, Paper TuG2 (Monterey, CA, 1991).

66. K. Kawano, T. Kitoh, H. Junmoji, T. Nozawa, M. Yanagibashi, and T. Suzuki, "Spectral-domain analysis of coplanar waveguide traveling-wave electrodes and their applications to Ti:LiNbO$_3$ Mach-Zehnder optical modulators," IEEE Trans. Microwave Theory Tech., MT-39 (1991) 1595–1601.

67. H. Chung, W. S. C. Chang, and E. L. Adler, "Modeling and optimization of traveling wave LiNbO$_3$ interferometric modulators," IEEE J. Quantum Electron., QE-27 (1991) 608–617.

68. M. Rangaraj, T. Hosoi, and M. Kondo, "A wide-band Ti:LiNbO$_3$ optical modulator with a conventional coplanar waveguide type electrode," IEEE Photonic. Technol. Lett., 4 (1992) 1020–1022.

69. G. K. Gopalakrishna, C. H. Bulmer, W. K. Burns, R. W. McElhanon, and A. S. Greenbelt, "40 GHz, low half-voltage Ti:LiNbO$_3$ intensity modulator," Electron. Lett., 28 (1992) 826–827.
70. M. Seino, T. Nakazawa, Y. Kubota, M. Doi, T. Yamane, and H. Hakogi, "A low DC drift Ti:LiNbO$_3$ modulator assured over 15 years," Proc. OFC'92, Post Deadline papers, PD3 (1992).
71. D. W. Dolfi and T. R. Ranganath, "50 GHz velocity matched broad wavelength LiNbO$_3$ modulator with multimode active region," Electron. Lett., 28 (1992) 1197–1198.
72. W. K. Burns, M. M. Hoverton, and R. P. Moeller, "Performance and modeling of proton exchanged LiTaO$_3$ branching modulators," J. Lightwave Technol., 10 (1992) 1403–1408.
73. K. Noguchi, O. Mitomi, K. Kawano, and M. Yanagibashi, "Highly efficient 40-GHz bandwidth Ti:LiNbO$_3$ optical modulator employing ridge structure," IEEE Photonic. Technol. Lett., 5 (1993) 52–54.
74. K. Kawano, "High-speed Ti:LiNbO$_3$ and semiconductor optical modulators," IEICE Trans. Electron., E76-C (1993) 183–190.
75. S. K. Korotky and J. J. Vesalka, "RC circuit model of long term Ti:LiNbO$_3$ bias stability," technical digest topical meeting on Integrated Photonics Research, paper FB3 (San Francisco, 1994) 187–189.
76. G. K. Gopalakrishnan, W. K. Burns, R. W. McElhanon, C. H. Bulmer, and A. S. Greenbelt, "Performance and modeling of broad band LiNbO$_3$ traveling waveoptical intensity modulators," J. Lightwave Technol., (1994) 1807–1819.
77. R. Madabhushi and T. Miyakawa, "A wide band Ti:LiNbO3 optical modulator with a novel low microwave attenuation CPW electrode structure," Proc. IOOC'95, WD1-3, paper, (Hong Kong, 1995).
78. O. Mitomi, Proc. IOOC'95, Paper PD1-8 (Hong Kong, 1995) 15–16.
79. K. Noguchi, O. Mitomi, H. Miyazawa, and S. Seki, "Broadband Ti:LiNbO$_3$ optical modulator with a ridge structure," J. Lightwave Technol., 13 (1995) 1164–1168.
80. M. Seino, T. Nakazawa, M. Doi, and S. Taniguchi, "The long term reliability estimation of Ti:LiNbO$_3$ modulator for DC drift," Proc. IOOC'95, Paper PD1-8 (Hong Kong, 1995) 15–16.
81. K. Komatsu and R. Madabhushi, "Gb/s-Range semiconductor and Ti:LiNbO$_3$ guided-wave optical modulators," IEICE Trans. Electron., E79-C (1996) 3–13.
82. K. Noguchi, O. Mitomi, and H. Miyazawa, "Push-pull type ridged Ti:liNbO$_3$ optical Modulator," IEICE Trans. Electron., E79-C (1996) 27–31.
83. R. Madabhushi, "Wideband Ti:LiNbO3 optical modulator with a low driving voltage," Proc. OFC'96, Paper ThB3, San Jose.
84. K. Noguchi, O. Mitomi, and H. Miyazawa, "Low voltage and broad bandit; LiNbO$_3$ optical modulator operating in the millimeter wave region," Proc. OFC'96, Paper ThB2 (San Jose, USA, 1996) 205–206.

85. H. Nagata, N. Mitsugi, K. Kikuchi, and J. Minowa, "Lifetime estimation for hermatically packaged 10 Gb/s LiNbO$_3$ optical modulators," Optics & Photon. News, 7 (1996).

86. M. Doi, S. Taniguchi, M. Seino, G. Ishikawa, H. Ooi, and H. Nishimoto, "40Gn/s integrated OTDM Ti:LiNbO3 modulators," Technical Digest International Topical Meetings on Photonics in Switching, Paper PThB1 (Sendai, 1996).

87. T. Kambe, R. Madabhushi, Y. Uematsu, and M. Kitamura, "DC-drift suppressed LiNbO$_3$ optical modulator with a novel Si double slit structure," Paper 11C1-1, 2nd Opto-Electronics & Communications Conference, OECC'97 (Seoul, Korea, 1997).

88. R. Madabhushi, Y. Uematsu, and M. Kitamura, "Wide-band Ti:LiNbO$_3$ optical modulators with reduced microwave attenuation," Proc. IOOC'97/ ECOC'97, Tu1B (Edinburgh, UK, 1997).

89. R. Madabhushi, "Microwave atenuation reduction techniques for wideband Ti:LiNbO$_3$ optical modulators," IEICE Trans. Electron., E81-C (1998) 1321–1327.

90. R. Madabhushi, Y. Uematsu, K. Fukuchi, and A. Noda, "Wide-band Ti: LiNbO$_3$ optical modulators for 40 Gb/s applications," Proc. ECOC'98 (Madrid, Spain, 1998) 547–548.

91. T. Kambe, Y. Urino, R. Madabhushi, Y. Uematsu, and M. Kitamura, "Highly reliable & high performance Ti:LiNbO$_3$ optical modulators," Proc. LEOS'98, Paper ThI5 (Florida, USA, 1998) 87–88.

92. K. Noguchi, O. Mitomi, and H. Miyazawa, "Millimeter-wave Ti:LiNbO$_3$ optical modulator," J. Lightwave Technol., 16 (1998) 615–619.

93. W. K. Burns, M. M. Hoverton, R. P. Moeller, A. S. Greenblatt, and R. W. McElhanon, "Broadband reflection traveling wave LiNbO$_3$ modulators," IEEE Photonic. Technol. Lett., 10 (1998) 805–806.

94. K. Noguchi, O. Mitomi, and H. Miyazawa, "Frequency-dependent propagation characteristics of coplanar waveguide electrode on 100 GHz Ti:LiNbO$_3$ optical modulator," Electron. Lett., 34 (1998) 661–663.

95. R. Madabhushi, "High speed LiNbO3 optical modulators for telecommunications," Proc. IPR'99, Invited paper RTuK5-1, (Santa Barbara, USA, 1999).

96. W. K. Burns, M. M. Hoverton, R. P. Moeller, A. S. Greenblatt, and R. W. McElhanon, "Broad-band reflection traveling wave LiNbO$_3$ modulator," IEEE Photonic. Technol. Lett., 10 (1999) 805–806.

97. W. K. Burns, M. M. Hoverton, R. P. Moeller, R. Krahenbuhl, A. S. Greenblatt, and R. W. McElhanon, "Low drive voltage Broadband LiNbO$_3$ modulators with and without Etched ridges," J. Lightwave Technol., 17 (1999) 2552–2555.

Chapter 7 | Electroabsorption Modulators

T.G. Beck Mason

Agere Systems, 9999 Hamilton Boulevard,
Breinigsville, PA 18031, USA

7.1. Introduction

7.1.1. FIBER OPTIC COMMUNICATIONS

Electroabsorption modulated sources offer many potential advantages for current and next-generation fiber optic communication systems. EA modulators offer advantages over other modulator types in size, cost, drive voltage, and compatibility with monolithic integration. In order to better understand the role of EA modulated sources in fiber optic communication systems we can first examine a conceptual system to see where they fit in. A generic optical communication system is composed of three main parts, a transmitter, a communication channel, and a receiver. In the vast majority of telecommunications systems data is transmitted digitally in either unipolar NRZ or unipolar RZ format using amplitude shift keying (ASK). This is also sometimes referred to as intensity modulation or on off keying (OOK) because the light is turned on and off to represent either a one or a zero bit. A simple conceptual diagram of a fiber optic communication system is shown in Fig. 7.1. The function of the transmitter element is to convert an electrical data signal into optical form and couple this optical signal into the communication channel. The transmitter typically consists of an optical source and some means of modulating that source either directly or externally, to encode the data onto the transmitted lightwave.

249

WDM TECHNOLOGIES: ACTIVE
OPTICAL COMPONENTS
$35.00

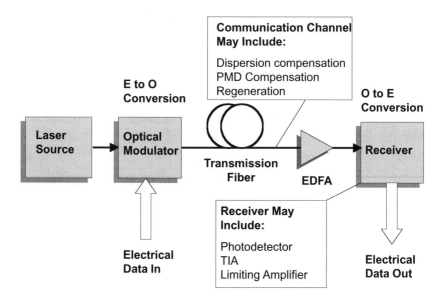

Fig. 7.1   A conceptual diagram of a fiber optic communication system showing the three basic elements of transmitter, communication channel, and receiver.

The communication channel provides a medium to transport the optical signal from the transmitter to the receiver without distorting it. In the most basic implementation this can be simply a single-mode optical fiber. However, for longer distance or higher capacity systems the channel can incorporate amplifiers, dispersion compensators, regenerators, polarization mode dispersion compensators, and other elements to maintain the fidelity of the transmitted optical signal.

The optical receiver converts an incoming optical signal at the output of the communication channel back into an electrical data signal. It usually consists of a photodiode, electrical amplifier, and a clock and data recovery circuit.

7.1.2.   MOTIVATION FOR EXTERNAL MODULATION

For short distance or low bit rate communication systems a directly modulated laser can be used as a transmitter. However, this technique has many limitations that preclude its use for longer distances or higher bit rates. The maximum modulation bandwidth for a semiconductor laser is limited by its relaxation resonance frequency, which can be approximated by Eq. (7.1) [2] for a laser above threshold. (where a is the differential gain, v_g is the optical group velocity, N_p, τ_p, and V_p are the photon density, photon effective lifetime, and optical cavity volume). From the second part of the equation we see that the bandwidth is proportional to the injection efficiency η_i, and the difference between the operating current and the threshold current.

$$\omega_R^2 \approx \frac{v_g a N_p}{\tau_p} = \frac{v_g a}{q V_p} \eta_i (I - I_{th}) \qquad (7.1)$$

This proportionality between the frequency response and the photon density places an upper limit on the extinction ratio for direct modulation. Typically, directly modulated lasers are only capable of achieving extinction ratios between 6 and 8 dB and are not practical for bit rates greater than 10 GB/s. There is a further limitation, which restricts the use of directly modulated sources in long haul or high bit rate systems. This is the undesirable frequency modulation response or frequency chirping of the laser output, which accompanies the amplitude modulation. Modulating the current in the active region of a semiconductor laser modulates both the photon density and the carrier density. The modulation of the carrier density changes the gain, and also changes the index of refraction of the active region. This causes a shift in the operating wavelength of the laser. This undesirable frequency shift broadens the modulated spectrum of the laser and increases the penalty for transmission through fiber links with nonzero dispersion. Directly modulated lasers are generally only used for low bit rate applications or in 1.31 um wavelength systems, which operate at the dispersion zero for standard single-mode fiber. This precludes their use in long haul or DWDM systems that depend on erbium-doped fiber amplifiers.

7.1.3. *PRINCIPLE OF EA MODULATORS VS. ELECTROOPTIC MODULATORS*

These fundamental limitations on direct modulation for fiber optic transmitters—limited bandwidth, large frequency chirp, and low extinction ratio—are the primary motivation behind the development of external optical modulators. Optical intensity modulators can be categorized into two main types based on the physical phenomenon that they use to modulate the light. One category contains modulators that rely on the electrooptic effect to change the effective index of a waveguide and modulate the phase of an optical signal. This type of modulator typically employs a Mach-Zehnder interferometer geometry to convert the phase change into

Mach-Zehnder Electroptic Modulator

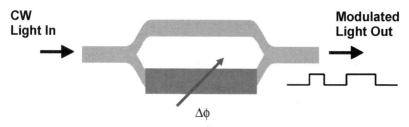

CW
Light In

Modulated
Light Out

$\Delta\phi$

Electroabsorption Modulator

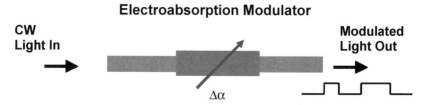

CW
Light In

Modulated
Light Out

$\Delta\alpha$

Fig. 7.2   Comparison of electrooptic and electroabsorption based optical modulators.

an intensity modulation. A common example of this type of modulator is a LiNbO$_3$ based Mach-Zehnder modulator, widely used in 2.5, 10, and 40 GB/s communication systems. The other category of modulators we will be discussing in this chapter are based on the electroabsorption effect, which changes the absorption in an optical waveguide to modulate the intensity of a lightwave passing through it. Schematic examples of these two different modulator types are shown in Fig. 7.2. There are a number of advantages and disadvantages associated with each of these classes of modulators, but in general they both offer substantial benefits over direct modulation. Among these benefits are low or negative chirp, high extinction ratio, and bandwidths that are high enough to support data rates up to 0C768 and beyond.

7.1.4.   ADVANTAGES AND BENEFITS OF EA MODULATORS

In the following sections of the chapter we will investigate the design, fabrication, characterization, and transmission performance of electroabsorption modulated sources. They are and have been key components in the evolution of optical communication systems. Compared with the LiNbO$_3$ alternative EA modulators are more compact, less expensive, compatible with monolithic integration, and offer lower drive voltages. However, fabrication complexity and open questions concerning the fidelity with which they transmit information make the exact role of EA modulators in advanced communication systems somewhat unclear. Later in the chapter we will describe more complex integrated devices that combine EA modulators with lasers, semiconductor optical amplifiers, and other EA modulators to perform more complex data modulation functions.

7.2. Electroabsorption

7.2.1. FRANZ-KELDYSH EFFECT

The electroabsorption effect for optical intensity modulators is based on one of two phenomenon: the Franz-Keldysh effect for bulk modulators or the quantum-confined Stark effect for multiple quantum well devices. The Franz-Keldysh effect allows electron-hole excitation with below band gap photons due to the possibility of lateral carrier tunneling under an applied electric field. All energies in principle are possible for these transitions along the electric field since the energy gap between the valence and conduction bands is a triangular potential well with a height h and width w given by Eq. (7.2) [2], where F is the applied electric field.

$$w = \frac{E_g}{eF}, \qquad h = \left(E_g - \frac{\hbar c}{\lambda} \right) \tag{7.2}$$

For a given nonzero field the tunneling probability depends exponentially on the barrier height, so even for a photon with zero energy there is still a finite probability of such transitions (Fig. 7.3). The reduction in the required photon energy for ionization is proportional to the product of the applied electric field and the effective tunneling length. For p-i-n based structures very high electrical fields on the order of 200 to 300 kV/cm can be generated with only a few volts of bias. This enables large changes in the absorption coefficient to be realized. For photon energies below the bandgap the absorption due to the Franz-Keldysh effect α_{FK} can be made to vary from a few cm^{-1} at low fields ($\sim$10 kV/cm), to almost 2000 cm^{-1} at high fields ($\sim$250 kV/cm) [1, 2]. This is very promising for modulators requiring high switching contrast and low insertion loss. One significant advantage of bulk active layer modulators that rely on the Franz-Keldysh effect is the large spectral width of the absorption change that this effect yields. This makes them less sensitive to temperature and more suitable for

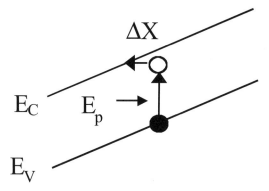

Fig. 7.3 Schematic illustration of below bandgap absorption of a photon in a semiconductor under an applied electric field.

use with widely tunable lasers than devices based on the quantum confined Stark effect [3].

The absorption coefficient for a bulk semiconductor in the presence of an electric field can is given by Eq. (7.3) [4] where n is the refractive index of the material, c is the velocity of light, e is the charge on an electron, m_e and m_h are the electron and hole effective mass, ω is the frequency of the light, E_c and E_v are the conduction and valence band energies, and P_{if} is the matrix element for photon absorption.

$$\Delta\alpha = R\theta_F^{\frac{1}{2}} F\left(\left[\frac{E_g - \hbar\omega}{\hbar\theta_F}\right]\right) \quad \text{where } \theta_F = \left[\frac{(eF)^2}{2\mu\hbar}\right]^{\frac{1}{3}}, \quad \frac{1}{\mu} = \frac{1}{m_e} + \frac{1}{m_h}$$

$$\text{and} \quad R = \frac{2e^2 C_o^2}{\hbar\omega n c m_e^2}\left(\frac{2\mu}{\hbar^2}\right)^{\frac{3}{2}} \tag{7.3}$$

The function $F(x)$ is given by Eq. (7.4), where $Ai(x)$ is the Airy function and $H(x)$ is the unit step function.

$$F(x) = |Ai'(x)|^2 - x|Ai(x)|^2 - \frac{1}{\pi}\sqrt{-x}H(-x) \tag{7.4}$$

For the case when the photon energy is below the bandgap the last part of Eq. (7.4) drops out, and if we replace the $Ai(x)$ and $Ai'(x)$ with their asymptotic expressions we get the following approximation for the absorption change:

$$\Delta\alpha = R\sqrt{\frac{e}{2\mu\hbar}}\left(\frac{\hbar}{E_g - \hbar\omega}\right) F e^{-\left(\frac{4}{3}\right)\left(\frac{E_g - \hbar W}{\hbar}\right)^{3/2}\left(\frac{2\mu\hbar}{e}\right)^{1/2}\left(\frac{1}{F}\right)} \quad (7.5)$$

This expression shows that the change in absorption due to the Franz-Keldysh effect is inversely proportional to the difference between the bandgap and the photon energy. Perhaps the most interesting part, however, is that it shows a newly linear dependence of the absorption change on the electric field F.

7.2.2. QUANTUM CONFINED STARK EFFECT

Excitonic effects have a very dramatic influence on the optical properties of semiconductors, particularly near the band edge. Below the band edge there is a strong excitonic resonance in the absorption and emission behavior. This causes a strong enhancement of the absorption process above the bandgap, especially in three-dimensional systems. Confining the exciton in a quantum well greatly increases the binding energy along with the oscillator strength. This increased binding energy allows the exciton to exist up to much higher temperatures than in a three-dimensional system. In a conventional three-dimensional semiconductor system an applied electric field ionizes the exciton state by pulling apart the electron–hole pair. However, in a quantum well when a field is applied in the transverse direction to the well the exciton is not ionized due to the confinement of the electron and hole states. Because of this exciton transitions can persist at electric fields as high as 100 kV/cm. In quantum well structures the dominant effect in the electroabsorption response is the quantum confined Stark effect. There are several effects that contribute to this. The most significant is the change in the intersubband energies. As the field is applied the bands become tilted and the electrons and holes no longer see a simple square potential well. The field pushes the electron and hole wavefunctions to opposite sides of the well, which reduces the intersubband separation or quantum well bandgap. This increased separation in the electron and hole wavefunctions also has the effect of reducing the exciton binding energy, which counters the shift in the intersubband energies. However, this effect is generally about ten times smaller than the shift in the bandgap although it does result in a significant increase in the exciton linewidth [8]. In Fig. 7.4 we show a schematic of a strain compensated quantum well structure with compressive strain in the well and tensile

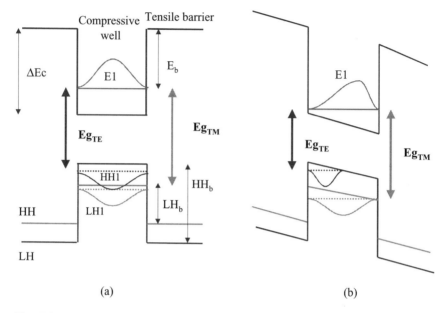

Fig. 7.4 Quantum Confined Stark Effect in a strained quantum well structure. (a) Quantum well with bound states and wavefunctions under zero field. (b) Effect of an applied electric field on the quantum well subband energies.

strain in the barriers. Note the splitting between the light-hole and heavy-hole subbands. This is due in part to the difference in their effective masses, which affects the bound state energy levels (E_n) in the well. These energies can be determined for a square potential well of width d and height ΔE_c by solving the characteristic equations for symmetric and anti-symmetric states (7.5) [2].

$$\tan\left[\frac{\pi}{2}\sqrt{\frac{2mw^2E_n}{\hbar^2\pi^2}}\right] = \left(\frac{\Delta E_c}{E_n} - 1\right)^{1/2} \quad (symmetric)$$

$$\tan\left[\frac{\pi}{2}\left(\sqrt{\frac{2mw^2E_n}{\hbar^2\pi^2}} - 1\right)\right] = \left(\frac{\Delta E_c}{E_n} - 1\right)^{1/2} \quad (antisymmetric)$$

$$(7.6)$$

A more significant effect is the fact that the light-hole and heavy-hole band energies are significantly different both in the well and in the barrier as a result of the deformation potentials. For strained material there is both a hydrostatic and a shear component to these. The hydrostatic strain causes a similar shift in the light-hole and heavy-hole valence band energies, however the shear component splits the degeneracy between the heavy-hole and light-hole bands. The net result of these effects is that we see a substantially different bandgap for absorption of light polarized with the electric field in the plane of the quantum well (TE) than for light polarized with the electric field perpendicular to the plane of the well (TM). Later in this chapter we will discuss how this phenomenon can be used advantageously in the design of multiple quantum well electroabsorption modulators. The effect of applying an electric field to the well can be seen in Fig. 7.4(b)—the electron and hole wavefunctions have been pushed to either side of the well and the effective bandgap has been reduced for both TE and TM polarized light. In principle these quantum well states are quasi-bound states in the presence of an electric field, because there is a finite probability that electrons and holes will eventually tunnel out of the well. This probability increases significantly with increased field strength.

An exact calculation of the intersubband separation for a quantum well in the presence of an electric field is beyond the scope of this text. There is a complete theoretical model by Debernardi and Fasano [5] that includes the effects of valence band mixing and coulomb effects. However, this problem generally is solved either with a variational approach or with the application of numerical techniques beyond the scope of this text. There are a number of efficient numerical techniques for handling this problem; among them is the transmission matrix method used by Johnson and Eng [6], and Gathak et al. [7]. The simulated absorption spectrum for a quantum well in the presence of an electric field is shown in Fig. 7.5. This calculation was performed using the transmission matrix method to solve for the changes in the intersubband energies. The effect of the electric field on the exciton peak can be clearly seen from this figure. As the field increases and the peak shifts to longer wavelengths, there is also a substantial increase in the broadening factor for the exciton caused by the reduction in the binding energy. Eventually at very high fields the exciton absorption peak disappears. For a typical operating wavelength that is detuned from the band edge by approximately 40 nm, a large part of the increase in the absorption is due to the broadening of the exciton linewidth.

While the exact calculation of the change in the quantum well bandgap under an applied electric field is quite difficult, an approximate solution can easily be derived using perturbation theory. This approach can give reasonable results for low electric fields. Adding a term to account for the electric field perturbation to the Hamiltonian for the unperturbed quantum

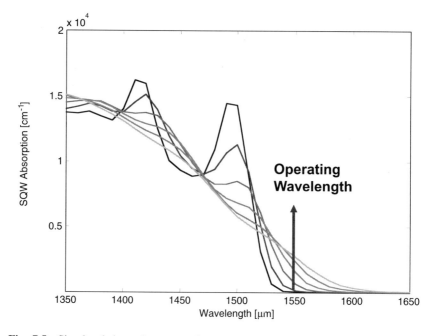

Fig. 7.5 Simulated absorption spectra for a quantum well structure in the presence of an applied electric field.

well yields

$$H = H_o + eF \tag{7.7}$$

The ground state eigenfunction for H_o has an even parity so the first-order correction is zero, thus the second-order correction must be used, which gives the following change in the ground state energy [8]:

$$\Delta E_1 = \frac{1}{24\pi^2}\left(\frac{15}{\pi^2} - 1\right)\frac{m^* e^2 F^2 w^4}{\hbar^2} \tag{7.8}$$

From this simple expression we can see that the change in the ground state subband energy has a quadratic dependence on the electric field F and increases strongly with increasing well width w. It also increases with increasing effective mass m^*. This would seem to suggest that for the highest modulation efficiency a wide well would be optimum. However, as the well width is increased the exciton binding energy is reduced and the overall change in absorption is not as great even though the change in the intersubband separation is large.

7.2.3. INDEX CHANGE KRAMERS KRONIG RELATION

One important property of electroabsorption modulators that must be considered is how the changes in the absorption spectrum for the material affect the index of refraction at the operating wavelength. The real and imaginary components of the index of refraction are not independent but are related to each other by a set of dispersion relations called Kramers-Kronig relations. These were first derived independently by H.A. Kramers (1927) and R. de L. Kronig (1926). They are general relations between the real and imaginary components of a response function. The dispersion relation for the index of refraction n in terms of the absorption coefficient α can be written as Eq. (7.9).

$$n(E) = 1 + \frac{hc}{2\pi^2} P \int_0^\infty \frac{\alpha(E')}{E'^2 - E^2} \, dE' \qquad (7.9)$$

In this equation P represents the Cauchy principle value of the integral defined as

$$P \int_0^\infty \equiv \lim_{a \to 0} \left[\int_0^{E-a} + \int_{E+a}^\infty \right] \qquad (7.10)$$

For optical intensity modulators it is more important to characterize the change in the index of refraction, which can be related directly to the change in the absorption spectrum using Eq. (7.11) [1]. Because the effect of the absorption at a given wavelength on the index of refraction at another wavelength is inversely proportional to the separation between these wavelengths, and also the effect of the electric field on the absorption is small for wavelengths far from the bandgap, the change in the index of refraction can be reliably estimated using the absorption spectrum over a limited range of wavelengths centered on the bandgap. This facilitates the use of Eq. (7.11) with either calculated or measured data for the absorption change in a semiconductor.

$$\Delta n(\lambda_0, V) = \frac{\lambda_0^2}{2\pi^2} \left(\lim_{\varepsilon \to 0} \int_0^{\lambda_0 - \varepsilon} \frac{\alpha(\lambda, V) - \alpha(\lambda, 0)}{\lambda_0^2 - \lambda^2} \, d\lambda \right.$$
$$\left. + \int_{\lambda_0 + \varepsilon}^\infty \frac{\alpha(\lambda, V) - \alpha(\lambda, 0)}{\lambda_0^2 - \lambda^2} \, d\lambda \right) \qquad (7.11)$$

The desired behavior for an electroabsorption modulator active region is to experience a significant change in absorption at the operating wavelength with a minimum change in the index of refraction. In particular it is often desirable to see a small reduction in the index of refraction at the operating wavelength as the absorption increases. This leads to optimal dispersion tolerance for transmission of signals over standard single-mode optical fiber. The relationship between the change in the index of refraction and the change in the absorption is typically characterized by the Henry alpha parameter (7.12) [9]; this is somewhat analogous to the linewidth enhancement factor for semiconductor lasers.

$$\alpha_H = \frac{\Delta n}{\Delta k} = (4\pi/\lambda_0)(\Delta n/\Delta \alpha)_{\lambda=\lambda_0} \qquad (7.12)$$

Some people refer to this as the chirp parameter, which can lead to confusion with the chirp parameter C for Gaussian pulses defined by Eq. (7.13) [9]. The chirp parameter C only applies to a Gaussian pulse whereas the alpha parameter is a more general parameter describing the relationship between the amplitude and phase modulation properties of a device.

$$\delta\omega = -\frac{\partial\phi}{\partial t} = \frac{C}{T_o^2}t \qquad (7.13)$$

The main source of confusion comes from the fact that a negative alpha parameter leads to positive chirp. That is an increase in the instantaneous frequency of the pulse from the leading to the trailing edge. For transmission of a pulse in an optical fiber the broadening depends on the relative signs of the group velocity dispersion parameter β_2 and the chirp parameter C. A Gaussian pulse will broaden monotonically with transmission distance if $\beta_2 C > 0$. However, if $\beta_2 C < 0$ it will undergo an initial narrowing stage before broadening monotonically. For standard single-mode fiber (SMF 28) β_2 is approximately -20 ps²/km in the 1550 nm wavelength range indicating anomalous dispersion, thus a positive chirp parameter or a negative alpha parameter is desirable for a modulator transmitting light through these fibers. The group velocity dispersion defined in Eq. (7.14) [9] is more typically specified by D, the dispersion parameter in units of [ps/(nm·km)].

$$\beta_2 = \frac{d^2\beta}{d\omega^2} = D\left(\frac{\lambda^2}{2\pi c}\right) \qquad (7.14)$$

If we reexamine the simulated absorption curves shown in Fig. 7.5 we can see that there is an increase in the absorption for wavelengths longer than the wavelength of operation. This will lower the index of refraction in the waveguide and contribute to positive chirp. However, the strong increase in the absorption for wavelengths below the operating wavelength will raise the index of refraction and contribute to negative chirp. This will be somewhat offset by the sharp reduction in the absorption of the exciton peaks at 1420 and 1500 nm, which will also contribute to positive chirp. However, the effect of the absorption change on the chirp diminishes as the separation from the operating wavelength increases so the effect will not be as strong as the changes in absorption that are closer in wavelength. It is clear from this analysis that as the detuning between the operating wavelength and the quantum well bandgap is reduced the chirp performance for a modulator will be improved.

7.2.4. DISPERSION PENALTY

In order to better understand the importance of the modulator chirp and its effect on transmission performance we can look at the pulse broadening of chirped Gaussian pulses propagating in an optical fiber. We can define an input Gaussian pulse with a chirp C as

$$A(0, t) = A_o \exp\left[-\frac{1 + iC}{2}\left(\frac{t}{T_o}\right)^2 \right] \qquad (7.15)$$

where A_o is the initial amplitude and T_o is the half-width of the pulse at the $1/e$ intensity point. It is related to the full width at half maximum by

$$T_{FWHM} = 2\sqrt{\ln(2)}T_o \qquad (7.16)$$

The linear propagation of pulses in an optical fiber can be described by Eq. (7.17) where t' is measured in a reference frame moving with the pulse at the group velocity, i.e. $(t' = t - z/v_g)$ [9].

$$i\frac{\partial A}{\partial z} = -\frac{i}{2}\alpha A + \left(\frac{1}{2}\beta_2\frac{\partial^2 A}{\partial t'^2} - \frac{1}{6}\beta_3\frac{\partial^3 A}{\partial t'^3}\right) - \gamma|A|^2 A \qquad (7.17)$$

The three different terms on the righthand side of the equation represent the absorption, dispersion, and nonlinearity respectively. If we consider operation at a wavelength far from the dispersion zero then the higher order dispersion effects can be neglected. Additionally we can assume that we are operating at a power level where the nonlinear dispersion is not significant.

Then the equation can be rewritten using the normalized amplitude $U(z, t)$ defined as

$$A(z, t) = U(z, t)A_o \exp\left(-\frac{\alpha}{2}z\right) \qquad (7.18)$$

Then the equation for the dispersion induced pulse broadening simply becomes [10]

$$i\frac{\partial U}{\partial z} = \frac{1}{2}\beta_2 \frac{\partial^2 U}{\partial t^2} \qquad (7.19)$$

This can be solved analytically to give Eq. (7.18), which defines the shape of a chirped Gaussian pulse after propagating a distance z in an optical fiber.

$$U(z, t) = \left(\frac{T_o}{\sqrt{T_o^2 - i\beta_2 z(1 + iC)}}\right) \exp\left(\frac{(1 + iC)t'^2}{2[T_o^2 - i\beta_2 z(1 + iC)]}\right) \qquad (7.20)$$

The benefit of considering Gaussian pulses is evident because the pulse retains its Gaussian shape at the output enabling the output pulse width to be written in a simple form (7.21), which clearly indicates the effect of the relative signs of the chirp and group velocity dispersion on the pulse width.

$$\frac{T_1}{T_o} = \left[\left(1 + \frac{C\beta_2 z}{T_o^2}\right)^2 + \left(\frac{\beta_2 z}{T_o^2}\right)^2\right]^{\frac{1}{2}} \qquad (7.21)$$

We can now use the broadening factor defined in Eq. (7.21) to estimate the dispersion penalty for a transmission link. The width of the Gaussian pulse can be defined by its root mean square (RMS) width σ, which is related to the $1/e$ width by $(\sigma = T_o/\sqrt{2})$. This factor can then be used as a criterion on the maximum bit rate. A commonly used criteria is that the limiting bit rate must be given by $(4B\sigma \leq 1)$. This ensures that 95% of the pulse energy remains within the bit slot. From Eq. (7.20) we see that the transmitted pulse remains Gaussian but that its peak power decreases due to the dispersion induced pulse broadening. We can define the power penalty for the transmission as the required increase in received power to compensate this reduction. This is given by

$$P = 10\log(T_1/T_o) \qquad (7.22)$$

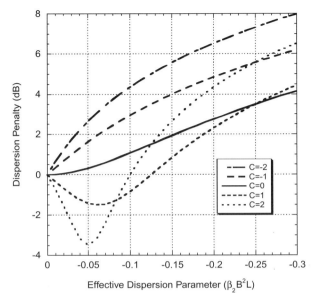

Fig. 7.6 The effect of chirp parameter on the dispersion-induced power penalty for optical fiber transmission systems.

If we now plug in the relationship for the pulse broadening (7.21) and perform a substitution to replace T_o with the bit rate we have an expression for the chirp induced power penalty.

$$P = 5\log\lfloor(1 + 8C\beta_2 B^2 L)^2 + (8\beta_2 B^2 L)^2\rfloor \qquad (7.23)$$

In Fig. 7.6 we have plotted the dispersion penalty as a function of the effective dispersion parameter $\beta_2 B^2 L$ for positive values of the group velocity dispersion. For fibers with normal dispersion (i.e., positive β_2) the plot would be the same, however the sign of the C parameter would be reversed. This plot shows that there is an initial reduction in the dispersion penalty for positively chirped pulses. However, if the chirp is too high then the penalty quickly increases beyond the penalty for unchirped pulses. Regardless of the magnitude of the positive chirp there will always be a distance beyond which the zero chirp case has a smaller dispersion penalty. This distance is defined as

$$L = \frac{-1}{4C\beta_2 B^2} \qquad (7.24)$$

As an example if we consider a fiber optic transmission system with an 80 km span operating at 10 GB/s and the effective dispersion parameter is 0.16 corresponding to a β_2 of $-20\,\mathrm{ps^2/km}$ or a dispersion D of 17 ps/nm·km, then the minimum dispersion penalty would be 1 dB for a chirp factor of 0.8. For an equivalent transmission penalty with a negative chirp of -0.8 the transmission distance would be only 20 km. This shows the critical nature of the source chirp for 10 GB/s systems. Ironically, at 40 GB/s the effect is not as important. Because the effective dispersion parameter varies with the square of the bit rate, the equivalent distances at 40 GB/s would be only 5 km for $C = 0.8$ and 1.25 km for $C = -0.8$. For these higher bit rate systems even with optimized chirp parameters only very short span lengths are possible for low dispersion penalties. Because of this most 40 GB/s systems incorporate dispersion-managed transmission links. For these applications the residual fiber dispersion can be either positive or negative so a near zero modulator chirp is desirable.

It is important to remember that this result is based on the assumption that we have Gaussian shaped pulses, which is not generally correct for real-world transmission systems. These systems have pulse shapes with steeper leading and trailing edges that are better characterized by higher order functions. This condition is more difficult to analyze and will generally result in an increased dispersion penalty when compared to the Gaussian pulse shape.

7.3.   EA Modulator Design

The design of an electroabsorption modulator involves a large number of optical and electrical considerations. In its simplest form an EA modulator consists of a semiconductor optical waveguide, with a PIN diode structure. The optical waveguide is formed by sandwiching a higher index of refraction (lower bandgap) core layer between two lower index of refraction (higher bandgap) cladding layers (Fig. 7.7). Lateral confinement for the waveguide can be achieved in a number of different manners, forming a shallow ridge in the upper cladding layer, etching a deep ridge all the way through the core layer, or regrowing around a deep etched ridge to form a buried heterostructure (Fig. 7.8). Typically the upper cladding layer is acceptor or p-type doped, the core layer is an undoped or intrinsic semi-conductor, and the lower cladding layer is donor doped making it n-type. This gives the device a basic PIN diode structure where the active layer is in the intrinsic or I region. This active layer is composed of a material

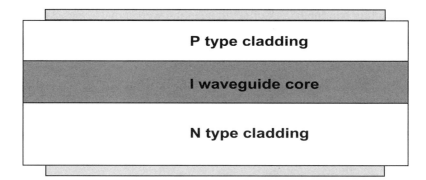

Fig. 7.7   Simple conceptual longitudinal cross section of an EA modulator.

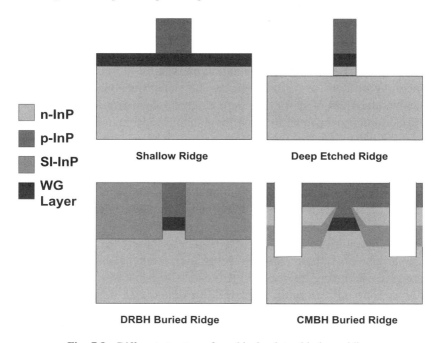

Fig. 7.8   Different structures for achieving lateral index guiding.

that has a bandgap energy that is slightly greater than the photon energy for the wavelength of light at which the device is intended to operate. This difference between the nominal bandgap energy and the operating wavelength is referred to as the detuning energy and is typically expressed in units of nm. Applying a reverse bias to the device creates a strong field in the intrinsic layer that shifts the absorption edge in the material to lower energies via the Franz-Keldysh effect in bulk semiconductor active layers, or the quantum confined Stark effect in quantum well active layers. This shift in the absorption edge is used to modulate the intensity of a lightwave propagating in the optical waveguide. The design of the optical waveguide properties strongly affects the insertion loss and extinction ratio for the device, whereas the design of the electronic properties of the device most critically affects the bandwidth, and saturation power. These two aspects of the design are intimately connected to each other and also to the fabrication processes available for the creation of the device. Each has a significant impact on the other and a careful methodology is necessary in order to achieve an efficient design.

The first step in the design process is to identify the desired performance parameters for the device. The selection of these parameters will depend critically on the application for which the modulator is intended. The main application space for EA modulators is for high-speed long-distance digital transmission in fiber optic communication systems. The first step in designing an EA modulator for this type of application is to select the operating bit rate for the transmitter. This determines the minimum 3dB bandwidth requirement for the EA modulator, which sets an upper limit on the junction capacitance. Because the junction capacitance scales with the area of the junction, the bandwidth requirement ultimately limits the device size. Because the width of the device is generally limited to a fairly narrow range by the requirement that the waveguide only support a single mode, the limit on the device size can be thought of as a limit on the total device length. Thus as the bandwidth requirement for the device increases, the size decreases. For shorter modulators a greater change in the absorption coefficient for the device is required in order to achieve the same extinction ratio as in a longer modulator. However, increasing the absorption per unit length and reducing the overall modulator size decreases the power level at which it begins to saturate. Despite this the output power requirement increases for systems operating at higher bit rates. The result of this is that it gets increasingly difficult to scale the EA modulator performance to higher and higher bit rates. Oftentimes this necessitates choosing fundamentally different technologies as the bandwidth requirements increase.

7.3.1. WAVEGUIDE DESIGN

The design of the optical waveguide for an EA modulator is coupled to the selection of the active absorbing layer. This layer also forms a major component of the waveguide core. It is generally good practice to limit the size of the waveguide such that it only supports a single transverse mode.

This places an upper boundary on the thickness and the width of the waveguide core. Because it is also desirable to limit the insertion loss of the device in the on state, the waveguide must be designed to have a mode size that is compatible with low loss coupling to optical fiber. To begin with we will derive a simple procedure for calculating the effective index and the transverse mode function for a dielectric waveguide. Full calculation of the transverse amplitude function for an arbitrary two-dimensional structure requires a numerical solution. However, a good approximate solution can be found using the effective index technique. In this technique the 1D effective index is solved for each of the three lateral regions of the waveguide as if they were infinite slab waveguides. In the case of the buried ridge the effective index of the regions on either side of the core is just the index of the regrown semiconductor. For clarity we will first set the conventions for describing the waveguide with reference to Fig. 7.9. The transverse direction will be the x direction normal to the plane of the surface of the device. The lateral direction or y direction will be in the plane perpendicular to the propagation direction of the light z. For the 1D slab approximation we will refer to the core region as region 2 and the cladding region as region 1. Light polarized with its e-field in the x direction perpendicular to the interface between the core and the cladding is referred to as transverse magnetic or TM, and light polarized with its e-field parallel to the interface is referred to as transverse electric or TE. For most devices the waveguide is asymmetric and the confinement in the vertical direction is greater than

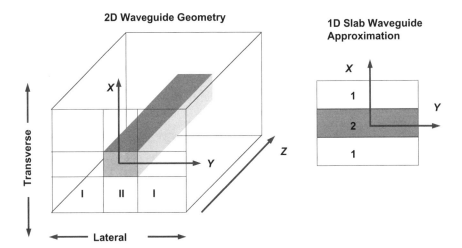

Fig. 7.9 Schematic waveguide cross section showing the transverse (x) and lateral (y) directions and the core and cladding regions.

in the lateral direction. For this reason we typically solve for the effective index in the vertical or x direction first and then use this to solve for the combined effective index in the lateral or y direction.

We can quickly derive the procedure for solving the 1D effective index of a three-layer slab waveguide. First we begin by writing the wave equation, which can be derived from Maxwell's equations.

$$\nabla^2 E = \mu\varepsilon\frac{\partial^2 E}{\partial t^2} \tag{7.25}$$

Here E is the electric field, μ is the magnetic permeability which is equal to μ_o for most semiconductor materials of interest, and ε is the dielectric constant, which can be complex. The imaginary component of ε represents the gain or loss in the material. We can assume a solution of the form

$$E(x, y, z, t) = \hat{e}_i E_o U(x, y)e^{j(\omega t - \tilde{\beta}z)} \tag{7.26}$$

where $\hat{e}_i$ is the unit vector that defines the polarization and $U(x, y)$ is the transverse wave function. Inserting Eq. (7.25) into (7.26) we find that the transverse wave function must satisfy the equation

$$\nabla^2 U(x, y) + \left(\tilde{n}^2 k_o^2 - \tilde{\beta}^2\right)U(x, y) = 0 \tag{7.27}$$

Now taking a simple three-layer slab waveguide of the form shown in Fig. 7.9, we can solve Eq. (7.27) in each of the three layers subject to the boundary conditions matching at the interfaces. We can simplify this process somewhat by replacing the complex index of refraction $\tilde{n}$ and propagation constant $\tilde{\beta}$ with their real components. This will not significantly reduce the accuracy of the solution because the imaginary components are typically very small compared to the real components for semiconductors. The general solution to Eq. (7.27) for region m with index n_m is given by

$$U_m(x) = A_1 e^{\left(k_o^2 n_m^2 - \beta^2\right)^{\frac{1}{2}}x} + A_2 e^{-\left(k_o^2 n_m^2 - \beta^2\right)^{\frac{1}{2}}x} \tag{7.28}$$

In the core region the index of refraction is greater than the effective index, which leads to sinusoidal solutions of the form

$$U(x) = \begin{cases} A\cos\left[\left(k_o^2 n_2^2 - \beta^2\right)^{\frac{1}{2}}x\right] & \text{(symmetric solutions)} \\ A\sin\left[\left(k_o^2 n_2^2 - \beta^2\right)^{\frac{1}{2}}x\right] & \text{(antisymmetric solutions)} \end{cases} \tag{7.29}$$

where n_2 represents the index of refraction in the core layer. In the cladding the effective index is greater than the material index, leading to exponential solutions. We consider only the exponentially decaying solutions, which correspond to guided modes defined by

$$U(x) = B e^{-\left(\beta^2 - k_o^2 n_1^2\right)^{\frac{1}{2}} x} \tag{7.30}$$

where n_1 represents the index of refraction in the cladding layers. If we define the width of the waveguide core layer as d and place the origin at the center of the waveguide, then we must match the boundary conditions for the two wavefunctions at the interface $d/2$. In the case of a symmetric guide we only need solve for a single set of boundary conditions. For TE modes at the boundary we have the condition that both the field and its first derivative must be continuous across the boundary. Thus $U_1 = U_2$ and $\partial U_1 / \partial x = \partial U_2 / \partial x$ at $x = d/2$. This allows us to derive a characteristic equation for the fundamental and higher order TE modes.

$$\frac{k_o d}{2} \left(n_2^2 - \bar{n}^2\right)^{\frac{1}{2}} = \arctan\left(\frac{\bar{n}^2 - n_1^2}{n_2^2 - \bar{n}^2}\right) + \frac{(m-1)\pi}{2} \tag{7.31}$$

Here $\bar{n}$ is the effective index for the mode defined by $\beta = k_o \bar{n}$. For TM modes the boundary conditions are somewhat different and $n_2^2 U_2 = n_1^2 U_1$; the characteristic equation is then given by

$$\frac{k_o d}{2} \left(n_2^2 - \bar{n}^2\right)^{\frac{1}{2}} = \arctan\left(\left(\frac{\bar{n}^2 - n_1^2}{n_2^2 - \bar{n}^2}\right)\left(\frac{n_2^2}{n_1^2}\right)\right) + \frac{(m-1)\pi}{2} \tag{7.32}$$

Solving these equations for the effective index and then inserting this into Eqs. (7.29) and (7.30) enables a piecewise solution of the transverse wave function to be found. Repeating this process in the lateral direction using the effective index for each layer the piecewise solution to the lateral wavefunction $U(y)$ can be found. This also gives the total effective index for the mode. Then the total transverse wavefunction is simply given by

$$U(x, y) = U(x) \cdot U(y) \tag{7.33}$$

In some instances there is no confined mode in the transverse direction for the lateral guiding regions. In this case an effective index can be calculated by computing the normalized index for the guided mode in the waveguide core section II from Fig. 7.9, with the index distribution of the layer structure in the cladding section I.

$$\bar{n}_I = \frac{\int |U_{II}(x)|^2 n_I(x)\,dx}{\int |U_{II}(x)|^2\,dx} \tag{7.34}$$

It is important to remember that if the characteristic equation for the TE modes is used in the transverse direction for the effective index method, then the characteristic equation for the TM modes must be used to solve for the lateral direction and vice versa.

7.3.2. CONFINEMENT FACTOR

This integral in Eq. (7.34) gives a weighted average of the index of refraction in the cladding layers of the waveguide where the weighting function is the square of the transverse mode field. This concept of using an overlap between the wavefunction and the material to measure a property of the mode is called an overlap integral. We can extend this concept to calculate the effect of the waveguide design on the change in the absorption in an EA modulator. The effective absorption coefficient seen by the optical mode as it propagates in the waveguide can be written as

$$\langle \alpha \rangle = \frac{\iint |U(x,y)|^2 \alpha(x,y)\,dx\,dy}{\iint |U(x,y)|^2\,dx\,dy} \tag{7.35}$$

where $\alpha(x,y)$ is the material absorption coefficient. In general the change in the absorption coefficient under an applied bias is limited to the active layer, which enables us to simplify this equation and rewrite it in terms of the differential absorption in the active layer $\Delta\alpha$.

$$\Delta\langle \alpha \rangle = (\Delta\alpha_a) \frac{\int_{-w/2}^{w/2} \int_{-d/2}^{d/2} |U(x,y)|^2\,dx\,dy}{\int_{-\infty}^{\infty} \int_{-\infty}^{\infty} |U(x,y)|^2\,dx\,dy} = \Gamma_{xy}\Delta\alpha_a \tag{7.36}$$

The integral on the righthand side of this equation is simply an overlap integral for the mode with the active layer. This overlap represents a very convenient concept called the confinement factor Γ, which can be used to greatly simplify the analysis of the differential absorption in an EA modulator. The extinction ratio for a modulator with a length L can now be written in the following simple form:

$$ER(dB) = 10\log_{10}(e^{-\Gamma_{xy}\Delta\alpha L}) \tag{7.37}$$

As a general point of reference the absorption coefficient for InGaAs at 1550 nm is approximately 6000 cm^{-1}. For a quantum well EA modulator under a strong electric field the change in the material absorption can be as high as 4500 cm^{-1}. Thus for a 100 μm long modulator with a confinement factor of 0.10 the maximum extinction ratio is approximately 20 dB. If the confinement factor is increased to 0.15, for example, by increasing the number of quantum wells in the active region from 8 to 12 then the extinction ratio increases to almost 30 dB. There is an equivalent scaling factor if the length is increased to 150 μm. However, increasing the length also increases the junction area and reduces the maximum bandwidth of the device. There is also a drawback to consider when increasing the confinement factor, which becomes evident when we consider the coupling efficiency between the device and an optical fiber.

7.3.3. INSERTION LOSS

The confinement factor for the active region of an EA modulator waveguide can be increased by either increasing the thickness of the waveguide or increasing the index difference between the waveguide core and the cladding. Because for most EA modulators of interest we are restricted to operating in the InP material system, the cladding material is generally InP. The core index of refraction depends on the bandgap of the active material, which is fixed by the operating wavelength for the device. The use of a higher bandgap separate confinement heterostructure can be used to increase the overall confinement factor for the waveguide but this has only a secondary effect on the active layer Γ. The only remaining method is to increase the thickness of the active layer itself. Increasing the confinement of the optical mode in the active layer increases the effective index of the mode and shrinks the effective size of the mode.

The insertion loss for an EA modulator is made up of two major components. One of these is the absorption, and scattering losses for light in the guided mode of the waveguide, and the other is made up of the coupling loss for the optical system used to couple light into and out of the device to single-mode optical fiber. The waveguide absorption is made up of residual absorption in the active layer, and interband carrier scattering, primarily intervalence band absorption in the doped materials. Careful design of the quantum wells and optimization of the doping profiles in the cladding layers can minimize these effects but they are inherent to the structure and cannot be eliminated entirely. Optical scattering loss is caused mainly by defects in the material or roughness in the sidewalls of the waveguide. It is a particular problem for ridge and deep etched ridge waveguide structures where there is a greater index contrast in the lateral wave guiding. However, in buried structures this factor contributes only a small amount of excess insertion loss. Scattered light can be particularly problematical in EA modulators because it not only increases the insertion loss, it can also reduce the extinction ratio. The small size of most EA modulators means that scattered light can easily be coupled into the output fiber leading to a substantial reduction in the extinction ratio for the device.

The other primary source of insertion loss is the fiber coupling loss. The smaller and more tightly confined the optical mode is in the modulator, the more difficult it is to achieve low coupling loss to the device. The effective mode size in an EA waveguide is only 2 to 3 μm while in a single-mode optical fiber it is approximately 8 μm. Even for perfect positioning of the fiber the coupling loss between it and the EA will be substantial. Furthermore, maximum coupling efficiency will occur when the fiber is in direct contact with the EA facet. To minimize the coupling loss either a micro optic lens system is used or a small lens shape is fabricated directly on the tip of the optical fiber. Lensed fibers are particularly effective and can produce spot sizes on the order of 3.5 μm at their focal point, which is around 20 μm from the fiber tip. The coupling efficiency between a lensed fiber and an EA modulator can be approximated by calculating the coupling efficiency for a Gaussian mode with a width of 3.5 μm to the EA modulator transverse mode profile calculated numerically or by using the effective index method described previously. This is done with a simple overlap integral, e.g. 7.38 [2], where the mode functions are normalized such that their integral over all space is unity. The factor t accounts for the impedance discontinuity between the free space region and the waveguide. If a proper antireflection coating is applied to the surface of the device this factor will be very close to unity [2].

$$\frac{P_g}{P_f} = \left| t \int\int U_g^*(x, y) \cdot U_f(x, y)\, dx\, dy \right|^2 \tag{7.38}$$

One way to overcome this tradeoff between the desire for a tightly confined mode in the active portion of the device and a more loosely confined mode at the input and output facet is with the use of beam expanders.

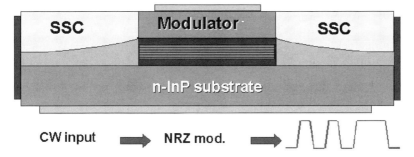

Fig. 7.10   EA modulator with integrated spot size converter waveguides.

7.3.4. BEAM EXPANDERS

A spot size converter or a beam expander (as they are sometimes called) is a section of waveguide wherein the thickness and/or the index of refraction of the guide are tapered so as to effect a transformation of the effective mode size as it propagates from one end to the other. An example of an EA modulator with integrated spot size converters is shown in Fig. 7.10. The spot size converters can serve two purposes, the first being to transform the mode from the small tightly confined state in the active region of the EA modulator into a larger size at the facet for easier coupling to optical fiber. The other reason for adding these long passive sections to an EA modulator chip is to increase the overall length of the device. For high-speed modulators the length of the active region is typically very short, sometimes less than 100 μm. It is extremely difficult to cleave and handle devices that are this short. The spot size converters eliminate this problem, making the chip length somewhat independent of the length of the active waveguide. Increasing the chip length also helps to reduce the amount of scattered light that is coupled from the input fiber to the output fiber by a path other than through the waveguide.

The majority of spot size converter designs employ tapered waveguide structures operating close to the modal cutoff to expand their spot size. These designs have been shown to provide improved coupling efficiency and more relaxed alignment tolerances. Both vertical and lateral waveguide tapers can be effective in transforming the mode size. For vertical tapers such as those shown in Fig. 7.10 the waveguide tapers can be produced by selective area growth [11] or selective area etching techniques [12]. Lateral tapers have also been used effectively for spot size converters in 1.3 μm lasers [13] and in polarization insensitive semiconductor optical amplifiers at 1.55 μm [14]. These are somewhat easier to fabricate than vertical tapers and can be very effective, particularly for 1.3 μm wavelength devices. A particularly important aspect of the spot size converter fabrication is the interface between the active waveguide in the modulator and the passive waveguide of the spot size converter section. Etching away the active waveguide layer and regrowing a butt-jointed passive spot size converter waveguide is a technologically challenging but highly effective method for performing this integration.

Now that we have a more thorough understanding for the optical waveguide parameters and how they impact the design of an EA modulator, we can begin to examine the electrical characteristics of the modulator itself and how they affect its performance.

7.3.5. FREQUENCY RESPONSE

In operation, an EA modulator is to first order a simple reverse-biased PIN diode. The frequency response can be accurately predicted from a simple equivalent circuit model based on extracting the series resistance, junction capacitance, and parasitics from the device structure. For low optical power levels we can neglect the effects of carrier transport and concentrate solely on the RC limited response. Examining the cross section for a typical device shown in Fig. 7.11 we can see that the electrical equivalent circuit is made up of a number of elements, which include the series resistance and junction

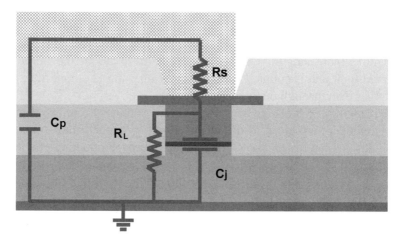

Fig. 7.11   Cross section of EA modulator with equivalent circuit model superimposed.

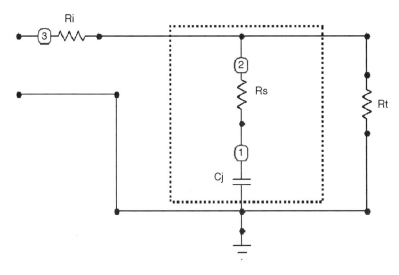

Fig. 7.12   Simplified equivalent circuit for EA modulator.

capacitance for the device, a parasitic capacitance associated primarily with the contact pad, and a leakage resistance in parallel with the junction. In reality the operation of the device is far more complicated but this simple equivalent circuit effectively captures the dominant factors in the device performance. For buried devices with a PNIN blocking structure there is also a significant parasitic capacitance associated with the blocking junction that can be included in the C_p term. Generally the leakage resistance is high enough that it can be approximated with a simple open circuit. If we assume also that the parasitic capacitance is much smaller than the junction capacitance and we neglect the voltage dependent current source associated with the absorption we have the simplified equivalent circuit shown in Fig. 7.12.

The bandwidth for the device is defined as the frequency range over which its response function remains within 3dB of the peak. This is equivalent in this case to the frequency at which the voltage across the junction falls to 50% of its DC value. If we assume that the device is being driven by a source with an impedance R_i, and has a termination resistor R_t, it is easily shown that the bandwidth is given by

$$BW = \frac{1}{2\pi \left(R_s + \left(\frac{1}{R_i} + \frac{1}{R_t} \right)^{-1} \right) C} \qquad (7.39)$$

The use of a termination resistor matched to the impedance of the source and the transmission line minimizes the reflection of RF power from the device.

Increasing the termination resistance above the impedance of the source reduces the bandwidth of the device, while decreasing the termination resistance increases the bandwidth but reduces the voltage across the junction, and thus the RF power coupled to the device since the voltage across the modulator scales with

$$V_j = V_{source}\left(\frac{R_t}{R_i + R_t}\right) \tag{7.40}$$

These simple models are very helpful in understanding the basic limitations on the EA modulator design. Now that the key parameters have been identified we can develop a simple procedure for estimating the bandwidth of a given structure. As an example we can calculate the maximum length for a modulator with a 1.5 μm wide waveguide and a 0.2 μm thick depletion region to achieve a bandwidth of 40 GHz. The series resistance and the junction capacitance are the dominant elements so we will consider these first. The junction capacitance scales with the device area and is given by Eq. (7.41), where A is the area of the junction, d is the depletion width, and ε is the dielectric constant for the material. For devices with well-defined doping profiles, d is approximately equal to the thickness of the intrinsic layer.

$$C_j = \frac{\varepsilon A}{d} \tag{7.41}$$

The series resistance is composed of several different components, which include the contact resistance for the n and p contacts, the series resistance of the p layer, and the series resistance of the n layer. In InP the mobility for holes is approximately 30 times lower than the mobility for electrons, also the specific contact resistance for the p contact is typically 10 times larger than for the n contact. In most device structures the n contacts are very large covering the entire back side of the wafer, while the p-contact area is limited to the width and length of the active region. To simplify the problem of calculating the series resistance we can then ignore the contribution from the n-contact and n-semiconductor layers, which will be on the order of 1 to 2 Ω, and just consider the p-contact resistance and the p cladding layer resistance, which will scale in inverse proportion to the device area. The specific contact resistance for an EA modulator can be found from transmission line measurements and is generally between 1E-5 and 1E-6 (Ωcm^2). A reasonably conservative number would be $r_c = $ 5E-6 (Ωcm^2). The mobility μ in p-type InP depends on the acceptor doping concentration n_a and is given by Eq. (7.42).

$$\mu(n_a) = 65 - 20\frac{n_a - 10^{18}}{10^{18}} \qquad (7.42)$$

A reasonable activated doping level for the InP cladding layer is $n_a = 1 \times 10^{18}$ cm^{-3}. The resistivity ρ and the conductivity σ is then given by

$$\frac{1}{\rho} = \sigma = q n_a \mu \qquad (7.43)$$

The total series resistance for the device can then be calculated using Eq. (7.44), where w is the width of the device, l is the length, and t is the thickness of the p cladding layer, which we can set to 2 μm.

$$R_s = \rho\left(\frac{t}{wl}\right) + \frac{r_c}{wl} \qquad (7.44)$$

The resistance and capacitance per unit length for the example can now be calculated: they are 1.6 Ω/mm and 931 pF · mm respectively. Solving Eq. (7.39) for the length we find that a 100-μm-long modulator will have approximately 40 GHz bandwidth.

A more complete model, which includes the parasitics of the device and termination including the effective inductance of the wire bonds used to connect to the device, is shown in Fig. 7.13. This type of model is best analyzed using a spice simulation tool. It is still approximate because it does not include distributed effects, but it will be more accurate than the model shown in Fig. 7.12.

Given the understanding we have developed so far of the design issues for EA modulators, it is useful to review the basic tradeoffs that exist between the different device performance requirements. The most fundamental of these is between the bandwidth and the extinction ratio. We have seen that the extinction ratio can be increased by increasing the length of the device or by increasing the absorption change per unit length. The change in the absorption per unit length is proportional to the confinement factor for the optical mode in the active layer and to the applied electric field. The maximum confinement factor, however, is limited by coupling considerations and the desire to maintain a single mode waveguide. Higher field strengths in the active layer can be achieved by thinning the depletion thickness,

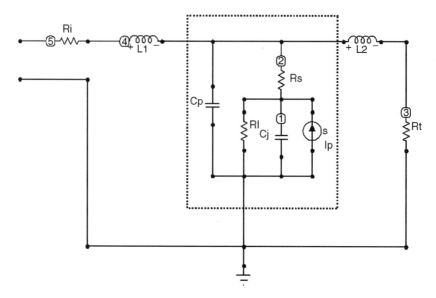

Fig. 7.13 More detailed lumped element equivalent circuit model for an EA modulator.

which increases the junction capacitance and reduces the bandwidth. Also the absorption change from the quantum confined Stark effect eventually saturates so thinning the depletion thickness can be used to trade off drive voltage with bandwidth for a given extinction ratio but not to increase the maximum absorption change. Some benefit can be had from reducing the waveguide width, but eventually the confinement factor begins to drop and this results in a decrease in the absorption per unit length. This brings us back to increasing the overall device length to increase the extinction ratio, which increases the junction capacitance and reduces the bandwidth. This fundamental tradeoff between the extinction ratio and the bandwidth provides an upper limit on the performance of lumped element EA modulators. Later in this chapter we will discuss how to overcome this limitation with a distributed design.

7.4.   EA Modulator Characterization

Evaluating the performance characteristics of EA modulators involves the application of a wide variety of different measurement techniques. There is a basic set of measurements, which can be used to characterize the main properties of the device. These include the insertion loss, extinction ratio, and frequency response. In addition to these basic measurements, additional tests are frequently performed to evaluate the scattering parameters, junction capacitance, saturation power, timing jitter, and optical coupling loss.

7.4.1. STATIC EA CHARACTERIZATION

The extinction ratio and insertion loss for an EA modulator must be characterized as a function of wavelength for the device. This is accomplished by coupling light from a tunable laser into the EA modulator and monitoring the output power as a function of bias voltage for several different wavelengths across the operating range of the device. A typical setup for measuring the insertion loss and DC extinction ratio for a device is shown in Fig. 7.14. First a lead-in cable is used to connect the tunable laser to the power meter. Then the transmitted power is measured as a function of wavelength. This reference measurement characterizes the source and allows the insertion loss of the connectors and the lead-in cable to be removed from the real measurement. The next step is to insert the EA modulator into the measurement system. A temperature controller is used to maintain the device at a constant temperature throughout the measurement, and a voltage source is added to bias the device.

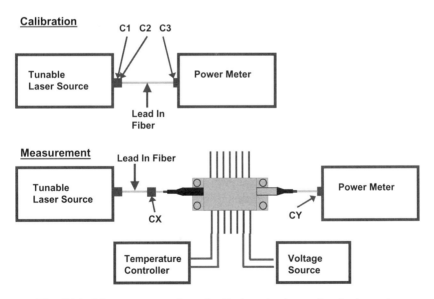

Fig. 7.14   Measurement configuration for insertion loss and extinction ratio.

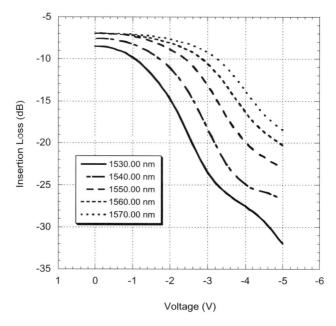

Fig. 7.15   Extinction curves of an electroabsorption modulator for a range of wavelengths.

A typical set of extinction ratio curves for an EA modulator is shown in Fig. 7.15. The curves are normalized to the input power to the device so the nominal insertion loss at a particular wavelength is given by the value at zero volts. As expected, the insertion loss increases as the de-tuning between the multiple quantum well active region bandgap and the operating wavelength is reduced. For this particular device the insertion loss varies from −9 dB at 1520 nm to −6 dB at 1570 nm. The bandgap of the material in this case is 1495 nm. For wavelengths that have a large amount of detuning from the bandgap the insertion loss is dominated by the fiber coupling loss. This is evident from the small amount of variation in the insertion loss for wavelengths from 1540 to 1570 nm when com-pared to the variation for wavelengths from 1520 nm to 1540 nm. There are other contributions to the insertion loss from interband absorption in the doped cladding materials and optical scattering caused by waveguide nonuniformities.

The shape of these curves is characteristic of multi-quantum well electro-absorption modulators that exhibit significant nonlinearity in their transfer function. The absorption curve is initially flat with bias then increases rapidly in slope before passing through an inflection point and finally saturating. The extinction ratio for a given voltage range is taken as the ratio of the transmitted power in the on state to the transmitted power in the off state. This can be read off the graph as the separation in dB between the on and off state voltages. For this particular device the extinction ratio at 1530 nm for a 5 V swing is approximately 20 dB.

There are four main sources of uncertainty in this measurement; these are caused by connector variations, power meter errors, polarization dependence of the device, and optical interference effects. For the test described previously, the calibration measurement includes the loss of the connection between the laser source and the lead-in fiber, and we assume that exchanging the connectors at the power meter does not significantly affect the measurement. However, in the device measurement there is an additional connector pair in the optical path so that the insertion loss measurement includes the device under test plus the loss of this additional connector pair, which can be anywhere from 0.5 to 1 dB [15]. Insertion loss is always calculated as the ratio of two power levels so absolute accuracy in power measurements is not required; however, the accuracy of the ratio is important. Uncertainty in the power ratio can result from nonlinearity in the response, polarization dependence, spatial inhomogeneity, and numerical aperture limitations in the detector head. The use of a high-quality commercial power meter is essential for performing accurate insertion loss and extinction ratio measurements. Because most EA modulators are highly polarization sensitive, the uncertainty caused by polarization dependent loss can be the dominant factor in device measurement error. For this reason it is critical to have precise control over the input polarization state of the light in the measurement setup. Figure 7.16 illustrates the effect on the extinction curves for varying degrees of polarization of the input light. There is little effect on the insertion loss at 0 V bias for wavelengths sufficiently detuned from the band edge that the nominal absorption is small. However, even for the case when the TE to TM ratio of the launched light is 20 dB there is approximately 1 dB of degradation in the extinction ratio of the device at −4 V. To overcome this sensitivity most polarization dependent EA modulators are pigtailed with a polarization maintaining fiber that is carefully aligned to the polarization axis of the device.

Reflections in the test system can cause two different types of measurement uncertainty. First, reflections in the test system can lead to optical interference that can result in a wavelength-dependent transmission function. This interference effect can magnify the impact of small internal

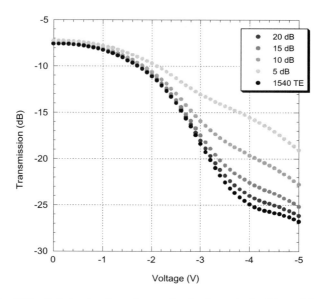

Fig. 7.16 Polarization dependence of extinction ratio for EA modulator.

reflections. If we consider just two reflections in the transmission path, r_1 and r_2, separated by a distance L, the amplitude of the transmitted light can be represented by Eq. (7.45) [2].

$$|S_{21}|^2 = \frac{(1 - R)^2}{(1 - R)^2 + 4R \sin^2 \beta L} \quad \text{where } R = r_1 r_2 e^{\alpha L} \text{ and } \tilde{\beta} = \beta + j\frac{\alpha}{2}$$

(7.45)

For the case when we have two 1% reflections and assume that there is no internal loss so that α is zero, the transmission loss uncertainty can be as high as 4% when βL is a multiple of π. Perhaps more significantly, if we have a situation when the EA modulator is bounded by two reflections, the path length may change with the change in the absorption within the modulator, resulting in a perturbation of the extinction curve. Reflections in the system can be minimized by using APC connectors wherever possible and antireflection (AR) coatings on the device facets. APC connectors have an angled interface, which typically reduces the back reflection below -60 dB. Another technique that can eliminate resonances in long patch cords is to decrease the coherence length of the optical source used in the measurement system. Many tunable laser sources have a coherence control that introduces a modulation to the source, which broadens the spectral line width. Without this coherence control external cavity tunable lasers can have linewidths in the 100 kHz range. To reduce the coherence length to less than 1 m, a typical length for fiber patch cords, a spectral width of 240 MHz is recommended [15].

7.4.2. DYNAMIC CHARACTERIZATION

A fundamental requirement of an EA modulator in a transmission system is that it have sufficient modulation bandwidth to allow the transmission and reception of the intended information. For digital systems using NRZ format the total system bandwidth needs to be greater than one-half the bit rate. In practice for optimum NRZ data transmission performance it has been shown that the rise and fall time at the transmitter should be 40% of the bit time, which is equal to $1/B$ where B is the bit rate [16]. This corresponds to a small signal bandwidth on the order of 90% of the bit rate. For the receiver the optimum bandwidth depends on the noise characteristics of the system. For optically amplified systems where the input power is far above the receiver sensitivity the optimum bandwidth is equal to the bit rate B. In this case thermal noise is negligible and the system is dominated by the intersymbol interference (ISI), which can be reduced by increasing the receiver bandwidth. However, when the thermal noise of the receiver dominates, the optimum bandwidth is approximately 60% of the bit rate. It is also important for the phase response to vary linearly with frequency over this bandwidth. Deviations from linear phase are indicative of group delay problems, which lead to increased intersymbol interference (ISI) penalties.

The bandwidth for an EA modulator is typically measured using a light-wave component analyzer or a vector network analyzer with a calibrated detector because these instruments are capable of measuring both the magnitude and the phase of the frequency response. A lightwave component analyzer is used to measure the small-signal linear transmission and reflection characteristics of a device as a function of frequency. It operates by injecting a modulated signal into a test device and comparing the modulated input signal to the signal that is transmitted or reflected by the test device. This comparison of the transmitted signal to the incident signal results in a ratio measurement. The concept of making ratio measurements to test the response of electrical devices and systems originated in the RF and microwave industry. For measuring an EA modulator a swept frequency RF source is applied to the device through a bias T. The bias T enables a DC bias voltage to be applied to the device to control the operating point about which the RF voltage will swing. Light coupled through the device is modulated by this RF signal. The light is converted back into an electrical signal in a reference receiver, where it is compared to the initial modulation signal. RF energy that is reflected back from the device is also compared to the initial input signal to determine the reflection coefficient for the device. These are generally referred to as the scattering or S parameters for the device. The frequency range over which the magnitude of the S_{21} response is within 3 dB of its peak magnitude is referred to as the 3 dB bandwidth of the device. A plot of the frequency response for an Agere Systems EA modulator is shown in Fig. 7.17. The input reflection coefficient for the same device is shown in Fig. 7.18. The electrical equivalent circuit for the device can be derived from the magnitude and phase of the S_{11} response, which is useful for evaluating the design of the EA.

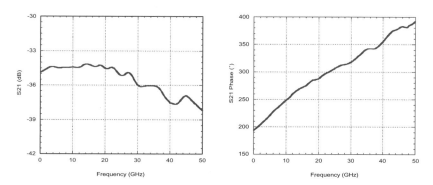

Fig. 7.17   Frequency response measurement (S_{21}), magnitude, and phase for Agere Systems EA modulator with 48 GHz bandwidth.

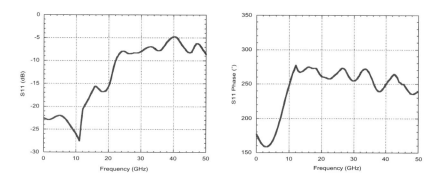

Fig. 7.18   Input reflection (S_{11}), magnitude, and phase for Agere Systems EA modulator.

7.4.3. *CHIRP MEASUREMENTS*

Lightwave component analyzer or network analyzer measurements can also be used to characterize the small signal alpha parameter for an EA modulator defined in Eq. (7.46), where n and k are respectively the real and imaginary parts of the modal index of the electroabsorption waveguide [17]:

$$\alpha_H = \frac{\partial n}{\partial k} \qquad (7.46)$$

For external modulators the variation in the phase ϕ with intensity I is related to the alpha parameter by Eq. (7.47).

$$\delta\phi = \frac{\alpha_H \delta I}{2I} \qquad (7.47)$$

A simple and accurate method for measuring the alpha parameter for an EA modulator, which can also be used to measure fiber dispersion, has been developed by Devaux et al. [18]. This technique uses small signal measurements in the frequency domain to analyze the chirp characteristics for light propagating in a dispersive medium. In these conditions they observe sharp resonance frequencies that originate from interferences between carrier and sideband wavelengths. By analyzing the frequencies at which these resonances occur it is possible to obtain accurate and reproducible values for the chirp parameter of the transmitting source.

The measurement is performed using a network analyzer with a calibrated optical receiver, the device to be measured, and a length of single-mode optical fiber with a nonzero dispersion (Fig. 7.19). The length of fiber required depends on the dispersion of the fiber and the maximum frequency of the measuring system. A general rule of thumb is for the total dispersion DL in ps/nm to be greater than $7E5/f_{max}^2$ where f_{max} is in GHz. To first calibrate out the frequency response of the modulator and the detector an S_{21} measurement is taken with the transmission fiber removed; this is used as a baseline. Then the transmission fiber is reinserted and the measurement is repeated. This measurement is divided by the original result to remove the bandwidth dependence of the modulator and receiver. The frequency response for the transmission through the dispersive fiber shows a number of resonances, which appear as sharp peaks in the frequency response (Fig. 7.20).

We can represent the electric field by

$$E = \sqrt{I} e^{j\phi(I)} \qquad (7.48)$$

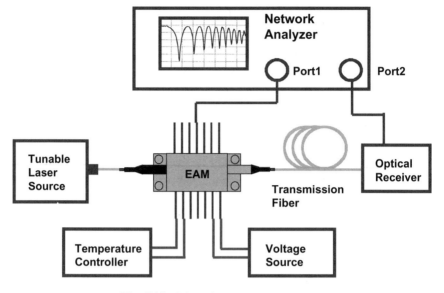

Fig. 7.19 Dispersion measurement setup.

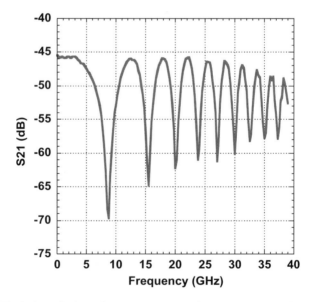

Fig. 7.20 Typical small signal frequency response for EA modulator after transmission through 50 km of single-mode fiber with $D = 17$ ps/(nm · km).

and assume that the transmitted intensity is given by Eq. (7.49) for small signal modulation with a frequency f and a modulation depth m being much less than 1.

$$I = I_0(1 + m\cos(2\pi ft)) \qquad (7.49)$$

After propagation through the fiber the frequency response, which is measured in Fig. 7.20, is given by Eq. (7.50) [18]. The resonances in this equation are the result of two simultaneous interferences between the carrier and the two sidebands. The resonance frequencies f_u shown correspond to the zeros of the cosine term and are given by Eq. (7.51).

$$I_f = I_0 m\sqrt{1 + \alpha_H^2}\left|\cos\left(\frac{\pi\lambda^2 DLf^2}{c} + \arctan(\alpha_H)\right)\right| \qquad (7.50)$$

$$f_u^2 L = \frac{c}{2D\lambda^2}\left(1 + 2u - \frac{2}{\pi}\arctan(\alpha_H)\right) \qquad (7.51)$$

Extracting the resonance frequencies from Fig. 7.20 and plotting them versus the parameter $2u$, which is twice the order of the resonance, we get a straight line shown in Fig. 7.21. The slope of this line is inversely proportional to the dispersion parameter for the fiber D, and the intercept

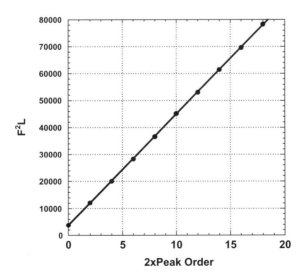

Fig. 7.21 Plot of the resonance frequencies squared times the fiber length versus twice the peak order.

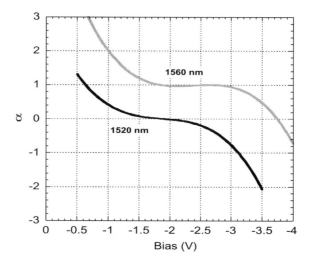

Fig. 7.22   Small signal alpha parameter as a function of bias voltage for the modulator.

can be used to calculate the small signal alpha parameter for the given operating conditions.

Repeating this process over a range of bias voltages and wavelengths enables us to plot the small signal alpha parameter characteristics for the modulator (Fig. 7.22).

From this plot we can see that the alpha parameter varies with the bias voltage on the modulator. For increasing bias voltage it becomes progressively more negative. It also shows a strong wavelength dependence. The alpha parameter is a useful technique for comparing the chirp performance of different active layers in electroabsorption modulators. However, the variation in the alpha parameter with voltage leads to a complicated dependence of the chirp on the drive signal. One parameter that can be used as a figure of merit for the device is the voltage at which the small signal alpha parameter becomes negative. This again is a good means of comparing two different quantum well structures. However, it is not as useful in predicting the transmission performance for the device. It is sometimes helpful to define an effective alpha parameter, which is representative of the average or cumulative chirp performance for the device. This can be done by first calculating the change in the imaginary index Δk as a function of bias. This can be found from DC extinction ratio measurements like those shown in Fig. 7.15 by using Eq. (7.52).

$$\frac{I(V)}{I(V_o)} = e^{-\frac{4\pi \Delta k L}{\lambda}} \tag{7.52}$$

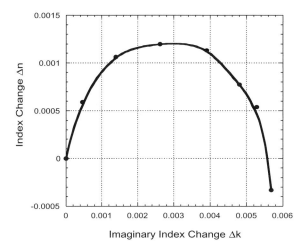

Fig. 7.23   Variation in the real component of the index of refraction plotted versus the variation in the imaginary component for 1520 nm.

Then the real part of the index change Δn can be calculated as a function of voltage using the alpha parameter, which is the derivative of the n versus k curve (Fig. 7.23). From this curve we can compute an effective large signal chirp parameter $\bar{\alpha}_H$ using Eq. (7.53).

$$\bar{\alpha}_H = \frac{n(V) - n(V_o)}{k(V) - k(V_o)} \tag{7.53}$$

This technique for calculating an effective alpha can be reasonably accurate for cases when the small signal alpha parameter does not change sign. However, in the case demonstrated here, the effective alpha calculated from Eq. (7.53) would be -0.06, which is clearly not representative of the chirp characteristics of this device.

A better estimate for the effective alpha parameter was proposed by Dorgeuille and Devaux [19] in 1994, based on the so-called 3 dB rule. This technique takes the average of the small signal alpha parameter over the first 3 dB of the extinction curve.

$$\bar{\alpha}_H = \frac{\int_{k_{on}}^{k_{-3dB}} \alpha_H(k)dk}{\int_{k_{on}}^{k_{-3dB}} dk} \tag{7.54}$$

Using this approach we can calculate the effective alpha parameter for the curve shown in Fig. 7.24 over the same range of absorption, and we get

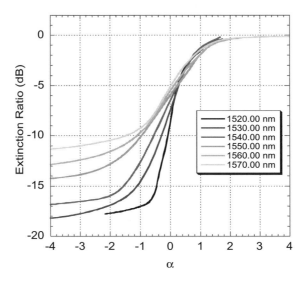

Fig. 7.24   Small signal alpha parameter plotted versus extinction ratio for a 50 nm range in wavelength.

an effective alpha of 0.9. If we bias the device with 3 dB of insertion loss then the effective alpha is reduced to 0.3, which illustrates quite clearly the tradeoff between insertion loss and chirp performance that is experienced by most practical modulators. Plotting the alpha parameter versus drive voltage can be misleading because there is no direct correlation to the transmitted light intensity, and it tends to show a greater wavelength dependence than exists in reality. If we instead plot the small signal alpha parameter as a function of the extinction ratio in the device for several wavelengths we see that the wavelength dependence is not as great as would be assumed from the previous plot.

It is important to note that for an EA modulator there is no adiabatic component to the chirp. This means that there is no inherent frequency shift associated with the on state versus the off state in the device; only during the rising and falling edges when the absorption is changing do we see a shift in the frequency of the light. This shift will be roughly equal and opposite for the rising and falling edges respectively. The large signal dynamic chirp for an EA modulator can best be observed using time resolved spectroscopy (TRS). This technique is better for illustrating the temporal distribution of the frequency shifts under large signal operation [20]. The basic setup for a TRS measurement is shown in Fig. 7.25. The modulator is

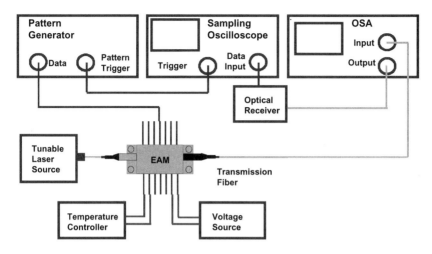

Fig. 7.25   TRS measurement setup.

driven with a pseudorandom binary sequence (PRBS) that has a relatively short word length (typically $2^7 - 1$). The transmitted light is coupled into a monochromator or an optical spectrum analyzer configured as a narrow-band tunable optical filter. The output light from the monochromator is coupled into a high-speed detector connected to a fast sampling oscillo-scope. The scope is triggered synchronously with the word pattern used to drive the modulator. The measurement is made by setting the resolution of the monochromator to as narrow a range as possible (<0.1 nm), and scanning over a frequency range, which is wide enough to contain more than 95% of the transmitted energy. For a 10 GB/s PRBS sequence this would correspond to approximately a 40 GHz range. For each wavelength an av-eraged trace is measured on the sampling oscilloscope. Typically a large number of averages are necessary for each trace due to the small amount of light transmitted through the monochromator. After the wavelength scan is complete the set of time-dependent responses for each wavelength is converted into a wavelength scan for each time interval. From this the mean wavelength at every point in time can be calculated. We can also recover the transmitted data pattern by summing the traces for every wave-length scan. Thus both chirp and extinction ratio can be simultaneously measured.

The optical intensity modulation and the corresponding frequency shift for a TRS measurement on an EA modulator operating at 10 GB/s is shown in Fig. 7.26. In this case the device was deliberately operated at a reduced

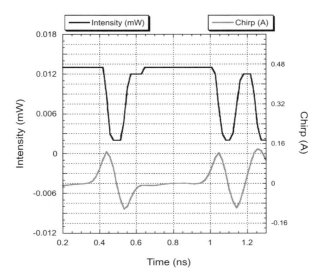

Fig. 7.26 Time resolved spectroscopy measurement for an EA modulator driven with a 10 GB/s pseudorandom binary sequence (PRBS).

extinction ratio so both the on and off state chirp could be measured. As discussed previously there is no inherent frequency shift associated with the on or off state and all the chirp is confined to the rising and falling edges of the data stream. For EA modulators integrated with semiconductor lasers there is often an additional adiabatic component to the chirp, which results in a shift in the wavelength for the on state relative to the off state. For an isolated pulse in the on state the wavelength increases linearly across the pulse. This corresponds to a decrease in the frequency across the pulse, which is characteristic of a negative chirp parameter C or a positive effective alpha parameter α_H. Conversely for the off state light the sign of the chirp is approximately equal and opposite. This is not particularly important for reasonable extinction ratios because there is very little light in the off state of the pulse. The TRS measurement is very useful for characterizing the overall performance of an EA modulator, because it can give us both chirp and dynamic extinction ratio in a single measurement. Here the dynamic extinction can be measured by taking the ratio of the power in the on and the off state, which is only 7.2 dB for this case. A further benefit of this measurement over the small signal alpha measurement is that it can distinguish between the dynamic and adiabatic components of the chirp in devices where both of these components are significant. The transmission performance for the EA measured in Fig. 7.26 can be predicted by extracting the Gaussian chirp parameter C from the peak-to-peak wavelength shift. First the wavelength chirp is converted into an angular frequency chirp using Eq. (7.55). Here the center wavelength was 1550 nm, the peak-to-peak wavelength shift from Fig. 7.26 was 0.223 Å, so the angular frequency shift across the pulse is approximately 17.5 GHz. If we assume a 50% eye crossing then the T_{FWHM} is equal to 1/Bit Rate or 100 ps.

$$\delta\omega = 2\pi c\left(\frac{1}{\lambda_1} - \frac{1}{\lambda_2}\right) \approx \frac{2\pi c\Delta\lambda}{\lambda_0^2} \qquad (7.55)$$

The Gaussian pulse width T_o can be calculated from Eq. (7.16), which gives a value of 60 ps. Then using Eq. (7.13) with t set to the bit time we can obtain an effective chirp parameter for the transmitted pulses of $C = 0.63$ assuming a Gaussian shape. Inserting this value into Eq. (7.23) we can solve for the maximum transmission distance for a 2 dB path penalty. Letting $\beta_2 = -20$ ps^2/km for standard single-mode fiber, we obtain a transmission distance of 45 km at 10 GB/s. This calculation relied on a number of approximations, the most important of which is that the shape of the transmitted pulses is Gaussian. However, as we will see in the next section it is remarkably consistent with the measured dispersion penalties from actual transmission experiments.

7.4.4. BIT ERROR RATIO TESTING

In the simplest case the fundamental measure for the transmission performance of an electroabsorption modulator is how accurately a receiver can determine the logic state of each transmitted bit. This figure of merit is called the bit error ratio. It is defined as the number of errors in a given time interval t divided by the total number of bits transmitted in that same time interval. The equipment used to measure this is known as a bit error ratio tester or BERT. It consists of three main components—a clock source, a pattern generator, and an error detector. The pattern generator creates a test pattern based on a pseudorandom binary sequence or PRBS. This is a repetitive pattern whose pattern length is of the form $2^N - 1$, where N is an integer. This ensures that the pattern repetition rate is not harmonically related to the data rate. Typical values of N available on a commercial BERT are 7, 15, 23, and 31. The bit sequence within the pattern is designed to simulate random data, particularly for longer word lengths. The frequency spectrum for a PRBS pattern consists of a series of discrete lines with a spacing defined by Eq. (7.56) that follows a $\sin(x)/x$ envelope function

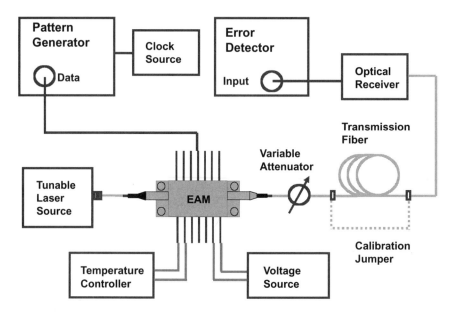

Fig. 7.27   Test setup for measuring the dispersion penalty for an EA modulator.

with nulls at integer multiples of the bit rate f_b.

$$\Delta f = \frac{f_b}{2^N - 1} \tag{7.56}$$

A typical test setup for measuring the chromatic dispersion power penalty for an EA modulator is shown in Fig. 7.27. The light from a tunable laser is modulated with the EA using a data stream from the pattern generator. This light is passed through a variable optical attenuator and then a length of optical fiber to a receiver. The receiver converts the optical signal back into an electrical data pattern, recovers the clock, and passes the data to an error detector. If a clock recovery circuit is not available, the error detector can be connected to the source clock for the pattern generator. However, this is not as stable because the phase shift between the transmitter clock and the received data can drift over time, particularly for long transmission distances. Inside the error detector is a decision circuit, which compares the incoming data signal I with a decision level I_d at a decision time t_d determined by the received clock. Ideally t_d is in the center of the bit time. The sampled value I fluctuates randomly about one of two values I_1 or I_0 depending on whether the received bit is a one or a zero. If I is greater than I_d the error detector records the bit as a one. If it is less than I_d it is recorded as a zero. If we assume that the fluctuations in the levels follow a Gaussian probability distribution with an rms width σ centered at the average signal level, then the bit error ratio can be calculated analytically by summing the probability for I_1 less than I_d and the probability for I_0 greater than I_d. Then the BER can be written as

$$BER = \frac{1}{4}\left[erfc\left(\frac{I_1 - I_D}{\sigma_1\sqrt{2}}\right) + erfc\left(\frac{I_D - I_0}{\sigma_0\sqrt{2}}\right)\right] \qquad (7.57)$$

where *erfc* stands for the complimentary error function defined as

$$erfc(x) = \frac{2}{\sqrt{\pi}}\int_x^\infty e^{-y^2}dy \qquad (7.58)$$

The minimum bit error rate is achieved by setting the decision level such that the contribution from the two error sources is equal. Then Eq. (7.57) can be rewritten as

$$BER = \frac{1}{2}erfc\left(\frac{Q}{\sqrt{2}}\right) \qquad \text{where } Q = \frac{I_1 - I_0}{\sigma_1 + \sigma_0} \qquad (7.59)$$

This form is used for plotting BER testing results where the bit error ratio is plotted against the received power using a vertical scale defined by the complimentary error function. Plotted in this way the bit error rate versus received power follows a straight line.

The first step in measuring the dispersion penalty is to connect a calibration jumper consisting of a short length of optical fiber between the transmitter and the optical receiver. Then the error ratio is measured as a function of the received power. This is plotted in Fig. 7.28 using solid circles, and is sometimes referred to as a back-to-back or baseline measurement. The next step is to remove the jumper and replace it with a length of transmission fiber. Then the BER is again measured as a function of received power. This is plotted in Fig. 7.28 with open circles. If we fit a straight line to the data the receiver sensitivity is typically defined as the received power level where the line intersects a bit error ratio of 10^{-10}, that is one error for every 10 billion bits of data. The difference between the receiver sensitivity for the back-to-back measurement and the measurement over the transmission fiber is the dispersion penalty.

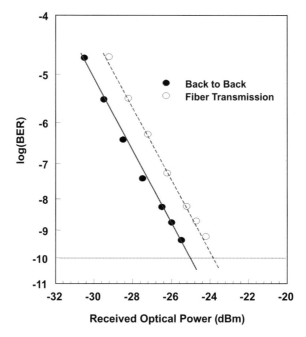

Fig. 7.28 Dispersion penalty test with bit error ratio curves for back-to-back and transmission through a dispersive fiber link.

7.5. Electroabsorption Modulators Integrated with Lasers

One of the biggest advantages of the EA modulator is its potential for integration with other semiconductor-based photonic devices. In fact some of the earliest examples of photonic integrated circuits are electroabsorption modulated lasers or EMLs which combine a conventional distributed feedback (DFB) telecommunication laser with a monolithically integrated electroabsorption modulator [21]. These devices have seen widespread application in a large number of 2.5 and 10 GB/s fiber optic communication systems. Their compact size and low cost make them an extremely attractive alternative to the more conventional approach of combining a fixed wavelength DFB laser with an external lithium niobate modulator. The basic elements of an EML are a DFB laser source, and an EA modulator. They also generally include a passive waveguide section linking the two devices to improve the electrical isolation between them. The operation of the electroabsorption modulator requires that the operating wavelength

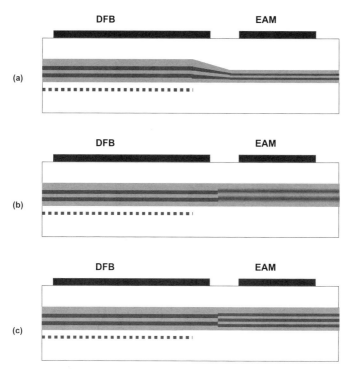

Fig. 7.29   Integrated EADFB fabricated with (a) selective area growth, (b) quantum well intermixing, (c) butt joint regrowth.

of the laser be at an energy that is below the bandgap of the active layer of the modulator. For this reason EMLs are typically fabricated with different active layer bandgaps in the modulator and the laser, although this is not always the case [22]. For this discussion the active layer is the portion of the waveguide in the laser that provides optical gain and in the modulator it is the portion that provides a voltage dependent absorption characteristic. For most EMLs a multiple quantum well layer is used for the active section both in the laser and in the modulator. There are several different techniques that can be used to integrate the different active regions, the most important are selective area growth or SAG [23], quantum well intermixing or disordering [24], and butt joint epitaxy [25] (Fig. 7.29). Selective area growth techniques use oxide masks patterned on the surface of the wafer to enhance the growth rate in certain localized regions during the epitaxial steps used to form the active layers of the device. This enables the thickness of the quantum wells in the laser section to be increased, which shifts their bandgap to lower energies relative to the quantum wells in the modulator section that do not experience the SAG effect. This is one of the most successful and widely used techniques for growing EML structures because of the ease with which it can be implemented and the high-quality modulator structures that are produced. One drawback of this approach is that it restricts the modulator and the laser to using the same number of quantum wells, and there is little room to optimize the different structures independently.

Quantum well intermixing is performed using a wide variety of techniques that introduce vacancies into the material. High-temperature annealing steps are used to diffuse these vacancies through the quantum well structure. The effect of diffusing the vacancies through the large concentration gradient in the quantum structure is to force an intermixing between the well and the barrier material. For properly designed wells this produces an increase in the effective bandgap of the structure. For EMLs the laser active layer would be kept as grown and the modulator active layer would be intermixed to shift it to a higher bandgap. This approach is also easily implemented but it has the disadvantage that the intermixed quantum wells in the modulator are less abrupt than they are in the case of a modulator grown by SAG. This reduction in the abruptness of the heterointerfaces reduces the confinement of the carriers in the well and broadens the exciton linewidth, which can lead to reduced extinction ratios and increased absorption loss in the device.

The third approach for active layer integration is the butt joint regrowth technique. In this method the active layer for the modulator is grown first and then etched off in the area where the laser is to be fabricated. Then the laser active layer is grown butt jointed to the original modulator active layer. This is the most difficult technique to implement but it provides the greatest flexibility, enabling any two waveguide layer types to be integrated [26]. The designer is then free to independently optimize the number, thickness, and strain of the quantum wells in the laser and modulator active layers.

The chirp behavior is somewhat different for an EML than it is for an isolated electroabsorption modulator. This is because in addition to the dynamic component of the chirp that comes from the EA itself there is also an adiabatic component that is caused by crosstalk between the EA and the DFB laser. The adiabatic chirp leads to a steady state difference between the wavelength of the device in the off-state and the wavelength in the on-state. The crosstalk that leads to adiabatic chirp has an electrical component which is a result of imperfect isolation between the EAM and the DFB laser. This is kept low by proper filtering of the drive circuitry to the laser and by providing a long isolation region between the EAM and the DFB. The other main source for the adiabatic chirp is the optical crosstalk caused by residual facet reflections. The operating wavelength for an index-guided DFB laser is extremely sensitive to the phase of the facet reflections. In the case of an EML the intensity of the reflected light from the front facet is modulated by twice the amount that the transmitted light is. This produces a large change in the intensity of the reflected light that is coupled back into the DFB laser between the on and off states for the modulator, which can lead to substantial shifts in the device wavelength. To mitigate this effect the facet reflectivity must be kept below 1E-4, which requires a very high-quality AR coating. Typical values for the peak-to-peak chirp in a good-quality EML are on the order of 0.1 Å, while the adiabatic chirp is generally smaller than 0.05 Å. This is generally more than an order of magnitude lower than for a directly modulated laser, which enables 1550 nm EMLs to achieve much longer transmission spans.

The high bandwidth, low chirp, and large extinction ratio of electroabsorption modulated DFB lasers has led to their widespread use in transmitters for 2.5 Gb/s long haul DWDM systems. Their compact size, low cost, and excellent transmission characteristics make them ideally suited to this application. At 2.5 GB/s error-free transmission has been demonstrated over more than 1000 km of standard single-mode optical fiber, while at 10 Gb/s dispersion penalty-free transmission has been achieved at up to 130 km [27], although most commercial devices are only suitable for 40 to 50 km transmission spans. For 10 GB/s systems the dispersion characteristics of the fiber are such that the importance of the modulator chirp is paramount, and thus lithium niobate Mach-Zehnder modulators compete effectively with EMLs. For 40 GB/s systems the dispersion limits are so short that even for optimized chirp, transmission distances are limited to a few kilometers. In these systems dispersion compensation is widely used and chirp parameters are actually more relaxed than at 10 GB/s.

The new advances in EML technology are following two directions. One is the push for higher and higher bit rates, which has lead to the development of 40 GB/s devices [28]. The other direction is the push for increased functionality, which has led to the development of tunable lasers integrated with EA modulators.

7.5.1. TUNABLE EMLS

The integration of an EA modulator with a tunable laser represents an increased design challenge both for the modulator design and for the overall device integration process. For tunable devices the EA must be capable of achieving low insertion loss, high extinction ratio, and minimal chirp over the entire tuning range of the device. For narrow tuning ranges between 8 and 10 nm, EA modulators can be integrated with tunable distributed Bragg reflector lasers. These lasers combine an active section, which provides optical gain, with a passive tuning element composed of a Bragg grating. The Bragg grating acts as a narrowband reflector, which controls the operating wavelength of the laser. Injecting current into the Bragg grating section lowers the effective index of the waveguide and tunes the laser to shorter wavelengths. The maximum tuning range is limited by the maximum index shift that can be achieved in the tuning section. A device of this type is shown in Fig. 7.30. This device also has an integrated semiconductor optical amplifier and an in-line tap for monitoring the output power. This device was capable of transmission at 2.5 GB/s over a distance of 680 km using any one of 20 fully stabilized wavelength channels spaced at 50 GHz [29].

Over time the enhanced functionality available with the wavelength tunable EMLs will enable them to displace fixed wavelength EMLs in many applications. The reduction in inventory, and the increased network flexibility that results from having tunable EMLs is enhanced for lasers with wider tuning ranges. The ultimate solution will be a widely tunable EML that can cover a wavelength range of more than 30 nm. One early example of a device that is capable of this is the sampled grating distributed Bragg reflector with integrated electroabsorption modulator [30]. The sampled grating DBR is similar to a DBR laser except that it has two grating

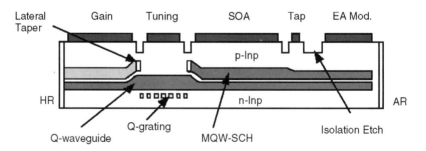

Fig. 7.30 Tunable DBR laser with integrated electroabsorption modulator.

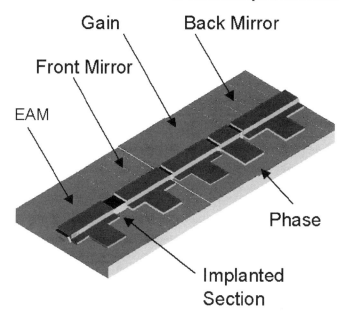

Fig. 7.31   Sampled grating DBR laser with integrated electroabsorption modulator.

mirrors instead of one. The SGDBR laser uses these two mirrors to employ a Vernier tuning mechanism that enables it to cover a much wider
wavelength range than the DBR. A schematic of the device is shown in
Fig. 7.31. The laser has four sections, front mirror, back mirror, gain and
phase control. Each of the front and back mirrors has multiple reflection
peaks, which are spaced so that only a single peak in the front and back
mirror can be aligned at one time. The laser operates at the wavelength
where the reflection peaks from both mirrors are aligned. Tuning any pair
of mirror peaks in tandem enables the laser wavelength to be tuned over
a range equivalent to that of a DBR. However, by differentially tuning the
mirrors, a new set of reflection peaks can be selected that covers an entirely
different wavelength range. Using this approach very wide tuning ranges
in excess of 60 nm have been demonstrated [31].

It is extremely difficult to design an EA modulator that can cover the
very wide tuning range of these lasers. Most multiple quantum well EA
modulators are useful over a range of only 15 to 20 nm. For a quantum
well the shift in the band edge absorption has a quadratic dependence on
the applied electric field. However, once a field is applied to a quantum
well the electron and hole states in the well are no longer truly confined

states because on one side of the well they see only a triangular barrier. The width of this barrier decreases with increasing electric field and the tunneling probability increases substantially, particularly for the electron states. For high fields the exciton resonance will be completely washed out and the absorption change due to the quantum confined Stark effect will saturate. This effect fundamentally limits the wavelength range for QCSE-based EA modulators. Using wider wells, or a greater barrier height in the conduction band, can help to extend the wavelength range, but this usually comes at the cost of reduced performance. For the sampled grating DBR device a bulk electroabsorption modulator was used instead of a multiple quantum well design. The bulk EA modulator is based on the Franz-Keldysh effect, which does not have the same saturation behavior that the QCSE effect does. The application of an electric field to the bulk active region introduces a finite absorption probability due to lateral tunneling at all wavelengths below the band edge. The probability increases with increasing photon energy and also with increasing electric field. DC extinction ratios of greater than 20 dB were demonstrated across the entire 40 nm tuning range of the laser at a −4.0 V bias for wavelengths from 1525 nm to 1565 nm (Fig. 7.32). This covers substantially more than the entire C-band for optical fiber transmission. As would be expected, the extinction is much higher for the shorter wavelengths, but the variation is unimportant because the typical

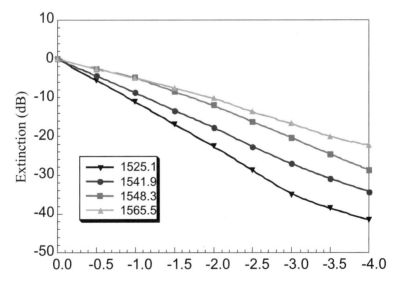

Fig. 7.32   Extinction curves for SGDBR with integrated bulk active EA modulator.

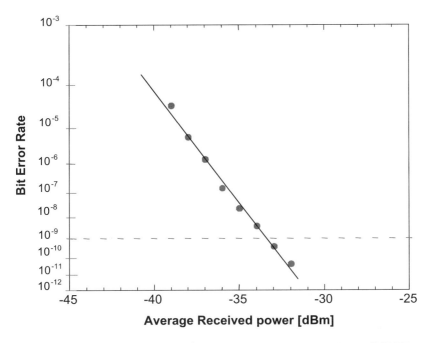

Fig. 7.33 Back-to-back bit error rate for data transmission with the EA/SGDBR at 2.5 GB/s.

requirement for the dynamic extinction ratio is less than 12 dB. The shape of the extinction curves is very different from those shown in Fig. 7.15 for a QCSE device. The Franz-Keldysh effect is generally not as efficient as the QCSE in dB/V, so an increased device length is generally required to achieve high extinction ratio at low drive voltages. The device in this example had a 200-μm-long modulator, which limited its maximum bit rate to 2.5 GB/s.

Bit error rate curves were measured for the combined EA/SGDBR at 2.5 GB/s and 1540 nm with a 2.5Vpp drive signal. The results, which indicate a receiver sensitivity for back-to-back measurements of −34 dBm, are plotted in Fig. 7.33. The receiver input eye diagram corresponding to a BER of 1E-9 is plotted in Fig. 7.34. This signal comes from the output of a limiting amplifier in the receiver path placed before the decision circuit. The limiting amplifier flattens the top and bottom rails, which improves the eye appearance, but it converts the amplitude variation to edge jitter. This explains the ∼20 ps of RMS jitter apparent in the crossing points.

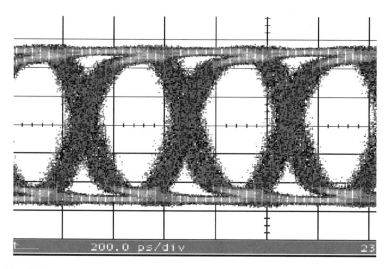

Fig. 7.34   Measured 2.5 GB/s eye diagram for the EA/SGDBR at a BER of 1E-9.

The tremendous potential of this device, which combines wide range tunability and high-speed data modulation capabilities integrated within a single semiconductor chip, represents an important future technology for electroabsorption modulated lasers, which in turn represent the future for electroabsorption modulators. The EML is perhaps the most commercially successful application for EA modulators. Fixed wavelength devices have been demonstrated for 2.5, 10 and 40 GB/s applications, and tunable devices capable of operating over 40 nm have been achieved. This will continue to be the largest volume application for EA modulators in the future, with tunable EMLs gradually replacing fixed wavelength EMLs in many system applications. Only in the most demanding long-haul applications where the highest levels of performance are required will external modulators continue to play a significant role.

7.6.   Advanced EA Modulator Designs

As the information-carrying capacity and span of optical networks continues to increase, the demands on the transmitter design become continuously more stringent. The push for ever-greater bandwidth, increased saturation power, and lower chirp stretches the limits of the physics that govern the performance of EA devices. To meet these requirements new approaches to the device design have been undertaken to overcome the traditional limits to EA performance, and to meet the needs of novel applications such as short pulse generation, optical demultiplexing, and clock recovery. In this next section we will examine some advanced EA modulator designs that are expanding the application space for these devices.

7.6.1. TRAVELING WAVE EA MODULATORS

The fundamental limitations on the bandwidth for conventional lumped element EA modulators are determined by the electrical time constant associated with the RC product of the device series resistance and the junction capacitance according to Eq. (7.39). Scaling of the device area indefinitely in order to achieve greater and greater bandwidths is impractical due to the tradeoff that exists between the device length and the optical extinction ratio. Reducing the impedance of the drive signal and the termination can significantly increase the bandwidth for a given junction capacitance. For example, operating with a 25 Ω input impedance and termination resistance can increase the bandwidth significantly. In the case of a device with a 10 Ω series resistance and a junction capacitance of 100 fF changing from a 50 Ω system to a 25 Ω system will increase the 3 dB bandwidth from 45 to 70 GHz. This represents about a 55% increase in bandwidth but reducing the impedance of the system doubles the drive power. Assuming one was willing to pay the penalty of increased drive power it is still substantially more difficult to design 25 Ω impedance transmission lines and the microwave losses associated with them are significantly higher than for standard 50 Ω lines.

A better alternative to breaking through the RC limitation on the device bandwidth is to use a distributed design where the junction capacitance is loaded with a series inductance to form a transmission line. In the ideal case the bandwidth is then limited only by the microwave losses in the line and the device length can be chosen quasi-independently of the bandwidth requirement. This approach is called a traveling wave design because the electrical wave traveling along the transmission line modulates the light as it travels along the optical waveguide. The critical problem with this approach is the need to match the electrical phase velocity on the transmission line to the optical group velocity in the waveguide. We can model the traveling wave device as a loaded transmission line using the equivalent circuit model shown in Fig. 7.35.

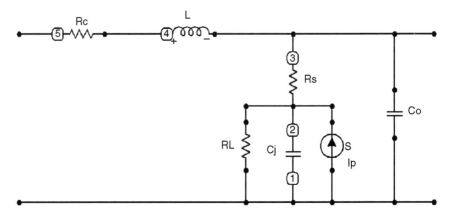

Fig. 7.35   Circuit model for a unit length of traveling wave EA modulator.

The impedance for the device Z_m and the propagation constant γ for the electrical signal are given by

$$Z_m = \sqrt{\frac{R_c + j\omega L}{\left(R_s + \left(\frac{R_L}{1+j\omega C_j R_L}\right)\right)^{-1} + j\omega C_o}} \qquad (7.60)$$

$$\gamma = \alpha + j\beta = \sqrt{(R_c + j\omega L)\left(\left(R_s + \left(\frac{R_L}{1 + j\omega C_j R_L}\right)\right)^{-1} + j\omega C_o\right)} \qquad (7.61)$$

where L and C_o are the inductance and capacitance per unit length of the unloaded transmission line, and C_j is the device capacitance per unit length. This formula treats the voltage controlled current source representing the photocurrent as an equivalent resistance and combines it with the leakage resistance R_L. Now that we have the formulas for the characteristic impedance and the propagation constant for the transmission line we can begin the much more complicated task of calculating the electroabsorption bandwidth response for a distributed transmission line with both microwave attenuation and velocity mismatch. The optical output power for the modulator can be expressed as

$$I_T = I_0 C^2 e^{-\alpha_o L} e^{-\Gamma \int_0^L \alpha(x)dx} \qquad (7.62)$$

where C is the coupling coefficient at the input and output of the device and α_o is the optical loss coefficient in the waveguide independent of the applied voltage bias and Γ is the confinement factor. If we rewrite the absorption coefficient in terms of a DC and an AC component we can approximate the small signal transmission response as

$$I_{AC} = -I_0 C^2 e^{-(\alpha_o + \Gamma \alpha_b)L} \left\{ \Gamma \frac{d\alpha}{dV} \int_0^L V_{ac}(x) dx \right\} \qquad (7.63)$$

From this we see that the frequency response for the modulator is governed by the integral of the voltage response function along the length of the device. The generalized formula for the AC voltage on the transmission line can be written as shown in Eq. (7.64) [32].

$$V_{AC}(x,t) = \frac{V_o T e^{j\omega t}}{1 - R_L R_S e^{-2\gamma_\mu L}} \left(e^{-\gamma_\mu x} + R_L e^{(\gamma_\mu x - 2\gamma_\mu L)} \right) \qquad (7.64)$$

The parameters T, R_S, and R_L represent the microwave transmission coefficient at the source and the reflection coefficients at the source and load respectively. The microwave propagation coefficient γ_μ is given by Eq. (7.61) and the microwave frequency is represented by ω. We can rewrite this formula in a reference frame propagating with the optical group velocity v_o by replacing t with $t_o + x/v_o$. This enables us to write the normalized frequency response in terms of the optical and microwave properties of the device using $\beta_o = \omega/v_o$ [32]. For InP based EA modulators the optical group velocity is around 3.6^2.

$$R = \left| \frac{T}{1 - R_L R_S e^{-2\gamma_\mu L}} \left(\frac{e^{(j\beta_o - \gamma_\mu)L} - 1}{(j\beta_o - \gamma_\mu)L} + R_L e^{-2\gamma_\mu L} \frac{e^{(j\beta_o + \gamma_\mu)L} - 1}{(j\beta_o + \gamma_\mu)L} \right) \right|^2$$

$$(7.65)$$

The reflection coefficients at the source and the load can be calculated from the impedance values for the input and output transmission lines and the impedance of the modulator transmission line.

$$R_S = \frac{Z_S - Z_m}{Z_S + Z_m}, \quad R_L = \frac{Z_L - Z_m}{Z_L + Z_m}, \quad T = 1 - R_S \qquad (7.66)$$

In the case of a lossless impedance matched line the response function R is equal to unity. Otherwise the frequency at which it drops to 0.5 can be used to calculate the traveling wave EA modulator bandwidth. For the case of an impedance matched device with no source or load reflections, the frequency response depends only on the microwave attenuation and the velocity mismatch between the electrical and optical waves in the modulator, then Eq. (7.65) can be rewritten as Eq. (7.67).

$$R = \left| \left(\frac{e^{(j\beta_o - \gamma_\mu)L} - 1}{(j\beta_o - \gamma_\mu)L} \right) \right|^2 \qquad (7.67)$$

In general the length of a traveling wave EA modulator is not substantially greater than a few hundred microns and so the microwave attenuation is generally relatively small even for high frequency operation. The bandwidth is then more fundamentally limited by the velocity mismatch between the electrical and optical waves. The traveling wave EA is one important class of devices that will become increasingly important as system requirements push the envelope of EA performance.

7.6.2. TANDEM EA MODULATORS

The growing interest in short pulse return to zero or RZ transmission as a means to increase the span length in optical fiber communication systems has led to the widespread use of EA modulators for short pulse generation. The nonlinear modulation transfer function of an EA modulator makes it well suited for use as a short pulse source. In 1992 Suzuki et al. demonstrated the generation of 14 ps optical pulses at a 15 GHz repetition rate by driving an EA modulator with a high power sine wave signal [33]. A second modulator can be added in series with the pulse generator to encode data onto the stream of short optical pulses, for short pulse RZ data transmission. When these two modulators are monolithically integrated the total insertion loss is substantially better than for two isolated devices in series, but it is still around 12 dB [34]. To overcome this insertion loss a semiconductor optical amplifier can be integrated on chip with the tandem EA modulators. In this case up to 14 dB of fiber-to-fiber gain has been demonstrated [35]. An example of a tandem EA with an integrated SOA from Agere Systems is shown in Fig. 7.36 [36]. This device has spot size converters on the input and output to reduce the coupling loss. The input light is coupled into the pulse carver first, which is driven with a single frequency sine wave at the bit rate. This carves out a periodic train of short optical pulses. The pulses are passed through the SOA and then to the data encoder where the data modulation is imposed on the pulse stream. The SOA is placed in between the carver and the encoder because the pulse carver has a high insertion

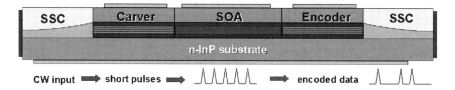

Fig. 7.36 Cross section of tandem EA modulator with integrated SOA and spot size converters.

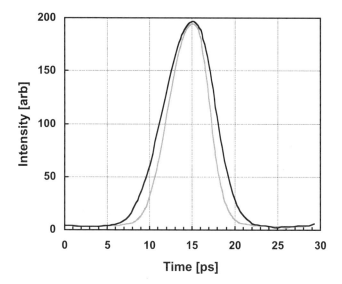

Fig. 7.37 Streak camera measurement of short pulse generation using a SGDBR.

loss and the periodic pulse stream can be amplified by the SOA without substantial distortion. In fact, pulses as short as 5.5 ps with extinction ratios greater than 20 dB can be produced by this device [37].

A streak camera measurement of the pulse shape for a tandem EA driven with a 40 GHz sine wave is shown in Fig. 7.37. The two pulses shown in this graph correspond to pulse widths of 7 ps and 5.5 ps. Both pulses have more than 20 dB of extinction. The wider pulse width is obtained with a −3 V bias and a 4Vpp sine wave drive signal. The shorter pulse required a −4 V bias and a 6Vpp drive signal. This ability to tune the pulse width by adjusting the drive conditions is a substantial advantage of EA modulators over other types of short pulse sources.

The large drive voltage swing required for the generation of very short pulses would seem to be a substantial problem for 40 GB/s applications.

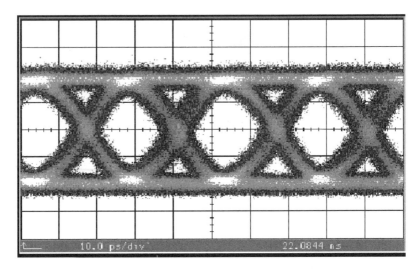

Fig. 7.38   40 GB/s NRZ eye diagram.

However it is important to remember that the drive signal for the pulse carver is a single frequency, and narrowband high power amplifiers are readily available even at 40 GHz. The broadband drive is more difficult and typical 40 GB/s drivers are only capable of 2.5 to 3Vpp swing. The data encoder side of the tandem EA discussed here is able to achieve more than 12 dB of dynamic extinction at this drive level, which is sufficient for most transmission applications. An NRZ eye diagram for the Tandem EA with just the data encoder being driven is shown in Fig. 7.38.

When the narrowband drive signal is applied and the phase is adjusted so that the pulse is centered at the middle of the data encoder eye the result is the RZ eye pattern shown in Fig. 7.39. The ringing in the bottom rail is caused by the limited bandwidth response of the photodetector. The advantage of the RZ transmission format is that it has a shorter dispersion length. This enables higher launch powers in the fiber without introducing nonlinear effects. The higher launch powers enable the signal to tolerate more optical signal-to-noise ratio degradation and thereby traverse a longer transmission span.

One drawback of the RZ transmission format is that it requires a larger frequency spectrum when compared to an NRZ transmission format. This limits the total fiber capacity by requiring increased spacing between adjacent channels in a DWDM system. A comparison of the optical spectrum for an RZ data stream and an NRZ data stream in shown in Fig. 7.40.

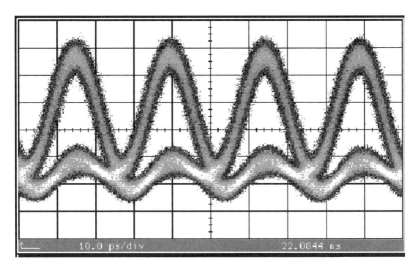

Fig. 7.39 RZ Eye diagram at 40 GB/s.

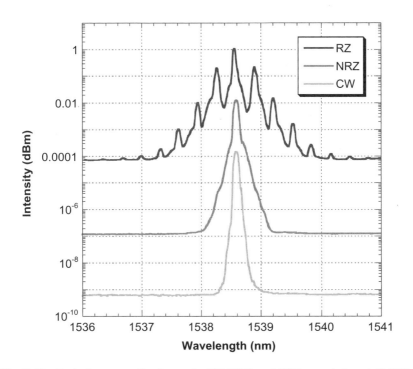

Fig. 7.40 Optical spectrum for short pulse RZ, NRZ, and CW transmission at 40 GB/s.

Note the strong sidebands in the RZ signal spaced at 40 GHz intervals. For the NRZ signal there is no energy in the spectrum at the frequencies corresponding to multiples of the bit rate.

The ability to generate very short optical switching windows is not only useful for RZ data transmission at a bit rate corresponding to the pulse interval, it also can be used as an enabling technology for optical time domain multiplexing and demultiplexing of signals to much higher bit rates than could otherwise be obtained with electronics. At the transmitter end 40 GB/s short pulse RZ sources can be optically multiplexed up to 160 GB/s or higher with each successive bit delayed by 6.25 ps so that it occupies a different time slot [38]. The higher the bit rate the shorter the pulse width required to prevent interference between adjacent bits. At the receiver end the signal can be demultiplexed back down to a desired lower bit rate using two cascaded EA modulators as gating elements. One EA is driven at the bit rate and the other is driven at twice the bit rate to create a narrower switching window. The EA drive signals are synchronized with the recovered clock at the receiver and a phase shifter can be used to select alternate bit streams. A schematic of a demultiplexer is shown in Fig. 7.41.

Because electronic clock recovery can be very difficult at these frequencies, the EA modulators can also be used in conjunction with a narrowband filter element to provide an optoelectronic clock recovery circuit. For the clock recovery circuit the output from the receiver would be fed to a narrow band high Q filter [39]. The output signal from this would be amplified to drive the first EA, and then doubled and amplified again to drive the second. When the narrowband drive signals are in phase with the incoming data stream the signal amplitude is maximized and the clock remains locked.

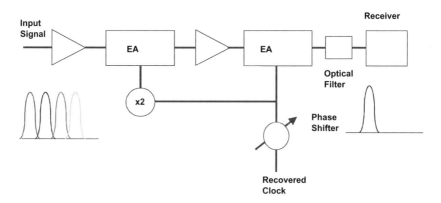

Fig. 7.41   EA modulator based OTDM demultiplexer.

7.7. Summary

The wide variety of research on current and future applications for electroabsorption modulators is indicative of the critical role that they play in optical communication systems. Their ability to transmit information with high fidelity over long spans of optical fiber has led to their widespread use in high bit rate fiber links. One of the greatest benefits of EA modulators is the ease with which they can be integrated with other components. This has played a large part in the commercial success of EA modulators. EA devices have been integrated with both fixed wavelength and tunable lasers for compact and efficient transmission sources. They have also been integrated with other EA modulators and with semiconductor optical amplifiers to perform more complicated functions such as short pulse RZ transmission, clock recovery, and optical time division demultiplexing. The compact size, low drive voltage requirement, and capability for ultrahigh bandwidth are making EAs a popular choice for many next-generation high-speed systems. EA modulator performance is constantly pushing the envelope with lumped element devices showing bandwidths greater than 50 GHz and traveling wave devices having the capability to push well beyond this. The future for EA modulators is clear, as bandwidth and extinction ratio requirements increase, traveling wave devices will become more prolific, and as network functionality becomes more complicated the focus on devices will increasingly shift to higher and higher levels of monolithic integration. The EA modulator will be a key building block in this integration, just as it has been a key building block in the evolution of the global photonic networks that exist today.

References

1. B. Knupfer *et al.*, IEEE Photonic. Technol. Lett., 5:12 (1993) 1386.
2. L. Coldren and S. Corzine, Diode Lasers and Photonic Integrated Circuits, (John Wiley and Sons, 1995).
3. B. Mason, European Conference on Optical Fiber Communication, ECOC 2000, Munich (2000).
4. C. W. Clark, Airy Functions: Physics Applications (NIST Digital Mathematical Library).
5. P. Debernardi and P. Fasano, IEEE J. Quantum Electron., 29:11 (1993) 2741–2755.
6. B. Jonnson and S. T. Eng, IEEE J. Quantum Electron., 26:11 (1990) 2025–2035.

7. A. K. Gathak *et al.*, IEEE J. Quantum Electron., 24:8 (1988) 1524–1531.
8. J. Singh, Physics of Semiconductors and Their Heterostructures (McGraw-Hill, 1993).
9. G. Agrawal, Fiber Optic Communication Systems (John Wiley and Sons, 1992).
10. A. F. Elrefaie, IEEE J. Lightwave Technol., 6:5 (1988) 704–709.
11. N. Yoshimoto *et al.*, Electron. Lett., 33:24 (1997) 2045–2046, 20.
12. G. Fish *et al.*, Topical Meeting OSA Trends in Optics and Photonics Series, 32 (2000) 17–19.
13. Y. Furushima *et al.*, Electron. Lett., 34:8 (1998) 767–768.
14. M. Bachmann *et al.*, Electron. Lett., 32:22 (1996) 2076–2078.
15. D. Derikson, ed., Fiber Optic Test and Measurement (Prentice Hall, 1998).
16. R. J. Nuyts *et al.*, IEEE Photonic. Technol. Lett., 9:4 (1997) 532–534.
17. Koyoma and Iga, J. Lightwave Technol., LT-6 (1988).
18. Devaux *et al.*, J. Lightwave Technol., 11:12 (Dec. 1993).
19. Dorgeuille and Devaux, J. Quantum Electron., 30:11 (Nov. 1994).
20. R. A. Linke, IEEE J. Quantum Electron., QE21 (1985) 593–597.
21. Y. Kawamura *et al.*, IEEE J. Quantum Electron., QE-23:6 (1987) 915–918.
22. A. Ramdane *et al.*, Electron. Lett., 30:23 (1994) 1980–1981.
23. T. Tanbun-Ek *et al.*, J. Crystal Growth, 145 (1994) 902–906.
24. S. O'Brien *et al.*, Appl. Phys. Lett., 58 (1991) 1363–1365.
25. N. Soda *et al.*, Electron. Lett., 26 (1990) 9–10.
26. S. Oshiba *et al.*, Electron. Lett., 29 (1993) 1528.
27. Y. K. Park *et al.*, IEEE Photonic. Technol. Lett., 8:9 (1996) 1255–1257.
28. H. Takeuchi, Indium Phosphide and Related Materials, 13th IPRM, Nara, Japan, paper WA3-1 (2001).
29. L. J. P. Ketelsen *et al.*, OFC 2000, Baltimore, paper PD14 (2000).
30. B. Mason *et al.*, IEEE Photonic. Technol. Lett., 12:7 (2000) 762–764.
31. B. Mason *et al.*, IEEE Photonic. Technol. Lett., 11:6 (1999) 638–640.
32. G. Li *et al.*, IEEE Trans. Microwave Theory and Techniques, 47:7 (1999) 1177–1183.
33. M. Suzuki *et al.*, Electron. Lett., 28 (1992) 1007–1008.
34. H. Tanaka *et al.*, Electron. Lett., 29:11 (1993) 1002–1004.
35. F. Devaux *et al.*, IEEE Photonic. Technol. Lett., 8:2 (1996) 1–3.
36. A. Ougazzaden *et al.*, Optical Fiber Communication Conference, OFC 2001 (2001).
37. B. Mason *et al.*, IEEE Photonic. Technol. Lett., (Jan. 2002).
38. B. Mikkelsen *et al.*, European Conference on Optical Fiber Communication, ECOC 1999, Nice (1999).
39. D. T. K. Tong *et al.*, Electron. Lett., 36:23 (2000) 1951–1952.
40. V. Swaminathay and A. T. Macrander, Materials Aspects of GaAs and InP Based Structures (Prentice Hall, 1991).

Part 3 | Photodetectors

Chapter 8 | P-I-N Photodiodes

Kenko Taguchi

Optoelectronic Industry and Technology Development Association,
Sumitomo Edogawabashiekimae Bldg., 7F,
20-10, Sekiguchi 1-Chome, Bunkyo-ku, Tokyo, 112-0014, Japan

8.1. Introduction

Higher transmission capacity in both trunk lines and access networks based on silica optical fiber is increasingly needed. To meet this need, high-performance photoreceivers and light sources must be developed, especially for use in actual WDM systems. The development of high-performance Er-doped fiber has enabled achievement of the highest over-all receiver sensitivity ever reported in the 1.55-μm wavelength region in systems using InGaAs PIN photodiodes with Er-doped fiber amplifiers (EDFAs).

Semiconductor photoreceivers based on InP materials are the device of choice for use in long-wavelength optical-fiber communication systems. Mixed compounds, such as InGaAs(P) and In(Al)GaAs lattice-matched to InP, can detect long-wavelength light, especially that with a nondispersion wavelength of 1.3 μm and a loss-minimum wavelength of 1.55 μm in silica optical fibers. Because $In_{0.53}Ga_{0.47}As$ (hereinafter referred to as InGaAs) lattice-matched to InP can detect all light emitted by InGaAs(P) and In(Al)GaAs materials lattice-matched to InP, this material is most widely used as a light absorber in optical-fiber communications. The characteristics of InP-based photodetectors are superior to those of conventional photodiodes composed of elemental Ge, which was the only material used for wavelengths below 1.5 μm. By using a heterostructure that

317

WDM TECHNOLOGIES: ACTIVE
OPTICAL COMPONENTS
$35.00

hasn't been expected until recently in group-IV elemental semiconductors, such as Si and Ge, new concepts and new device design for high-performance photodetectors have been developed. It has been found, for example, that the absorption region can be confined to a limited layer, and the wide-bandgap layer can serve as a transparent layer for specific communication wavelengths. Also, a heterojunction layer structure enabling a high speed with a high optical-to-electronic conversion efficiency, or quantum efficiency, and easier coupling to optical fiber has been analyzed experimentally.

Because photodiodes operate under a reverse bias, high-quality semiconductor layers are needed. To obtain photodiodes that operate at a low bias and have a low dark current, it is necessary to produce epitaxial layers that are pure and that have few defects, such as dislocations, point defects, and impurity precipitates. Fabrication and processing technologies, such as impurity diffusion, passivation, and metalization of ohmic contacts, will play an important role in the production of reliable high-performance photodetectors.

The performance of photodetectors can often be evaluated in terms of three main characteristics: responsivity, noise, and bandwidth. For practical use in systems, photodetectors must also be highly reliable and inexpensive. With regard to noise, there is a limitation on the minimum signal level needed to achieve a signal error rate that can be connected to the ratio of signal-to-noise (S/N). This chapter describes PIN-type photodiodes (PDs) composed mainly of InGaAs as a light-absorption layer with no internal gain. Avalanche photodiodes (APDs) with an internal gain are described in the next chapter. Section 8.2 describes the basic device parameters needed to obtain a large S/N ratio in receiver circuits and discusses the concept, design, and expected performance of photodetectors. Section 8.3 describes frequency response measured for different layer structures and light penetrations. Response limitations in InGaAs/InP are also evaluated. In Section 8.4, dark-current characteristics both theoretically estimated and experimentally obtained for an InGaAs p^+n junction are discussed. The dark-current reduction is one of the most important factors responsible for a large S/N. In Section 8.5, we introduce several typical detectors that are applicable to WDM systems and that are important to future systems. These include matured basic InGaAs PIN-PDs, highly efficient high-speed waveguide PIN-PDs, uni-traveling carrier PDs, highly efficient PDs easily coupled to optical fibers, and PDs responsible for high-input power handling. A summary is provided in Section 8.6.

8.2. Basic Photodiode Concepts, Design, and Requirements for Use in Optical Fiber Communications

Photodiodes used in receiver circuits must be reliable, must efficiently translate optical signals into electrical signals, and must be able to receive data transmitted through lightwave systems. In this section, we describe the concepts and operation of photodiodes. The photodiode design and device parameters required for practical use in fiber communications are discussed on the basis of a simplified theoretical analysis of receiver sensitivity.

8.2.1. ABSORPTION COEFFICIENT

The fundamental mechanism behind the photodetection process is light absorption. Light absorption can be expressed using an absorption coefficient. The absorption coefficient is defined as follows. When an incident light with optical power P_{i0} penetrates the surface of an absorbing media, as shown in Fig. 8.1, lost optical power $-dP_i(x)$ in region dx at position x from the absorbing media surface can be expressed as being proportional to optical power $P_i(x)$ at both x and dx by using proportional constant $\alpha(\text{cm}^{-1})$, which is the absorption coefficient.

$$-dP_i(x) = \alpha P_i(x)dx,$$
$$P_i(x) = P_{i0}e^{-\alpha x}. \tag{8.1}$$

The absorption length, or the penetration depth, is equal to $1/\alpha$, which is the optical power level position decreased to $1/e$ (about 37%) of the input power of P_{i0} or the light penetration depth point absorbed with a constant P_{i0}.

Near the fundamental absorption edge (band-to-band transition) or the bandgap wavelength of $\lambda_g = hc/Eg$, of a semiconductor material, the absorption coefficient can be expressed as $\alpha \sim (h\nu - Eg)^\gamma$, where h is the Planck's constant, c is the velocity of light in a vacuum, $h\nu$ is the photon energy, Eg is the bandgap of the material, and γ is a constant, that is, respectively, $1/2$ and 2 for the allowed direct and indirect transition [1], as shown in Fig. 8.2, where an exaggerated band structure and transmissions are illustrated for the different mechanisms. In the indirect transition, lattice phonons are needed to conserve the momentum. Therefore, in general, direct-transition materials have steeper and larger absorption coefficients

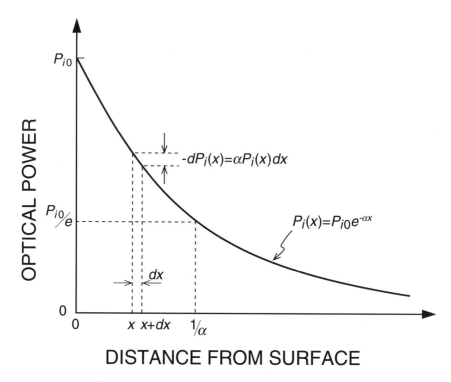

Fig. 8.1 Light attenuation within an absorbing media.

near the absorption edge than do indirect materials. As a result, from the design point of view, direct-transition materials with a bandgap narrower than the bandgap corresponding to the objective wavelength are more applicable to photodetectors.

Figure 8.3 shows the dependence of the absorption coefficient on the wavelength for different semiconductors used in photodetectors. Absorption has asymptotic behavior near bandgap wavelength λ_g where the material is transparent, and there is a strong dependence of the absorption on the wavelength shorter than λ_g. Silicon and Ge have indirect bandgaps. The absorption-coefficient dependence of Ge on the wavelength is fairly similar to that of direct-transition materials because of the narrow bandgap and narrow displacement between Γ and L in the momentum space. However, Ge is not very effective in detecting light with a loss-minimum wavelength of 1.55 μm. This is because a depleted absorption region of at least a few tenths of micron is needed to obtain a 90% optical-to-electronic conversion

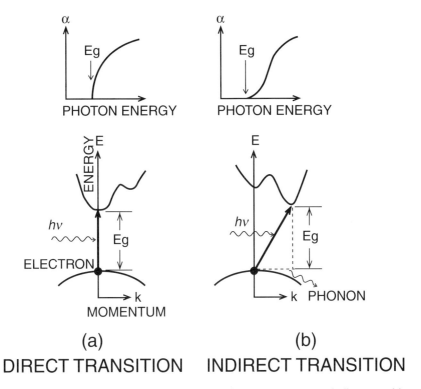

(a)
DIRECT TRANSITION

(b)
INDIRECT TRANSITION

Fig. 8.2   Optical transitions in semiconductors: (a) direct transition, (b) indirect transition including phonons.

efficiency. In contrast, a 3–4 μm InGaAs layer can enable high conversion efficiency in the 1.3–1.55 μm wavelength region.

8.2.2.   PHOTODIODE OPERATION

Photodiodes operate under a reverse bias to create a depleted region in which photogenerated electron-hole pairs are separated and swept across the semiconductor, generating a flow of electric current. Figure 8.4(a) shows a cross section of a photodiode with a p^+-n-n^+ structure. It also shows optical absorption, or photocarrier generation, which depends on the absorption coefficient, α, of the material for the incident light and decreases exponentially with an increase in the distance from the diode front p^+-region. Figure 8.4(b) and (c) show, respectively, the electric-field distribution and energy band. Most photocarriers are designed for use in a fully

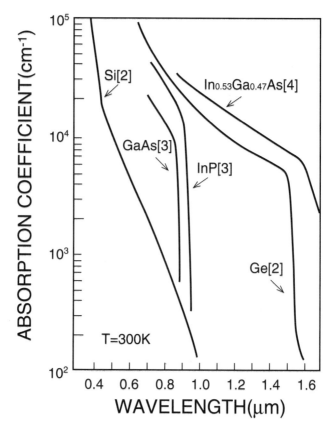

Fig. 8.3   Dependence of absorption coefficient on wavelength in different semiconductors [2–4].

depleted n-region so that they had a high-speed response: electrons and holes generated within the depleted region are instantaneously separated by an electric field and drift in the opposite direction, inducing a photocurrent in the external circuit. At the same time, minority carrier holes excited within the average diffusion length in the undepleted n^+(or n)-region adjacent to the depleted region diffuse into the edge of the depleted junction with some recombination and are collected across the high-field region, which results in a diffusion photocurrent in the external circuit. The diffusion photocurrent is generally characterized by a slow response to the optical signal, because the speed of the response depends on the time it takes the photogenerated minority carriers to diffuse from where they are generated in the neutral undepleted region into the edge of the depleted

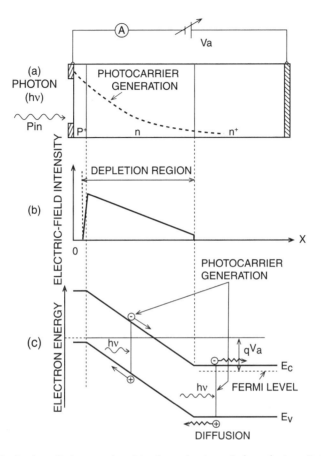

Fig. 8.4 Basic photodiode operation: (a) schematic view of p$^+$-n-n$^+$ photodiode under a reverse bias; (b) electric-field distribution; (c) energy-band diagram.

region. These photoresponse-frequency characteristics will be discussed in greater detail in the following section. Photodiodes should, therefore, be designed in such a way as to minimize optical absorption in the undepleted neutral region as much as possible. For the same reason as well as to reduce the recombination loss of photocarriers generated in the p$^+$-region on the front side of the diode, the p$^+$-region must be as thin as possible.

When the electric field of a diode is elevated to several hundreds of kilovolts per centimeter by increasing the reverse bias, an internal gain for the primary photocurrent can be obtained. This gain is a result of the electron-hole pair creation avalanche process initiated by the photogenerated carriers, which is governed by the relationship between the strength of the electric field and the electron-hole impact ionization rates of the material itself, which will be discussed in the next chapter.

Based on the preceding discussion, the depletion region must be as large as possible to suppress the slow photocurrent created in the neutral region and enable a high conversion efficiency. For instance, for an objective wavelength to enable a conversion efficiency as high as 95%, the depletion layer of inversely about three times the absorption coefficient must be 40–50 μm for Si for a 0.85 μm wavelength as well as 30 μm for Ge and 6 μm for InGaAs for a 1.55 μm wavelength. Figure 8.5 shows the relationship between the junction capacitance in the units of area C_j/S and donor concentration N_D as a function of bias voltage V_a including the built-in voltage for a p⁺n one-side abrupt junction having a dielectric constant, ε_0, of 12.

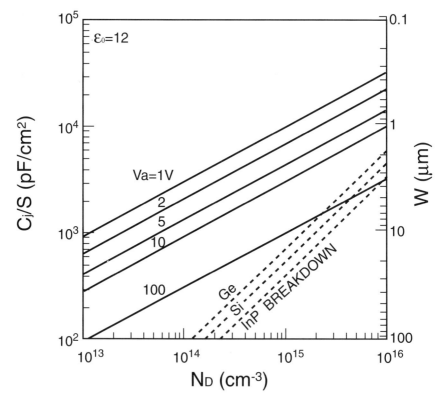

Fig. 8.5 p⁺n-junction capacitance vs. donor concentration at different bias voltages.

This is based on the following relationships:

$$V_a = \left(\frac{q}{2\varepsilon}\right) N_D W^2, \tag{8.2}$$

$$C_j = S\sqrt{q\varepsilon N_D/2V_a} \tag{8.3}$$

where q is the electric charge, ε is the permittivity, N_D is the donor concentration, W is the depletion-layer width, and S is the diode area. For instance, at an operating voltage of -5 volts, the concentration for Si must be smaller than 1×10^{13} cm^{-3} and that for InGaAs must be smaller than 2×10^{14} cm^{-3}. This is why a highly purified absorption layer is needed for photodiodes.

8.2.3. QUANTUM EFFICIENCY

Quantum efficiency η is defined as a ratio of the electron-hole pair-generation rate contributed to photocurrent I_p to the photon incidence rate.

$$\eta = (I_p/q)/(P_{in}/h\nu) \tag{8.4}$$

where P_{in} is the incident optical power. Responsivity defined as a ratio of I_p to P_{in} in the units of A/W is also usually used to evaluate the conversion efficiency. External overall efficiency depends on the reflection of incident light on the photodiode surface. By using reflection rate R, optical power P_{i0} in the photodetector can be expressed as

$$P_{i0} = (1 - R)P_{in} \tag{8.5}$$

To minimize the reflection, antireflection dielectric film is usually over-arrayed on the semiconductor surface. The antireflection coating film thickness is tailored to the objective wavelength where the thickness is set to the wavelength divided by four times the coating-film refractivity. When the absorption layer is fully depleted, the quantum efficiency can be approximated by

$$\eta = (1 - R)[1 - \exp(-\alpha W)] \tag{8.6}$$

where α is the absorption coefficient of the light and W is the thickness of the absorption layer.

Figure 8.6 shows the typical spectral external quantum efficiency of commercially available photodiodes. There is a quantum efficiency cut-off, or a decrease, at long wavelengths, corresponding to each absorption

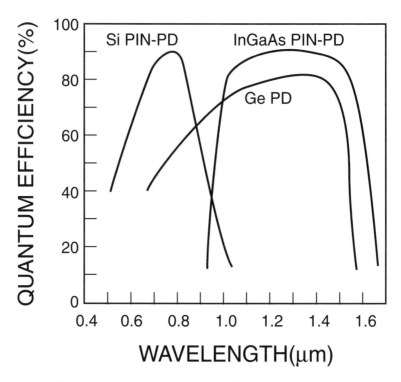

Fig. 8.6   Spectral external quantum efficiency of photodiodes.

edge, and the short wavelength side decrease in the efficiency is due to the recombination loss of the photogenerated carriers in the surface high-doped region in elemental photodiodes or in the wide-bandgap capping layer used in heterostructures. In InGaAs PIN photodiodes discussed following, a high efficiency can be obtained for a wide wavelength range by using a heterostructure with a wide bandgap cap and contact layers adjacent to the absorption layer.

8.2.4.   EQUIVALENT CIRCUIT AND RC TIME CONSTANT

Following the schematization commonly used in electronics, the small-signal behavior of a photodiode can be described by using an equivalent circuit as shown in Fig. 8.7, to which external load resistance R_L is connected. Here, C_L is the equivalent capacitance of the load including the output terminal parasitic capacitance. The signal is given by the current generator driven by photogenerated current $I_p(\omega)$, in parallel with the

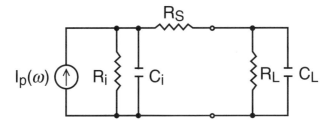

Fig. 8.7 Equivalent circuit of a photodiode and a load.

internal capacitance, C_i, that takes into account the junction capacitance and packaging capacitance. R_i is the internal resistance, or the dispersion resistance, that takes into account finite conductance dI/dV of the junction. The series resistance, R_s, is due to the ohmic contacts and the undepleted bulk resistance.

The output electrical power, $P_{out}(\omega)$, as a function of frequency obtained in the circuit can be expressed as

$$P_{out}(\omega) = \frac{I_p(\omega)^2 R_L}{1 + \omega^2 R_{eq}^2 (C_i + C_L)^2} \qquad (8.7)$$

where $R_{eq} = R_i R_L / (R_i + R_L)$ when $R_s \ll R_L$. Then, the electrical RC cut-off frequency (3-dB-down bandwidth) can be defined as

$$f_c(\text{RC}) = 1/(2\pi(C_i + C_L)R_{eq}). \qquad (8.8)$$

In Eq. (8.7), the maximum output power can be obtained under the condition of $R_L = R_i$. Because the internal resistance, R_i, is usually very high, typically 1–100 MΩ, R_{eq} can be approximated to be R_L. Then, for a low capacitance of $C_L + C_i$ and a high speed of $I_p(\omega)$, the load resistance, R_L, must be as large as possible to satisfy Eq. (8.8) in which $f_c(\text{RC})$ is equal to or slightly greater than the objective bandwidth, in order to enable a highly sensitive detection. This is because the receivers composed of a low-capacitance PIN-PD and high-impedance FET amplifiers are used for bit-rate systems of less than a few hundred Mb/s [5].

8.2.5. NOISE AND RECEIVER SENSITIVITY

Receiver sensitivity has been analyzed for a variety of signal waveforms by Personick [6] and Smith [7]. The present discussion of sinusoidal optical signal detection with a PD is simplified, focusing on the receiver circuit combined with load resistance R_L followed by a preamplifier with equivalent input noise F_{amp}. For a sinusoidal signal with a full modulation depth, the mean-square signal current is given by

$$\langle i_p^2 \rangle = I_p^2/2, \tag{8.9}$$

where I_p is the photocurrent transferred from the optical signal to the electrical signal. The relationship between the input optical signal and the photocurrent is give Eq. (8.4).

The noises in the circuit are shot noise and circuit noise (including the following preamplifier noise). The shot noise is due to the diode dark current, I_d, and photocurrent, I_p. The total mean-square shot-noise current is thus

$$\langle i_s^2 \rangle = 2q(I_p + I_d)B, \tag{8.10}$$

where B is the objective bandwidth. The receiver circuit noise can be simplified to the circuit thermal noise including the following preamplifier noise, F_{amp}, as follows:

$$\langle i_{sc}^2 \rangle = 4kT F_{amp} B / R_{eq}, \tag{8.11}$$

where R_{eq} is the equivalent circuit resistance usually represented by the load resistance, R_L, described previously, k is the Boltzman constant, and T is the absolute temperature. The thermal noise can be further described by using the shot noise due to the preamplifier FET gate leak current and its channel noise [8].

The signal-to-noise ratio (S/N) in the circuit can thus be expressed as

$$S/N = \langle i_p^2 \rangle / (\langle i_s^2 \rangle + \langle i_{sc}^2 \rangle)$$
$$= (I_p)^2 / \{2[2q(I_p + I_d)B + 4kT F_{amp} B / R_L]\} \tag{8.12}$$

The minimum optical signal power, P_{min}, required for a given S/N ratio can be calculated from Eq. (8.12). If the dark current, I_d, is negligibly small, P_{min} is limited by the thermal noise to

$$P_{min} = [h\nu/(q\eta)](S/N)^{1/2}(8kT F_{amp} B / R_L)^{1/2} \tag{8.13}$$

and $P_{min}(S/N = 1)$ is equal to the thermal noise power. Then, the noise equivalent power, NEP, can be expressed by $P_{min}(S/N = 1)$ for a unit of frequency as $P_{min}(S/N = 1)/B^{1/2}$ (W/Hz$^{1/2}$). This is one of the performance indices of photodetectors. Based on this, the dark-current reduction is one of the most important factors in high-performance photodetectors, especially at low bit rates.

8.3.   Frequency-Photoresponse Calculations

The main factors limiting the photodiode response speed are the carrier diffusion and drift (transit) time of the photogenerated carriers and the diode capacitance (RC time constant in the circuit). An AC analysis of the photogenerated diffusion current was carried out earlier by Sawyer and Rediker [9], and a response analysis of the photo-excited drift current for PIN-PD depletion layers was done by Lucovsky et al. [10]. In this section, the general expression of frequency response in the pn-junction is analyzed to clarify the photo-response phenomena.

The carrier behavior in semiconductors, especially the carrier deviation from the thermal-equilibrium condition under the influence of external conditions, is basically governed by the current-density conditions and continuity equations.

8.3.1.   FREQUENCY RESPONSE FOR PHOTOGENERATED DRIFT CURRENT

For a PIN-type photodiode shown in Fig. 8.8, the following conditions are assumed: the I-layer is fully depleted, resulting in a constant electric field in the I-region; the carrier recombination is neglected in the I-region; Φ_0 is the incident photon density in the unit area on the p-side surface.

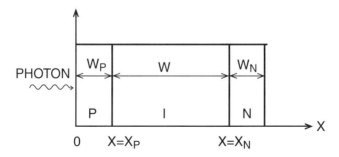

Fig. 8.8   Calculation model for a PIN photodiode.

Under these conditions, the continuity equations can be written as follows:

$$\frac{\partial n_p}{\partial t} = \frac{1}{q}\frac{\partial J_n}{\partial x} + g \tag{8.14}$$

$$\frac{\partial p_n}{\partial t} = -\frac{\partial J_p}{\partial x} + g \tag{8.15}$$

where n_p and p_n are, respectively, the minority electron- and the hole-carrier density. The normalized generation rate, g, of an electron-hole pair is given by

$$g = \alpha \Phi_0 \exp(-\alpha x) \tag{8.16}$$

The current density for the axis shown in Fig. 8.8 is a negative value so that the current is defined as $J_n = -q v_n n$ and $J_p = -q v_p p$. Here, v_n and v_p are, respectively, the electron drift velocity and the hole drift velocity. The continuity equations for an AC solution are as follows:

$$\frac{\partial J_n}{\partial x} + j\frac{\omega}{v_n}J_n = -q\alpha\Phi_0 m e^{-\alpha x}e^{j\omega t} \tag{8.17}$$

$$\frac{\partial J_p}{\partial x} - j\frac{\omega}{v_p}J_p = q\alpha\Phi_0 m e^{-\alpha x}e^{j\omega t} \tag{8.18}$$

where J_n and J_p are, respectively, the electron current density and the hole current density. It is assumed that the intensity of the incident light is modulated by a function of $\exp(\omega t)$ with modulation degree m. The electron current density can be obtained from Eq. (8.17) by using a boundary condition of $J_n = 0$ at $X = X_p$.

$$J_n(x,t) = \frac{q\Phi_0 m e^{j\omega t}}{\left(1 - j\frac{\omega}{\alpha v_n}\right)}\left(e^{-\alpha x} - e^{-j\frac{\omega}{v_n}(x-x_p)-\alpha x_p}\right) \tag{8.19}$$

The hole current can also be obtained from Eq. (8.18) by using a boundary condition of $J_h = 0$ at $X = X_n$.

$$J_p(x,t) = \frac{-q\Phi_0 m e^{j\omega t}}{\left(1 + j\frac{\omega}{\alpha v_p}\right)}\left(e^{-\alpha x} - e^{-\alpha x_n}e^{j\frac{\omega}{v_p}(x-x_n)}\right) \tag{8.20}$$

Based on Eqs. (8.19) and (8.20), the total current density is obtained by averaging the currents in the depletion region.

$$J_{ex}(drift) = \frac{1}{W}\int_{x_p}^{x_n}\{-(J_n + J_p)\}dx \tag{8.21}$$

Then, the total current density can be expressed as

$$J_{ex}(drift) = q\Phi_0 m e^{j\omega t} F_{drift}(\omega) \tag{8.22}$$

where $F_{drift}(\omega)$ is the transit-time frequency response. The electron-drift transit-time frequency response, $F_{n-drift}(\omega)$, can be expressed as

$$F_{n-drift(\omega)} = \frac{e^{-\alpha x_p}}{\left(1 - j\frac{\omega}{\alpha v_n}\right)} \left[\frac{1 - e^{-j\omega\frac{W}{v_n}}}{j\omega\frac{W}{v_n}} - \frac{(1 - e^{-\alpha W})}{\alpha W} \right] \tag{8.23}$$

In Eq. (8.23), the limit case of $\alpha \to \infty$ and $X_p = 0$ corresponds to the transit-time response of the electron carriers injected into the depletion edge. This can be expressed as

$$F(\omega)_{n-drift}_{\alpha \to \infty} = \frac{1 - e^{-j\omega\frac{W}{v_n}}}{j\frac{W}{v_n}} \tag{8.24}$$

Here, the injected electron carriers are electron carriers that have reached the depletion edge from the photogenerated position in the neutral p-region. This is also the frequency response added to the diffused-electron photocurrent-injected edge from the adjacent neutral region as an external circuit.

In photodiodes with a back-illuminated structure (incident on the n-type semiconductor surface), the frequency response can be expressed using $m\alpha\phi_0 e^{-\alpha(W_n+W+W_p)}e^{\alpha x}e^{j\omega t}$ instead of g.

8.3.2. DIFFUSION-CURRENT FREQUENCY RESPONSE

The frequency response of a photogenerated diffusion current is given by the minority-carrier continuity equations including photo-induced carrier-generation and recombination terms in the neutral semiconductor region. Because the electric field in the neutral region can be neglected assuming that there is a low carrier-injection level and no doped-impurity gradient, the equations can be expressed as

$$\frac{\partial n_p}{\partial t} = D_n \frac{\partial^2 n_p}{\partial x^2} - \frac{n_p - n_{p0}}{\tau_n} + \alpha\Phi_0 m e^{-\alpha x}e^{j\omega t} \tag{8.25}$$

$$\frac{\partial p_n}{\partial t} = D_p \frac{\partial^2 p_n}{\partial x^2} - \frac{p_n - P_{n0}}{\tau_p} + \alpha\Phi_0 m e^{-\alpha x}e^{j\omega t} \tag{8.26}$$

where D_n and D_p are, respectively, the carrier diffusion constant for the electron and that for the hole, which are the functions of mobility and are expressed using the Einstein relationships of $D_n = (kT/q)\mu_n$ and $D_p = (kT/q)\mu_p$ for semiconductors with a non-degradation condition. Here, n_p and p_n are, respectively, the minority electron density in p-type semiconductors and the minority hole density in n-type semiconductors; n_{p0} and p_{n0} are, respectively, the thermal equilibrium of the electron and hole density; and τ_n and τ_p are, respectively, the electron lifetime in p-type semiconductors and the hole lifetime in n-type semiconductors.

The minority-electron-carrier AC expression in a p-type semiconductor can be deduced from Eq. (8.25) using the boundary conditions as follows:

$$D_n \frac{\partial n_p(x,t)}{\partial x} = S_n n_p(x,t) \quad \text{at } x = 0, \tag{8.27}$$

$$n_p(x,t) = 0 \quad \text{at } x = x_p, \tag{8.28}$$

where S_n is the recombination velocity on the surface of the p-type semiconductor. The electron diffusion current induced into the depletion layer can thus be expressed taking into account the flow direction by the following equation:

$$J_n(diff) = -q D_n \frac{\partial n_p}{\partial x} \quad \text{at } x = x_p \tag{8.29}$$

As described previously, the external current caused by the diffusion current is also affected by the drift in the depletion region. Therefore, the external electron diffusion current can be expressed as

$$J_{ex}(n - diff) = \frac{qm\Phi_0 e^{j\omega t}}{\left(1 - \frac{1}{\alpha^2 L_n'^2}\right)} \left(\frac{1 - e^{-j\omega \frac{W}{v_n}}}{j\omega \frac{W}{v_n}} \right)$$

$$\times \left[\frac{\left(1 + \frac{S_n}{\alpha D_n}\right) - e^{-\alpha W_p}\left\{\left(1 + \frac{S_n}{\alpha D_n}\right)\cosh\left(\frac{W_p}{L_n'}\right) + \left(\frac{S_n L_n'}{D_n} + \frac{1}{\alpha L_n'}\right)\sinh\left(\frac{W_p}{L_n'}\right)\right\}}{\cosh\left(\frac{W_p}{L_n'}\right) + \frac{S_n L_n'}{D_n}\sinh\left(\frac{W_p}{L_n'}\right)} \right]$$

$$\tag{8.30}$$

where $L_n' \equiv L_n/(1 + j\omega\tau_n)^{1/2}$ and $L_n = (D_n\tau_n)^{1/2}$. For an ohmic contact on the surface ($x = 0$) of a p-side surface, S_n is an infinity, and for an ideal hetero-interface, S_n is zero.

The diffusion current due to the photo-excited hole carriers in the n-type neutral semiconductor region can be deduced by using the boundary conditions as follows:

$$D_p \frac{\partial p_n(x,t)}{\partial x} = -S_p p_n(x,t) \quad \text{at } X = W_p + W + W_n \quad (8.31)$$

$$p_n(x,t) = 0 \quad \text{at } X = W_p + W \quad (8.32)$$

8.3.3. FREQUENCY RESPONSE FOR InP/InGaAs/InP DOUBLE-HETEROSTRUCTURE PIN-PDs

In this section, we analyze the frequency response of the structure shown in Fig. 8.9 as a function of the bias voltage. We also analyze the incident light direction. In this section, our obtained results are summarized. In the calculations of the InP cap, buffer, and substrate, the photo-absorption was neglected and the drift effect was included only for the depleted region. The trap effect and the time-delay effect at the hetero-interfaces, which can

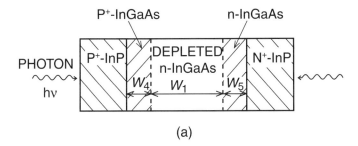

(a)

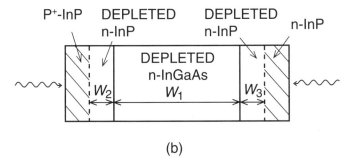

(b)

Fig. 8.9 Frequency-response calculation model for InP/InGaAs/InP double heterostructure: (a) depletion stayed in the absorption layer; (b) depletion spread out in the wide bandgap layers.

be easily evaluated by introducing a delay function, were not taken into account.

The light penetration on the surface of the p^+-InP side can be expressed as follows:

Hole drift current:

$$J_{ex}(p - drift) = \frac{mq\,\Phi_0 e^{j\omega t} e^{-\alpha W_4}\left(1 - e^{-j\omega\frac{W_2}{v_p(B)}}\right)}{j\omega\frac{W_2}{v_p(B)}}$$

$$\times \left\{ \frac{1}{\left(1 + j\frac{\omega}{\alpha v_p(T)}\right)} \left(\frac{e^{-\alpha W_1}\left(e^{-j\omega\frac{W_1}{v_p(T)}} - 1\right)}{j\omega\frac{W_1}{v_p(T)}} + \frac{(1 - e^{-\alpha W_1})}{\alpha W_1} \right) \right\}$$

(8.33)

where $v_p(B)$ and $v_p(T)$ are the hole drift velocities in the InP layer and InGaAs layer, and W_2 is zero when W_4 is not zero.

Electron drift current:

$$J_{ex}(n - drift) = mq\,\Phi_0 e^{j\omega t} e^{-\alpha W_4} \frac{\left(1 - e^{-j\omega\frac{W_3}{v_n(B)}}\right)}{j\omega\frac{W_3}{v_n(B)}}$$

$$\times \left\{ \frac{1}{\left(1 - j\frac{\omega}{\alpha v_n(T)}\right)} \left(\frac{\left(1 - e^{-j\omega\frac{W_1}{v_n(T)}}\right)}{j\omega\frac{W_1}{v_n(T)}} - \frac{(1 - e^{-\alpha W_1})}{\alpha W_1} \right) \right\}$$

(8.34)

where W_3 is zero when W_5 is not zero.

Hole diffusion current:

$$J_{ex}(p - diff) = mq\,\Phi_0 e^{j\omega t} e^{-\alpha(W_4 + W_1)} \left\{ \frac{1 - e^{-j\omega\frac{W_2}{v_p(B)}}}{j\omega\frac{W_2}{v_p(B)}} \right\}$$

$$\times \left\{ \frac{1 - e^{-j\omega\frac{W_1}{v_p(T)}}}{j\omega\frac{W_1}{v_p(T)}} \right\} \left\{ \frac{1}{\left(1 - \frac{1}{(\alpha L_p')^2}\right)} \right\}$$

$$\times \left\{ \frac{-\left(1 - \frac{S_p}{\alpha D_p}\right)e^{-\alpha W_5} + \left(1 - \frac{S_p}{\alpha D_p}\right)\cosh\left(\frac{W_5}{L_p'}\right) - \left(\frac{1}{\alpha L_p'} - \frac{S_p L_p'}{D_p}\right)\sinh\left(\frac{W_5}{L_p'}\right)}{\cosh\left(\frac{W_5}{L_p'}\right) + \frac{S_p L_p'}{D_p}\sinh\left(\frac{W_5}{L_p'}\right)} \right\}$$

(8.35)

where W_2 is zero when W_4 is not zero, and $L_p' \equiv L_p/(1 + j\omega\tau_p)^{1/2}$.

Electron diffusion current:

$$J_{ex}(n - diff)$$

$$= mq\,\Phi_0 e^{j\omega t} \left\{ \frac{1 - e^{-j\omega \frac{W_1}{v_n(T)}}}{j\omega \frac{W_1}{v_n(T)}} \right\} \left\{ \frac{1 - e^{-j\omega \frac{W_3}{v_n(B)}}}{j\omega \frac{W_3}{v_n(B)}} \right\} \left\{ \frac{1}{\left(1 - \frac{1}{(\alpha L_n')^2}\right)} \right\}$$

$$\times \left\{ \frac{\left(1 + \frac{S_n}{\alpha D_n}\right) - e^{-\alpha W_4}\left\{\left(1 + \frac{S_n}{\alpha D_n}\right)\cosh\left(\frac{W_4}{L_n'}\right) + \left(\frac{S_n L_n'}{D_n} + \frac{1}{\alpha L_n'}\right)\sinh\left(\frac{W_4}{L_n'}\right)\right\}}{\cosh\left(\frac{W_4}{L_n'}\right) + \frac{S_n L_n'}{D_n}\sinh\left(\frac{W_4}{L_n'}\right)} \right\}$$

$$(8.36)$$

where W_3 is zero when W_4 is not zero, and $L_n' \equiv L_n/(1 + j\omega\tau_n)^{1/2}$. From Eqs. (8.33–8.36), the external current is given by

$$J_{ex}(total) = J_{ex}(n - drift) + J_{ex}(n - diff) + J_{ex}(p - drift) + J_{ex}(p - diff)$$
$$= mq\,\Phi_0 e^{j\omega t} F_{ex}(\omega) \qquad (8.37)$$

Here, $F_{ex}(\omega)$ is the analytical solution of the objective frequency response, and $F_{ex}(0)$ is the quantum efficiency, which can be easily divided into a drift term and a diffusion term, based on Eqs. (8.33) + (8.34), and (8.35) + (8.36). The frequency response for the light penetration on an n^+-surface can be obtained in the same way.

8.3.4. FREQUENCY RESPONSE CALCULATIONS OF InGaAs PHOTODIODES AND ITS HIGH-SPEED LIMITATIONS

Photoresponse is usually calculated using a load resistance (usually $50\,\Omega$). As a result, the photo-induced current suffers from the RC time constant effect. The following equation is the expected frequency-response solution:

$$f_{ex}(\omega) = F_{ex}(\omega)/(1 + j\omega C_t R_{eq}) \qquad (8.38)$$

where C_t and R_{eq} are, respectively, the approximated diode capacitance and its load resistance.

The calculations were done for an InP/InGaAs/InP-structure diode with a p^+n junction diameter of $100\,\mu m$, an InGaAs absorption-layer concentration of $3 \times 10^{15}\,cm^{-3}$, and an absorption-layer depth of $3\,\mu m$ at a bias of -5 V including the built-in voltage and an additional stray capacitance of $C_{st} = 0.3$ pF (mounted on a TO18 can case). The junction front was formed on the InGaAs layer closely to the InP/InGaAs interface.

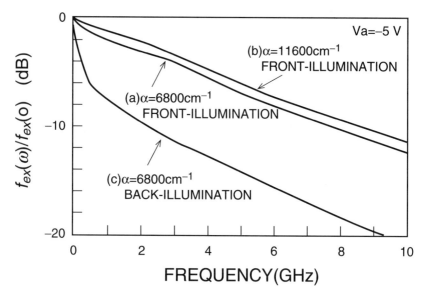

Fig. 8.10 Calculated frequency response: (a) for a 1.55 μm light; (b) for a 1.30 μm light; (c) n-side illumination with a 1.55 μm signal light.

The theoretical frequency response for the light penetration from the p^+-side with an absorption coefficient of $\alpha = 6800$ cm^{-1} [11] (equal to the wavelength of 1.55 μm) is shown in Fig. 8.10(a). In the calculations, we assumed that there was no p^+-InGaAs region, and a built-in voltage was added to the bias voltage. We found that the total capacitance was 0.9 pF and the 3-dB-down bandwidth was 2 GHz. The internal quantum efficiency was 87%, which consisted of a drift-term efficiency of 66% in the depleted 1.6 μm region and a diffusion-term efficiency of 21% from the neutral 1.4 μm region. The steep signal degradation in the low-frequency region was less than 0.3 GHz. Figure 8.10(b) shows the frequency response for a light with $\alpha = 116000$ cm^{-1} [11] (equal to a 1.3 μm light). Because of the large absorption coefficient compared to that of the 1.55 μm light, the effective hole drift distance decreased, resulting in a bandwidth of 2.8 GHz, which is higher than that for the 1.55 μm light. The expected quantum efficiency of 97% means that an InGaAs layer exceeding 3 μm is not needed for 1.3 μm wavelength signal detection. Figure 8.10(c) shows the frequency response of the back-illuminated (n^+-side illumination) structure for a signal with a 1.55 μm light. The response characteristics were strongly affected by the diffusion from the neutral region. However, the obtained quantum efficiency was the same as that for the front-illuminated structure, even though the diffusion-term efficiency was 63%. The diffusion length and lifetime used in the calculations were, respectively, 75 μm and 6×10^{-6} sec [12]. The drift velocities of the electrons and holes used were, respectively, $\upsilon_e = 6.5 \times 10^6$ cm/s and $\upsilon_h = 4.5 \times 10^6$ cm/s [13], and they were independent constants from the electric field.

8.3.5. BANDWIDTH LIMITATIONS IN InGaAs PHOTODIODES

Figure 8.11 shows the bandwidth calculated against the InGaAs absorption-layer thickness under full-depleted conditions in the InGaAs layer. The pn-junction diameter was treated as a parameter. The load resistance (equivalent-circuit resistance) was 50 Ω. For a given pn-junction diameter there was a maximum bandwidth. This is due to the effect of the diode capacitance, which is inversely proportional to the thickness of the depleted absorption layer. The cut-off frequencies limited solely by the transit time and free from the capacitance effect, for the back and front illumination, are shown in Fig. 8.11. At a given thickness, the speed was somewhat higher for the front illumination than it was for the back illumination. This is because the saturation velocity of electrons is higher than that of holes. It is obvious that a diode with a 100 μm diameter cannot operate at 10 GHz or above. To obtain a bandwidth exceeding 10 GHz for a 1.55 μm light, the InGaAs layer must be thinner than 3 μm for the front illumination and thinner than 2.5 μm for the back illumination.

Figure 8.12 shows the bandwidth calculated against the InGaAs donor concentration at a bias of −5 volts including the built-in voltage. The RC time-constant effect was neglected to observe the limitations. The InGaAs layer thickness, d_T, was treated as a parameter. For the InGaAs layers thinner than 1.5 μm, there was almost no bandwidth dependence on the concentration. In contrast, for $d_T > 2$ μm, the degradation became significant especially at the donor concentration higher than 5×10^{15} cm^{-3}. For $d_T = 3$ μm, a concentration lower than 3×10^{15} cm^{-3} is required to obtain a cutoff frequency of over 10 GHz. The figure shows the maximum bandwidth for a thin undepleted neutral InGaAs region. From the figure, we can conclude that when there is an undepleted thin region, the transit time decreases and there is almost no effect of the diffusion current due to the undepleted region on the overall bandwidth, compared to what happens under depleted conditions.

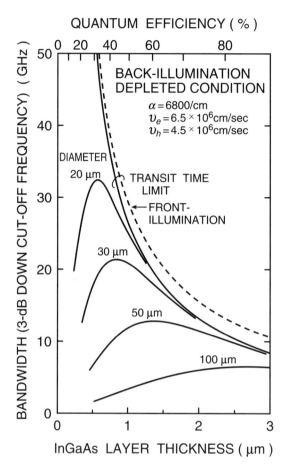

Fig. 8.11   Bandwidth vs. InGaAs absorption-layer thickness calculated for different p^+n-junction diameters.

8.4.   Current Transport in InGaAs p^+n-Junction

The dark current is one of the most important parameters in determining the receiver system S/N, as was discussed in Section 8.2. By evaluating the dark current and its temperature-dependent characteristics for an InGaAs p^+n junction, we can clarify the current transport process in the junction. The dark current can also be used to evaluate the composing crystallinity and processing technology employed. Based on this evaluation, the limitations on the dark current in InGaAs materials can be estimated. Forrest [14] and Takanashi [15] earlier attempted to analyze the exponentially increased

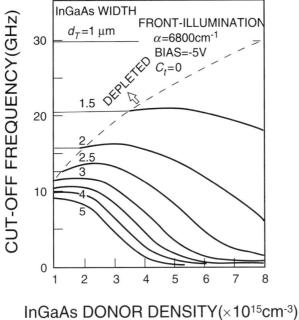

Fig. 8.12   Bandwidth vs. InGaAs donor concentration.

dark current in InGaAs p^+n-junctions as a band-to-band tunneling current. This exponentially increased current as a function of the reverse bias was characterized by a low mass of electrons in narrow bandgap materials. However, the obtained data were often explained qualitatively by using fitting parameters [16] or were deduced from the effective mass [17].

In this section, the tunneling current at a relatively high bias and the dark current under a low bias are analyzed separately. Based on the temperature-dependent characteristics, we attempted to separate the dark current into a generation and a diffusion current. We found that there is a residual dark current around the tunneling current in the explicitly increased region that cannot be divided into the two components.

8.4.1. InGaAs PIN-PD SAMPLE FABRICATION

A cross section of the diode we examined is shown in Fig. 8.13. The layer structure was fabricated by using hydride vapor phase epitaxy. The resultant structure was an InP-cap-layer/InGaAs-photo-absorption-layer/InP-buffer-layer structure on an InP substrate. The concentration of the InP cap layer

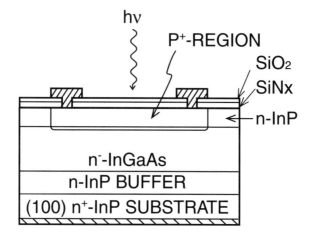

Fig. 8.13   Planar InGaAs-PIN-PD cross section.

was typically about 1×10^{16} cm^{-3} and that of the InP buffer layer was over 1×10^{17} cm^{-3}. The p$^+$n junction planar structure was obtained by selective diffusion of Zn (Zn$_3$P$_2$ source) and Cd (Cd$_3$P$_2$ source) [18]. The InP-cap-layer surface was coated with a passivated SiN$_x$/SiO$_2$ bilayer. For the contacts in the p- and n-layers, we used Ti/Pt/Au and AuGe alloys. The p$^+$n junction area was 95–100 μmϕ.

8.4.2.   TUNNELING BREAKDOWN CHARACTERISTICS UNDER A HIGH BIAS

The exponentially increased dark current is often explained by the Kane theory [19] that assumes a parabolic potential barrier and an uniform electric field. However, the resulting theoretical value is several orders-of-magnitude larger than the experimental values [14, 15]. The differential equation formula based on the dependence of the electric field on its position can better explain the data. The differential equation is as follows:

$$\frac{dJ_{tun}}{dx} = \frac{q^3 m^{*1/2} E^2(x)}{2\sqrt{2}\pi^3 \hbar^2 E_g^{1/2}} \exp\left(-\frac{\pi m^{*1/2} E_g^{3/2}}{2\sqrt{2}q\hbar E(x)}\right) \qquad (8.39)$$

where m^* is the effective tunneling mass. Figure 8.14(a) shows the tunneling current characteristics of the samples with an InGaAs donor concentration of about 5×10^{15} cm^{-3}. The solid lines in Fig. 8.14(a) show the theoretical curves based on Eq. (8.39), calculated for different

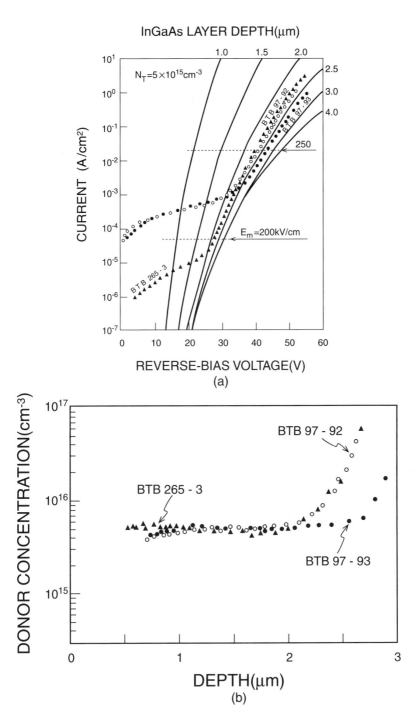

Fig. 8.14 (a) Tunneling dark current. Solid lines show theoretical results for different thicknesses. (b) Donor concentration profiles estimated by C-V measurements.

InGaAs-layer widths. In the calculations, $m^* = 0.04m_0$ [20], and $E_g = 0.75$ eV. Figure 8.14(b) shows the measured carrier profiles of the diodes used to obtain the tunneling current characteristics shown in Fig. 8.14(a). We can see that the tunneling dark current characteristics depend on the InGaAs layer concentration and thickness. With regard to the dark current in the high-electric field, Kane's differentiated theory explains the experimental results very well.

The temperature dependence of breakdown is usually evaluated by using the following equation:

$$V(T) = V(T_0)\{1 + \gamma(T - T_0)\} \qquad (8.40)$$

where the avalanche process is dominant for $\gamma > 0$, and the tunneling process is dominant for $\gamma < 0$. In the experiments with the InGaAs p^+n diodes, γ was around -6×10^{-4} (K^{-1}).

There are no temperature-dependent terms in Eq. (8.39). To explain this, the dependence of the energy gap on the temperature is used [21].

$$E_g(T) = 0.75\{1 - 3 \times 10^{-4}(T(^\circ C) - 20)\}. \qquad (8.41)$$

The calculated value of γ agreed well with the experimental one.

8.4.3. DARK-CURRENT CHARACTERISTICS AT A LOW BIAS AND EFFECTIVE LIFETIME

In this section, we discuss the dark-current characteristics at a low bias that did not appear in the tunneling current. By setting the $J \propto V^{1/2}$ line for the experimental dark current versus the bias curves, the effective lifetime, τ_{eff}, can be estimated. The generation-recombination (g-r) current due to the depleted region under a reverse bias is expressed as [22]

$$J_g = qn_i W/\tau_{eff} \qquad (8.42)$$

where n_i is the intrinsic carrier concentration (about 6×10^{11} cm^{-3} at 300 K for InGaAs) and W is the depleted layer width. In the p^+n junction, W equals $(2\varepsilon V/qN_T)^{1/2}$ where N_T is the donor concentration of InGaAs, and therefore, the g-r current can be evaluated by setting the line of $V^{1/2}$ for $|V| > 3k/q$. Based on this, τ_{eff} was obtained as a function of the InGaAs concentration for different photodiodes as shown in Fig. 8.15. From the leading edge that better fits the data, τ_{eff} can be approximated as

$$\tau_{eff}(\text{sec}) = 1.5 \times 10^{11}/N_T \text{ (cm}^{-3}) \qquad (8.43)$$

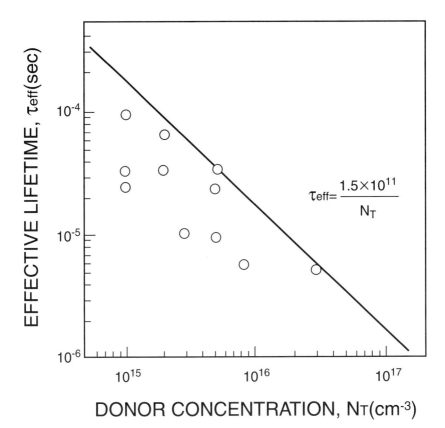

$$\tau_{eff}= \frac{1.5\times10^{11}}{N_T}$$

Fig. 8.15   Effective lifetime in InGaAs p^+n photodiodes.

For the equal cross sections of the electrons and holes in the g-r process, the theoretical lifetime is given by [23]

$$\tau_{eff} = 2\tau' \cosh\{(E_i - E_t)/kT\} \tag{8.44}$$

When E_i equals E_t, τ_{eff} is the minimum value, and when $|E_i - E_t|$ equals $(1/2)E_g$, it is the maximum value. Here, E_t is the g-r energy level and τ' is the same as the minority-carrier lifetime in the crystal, which is in the order of 10^{-9} seconds for most group III-VI compounds. Therefore,

$$10^{-3} \geq \tau_{eff}(\text{sec}) \geq 10^{-9} \tag{8.45}$$

Because τ' is inversely proportional to the impurity concentration, τ_{eff} should have the same tendency as does τ'. It is remarkable that the obtained τ_{eff} characteristics (see Fig. 8.15 and Eq. (8.43)) inversely proportional to the donor concentration coincided well with the extrapolated line of the lifetime dependence on a carrier concentration of over 10^{17} cm^{-3} reported by Henry [24]. Therefore, the maximum τ_{eff} of about 10^{-3} will be achieved at a high purified donor concentration lower than that in the order of 10^{14} cm^{-3}.

The diode dark current usually consists of a g-r current and a diffusion current. We attempted to divide the dark current into the two components: the g-r term proportional to the root-square of the bias voltage and the diffusion term with no dependence on the bias voltage. Figure 8.16 shows the component separation for a diode at different temperatures. The separation was successful. However, we found that there is a residual component

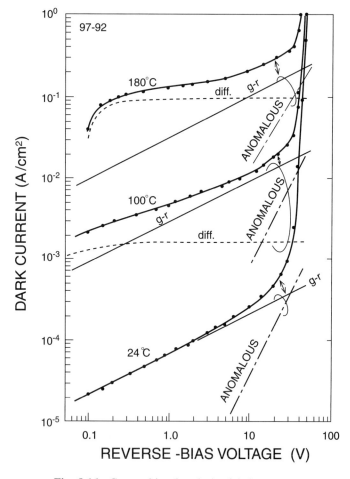

Fig. 8.16   Compositional analysis of dark current.

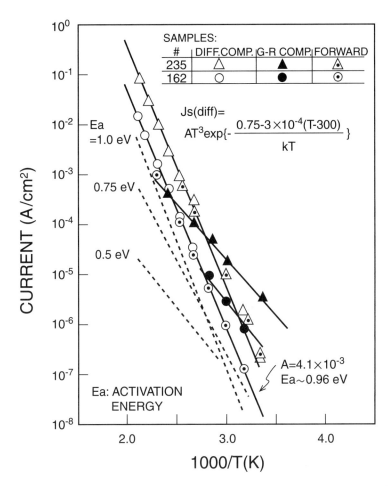

Fig. 8.17 Arrhenius plot of compositionally analyzed diffusion current and generation-recombination (g-r) current.

that cannot be divided into the two components. The characteristics of the divided components are discussed separately following.

Figure 8.17 shows the results obtained from the same process as that in the experiment shown in Fig. 8.16 including the diffusion-current component and the g-r current component at a bias of -5 V. Here, the diffusion current data were strengthened with forward saturation current data at low temperature to clarify the activation energy. The activation energy obtained in this experiment was greater than 0.75 eV in the InGaAs bandgap. The reason activation energy Ea for the InGaAs p^+n-diode was greater than the energy gap of InGaAs can be explained as follows: The diffusion current is equal to the forward saturation current and can be approximated as follows [25]:

$$J_s \sim q D_p p_{n0}/L_p = q(D_p/\tau_p)^{1/2} n_i^2/N_D \tag{8.46}$$

$$n_i = (N_c N_v)^{1/2} \exp(-E_g/2kT)$$

$$= 4.9 \times 10^{15} \left(m_{\mathrm{de}} m_{\mathrm{dh}}/m_0^2\right)^{3/4} T^{3/2} \exp(-E_g/2kT) \tag{8.47}$$

where L_p is $(D_p \tau_p)^{1/2}$, $N_c(N_v)$ is the effective carrier density in the conduction band (balance band), and $m_{\mathrm{de}}(m_{\mathrm{dh}})$ is the effective mass in the conduction band (valence band). Then, D_p/τ_p can be assumed to be proportional to T^γ, which gives us

$$J_s \propto T^{(3+\gamma/2)} \exp(-E_g/kT)/N_D \tag{8.48}$$

Thus, the diffusion current can be approximated to [26]

$$J_s \propto AT^3 \exp(-E_g/kT) \tag{8.49}$$

By using the temperature dependence of the InGaAs bandgap described previously, the value of Ea was found to be about 0.96 eV, which was in good agreement with the experimental value.

The temperature dependence of the divided g-r current at -5 V is shown in Fig. 8.17. Based on Eq. (8.42), the effective lifetime was re-estimated, resulting in

$$\tau_{\mathrm{eff}} \,(\mathrm{sec}) = 3 \times 10^{11}/N_T (\mathrm{cm}^{-3}) \tag{8.50}$$

The extrapolated value was twice as large as the value obtained in Eq. (8.43).

As can be seen in Fig. 8.16, there was a residual anomalous dark current J_{anom} at around the intercept region between the g-r current as well as the diffusion and the tunneling current that could not be divided into the g-r and diffusion components. In the experiment, a relationship of $J_{\mathrm{anom}} \propto V^{3/2}$ was obtained. From the Arrhenius plot, the activation energy was estimated to be approximately 0.5 eV. Here, $J_{\mathrm{anom}} \propto V^{3/2}$ is equal to the tunneling process in the energy bandgap of $E_g = 0.01$ eV in Eq. (8.39), and in the multi-step tunneling of a few tenths to one hundred steps, by assuming the multi-step tunneling process described by Riben [27]. The tendency in the anomalous components was the same as that in the planar devices reported by Hasegawa et al. [28] and Kagawa et al. [29]. However, more research is needed to clarify this phenomena and improve the growth and processing technologies. Finally, in the case of poor growth and poor processing, the dark-current characteristics were deteriorated more than the data ones reported here.

8.5. Photodiodes

The photodiodes (PDs) described in this section are mainly of the PIN-type. "PIN" means a layer structure in which an unintentionally doped high-purity layer is sandwiched between the p^+ and n^+ layers. The light-absorption layer described here is an InGaAs layer lattice-matched to InP. This is because InGaAs is responsive to all wavelengths of the WDM light sources based on InGaAs(P)/InP and In(Al)GaAs/InP material systems.

8.5.1. BASIC InGaAs PIN PHOTODIODES

To fabricate a simple and reliable planar-structure PD, we grew a double heterostructure consisting of InGaAs/InP with an InP capping layer. This step was followed by selective impurity diffusion to form a p^+n-junction. A window for the light to pass through was formed on the front (the grown-layer surface) and back surfaces of the InP substrate. The back-illuminated structure is the one often used to obtain a low capacitance for high-speed operation.

A cross-sectional view of the front-illuminated planar-structure InGaAs PIN-PD is shown in Fig. 8.18. The front of the pn-junction was formed on the InGaAs absorption layer close to the InGaAs/InP interface by using thermal diffusion with Zn and Cd [18]. The dark current characteristics of the diode with an effective junction diameter of $104\,\mu m$ (a light-receiving area diameter of $80\,\mu m$) and a 4-μm-thick InGaAs absorption layer with a carrier concentration of 2×10^{15} cm^{-3} measured at different temperatures are shown in Fig. 8.19. As was discussed in the previous section, the exponential dark-current increase with an increase in the bias, observed when the reverse bias was large, was due to the InGaAs band-to-band tunneling [14, 15]. The diode must be operated under a moderate reverse bias, where it is not affected by the tunneling current. At a bias lower than 10 V, the dark current of the well-fabricated diodes was less than sub-nA and it changed by about one order of magnitude when the diode temperature changed by $40°C$. This low dark current was obtained using a planar structure that terminated the pn-junction in the wide-bandgap InP capping layer. The dark current in a mesa-structure, which is controlled by the surface leakage current at the mesa wall, is generally two or more orders of magnitude higher than that in

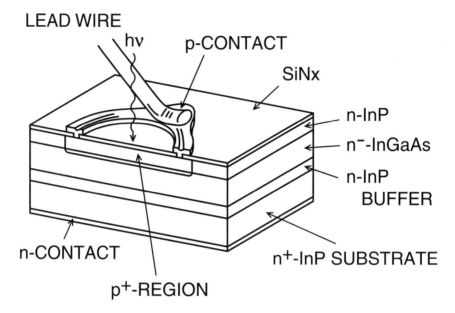

LEAD WIRE
hv   p-CONTACT
Sinx
n-InP
n⁻-InGaAs
n-InP
BUFFER
n-CONTACT
p⁺-REGION
n⁺-InP SUBSTRATE

Fig. 8.18   Planar InGaAs double-heterostructure photodiode.

a planar heterostructure. Figure 8.20 shows the spectral external quantum efficiency for a diode under a 5-V bias. The quantum efficiency was higher than 80% at wavelengths between 1.0 and 1.55 µm. This wide-range high efficiency was due not only to the antireflection coating of the surface but also to the thin p^+-InGaAs region, which acted as a recombination region for the photocarriers generated in it. The high efficiency was also due to the reduction in the interface recombination velocity of the photogenerated carriers at the p^+-InP/p^+-InGaAs heterojunction, the configuration of which cannot be expected in conventional homojunction photodiodes. The cut-off at the short wavelength (about 0.95 µm) was controlled by the bandgap of the InP ($E_g = 1.35$ eV) capping layer. The bandwidth (3-dB-down cut-off frequency) was about 5 GHz with a 50-Ω load resistance.

Planar double-heterostructure PDs are needed for use in practical transmission systems because of their stable operation and reliability. Accelerated life tests using diodes with a surface coated with a plasma-deposited SiN_x/CVD-SiO_2 double layer to prevent the formation of pin holes and reduce the film thermal stress in InGaAs/InP showed that there was no significant degradation after a 5500-h aging at 250°C at a reverse bias of 10 V when Ti/Pt and Ti/Au were used for the p-side contacts [30]. A similar reliability of layer-structure diodes with BeAu/Cr/Au as a p-metal in humid

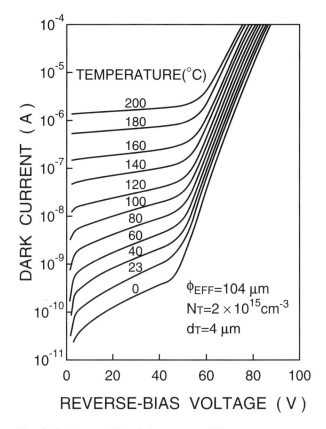

Fig. 8.19   Reverse-bias dark current at different temperatures.

ambients was reported. The diodes had a 20-year hazard level of less than 100 FITs in the devices operating at an ambient of 45°C/50% RH [31].

Cut-off frequencies higher than 100 GHz were earlier obtained in a thin small-area diode using a graded-bandgap layer to reduce the carrier trapping at the InGaAs/InP heterointerface [32, 33]. Figure 8.21 shows a mushroom-mesa-geometry PIN-PD with an air-bridge contact metal developed for ultra-high-speed operation [34]. The mushroom mesa was used to reduce the junction capacitance while maintaining a large contact area for the low series resistance [35], and the air-bridge contact was used to minimize the parasitic capacitance. A photodiode with a 2-μm diameter and a 0.18-μm-thick InGaAs layer had a response time of 2.7 ps (full width at half maximum) for Ti-Sapphire laser pulses with a wavelength of 0.98 μm. The cut-off frequency obtained by the fast-Fourier transform

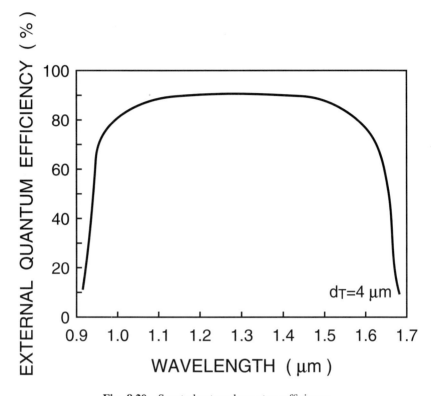

Fig. 8.20   Spectral external quantum efficiency.

of the measured pulse responses was 120 GHz. The estimated external quantum efficiency at a wavelength of 1.3 μm was 28%.

Other techniques have been developed to improve the low quantum efficiency in vertical-illumination-type high-speed PIN-PDs associated with the small junction area. These include the use of signal light reflection at the contact metal deposited on the rear surface of the InP substrate and a Bragg reflector to enhance the efficiency [36, 37], as well as monolithic lens integration [38, 39] on the substrate to magnify the effective receiving area.

8.5.2. MSM PHOTODIODES

A metal-semiconductor-metal (MSM) structure with an interdigitated contact has been used in photodetectors because this structure is easy to fabricate and the simplicity of the planar contact makes it suitable for integration with electrical circuits. MSM photodiodes have therefore often been used in receiver OEICs. They also have a low capacitance per unit area, which

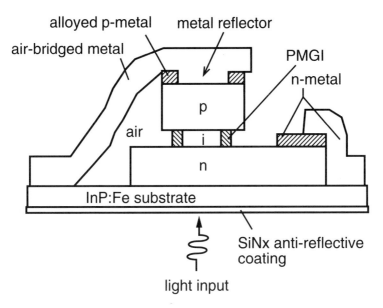

alloyed p-metal   metal reflector
air-bridged metal
PMGI
n-metal
p
air   i
n
InP:Fe substrate
SiNx anti-reflective coating
light input

Fig. 8.21 Mushroom-mesa PIN-PD with an air-bridge metal [34].

is advantageous and useful in photoreceivers with a large photoreceptive area. The main problems with these photodiodes are the dark current and quantum efficiency. The dark-current degradation in the InGaAs/InP material systems is mainly due to the low Schottky barrier heights of InP and InGaAs(P). This problem can be solved by using an InAlAs capping layer [40]. However, the degradation in quantum efficiency caused by electrode masking cannot be eliminated. There have been reports on transparent electrodes [41, 42] and back illumination [43, 44] for the substrate side of MSM-PDs to improve the efficiency. However, the use of these new structures results in response-speed degradation due to the low electric field underneath the electrodes. Electron-beam lithography has also been used to reduce the size of the contact-metal region. For high-speed applications, an MSM structure with a submicrometer line-and-space layout has been analyzed by using electron-beam lithography [45].

8.5.3. WAVEGUIDE PIN PHOTODIODES

There is a trade-off between the speed and efficiency in conventional vertical-type photodiodes. In contrast, waveguide(WG)-structure photodiodes are basically free from this trade-off problem because of the parallel penetration of signal light along the absorption layer. In this structure, the internal quantum efficiency is a function of the length, propagation mode, and mode-confinement factor. To improve coupling, several structures have been developed.

WG-PDs with a bandwidth of 50 GHz and a quantum efficiency of 40% at a wavelength of 1.53 µm were described by Wake *et al.* [46], who developed a PD with an asymmetric InGaAsP waveguide structure. A thin (0.13-µm-thick) InGaAs absorption layer was sandwiched between a 3-µm-thick n-doped InGaAsP layer (bandgap wavelength of $\lambda_g = 1.3\,\mu m$) and a 0.1-µm-thick undoped InGaAsP layer. The absorptive InGaAs and thick InGaAsP layers were designed to enable a high external efficiency: the thick InGaAsP layer largely determined the transverse waveguiding properties of the diode, ensuring a large mode size comparable to the mode size of lensed fiber. The InGaAs absorption layer was thin to enable a large mode size. The diodes were 5 µm wide and 10 µm long, and their capacitance was less than 0.1 pF.

A multimode waveguide structure with symmetric InGaAsP intermediate layers inserted between the InGaAs light-absorption and InP cladding layers was developed by Kato *et al.* [47]. It was shown that higher-order mode lights in the structure increased the efficiency of coupling between the waveguide PD and the fiber. Experiments using a structure with a 0.6-µm-thick InGaAs absorption layer sandwiched between 0.6-µm-thick InGaAsP ($\lambda_g = 1.3\,\mu m$) layers yielded a bandwidth of 50 GHz with an external quantum efficiency of 68%. Figure 8.22 shows the simulated coupling efficiency as a function of InGaAs thickness for multi-mode WG-PDs in which the total thickness of the InGaAs and InGaAsP layers was kept constant at 1.8 µm. Here, η_{cy0} and η_{cy2} show, respectively, the calculated fundamental and second-order mode contributions to the coupling efficiency. The calculations showed that the coupling efficiency depends on the total thickness of the InGaAs and two InGaAsP intermediate-bandgap layers. These results mean that the InGaAs absorption-layer thickness can be designed not only for the coupling but also for the objective speed. Figure 8.23 shows a multimode waveguide PIN-PD with a mushroom-mesa structure [35]. The mushroom mesa configuration was used to reduce the diode capacitance, and this configuration enabled leaving a wide area for the metal contact to minimize the series resistance. The layer structure consisted of 0.8-µm-thick p-doped and n-doped InGaAsP layers with a 0.2-µm-thick unintentionally doped InGaAs absorption layer. The mushroom mesa was made by forming 6-µm-wide cladding layers that were then selectively wet-etched to decrease the junction capacitance. The diode with a 1.5-µm-width

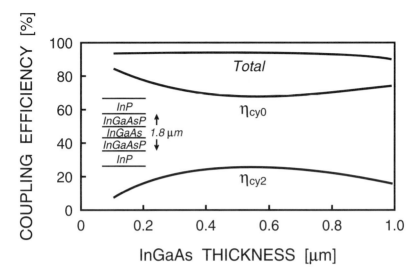

Fig. 8.22 Simulated coupling efficiency (total) as a function of InGaAs thickness for multi-mode WG-PDs [47].

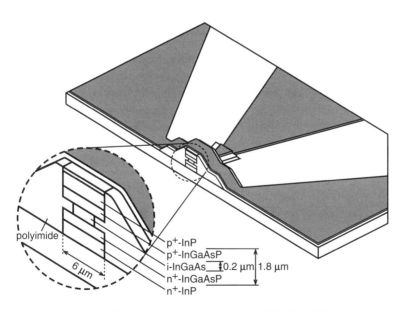

Fig. 8.23 Mushroom-mesa multimode WG-PD [35].

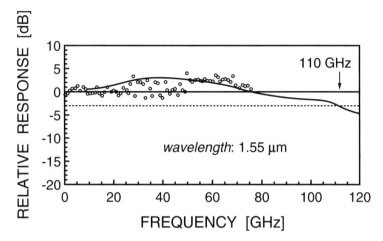

Fig. 8.24   Frequency response measured by a spectrum analyzer (circle) and deduced from the Fourier transform [35].

core showed an external efficiency of 50% at a wavelength of 1.55 μm and had a capacitance of 15 fF and a series resistance of 10 Ω. The frequency response for a circuit with an impedance of 50 Ω was measured and is shown in Fig. 8.24. The response was almost flat over the frequency range between 0 to 75 GHz. The Fourier transform of the measured short-pulse responses indicated a bandwidth of 110 GHz. The obtained bandwidth-efficiency product of 55 GHz was 1.6 times larger than that of the 120-GHz-bandwidth vertical PIN-PD described in the previous section.

Multimode waveguide structures have been attracting much attention because of their potential ability to be coupled easily to fiber and planar lightwave circuits (PLCs) without the use of a focusing lens, which lowers the cost of receiver modules [48]. The operating speed of receiver modules in access networks should not exceed a few gigahertz, so the key issue for this application is the alignment tolerance of the structures to fibers and optical waveguides. A WG-PD that can be used in 1.55-μm-wavelength-range access receivers was earlier designed and fabricated for receivers related to the input spot size and for waveguide layer structures [49]. Figure 8.25 shows the maximum external quantum efficiency, η_{ex}, for both symmetric and asymmetric waveguide structures as a function of the total thickness of the guide layers and InGaAs core layer calculated by using a beam-propagation method. Here, the InGaAs layer was 1.5 μm thick, which enabled a low-bias operation, and the guide layers were InAlGaAs with a 1.3-μm bandgap wavelength. An input spot size of 0.75 μm was obtained

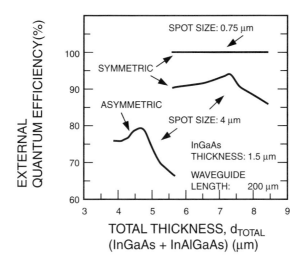

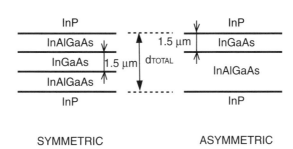

SYMMETRIC   ASYMMETRIC

Fig. 8.25 Calculated maximum external quantum efficiency as a function of total thickness of the guide layer (variable) and InGaAs layer (constant at 1.5 μm). The waveguide length was 200 μm.

by using a hemispherically ended fiber with a sufficiently small spherical radius. A typical value for a flat-ended fiber or a silica waveguide is 4 μm. The calculation results showed that the symmetric waveguides had a higher η_{ex} than did the asymmetric ones, which was a result of the optical confinement difference. Figure 8.26 shows the calculated 1-dB-down full widths of the vertical coupling tolerance curves as a function of the total thickness of the guide and core layers. Here, the waveguide length was fixed, because it is usually determined by the specifications of the junction capacitance, which is the objective speed. The vertical tolerance for the spot size of 0.75 μm decreased abruptly at a total thickness of around 7.5 μm. This means that the depth of the coupling dips, which is due to the coupling of

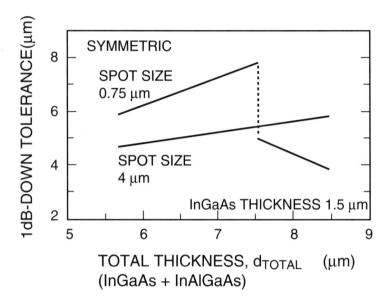

Fig. 8.26   Calculated 1-dB-down vertical tolerance for different total thicknesses and input spot sizes.

the input spot light with the weakly confined propagation modes, exceeded the 1-dB allowance at this point. From these calculations, we conclude that a high external efficiency and a large vertical tolerance can be achieved by using a symmetrical waveguide, an optimized guide-layer thickness, and an input spot of a small size. The device structure we examined consisted of a p^+-InP cladding layer, a 2.5-μm-thick p^+-InAlGaAs guide layer, a 1.4-μm-thick undoped InGaAs layer, a 2.5-μm-thick n^+-InAlGaAs layer, and an n^+-InP cladding layer on a semi-insulating InP substrate. The waveguide mesa of the WG-PD was 18 μm wide and 100 μm long. Figure 8.27 shows a schematic of the fabricated WG-PD with alignment markers for flip-chip mounting. We ensured that the alignment system recognized these markers by using an infrared camera to set the system on a Si submount bench with objective aligned markers and a V-grooved trench for the input optical fiber. Figure 8.28 shows the coupling tolerance curve measured in the vertical direction. The maximum quantum efficiency was over 95% at a wavelength of 1.55 μm and the 1-dB-down vertical tolerance was as large as 6.5 μm. The capacitance of the tested diode was 0.28 pF, and its bandwidth was higher than 10 GHz at a bias of −2 V.

Greatly simplified structures that enable easy coupling at a low bias were previously reported [50]. Their layer structure consists of a 3-μm-thick InGaAsP ($\lambda_g = 1.4$ μm) photoabsorbing core layer and two 2-μm-thick

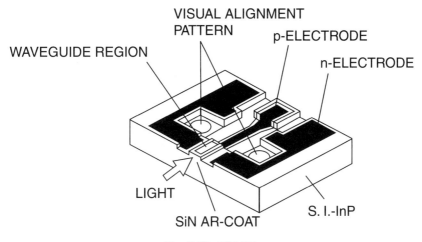

Fig. 8.27   WG-PD.

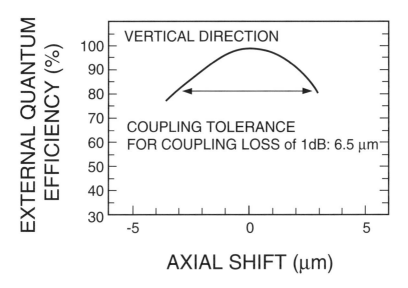

AXIAL SHIFT (μm)

Fig. 8.28   Measured coupling tolerance in the vertical direction. Hemispherically ended fiber had a spherical radius of 10 μm.

InGaAsP ($\lambda_g = 1.2\,\mu$m) intermediate layers for 1.3-μm-wavelength appli-cations. Selective impurity diffusion was used to form the pn-junction and slab waveguide, and the pn-junction front was designed to be deep in the light-absorption core layer. A tolerance of 5.5 μm in the vertical direction, a bandwidth of 500 MHz, and a responsivity of 0.87 A/W at a wavelength of 1.31 μm were achieved at a 1-V bias.

8.5.4. EVANESCENTLY COUPLED PHOTODIODES

For next-generation systems with a data transmission rate of 40 Gb/s, a waveguide structure can be used to obtain photodetectors with high-speed and high-efficiency characteristics, as discussed in the previous section [51]. However, the detectors must enable a stable operation at high power levels, because the use of EDF preamplifiers in such systems results in the optical input to the detectors being at a high level normally, for example, a few mW. High power inputs of several milliwatts often damage the input facet of simple waveguide photodiodes due to the absorption of input light. To improve these high-power handling capabilities, the characteristics of an evanescently coupled waveguide photodiode (EC-PD) were analyzed [52, 53]. Figure 8.29 compares the structure of a WG-PD with that of an EC-PD. The parameters of the layer structures are shown in the figure. From the photocurrent distribution curves calculated by a beam-propagation method for both the 0.5-μm InGaAs-core WG-PD and EC-PD, it can be seen that the photocurrent density near the input edge of the PD region in the EC-PD decreased by one half, compared to that in the WG-PD. This is because the input light in the EC-PD gradually penetrated the absorption layer

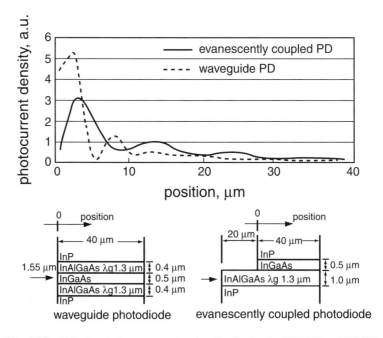

Fig. 8.29   Calculated photocurrent density distribution for WG-PD and EC-PD.

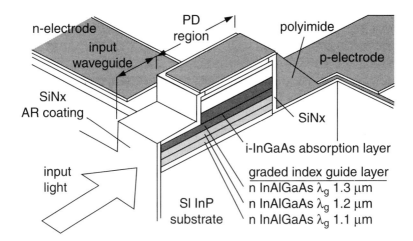

n-electrode

input
waveguide

PD
region

polyimide

p-electrode

SiNx
AR coating

SiNx

i-InGaAs absorption layer

input
light

graded index guide layer

SI InP
substrate

n InAlGaAs λ_g 1.3 μm
n InAlGaAs λ_g 1.2 μm
n InAlGaAs λ_g 1.1 μm

Fig. 8.30   EC-PD with three-layer graded index guide [53].

from the guide layer. This means that the EC-PD will be much more robust than the WG-PD under high-input-power conditions. Figure 8.30 shows the device structure of an EC-PD with a graded index guide. The critical photocurrent for the EC-PDs we examined was approximately twice as large as that for the WG-PDs. An external quantum efficiency of about 60% was obtained for 6-μm-wide and 30-μm-long devices and a bandwidth of 40.2 GHz at a bias of −5 V was obtained for the average photocurrent of 10 mA, which corresponds to an output peak voltage of 1 V for a return-to-zero signal.

8.5.5.   UNI-TRAVELING CARRIER PHOTODIODES

To achieve a high-speed response and a high saturation output by using only the electrons as drift carriers, thereby suppressing the space charge effect, a uni-traveling carrier (UTC) photodiode with an npn structure was developed by Ishibashi *et al.* [54]. This device structure shown in Fig. 8.31 was configured using a thin p-type neutral narrow-gap light-absorption layer and a depleted n-type wide-gap carrier-collector layer. A wide-gap n^+-layer anode was used to block the photo-generated electron diffusion into the anode. This configuration enables using only the electrons as active carriers. Minority electrons photo-generated in the absorption layer diffuse, or are enhanced by the internal field due to the potential gradient, into the carrier-collector n^--layer, while excess holes photo-generated in the absorption layer are swept out as a conduction current. Similar structures with

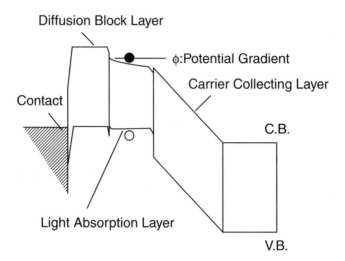

Fig. 8.31   Band diagram of uni-traveling carrier photodiode structure [54].

a photo-absorption in a neutral layer were developed to suppress the dark current, reduce the diode capacitance, and enable a high current. However, these structures failed to enable a high speed [55, 56]. The answer lies in the thin p-type absorption layer with a hopefully potential gradient and in the smooth injection of photo-generated electrons into the n^--collector layer. Simulation has shown that a potential gradient of more than 50 meV in a 0.2-μm-thick p-layer is needed to enable a speed higher than that enabled by a pin-structure photodiode with the same dimensions [54].

In the experiments shown in Fig. 8.32, two different absorption layers were tested and compared [57]. The type-I layer was a 140-nm-thick photo-absorption layer uniformly doped to a concentration of 2×10^{18} cm^{-3}. The type-II layer had different doping levels of 2×10^{18} and 2×10^{17} cm^{-3} in the absorption layer, as shown in Fig. 8.32(b). As shown in Fig. 8.32(a), the UTC-PD structures consisted of an $(n^+$-$)$InP-InGaAs-InP subcollector layer on a semi-insulating InP substrate, followed by a 208-nm-thick InP collector layer (Si $= 4 \times 10^{16}$ cm^{-3}), a 10-nm-thick InP cliff layer (Si $= 3 \times 10^{18}$ cm^{-3}), a 2-nm-thick undoped InP spacer layer, a 2-nm-thick undoped InGaAs spacer layer, a 140-nm-thick carbon-doped InGaAs absorption layer, a 15-nm-thick p^+-InGaAsP barrier layer, and a 60-nm-thick p^+-InGaAs cap-and-contact layer. The capacitance of the fabricated devices obtained by S-parameter measurements was 0.5 fF/μm^2, and the external quantum efficiency was 13% for both structures. Figure 8.33 shows the

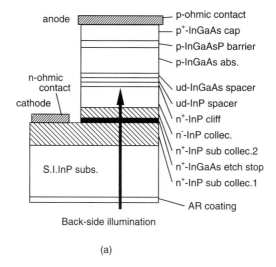

anode — p-ohmic contact
— p$^+$-InGaAs cap
— p-InGaAsP barrier
— p-InGaAs abs.

n-ohmic contact
cathode
— ud-InGaAs spacer
— ud-InP spacer
— n$^+$-InP cliff
— n$^-$-InP collec.
— n$^+$-InP sub collec.2
S.I.InP subs. — n$^+$-InGaAs etch stop
— n$^+$-InP sub collec.1
— AR coating

Back-side illumination

(a)

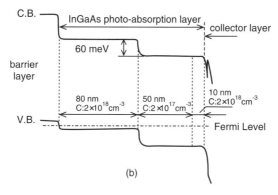

C.B.
InGaAs photo-absorption layer
collector layer
60 meV
barrier layer
80 nm C:2×10^{18}cm^{-3} 50 nm C:2×10^{17}cm^{-3} 10 nm C:2×10^{18}cm^{-3}
V.B.
Fermi Level

(b)

Fig. 8.32 (a) Cross section of UTC-PD. (b) Doping profile of InGaAs absorption layer for type-II devices [57].

relationship between the peak output voltage and the bandwidth at different input laser-power levels for devices with a 20-μm^2 active area at biases of −1.5 and −4.0 V. The measurements were done for laser pulses with a 1.55 μm wavelength by using an electrooptic (EO) sampling technique. The maximum bandwidths were, respectively, 125 and 152 GHz for the type-I and type-II devices. When the bias was −1.5 V for the type-II devices, the output voltage saturated at around 1 volt. At a bias of −4 V, the saturation voltages increased to around 1.9 V. The maximum bandwidth at a bias of −4 V was smaller than that at a bias of −1.5 V. This is because

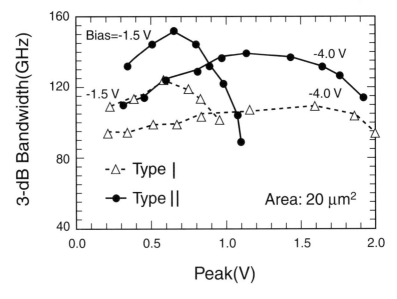

Fig. 8.33   Bandwidth vs. output peak voltage for fabricated devices [57].

the electron-velocity overshoot in the InP collector layer was clearer at the lower bias. It was found that a device based on this concept can directly drive a logic IC with 40-Gb/s optical signals [58]. A 40-Gb/s operation of a monolithically integrated digital OEIC composed of a UTC-PD and InP HEMTs was also reported [59]. Recently, a bandwidth of 310 GHz and a pulse width of 0.97 ps at a wavelength of 1.55 μm were obtained with a UTC-PD, in which the InGaAs absorption layer was made thin (30 nm) and the collector layer was made relatively thick (230 nm) to keep the junction capacitance low [60].

A UTC structure was used in the high-speed high-efficiency multi-mode waveguide photodiode described previously to improve the quantum efficiency of the vertically illuminated structure [61]. The UTC structure used in the WG-PD consisted of a 0.1-μm-thick InGaAs absorption layer and a 0.2-μm-thick InGaAsP layer. To form a multi-mode double-core waveguide, intermediate bandgap InGaAsP layers were used above and below the UTC structure. The 0.85-μm-thick upper layer also acted as a diffusion block layer, and the 0.63-μm-thick lower layer acted as a subcollector layer. The device showed a quantum efficiency of 32% at the 1.55-μm wavelength, an output voltage of 1.3 V with a bandwidth of 55 GHz, and an output voltage of 0.45 V with a bandwidth of 70 GHz.

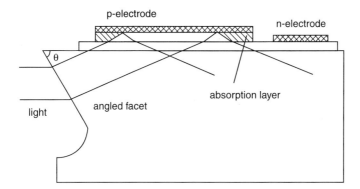

Fig. 8.34 Cross-sectional view of fabricated refracting-facet photodiode [63].

8.5.6. REFRACTING-FACET PHOTODIODES

To reduce the cost of optical modules, the optical axis alignment must be simplified and the number of optical components, such as lenses, must be reduced. For photodiodes, a large optical axis misalignment tolerance is the most important factor in reducing the cost. A low-cost edge-illuminated refracting-facet photodiode was previously developed, in which the incident light parallel to the device surface is refracted at an inwardly angled facet and absorbed by the absorption layer as shown in Fig. 8.34 [62, 63]. It improves the optical axis misalignment tolerance and responsivity. This is because the tolerance of the optical axis to the most severe misalignment in the vertical direction is determined not by the thickness of the absorption layer as in the waveguide photodiodes, but by its length with the incident light refraction. The angle facet was formed by anisotropic chemical etching to produce a (111)A plane with an angle of 54.7 degrees on InP (001) surface. In the $14 \times 20\text{-}\mu\text{m}^2$ diode with a 1-μm-thick absorption layer, the external quantum efficiency was found to be 80% for the 1.55-μm light, the bandwidth was 38 GHz, and the vertical misalignment tolerance was 3.3 μm for 1 dB down [62].

This structure with a thin absorption layer also enables a high output peak voltage [63]. This is because a high responsivity can be obtained even with a thin absorption layer, and this structure enables a decrease in the photo-generated carrier density. Raising the electric field by thinning the absorption layer is thought to be effective in improving the saturation output characteristics because the field modulation effect due to the space charge in the depleted region can be suppressed. In the experiments, the device structure consisted of a 0.4-μm-thick undoped InGaAs absorption

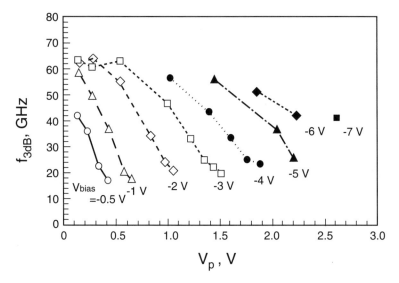

Fig. 8.35   Bandwidth f_{3dB} vs. output peak voltage V_p with changing input power at different bias voltages, V_{bias}, for refracting-facet PD with a junction of $4 \times 14\,\mu m^2$ [63]. $V_{bias} = -0.5\ V$ (○); $V_{bias} = -1\ V$ (△); $V_{bias} = -2\ V$ (◇); $V_{bias} = -3\ V$ (□); $V_{bias} = -4\ V$ (●); $V_{bias} = -5\ V$ (▲); $V_{bias} = -6\ V$ (◆); $V_{bias} = -7\ V$ (■).

layer and a 0.03-μm-thick p-doped InGaAs contact layer. A non-alloyed p-metal for the p-electrode acted as a reflector to double the absorption length. The external quantum efficiency measured at a 1.5-μm light and the bandwidth were, respectively, 49% and 66 GHz for a $6 \times 9\ \mu m^2$ diode at a bias of -2 V. Figure 8.35 shows the bandwidth against the output peak voltage with changing input power for a $4 \times 14\ \mu m^2$ diode. The bandwidth decreased with an increase in the output peak voltage (input power). A bandwidth of more than 40 GHz was obtained at an output peak voltage of 1 V when the diode was biased at -3 V and at an output peak power of over 2.5 V when the diode was based at -7 V. This structure using a selectively impurity-diffused planar pn-junction is believed to be as reliable as the planar devices described earlier in this chapter.

8.5.7.   RESONANCE-CAVITY-ENHANCED PHOTODIODES

To improve the quantum efficiency of surface-illuminated photodiodes with a thin absorption layer, mirrors for multiple optical passes through the absorption layer (similar to the resonances of Fabry-Perot microcavities) have been used [36, 37]. The schematic diagram of a resonance-cavity-enhanced (RCE) photodiode is shown in Fig. 8.36 [64]. A thin InGaAs

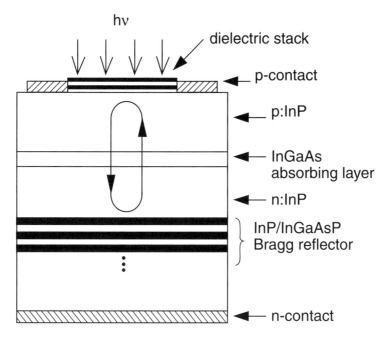

hν

dielectric stack

p-contact

p:InP

InGaAs
absorbing layer

n:InP

InP/InGaAsP
Bragg reflector

n-contact

Fig. 8.36   Cross section of InP/InGaAsP/InGaAs resonance-cavity-enhanced photodiode
[64].

absorption layer is sandwiched between two p- and n-doped InP spacer
layers. The top mirror can be the air/semiconductor interface. The external
quantum efficiency of a RCE for a wavelength of λ is given by

$$\eta = \frac{(1 + R_2 e^{-\alpha d})(1 - R_1)(1 - e^{-\alpha d})}{1 - 2\sqrt{R_1 R_2} e^{-\alpha d} \cos(4\pi n L_{CAV}/\lambda + \phi_1 + \phi_2) + R_1 R_2 e^{-2\alpha d}}$$

$$(8.51)$$

where R_1, R_2 are the reflectivities of the front and back mirrors, d is the
absorption-layer thickness, ϕ_1 and ϕ_2 are the phase shifts for the reflection
on the front and back mirrors, L_{CAV} is the distance between the mirrors, and
n is the refractive index of the cavity. The efficiency has its maximum when
$4\pi n L_{CAV}/\lambda + \phi_1 + \phi_2 = 2\pi m$ and $R_1 = R_2 e^{-2\alpha d}$. The RCE photodiode
also acts as a wavelength selector due to its operation principle. In an
InP-based photodiode that had a 0.2-μm-thick InGaAs absorption layer
embedded in the cavity consisting of InP spacer layers, a bottom mirror of
an InP/InGaAsP quarter-wave stack, and a top mirror of a single ZnSe/CaF$_2$
pair, the external quantum efficiency was 82% at a wavelength of 1.48 μm
[64]. However, even though there have been a number of theoretical studies

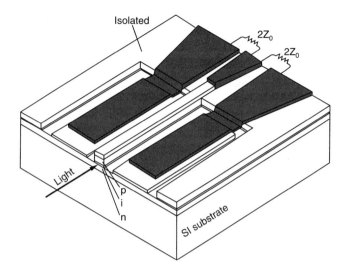

Fig. 8.37   Traveling-wave photodiode [65].

of and numerical simulations with RCE photodiodes, an RCE detector for transmission experiments could not be obtained because of the difficulties with its fabrication, material growth, and structure design.

8.5.8. TRAVELING-WAVE PHOTODIODES

To design high-speed photodetectors, a traveling-wave (TW) photodiode was developed [65, 66]. This is a waveguide photodiode with an electrode structure designed to support traveling electromagnetic waves with a phase velocity or a characteristic impedance matched to that of the external circuit, as shown in Fig. 8.37. In a TW photodetector, the optical dielectric waveguide is designed to also be an electrical waveguide for propagating electric wavefields. Because the TW photodetector is an electrically distributed structure, it is free from the RC limitation of the waveguide photodetector. As a result, larger bandwidths can be obtained compared to those obtained by using waveguide photodetectors. The bandwidth of TW photodetectors is limited by the optical absorption coefficient and the velocity mismatch between the optical and the electrical waves. The absorption contribution to the bandwidth is practically independent of the device length, which means that TW photodetectors with larger bandwidth efficiency than those of waveguide photodetectors are possible. A bandwidth of 172 GHz and a 76 GHz bandwidth-efficiency product were obtained for a GaAs/AlGaAs diode, 1 μm wide and 7 μm long [66]. A unique feature of TW photode-tectors is their ability to detect high-power light intensity. This is because of their large size compared to that of waveguide photodetectors in which the size and, as a result, the maximum power dissipation are limited by the requirements imposed on the detectors to achieve a small capacitance.

Photodetectors with high power-handling capabilities and a high speed are needed for receivers with a wide dynamic range. The power-handling capabilities of high-speed photodiodes are limited by the screening effects of the photo-generated carriers. When the carrier density becomes too large in the absorption region, the electric field is screened, and the carriers are no longer efficiently collected. Consequently, the transit time increases. In high-speed detectors with a small active area to minimize the capacitance, the power-handling capabilities are not very good because the detectors are designed to increase the bandwidth. To extend the saturation power to the 100-mW level, the absorbing volume must be enlarged. A new photodetec-tor structure was developed for simultaneously achieving a high saturation power and a high speed [67]. The detector, known as a velocity-matched (VM) distributed photodetector, achieves a high power-handling capability by combining the outputs of multiple photodetectors. The VM distributed photodetector, which is the same as a periodic traveling-wave photodetec-tor, consists of an array of discreet photodetectors serially connected on a passive optical waveguide, the output of which is collected by a separate velocity-matched electrical transmission line. In this structure, the optical waveguide, photodiodes, and transmission line can be independently op-timized at the cost of increased complexity and higher losses due to the reflection and scattering at each distributed photodiode. The overall device bandwidth is determined by that of the individual photodiode elements and by the velocity matching. Each photodiode along the transmission line is fed by an optical waveguide, which runs parallel to the transmission line. Making the electrical and optical wave velocities equal ensures that the photocurrents are added in-phase, which leads to efficient combining of the photo-generated signal. The above scheme was implemented using GaAa/AlGaAs, an integrated optical ridge waveguide, and a coplanar-strip transmission line. MSM detectors were placed across the coplanar strips, and the waveguide ran beneath them. The detectors, fabricated by using electron-beam lithography, had 0.3-μm finger widths and 0.2-μm finger spacings. By using five detectors spaced at 150-μm intervals, peak com-bined photocurrents of 66.5 mA (optical power of 98 mW) at a wavelength of 860 nm were obtained at 1-dB compression under pulsed operation.

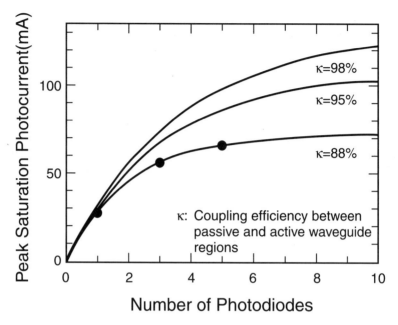

Fig. 8.38   Theoretical and measured peak saturation power of VM-PD versus the number of PDs [67].

This is much higher than 28 mW obtained for a similar structure with just one device. Because the waveguide feeds the detectors serially, a good coupling is needed to the sections of the waveguide that pass beneath each detector. This coupling between the passive and active sections of the waveguide was 88%. As shown in Fig. 8.38, further improving this coupling increases the saturation current. The saturation characteristics in the described device were affected mainly by the first of the serially connected diodes. Thus, to achieve a uniform illumination, the optical signal should be split evenly n ways before illuminating n detectors [68].

8.5.9.   PHOTONIC INTEGRATED CIRCUITS INCLUDING PHOTODIODES

One key element in WDM system is the wavelength-demultiplexing receiver that can resolve the wavelength channels and receive the signals. These functions were earlier integrated on a single chip [69]. Figure 8.39 shows a monolithic 8-wavelength demultiplexing receiver [70]. The waveguide grating router (WGR) consists of eight input waveguides, eight output

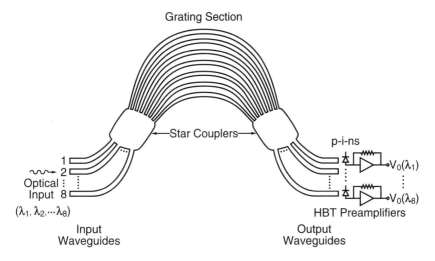

Fig. 8.39   Monolithic 8-wavelength demultiplexing receiver [70].

waveguides, two star couplers, and an arrayed-waveguide-grating (AWG) section. When eight wavelengths are launched simultaneously into any one input waveguide, the WGR spectrally resolves the eight signals, sending one into each of the eight output waveguides. Each signal is then coupled into a PIN-PD, and the photocurrent is amplified by an integrated amplifier. The WGR has a buried rib waveguide consisting of an n^--InP lower cladding layer, a 0.3-μm n^+-InGaAsP ($\lambda_g = 1.3\,\mu$m) waveguide, a 12-nm-thick n^--InP stop-etch layer, a 40-nm-thick n^--InGaAsP rib layer, and a 1.5-μm-thick undoped InP layer burying the rib waveguide. The chip can demultiplex 8 wavelengths spaced 0.81 nm apart with a nearest neighbor crosstalk of less than -15 dB. Polarization-independent operation was achieved using polarization-dispersion compensation, with two different sections for the array waveguide for an 8-channel high-speed (10 GHz) phase-array-demultiplexing receiver monolithically integrated with photodiodes [71].

To make a polarization-insensitive AWG section, nonbirefringent waveguides were developed and used [72]. A deep-ridge waveguide structure was used for this purpose and also to obtain a small bending radius. It was found that using an InGaAsP core layer close to the InP composition can increase the fabrication tolerance of nonbirefringent waveguides. A multiwavelength photodetector chip with 16 channels and a channel spacing of 100 GHz was developed by integrating a demultiplexer with a

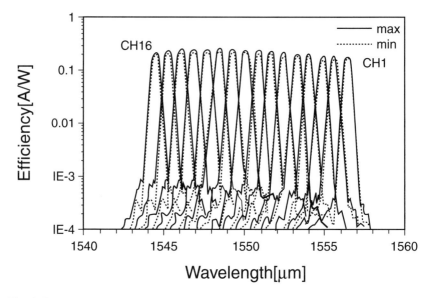

Fig. 8.40 Output spectra of multiwavelength photodetector with 16 channels and 100-GHz channel spacing [73].

photodiode array [73]. As can be seen in Fig. 8.40, the crosstalk among the neighboring cannels was lower than -20 dB with almost polarization-insensitive characteristics. The capacitance of the detector in this chip was less than 0.5 pF at a bias of -1 V. This means that the detector can be used as a multiwavelength detector for WDM applications.

With regard to access networks, bidirectional links for broadband integrated service digital networks and fiber-to-home architectures are receiving a lot of attention. Optical receivers with different circuit configurations have been studied and developed. For these devices, cost reduction and high performance are essential. There have been several studies of transceiver devices, both of a semiconductor photonic-integrated-circuit (PIC) type [74, 75] and a hybrid type with silica waveguides [76]. All optical components in PICs are monolithically integrated on a semiconductor substrate. Thus, the number of optical alignments can be reduced, and the assembly costs can be lower than they are for hybrid transceivers. The most common fabrication process for PICs, however, requires repetitive etching and regrowth, which in turn reduces uniformity and lowers the fabrication yield. The new technology allowing for selective-area metal-organic-vapor-phase epitaxy will overcome the limitations of conventional fabrication

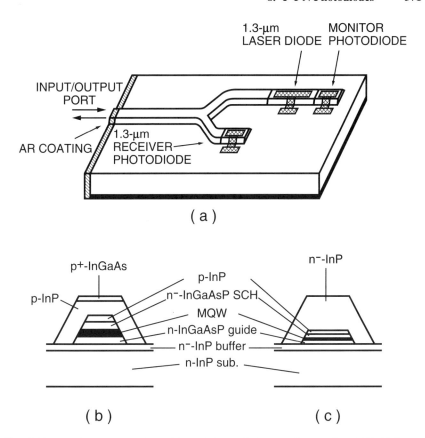

Fig. 8.41   Fabricated transceiver PIC. (a) Schematic layout of transceiver PIC. (b) Layer structure of laser diode and PDs. (c) Layer structure of passive waveguide.

techniques by enabling in-plane bandgap control [77], which will be described in greater detail in Chapter 10. The bandgap energy of InGaAs(P) layers can be controlled by varying the mask width. Figure 8.41 shows a conceptual diagram of a WDM transceiver PIC [78]. The active layer of the LDs, the absorption layer of the PDs, and the core layers of the passive waveguides can be grown simultaneously on a mask-patterned substrate, and this growth will be followed by InP regrowth over the layers. Active and passive components can be fabricated without complicated etching and partial regrowth and without forming waveguide discontinuities. Thus, the new technology will increase the fabrication yield at the same time as it lowers the cost of producing the devices.

8.6. Conclusion

This chapter described the design and performance of photodetectors for use in optical fiber communications, focusing in particular on WDM applications. Photodiodes based on a new concept with a heterostructure not expected in conventional elemental materials, such as Si and Ge, were developed.

High-speed and low-cost optical modules are essential for WDM systems. To reduce the cost of optical modules, the alignment of the optical axis must be simplified, and the number of optical components, such as lenses, must be reduced. For photodiodes, achieving a high tolerance to optical axis misalignment is one of the most important goals to minimize the cost, improve the wavelength-handling capability, and increase the power of the photodiodes.

References

1. R. A. Smith, Semiconductor (Cambridge University Press, Cambridge, England, 1959).
2. S. M. Sze, Physics of Semiconductor Devices (Wiley-Interscience, New York, 1969).
3. B. R. Bennett and R. A. Soref, "Electrorefractance and electroabsorption in InP, GaAs, GaSb, InAs, and InSb," IEEE J. Quantum Electron., QE-23 (1987) 2159–2166.
4. F. R. Bacher, J. S. Blakemore, J. T. Ebner, and J. R. Arthur, "Optical absorption coefficient of $In_{1-x}Ga_xAs/InP$," Phys. Rev., B37 (1988) 2551–2557.
5. D. R. Smith, R. C. Hooper, P. P. Smith, and D. Wake, "Experimental comparison of a germanium avalanche photodiode and InGaAs PINFET receiver for long wavelength optical communication systems," Electron. Lett., 18 (1982) 453–454.
6. S. D. Personick, "Receiver design for digital fiber optic communication systems, Parts I & II," Bell Syst. Tech. J., 52 (1973) 843–886.
7. R. G. Smith and S. D. Personick, Semiconductor Devices for Optical Communication, 2nd ed., H. Kressel, ed., (Springer Verlag, New York, 1982) 89–160.
8. D. R. Smith, M. C. Brain, and R. C. Cooper, "Receivers for optical fibre communication systems," Telecom. J., 48 (1981) 670–685.
9. D. E. Sawyer and R. H. Rediker, Proc. IRE., 46 (1958) 1122.
10. G. Lucovsky, R. F. Schwarz, and R. B. Emmons, "Transit-time consideration in p-i-n diodes," J. Appl. Phys., 35 (1964) 622–628.

11. D. A. Humphreys, R. J. King, D. Jenkins, and A. J. Moseley, "Measurement of absorption coefficients of $Ga_{0.47}In_{0.53}As$ over the wavelength range 1.0-1.7μm," Electron. Lett., 25/26 (1985) 1187–1189.

12. R. Torommer and L. Hoffman, "Large-hole diffusion length and lifetime in InGaAs/InP double-heterostructure photodiodes," Electron. Lett., 22 (1986) 360–361.

13. P. Hill, J. Schlafer, W. Powazinik, M. Urban, E. Eichen, and R. Olshansky, "Measurement of hole velocity in n-type InGaAs," Appl. Phys. Lett., 50 (1987) 1260–1262.

14. S. R. Forrest, R. F. Leheny, R. E. Nahory, and M. A. Pollack, "InGaAs photodiodes with dark current limited by generation-recombination and tunneling," Appl. Phys. Lett., 37 (1980) 322–325.

15. Y. Takanashi, M. Kawashima, and T. Horikoshi, "Required donor concentration of epitaxial layers for efficient InGaAsP avalanche photodiodes," Japan J. Appl. Phys., 19 (1980) 693–701.

16. S. R. Forrest, M. Didomenico, R. S. Smith, and H. J. Stocker, "Evidence for tunneling in reverse-biased III-V photodetector diodes," Appl. Phys. Lett., 36 (1980) 580–583.

17. M. Ito, T. Kaneda, K. Nakajima, Y. Toyama, and H. Ando, "Tunneling current in In0.53Ga0.47As homojunction diodes and design of InGaAs/InP heterostructure avalanche photodiodes," Solid-State Electron., 24 (1981) 421–424.

18. Y. Matsumoto, "Diffusion of Cd and Zn into InP and InGaAsP (Eg = 0.95 − 1.35 eV)," Japan J. Appl. Phys., 22 (1983) 1699–1704.

19. E. O. Kane, "Theory of tunneling," J. Appl. Phys., 32 (1961) 83–91.

20. R. J. Nicholas, J. C. Portal, C. Houlbert, P. Perrier, and T. P. Pearsall, "An experimental determination of the effective masses for $Ga_xIn_{1-x}As_yP_{1-y}$ alloys grown on InP," Appl. Phys. Lett., 34 (1979) 492–494.

21. S. R. Forrest, R. F. Leheny, R. E. Nahory, and M. A. Pollack, "InGaAs photodiodes with dark current limited by generation-recombination and tunneling," Appl. Phys. Lett., 37 (1980) 322–325.

22. S. M. Sze, "Physics of Semiconductor Devices," (Wiley-Interscience, New York, 1969) 102–104.

23. A. S. Grove. "Physics and Technology of Semiconductor Devices," (John Wiley and Sons, Inc., New York, 1967) 173–175.

24. C. H. Henry, R. A. Logan, F. R. Merritt, and C. G. Bethea, "Radiative and nonradiative lifetimes in n-type and p-type 1.6 μm InGaAs," Electron. Lett., 20 (1984) 358–359.

25. W. Shockley, "Electrons and holes in semiconductors," (Van Nostrand Book Co., 1950).

26. S. M. Sze, "Physics of Semiconductor Devices," (Wiley-Interscience, New York, 1969) 96–102.

27. A. R. Rieben and D. L. Feucht, "nGe-pGaAs heterojunctions," Solid-State Electron., 9 (1966) 1055–1065.

28. K. Hasegawa, K. Ohnaka, M. Kubo, Y. Hori, and H. Serizawa, "InGaAs PIN photodiode with low-dark current," Extended abstract of 16th conf. on Solid State Devices and materials, (Kobe, 1984) 579–582.

29. T. Mikawa, S. Kagawa, and T. Kaneda, "InP/InGaAs PIN photodiodes in the 1 μm wavelength regio," Fujitsu Sci. Tech. J., 20 (1984) 201–218.

30. H. Ishihara, K. Makita, Y. Sugimoto, T. Torikai, and K. Taguchi, "High-temperature aging tests on planar structure InGaAs/InP PIN photodiodes with Ti/Pt and Ti/Au contact," Electron. Lett., 20 (1984) 654–656.

31. J. W. Osenbach and T. L. Evanosky, "Temperature-humidity-bias-behavior and acceleratio model for InP planar PIN photodiodes," J. Lightwave Technol., 8 (1996) 1865–1881.

32. J. E. Bowers and C. A. Burrus, "Ultrawide-band long-wavelength p-i-n photodetectors," J. Lightwave Technol., LT-5 (1987) 1339–1350.

33. Y. G. Wey, D. L. Crawford, K. Giboney, J. E. Bowers, M. J. Rodwell, P. Silvestre, M. J. Hafich, and G. Y. Robinson, "Ultrafast graded double-heterostructure GaInAs/InP photodiode," Appl. Phys. Lett., 58 (1991) 2156–2158.

34. I.-H. Tan, C.-K. Sun, K. S. Giboney, J. E. Bowers, E. L. Hu, B. I. Miller, and R. J. Capik, "120-GHz long-wavelength low-capacitance photodetector with an air-bridge coplanar metal waveguide," IEEE Photonic. Technol. Lett., 7 (1995) 1477–1479.

35. K. Kato, A. Kozen, Y. Muramoto, Y. Itaya, T. Nagatsuma, and M. Yaita, "110-GHz, 50%-efficiency mushroom mesa waveguide p-i-n photodiode for a 1.55 μm wavelength," IEEE Photonic. Technol. Lett., 6 (1994) 719–721.

36. A. Chin and T. Y. Chang. "Enhancement of quantum efficiency in thin photodiodes through absorptive resonance," J. Lightwave Technol., 9 (1991) 321–328.

37. K. Kishino, M. S. Ünlü, J.-I. Chyi, J. Reed, L. Arsenault, H. Morkoç, "Resonant cavity-enhanced (RCE) photodetectors," IEEE J. Quantum Electron., 27 (1991) 2025–2033.

38. O. Wada, T. Kumai, H. Hamaguchi, M. Makiuchi, A. Kuramata, and T. Mikawa, "High-reliability flip-chip GaInAs/InP photodiode," Electron. Lett., 26 (1990) 1484–1486.

39. M. Makiuchi, H. Hamaguchi, O. Wada, and T. Mikawa, "Monolithic GaInAs quad-pin photodiodes for polarization-diversity optical receivers," IEEE Photonic. Technol. Lett., 6 (1991) 536–537.

40. J. B. D. Soole and H. Schumacher, "InGaAs metal-semiconductor-metal photodetectors for long wavelength optical communications," IEEE J. Quantum Electron., 27 (1991) 737–752.

41. J.-W. Soe, C. Caneau, R. Bhat, and I. Adesida, "Application of indium-tin-oxide with improved transmittance at 1.3 μm for MSM photodetectors," IEEE Photonic. Technol. Lett., 5 (1993) 1313–1315.

42. R.-H. Yuang, J.-I. Chyi, Y.-J. Chan, W. Lin, and Y.-K. Tu, "High-responsivity InGaAs MSM photodetectors with semi-transparent schottky contacts," IEEE Photonic. Technol. Lett., 7 (1995) 1333–1335.
43. J. H. Kim, H. T. Griem, R. A. Friedman, E. Y. Chan, and S. Ray, "High-performance back-illuminated InGaAs/InAlAs MSM photodetector with a record responsivity of 0.98 A/W," IEEE Photonic. Technol. Lett., 4 (1992) 1241–1244.
44. O. Vendier, N. M. Jokerst, and R. P. Leavitt, "Thin-film inverted MSM photodetectors," IEEE Photonic. Technol. Lett., 8 (1996) 266–268.
45. D. Kuhl, E. H. Böttcher, F. Hieronymi, E. Dröge, and D. Bimberg, "Inductive bandwidth enhancement of sub-μm InAlAs-InGaAs MSM photodetectors," IEEE Photonic. Technol. Lett., 7 (1995) 421–423.
46. D. Wake, T. P. Spooner, S. D. Perrin, and I. D. Henning, "50 GHz InGaAs edge-coupled pin photodetector," Electron. Lett., 27 (1991) 1073–1075.
47. K. Kato, S. Hata, K. Kawano, J. Yoshida, and A. Kozen, "A high-efficiency 50 GHz InGaAs multimode waveguide photodetector," IEEE J. Quantum Electron., 28 (1992) 2728–2735.
48. Y. Akatsu, Y. Muramoto, K. Kato, M. Ikeda, M. Ueki, A. Kozen, T. Kurosaki, K. Kawano, and J. Yoshida, "Long-wavelength multimode waveguide photodiodes suitable for hybrid optical module integrated with planar lightwave circuit," Electron. Lett., 31 (1995) 2098–2100.
49. T. Takeuchi, T. Nakata, M. Tachigori, K. Makita, and K. Taguchi, "Design and fabrication of a waveguide photodiode for 1.55-μm-band access receivers," Japan J. Appl. Phys., 38 (1999) 1211–1214.
50. K. Kato, M. Yuda, A. Kozen, Y. Muramoto, K. Noguchi, and O. Nakajima, "Selective-area impurity-doped planar edge-coupled waveguide photodiode (SIMPLE-WGPD) for low-cost, low-power-consumption optical hybrid modules," Electron. Lett., 32 (1996) 2078–2079.
51. T. Takeuchi, T. Nakata, K. Fukuchi, K. Makita, and K. Taguchi, "A high-efficiency waveguide photodiode for 40-Gb/s optical receivers," IEICE Trans. Electron., E82-C (1999) 1502–1508.
52. T. Takeuchi, T. Nakata, K. Makita, and K. Taguchi, "Reliable input-facet structure of waveguide photodiodes for high-power input light," Extended Abstracts (The 60th autumn meeting), The Japan Society of Applied Physics, (1999) 985.
53. T. Takeuchi, T. Nakata, K. Makita, and M. Yamaguchi, "High-speed, high-power and high-efficiency photodiodes with evanescently coupled graded-index waveguide," Electron. Lett., 36 (2000) 1–2.
54. T. Ishibashi, S. Kodama, N. Shimizu, and T. Furuta, "High-speed response of uni-traveling-carrier photodiodes," Japan J. Appl. Phys., 36 (1997) 6263–6268.

55. T. P. Pearsall, M. Piskorski, A. Brochet, and J. Chevrier, "A $Ga_{0.47}In_{0.53}As/InP$ heterophotodiode with reduced dark current," IEEE J. Quantum Electron., QE-17 (1981) 255–259.

56. G. A. Davis, R. E. Weiss, R. E. LaRue, K. J. Williams, and R. D. Esman, "A 920-2650-nm high-current photodetector," IEEE Photonic. Technol. Lett., 8 (1996) 1373–1375.

57. N. Shimizu, N. Watanabe, T. Furuta, and T. Ishibashi, "InP-InGaAs uni-traveling-carrier photodiode with improved 3-dB bandwidth of over 150 GHz," IEEE Photonic. Technol. Lett., 10 (1998) 412–414.

58. Y. Miyamoto, M. Yoneyama, K. Hagimoto, T. Ishibashi, and N. Shimizu, "40 Gbit/s high sensitivity optical receiver with uni-traveling-carrier photodiode acting as decision IC driver," Electron. Lett., 34 (1998) 214–215.

59. N. Shimizu, K. Murata, A. Hirano, Y. Miyamoto, H. Kitabayashi, Y. Umeda, T. Akeyoshi, T. Furuta, and N. Watanabe, "40 Gbit/s monolithic digital OEIC composed of unitraveling-carrier photodiode and InP HEMTs," Electron. Lett., 36 (2000) 1220–1221.

60. H. Ito, T. Furuta, S. Kodama, and T. Ishibashi, "InP/InGaAs uni-traveling-carrier photodiode with 310 GHz bandwidth," Electron. Lett., 36 (2000) 1809–1810.

61. Y. Muramoto, K. Kato, M. Mitsuhara, O. Nakajima, Y. Matsuoka, N. Shimizu, and T. Ishibashi, "High-output-voltage, high speed, high efficiency uni-traveling-carrier waveguide photodiode," Electron. Lett., 34 (1998) 122–123.

62. H. Fukano, K. Kato, O. Nakajima, and Y. Matsuoka, "Low-cost, high-speed and high-responsivity photodiode module employing edge-illuminated refracting-facet photodiode," Electron. Lett., 35 (1999) 842–843.

63. H. Fukano, Y. Muramoto, and Y. Matsuoka, "High-speed and high-output voltage edge-illuminated refracting-facet photodiode," Electron. Lett., 35 (1999) 1581–1582.

64. A. G. Dentai, R. Kuchibhotla, J. C. Campbell, C. Tsai, and C. Lei, "High quantum efficiency, long wavelength InP/InGaAa microcavity photodiode," Electron. Lett., 27 (1991) 2125–2127.

65. K. S. Giboney, M. J. W. Rodwell, and J. E. Bowers, "Traveling-wave photodetectors," IEEE Photonic. Technol. Lett., 4 (1992) 1363–1365.

66. K. S. Giboney, R. L. Nagarajan, T. E. Reynolds, S. T. Allen, R. P. Mirin, M. J. W. Rodwell, and J. E. Bowers, "Travelling-wave photodetectors with 172-GHz bandwidth and 76-GHz bandwidth-efficiency product," IEEE Photonic. Technol. Lett., 7 (1995) 412–414.

67. L. Y. Lin, M. C. Wu, T. Itoh, T. A. Vang, R. E. Muller, D. L. Sivco, and A. Y. Cho, "High-power high-speed photodetectors—design, analysis, and experimental demonstration," IEEE Trans. Microwave Theory Techniques, 45 (1997) 1320–1331.

68. C. L. Goldsmith, G. A. Magel, B. M. Kanack, and R. J. Baca, "Coherent combining of RF signals in a traveling-wave photodetector array," IEEE Photonic. Technol. Lett., 9 (1997) 988–990.

69. H. Takahashi, Y. Hibino, Y. Ohmori, and M. Kawachi, "Polarization-insensitive arrayed-waveguide wavelength multiplexer with birefringence compensating film," IEEE Photonic. Technol. Lett., 5 (1993) 707–709.

70. S. Chandrasekhar, M. Zirngibl, A. G. Dentai, C. H. Joyner, F. Storz, C. A. Burrus, and L. M. Lunardi, "Monolithic eight-wavelength demultiplexed receiver for dense WDM applications," IEEE Photonic. Technol. Lett., 7 (1995) 1342–1344.

71. C. A. M. Steenbergen, C. Dam, A. Looijenl, C. G. P. Herben, M. Kok, M. K. Smith, J. W. Pedersen, I. Moerman, R. G. Baets, and B. H. Verbeek, "Compact low loss 8 × 10 GHz polarisation independent WDM receiver," Proc. 22th European Conf. Optical Communication, 1 (Oslo, 1996) 129–132.

72. M. Kohtoku, H. Sanjoh, S. Oku, Y. Kadota, and Y. Yoshikuni, "Polarization independent semiconductor arrayed waveguide gratings using a deep-ridge waveguide structure," IEICE Trans. Electron., E81-C (1998) 1195–1204.

73. R. Yoshimura, M. Kohtoku, H. Sanjoh, T. Kurosaki, S. Oku, H. Okamoto, and Y. Yoshikuni, "2.5Gb/s-16ch-100GHz receiver module using a semiconductor AWG-PD," Extended Abstract (The 2001 IEICE General Conference), The Institute of Electronics, Information and Communication Engineers (2001) 258.

74. R. Matz, J. G. Bauer, P. Clemens, G. Heise, H. F. Mahlein, W. Metzger, H. Michel, and G. Schulte-Roth, "Development of a photonic integrated transceiver chip for WDM transmission," IEEE Photonic. Technol. Lett., 6 (1994) 1327–1329.

75. G. H. Foster, J. R. Rawsthorne, J. P. Hall, M. Q. Kearley, and P. J. Williams, "OEIC WDM transceiver modules for local access networks," Electron. Lett., 31 (1995) 132–133.

76. Y. Yamada, S. Suzuki, K. Morikawa, Y. Hibino, Y. Tohmori, Y. Akatsu, Y. Nakasuga, T. Hashimoto, H. Terui, M. Yanagisawa, Y. Inoue, Y. Akahori, and R. Nagase, "Application of planar lightwave circuits platform to hybrid integrated optical WDM transmitter/receiver module," Electron. Lett., 31 (1995) 1366–1367.

77. T. Sasaki, M. Kitamura, and I. Mito, "Selective metalorganic vapor phase epitaxial growth of InGaAsP/InP layers with bandgap energy control in InGaAs/InGaAsP multiple-quantum well structures," J. Crystal Growth, 132 (1993) 435–443.

78. T. Takeuchi, T. Sasaki, M. Hayashi, K. Hamamoto, K. Makita, K. Taguchi, and K. Komatsu, "A transceiver PIC for bidirectional optical communication fabricated by bandgap energy controlled selective MOVPE," IEEE Photonic. Technol. Lett., 8 (1996) 361–363.

Chapter 9 | Avalanche Photodiodes

Masahiro Kobayashi and Takashi Mikawa

Fujitsu Quantum Devices, Limited
Kokubo Kogyo Danchi, Showa-cho, Nakakoma-gun,
Yamanashi-ken 409-3883, Japan

9.1. Introduction

Explosive growth of information due to the expanding Internet and mul-
timedia society has accelerated the further increase in transmission capacity and data rate in all the photonic network systems from long haul
to metropolitan and local access area. In the time-division multiplexing
(TDM) systems, 10 gigabit per second (Gbps) transmission has already
been established, and development of ultra high-speed transmissions of
40 Gbps has become an urgent issue. Rapid increase in transmission capacities of wavelength-division multiplexing (WDM) systems using fiber
amplifiers has also required widening of the usable spectral range beyond
1.6 μm wavelength (L-band region) [1, 2].

The heart of a receiver for any lightwave transmission system is the
optoelectric component that is used as a photodetector. In high-speed
and large-capacity TDM and WDM systems for 2.5 Gbps and 10 Gbps,
InP/InGaAs avalanche photodiodes (APDs) have played an especially important role as the key component satisfying system requirements. In
short- and/or medium-distance transmission systems such as metropolitan networks, the benefits of using APDs—cost effectiveness, low power
consumption, and compact size—have made them a very attractive solution for system design as well as very high-speed and high-sensitivity performances without fiber pre-amplifier [3]. Internal gain of the avalanche

379

WDM TECHNOLOGIES: ACTIVE
OPTICAL COMPONENTS
$35.00

photodiode improves the receiver sensitivity, which increases the maximum allowable loss design in the high-speed signal transmission channel. The first avalanche photodiode detector that had been made commercially available for long-wavelength ligthwave transmission systems (1.3 μm wavelength window) was the Germanium avalanche photodiode (Ge-APDs) [4]. It was the most useful avalanche photodiode detector in 1980s. Limitation of Ge-APD performance for dark current, multiplication noise, and sensitivity at longer wavelength window at 1.55 μm comes from its material parameters. To respond to the needs for higher sensitivity APDs both at 1.3 μm and 1.55 μm wavelength windows, compound semiconductor materials were widely utilized. InP has a larger bandgap energy and a smaller intrinsic carrier concentration than Ge, so low dark current can be natively expected, as well as higher speed operation by its k value (ionization coefficient ratio). InGaAs is a material lattice-matched to InP and has a narrow bandgap energy suitable for 1.55 μm light absorption.

To realize high sensitivity InP/InGaAs APDs, the key issues to be studied were the influence of the InP/InGaAs hereto-interface to the dark current and the frequency response and the design of a useful guard ring structure. Many studies have also been done to reduce and eliminate crystalline imperfections, which were present in the semiconductor starting materials or introduced during device fabrication of epitaxial growth and wafer process. Defects and imperfections in the high electric field active region resulted in unexpected generation of dark current and occurrence of localized multiplication. They often caused avalanche photodiodes to fail in the early stage of their operation lives. High quality InP, InGaAsP, and InGaAs alloy semiconductors have now been realized by several epitaxial growth methods, and device fabrication technologies have been refined as well. After many efforts to put InP/InGaAs avalanche photodiodes (InP/InGaAs-APDs) into practical use, they have been made commercially available since the mid to late 1980s [5–7]. Then, they shortly became indispensable components for optical front ends of receivers, to realize large-capacity and long-distance lightwave transmission systems. In a typical long-wavelength receiver design using a low noise pre-amplifier integrated circuit of GaAs FETs for 2.5 Gb/s operation, a receiver with an InP/InGaAs APD gives about 10 dB better sensitivity than that with an InGaAs PIN photodiode. No alternative to the InP/InGaAs APDs has become available presently for the long-wavelength and high-speed systems because manufacturing techniques, productivity, and high reliability have been refined and established for volume.

In this chapter we deal with APDs for long-wavelength lightwave transmission system applications. Fundamental design expressions for important characteristics will be derived in Section 9.2. Germanium APDs are briefly explained in Section 9.3 because they support long-wavelength optical communication systems in their early stages. In the following section, Section 9.4, the main subject will move to InP/InGaAs APDs. The most important structural concept for these devices, SAM (Separated Absorption and Multiplication region) structure, is first introduced. The key issue for realizing practical APDs is to form effective planer guard ring structure, and several examples are discussed. Then, the conventional InP/InGaAs APDs and APD/pre-amplifier-IC hybrid integration receivers for 10 Gbps system applications are reviewed. Reliability of InP/InGaAs APDs is also addressed in this section. APDs with a strain-compensated multiple quantum well (MQW) absorption layer to enhance the responsivity above 1.6 um wavelength are described. Recent studies of novel APDs meant to improve sensitivities for ultra-high-speed systems beyond 10 Gbps are surveyed in Section 9.5. These novel devices include structures such as the superlattice avalanche multiplication region, the thin multiplication region, and the Si/InGaAs hetero-interface (wafer fusion) type.

9.2.   Basic Design and Operation of Avalanche Photodiodes

Avalanche photodiodes used in the lightwave transmission systems are required to satisfy several performance criteria, such as high responsivity (quantum efficiency), high multiplication gain, wide bandwidth (fast response speed), and low noise generation to achieve better receiver sensitivity.

This section proceeds from addressing the avalanche multiplication process to the fundamental operation mechanism that influences device performance. The carrier impact-ionization process will be reviewed and analytical expressions that describe the basic operation of the avalanche photodiodes will be derived. These expressions are rather complicated but should be useful when considering the design of the practical avalanche photodiodes. Sensitivity of the APD receiver will also be explained in the last part of this section.

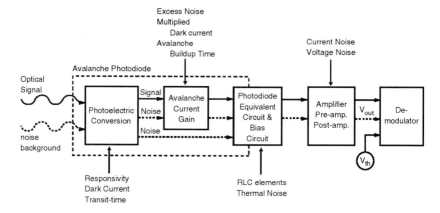

Fig. 9.1   Signal and noise flow along with photodetection, avalanche multiplication, and electrical amplification process in an avalanche photodiode and a receiver.

9.2.1.   DETECTION AND GAIN PROCESS OF AVALANCHE PHOTODIODE AND RECEIVER

Functional Diagram of APD Receiver

Figure 9.1 is a functional block diagram for a receiver with an avalanche photodiode at the front end. The system shows signal and noise flow through the receiver to the demodulation circuit as well as key parameters that influence the total performance [8–10]. The input light with signal and noise components is absorbed in the photodiode and converted into the current (primary photocurrent). The photocurrent may contain undesired excess optical noise (due to cross-coupling of laser frequency noise and amplitude noise, mixing of polarization mode, optical feedback-induced laser instability, and so on) and a background noise component in addition to the desired signal and quantum shot-noise. The excess optical noise, however, is not considered further here. The current gain mechanism of the photodiode multiplies both the signal current and the noise current. The multiplied noise current is composed of that from the noise of the input light and from a part of dark current that flows through the region in the photodiode where avalanche multiplication takes place. Then, the dark current is distinguished into the multiplied dark current and the unmultiplied dark current. The avalanche gain mechanism also generates an additional noise component, called excess noise. The APD includes equivalent electrical elements resulting from its physical structure. These elements could influence the signal and the noise as well as the bias circuit and the electrical elements from packaging for the APD. The APD is connected to the external circuit, which consists of an external load resistor in the simplest case, to deliver a signal and noise output. The resulting signal is amplified by electric amplifiers that introduce additional voltage noise and current noise, and finally the system delivers a necessary amplitude of signal voltage to the demodulator to extract the desired information. The receiver sensitivity will be discussed later in this section.

Avalanche Multiplication Process

The most important physical process in an avalanche photodiode is the avalanche multiplication process that appears in the functional diagram of Fig. 9.1. In a semiconductor material, carriers traveling through the high electric field region ($>1 \times 10^5$ V/cm) gain energy from the field and generate additional free electron–hole pairs by impact ionization of valence electrons into the conduction band, leaving free holes in the valence band. The secondary electrons can be accelerated by the electric field and generate more electron–hole pairs when they ionize other valence electrons. This process, with increasing number of carriers in the high electric field region, is called avalanche multiplication. The generation of electron–hole pairs can be described in terms of the impact-ionization coefficient or the ionization ratio, and these coefficients increase with applied electric field and decrease with elevating the material and/or device temperature. They correspond to the probability that the carrier will cause an impact-ionization in a unit length or the average distance a carrier will travel before impact-ionization. Their dimensions are in cm^{-1}. Here, we present the impact-ionization coefficient for electrons and holes as α and β, respectively.

The feedback by electrons and holes traveling opposite directions causes statistical fluctuation in the avalanche multiplication process, which is schematically explained in Fig. 9.2 [11]. The figure shows the multiplication gain of 8 (eight) both for the case of $\beta = 0$ (only electrons make impact-ionization) and the case of $\alpha = \beta$ (equal ionization coefficients). An electron that enters the high electric field region in Fig. 9.2(a) is rapidly accelerated and the avalanche proceeds from left to right. In Fig. 9.2(b), an electron moving to the right can generate a hole that moves to the left and create an electron that moves to the right, etc. This is positive feedback, and it is the cause of the excess shot-noise due to the avalanche multiplication process. From the figures, all the carriers are collected in shorter period

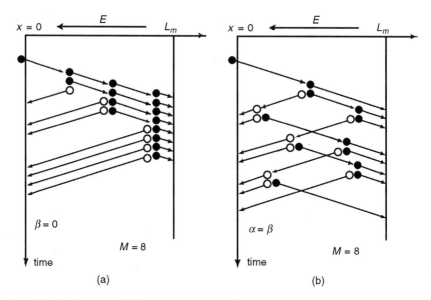

Fig. 9.2 Avalanche multiplication process for $M = 8$ (a) in the case that only electrons contribute to impact-ionization ($\beta = 0$) and (b) in the case that both electrons and holes impact-ionize equally ($\alpha = \beta$).

of time and fluctuation becomes smaller when only electrons contribute to the ionization than when both the carriers do. The uncertainty in the multiplication process and the extra time involved to clean all the ionized carriers would reduce and restrict performance of APDs. The actual characteristics depend strongly on the exact location at which the carrier enters the multiplication region and the spatial variation of α and β within the region.

Electron and hole ionization coefficients, α and β, greatly influence APD performance such as current multiplication gain, multiplication noise, and response speed characteristics. For a given temperature, the ionization coefficients are exponentially dependent on the electric field and have a functional form of

$$\alpha(E) = A_e \exp(-B_e/E) \quad \text{and} \quad \beta(E) = A_h \exp(-B_h/E) \quad (9.1)$$

where A_e, A_h, B_e, and B_h are experimentally determined constants, and E is the magnitude of the electric field. Measurement data for InP and InGaAs and other key materials are summarized in Fig. 9.3 [12–16]. An important factor for designing APD performance is the ionization coefficient ratio, $k = \alpha/\beta$ (or $k = \beta/\alpha$ when α is larger). Semiconductor material with a

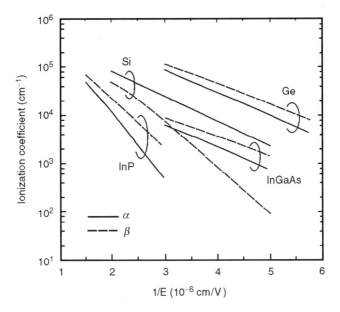

Fig. 9.3 Electron and hole ionization coefficients of various semiconductor materials used for long wavelength avalanche photodiodes.

large difference in the ionization coefficients is desirable to achieve better performance for lightwave transmission use.

9.2.2. BASIC PERFORMANCE EXPRESSIONS OF APD

The functional block diagram could be linked with a typical avalanche photodiode structure and made more applicable to a description of its basic performance expressions based on the structure. Because the current DWDM systems usually employ InP/InGaAs avalanche photodiodes, the structural modeling is based on and resembles APDs.

Expression for Multiplication

An avalanche photodiode composed of a p^+-i-n^+ multiplication region extending from $x = 0$ to $x = L_m$ and a depletion region extending to $x = w_d$ is considered and modeled as shown in Fig. 9.4. The stack of multiple layers with p^+-i-n^+-n^--n^+ (substrate) impurity doping profile is shown in Fig. 9.4(a) and (b). The electric field distribution when a reverse bias is applied to the device is drawn in Fig. 9.4(c). Figure 9.4(d) indicates

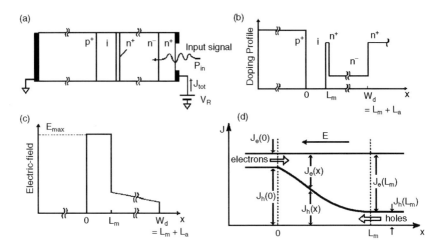

Fig. 9.4 Simplified model of avalanche photodiode: (a) structural concept by stack of multiple layers, (b) image of the doping profile, (c) image of electric field, and (d) schematic diagram of avalanche multiplication region. (The electric field is applied between $x = 0$ to L_m. Definition of direction of electric field and electron and hole currents are also defined.)

the definitions of current flows. The direction of the field is assumed so that the electrons travel in the positive x direction and the holes drift to the negative x direction with velocities of v_e and v_h, respectively. $J_e(x)$ and $J_h(x)$ are the electron and hole current densities and defined as in the figure. The equations that describe the electron and hole current transportation through the avalanche multiplication region are as follows [9, 17, 18]:

$$\frac{1}{v_h}\frac{\partial J_h(x,t)}{\partial t} - \frac{\partial J_h(x,t)}{\partial x} = \alpha(x)J_e(x,t) + \beta(x)J_h(x,t) + qG(x,t)$$

(9.2a)

$$-\frac{1}{v_e}\frac{\partial J_e(x,t)}{\partial t} + \frac{\partial J_e(x,t)}{\partial x} = \alpha(x)J_e(x,t) + \beta(x)J_h(x,t) + qG(x,t)$$

(9.2b)

where q is the electronic charge, $qG(x,t)$ is the sum of the thermal and optical current generation rates. The electron and hole ionization coefficients $\alpha(x)$ and $\beta(x)$ are functions of the electric field at x. For the steady-state condition ($\partial/\partial t = 0$), the expression for the position-dependent multiplication gain at the position $x = x_0$, where an electron–hole pair is generated in or injected into the multiplication region is obtained from Eq. (9.2)

$$M(x_0) = \frac{\exp\left[-\int_0^{x_0}(\alpha(x) - \beta(x))\,dx'\right]}{1 - \int_0^{L_m}\alpha(x)\exp\left[-\int_0^x(\alpha(x) - \beta(x))\,dx'\right]dx} \quad (9.3)$$

For $\beta > \alpha$, hole injection from the n^+-side edge of the multiplication region makes the holes travel the entire width of the multiplication region, and optimum multiplication could be achieved. In the case of $\alpha > \beta$, the electron should be injected from the p^+-side edge of the multiplication region. An injection scheme with the higher ionization coefficient carrier is important to obtain the lowest possible excess-noise factor for the given material. The multiplication factor M of the specific device can be defined as the ratio of the total current density that flows through the avalanche photodiode J_{tot} to the total injection current density into the multiplication region J_0

$$M = \frac{J_{tot}}{J_0} = \frac{J_e(0)M(0) + J_h(L_m)M(L_m) + q\int_0^{L_m}G(x)M(x)\,dx}{J_e(0) + J_h(L_m) + q\int_0^{L_m}G(x)\,dx} \quad (9.4)$$

where $J_e(0)$ is the electron injection current at $x = 0$ (p^+-side edge of the multiplication region), and $J_h(L_m)$ is the hole injection current at $x = L_m$ (another edge of the multiplication region). The avalanche breakdown voltage, which is not influenced by the type of carrier injection, is defined as that voltage where the multiplication factor becomes infinite, and it can be expressed as [19]

$$\int_0^{L_m}\alpha(x)\exp\left[-\int_0^x(\alpha(x) - \beta(x))\,dx'\right]dx = 1 \quad (9.5)$$

In the actual device, the multiplication factor is dependent on current levels that flow through the device. There are three factors that contribute to this characteristic. Voltage drop across device series resistance (contact resistances and undepleted bulk semiconductor regions) lowers the actual applied voltage to the p-n junction. Carriers drifting through the depletion region reduce the electric field (space-charge effects). The electron and hole ionization coefficients are reduced by elevation of junction temperature due to device thermal resistance. The multiplication factor can be empirically expressed as

$$M = \frac{1}{1 - [(V - I_{tot} \cdot R_s)/V_B]^n} \qquad (9.6)$$

where I_{tot} is the total current, R_s is the device series resistance, V_B is the breakdown voltage, and the exponent n is an adjustable parameter that could be obtained by fitting Eq. (9.6) to the experimental data [19]. This exponent depends on the material parameters and device structure. The maximum achievable multiplication factor for a given primary photocurrent I_0 is then [10]

$$M_{max} = \sqrt{V_B/nI_0R_s} \propto 1/\sqrt{I_0} \qquad (9.7)$$

Localization of avalanche multiplication also limits the electric field that can be maintained over the entire active area of the device (micro-plasma effect). This is mainly associated with crystalline defects and imperfections and/or doping fluctuations introduced during the device fabrication process.

Response Speed

The response speed of avalanche photodiodes can be an important factor in determining the performance of high-bit-rate optical front-end receivers. The response speed is generally limited by a carrier transit-time effect and a RC parasitic effect (RC time-constant) when carriers are well absorbed in the depletion region and carrier diffusion effects can be neglected.

When holes are the carriers of interest (for example, $\beta > \alpha$), the transit-time of the avalanche photodiode shown in Fig. 9.4 consists of three components [10]. They are the time for holes to travel the absorption region, the time for holes to transit the multiplication region, and the time for electrons generated by the avalanche multiplication process to travel both the multiplication and absorption regions to the n-type neutral region. The transit-time τ_{tr} could be approximately written as

$$\tau_{tr} = (L_m + L_a)\frac{v_h + v_e}{v_h v_e} \qquad (9.8)$$

where $L_m + L_a$ is the total depletion width, v_e and v_h are the electron and hole mean velocities, respectively. For the assumption that all primary carriers (holes) are generated at the end of the depletion region and they drift the entire depletion layer to the multiplication region, the normalized frequency response determined by the carrier transit-time $F_{tr}(\omega)$ could be expressed as

$$F_{tr}(\omega) = \left| \frac{1 - \exp(i\omega\tau_{tr})}{i\omega\tau_{tr}} \right| = \frac{\sin(\omega\tau_{tr}/2)}{\omega\tau_{tr}/2}. \tag{9.9}$$

This should be the limited case of response, and the 3-dB cut-off frequency is $0.446/\tau_{tr}$.

The ultimate speed of an APD does not equal that of a p-i-n photodiode with the same traveling distance of photo-generated carriers. There is an inherent time delay because the regenerative nature of the impact ionization process requires longer periods to build up the avalanche multiplication than a single path transit across the multiplication region. The relation of multiplication factor and bandwidth for a p-i-n structure APD was examined by Emmons [17] and Chang [18]. The current transportation equations are given in Eq. (9.2), and they can be solved to determine the frequency response of the multiplication factor $M(\omega)$ under the simplification of constant electron and hole velocities (at the saturated velocities) and constant electron and hole ionization coefficients in the multiplication region. The frequency variation of the multiplication for the hole injection scheme can be represented by an equation of the form

$$M(\omega) = M_0/\sqrt{1 + (\omega M_0 \tau_m)^2} \tag{9.10}$$

where ω is the angular frequency, M_0 is the dc (low frequency at $\omega \to 0$) multiplication factor, and τ_m is the effective transit-time. The avalanche buildup time corresponding to the dc multiplication factor is

$$\tau_{av} = \tau_m M_0 \tag{9.11}$$

The effective transit time depends on the reduced effective carrier ionization coefficient ratio $k_{eff} = (\alpha/\beta)_{eff}$ (which is different from α/β as a function of electric field) and the effective carrier drift velocity v_{eff} in the multiplication region, and it is expressed by the formula of

$$\tau_m = N\left(\frac{L_m}{v_{eff}}\right)k_{eff} \tag{9.12}$$

where N is a slowly varying number between $1/3$ (for $\alpha = \beta$) and 2 (for $\alpha/\beta = 10^{-3}$). To achieve large gain-bandwidth products, L_m and k_{eff} should be small. Kaneda [20] experimentally derived $\tau_m = 5 \times 10^{-12}$ sec for Ge. In the case of InP, $\tau_m = 1.5 \times 10^{-12}$ sec can be expected from the gain-bandwidth products of InP/InGaAs APDs reported so far [21–25].

An approximate frequency response of an APD could be written by the three factors including the RC-time limitation, and the normalized expression is

$$F_{APD}(\omega) = F_{tr}(\omega) \cdot \frac{M(\omega)}{M_0} \cdot F_{RC}(\omega)$$

$$= \frac{\sin(\omega\tau_{tr}/2)}{\omega\tau_{tr}/2} \cdot \frac{1}{\sqrt{1 + (\omega M_0 \tau_m)^2}} \cdot \frac{1}{\sqrt{1 + (\omega\tau_{RC})^2}} \qquad (9.13)$$

where $\tau_{RC} = (R_s + R_L)C_t$ is the RC-time constant for the device capacitance, C_t, the series resistance of the device, R_s, and the load resistor of the measurement circuit (or receiver circuit), R_L. C_t is the capacitance measured between the anode and the cathode terminals of the device and usually includes parasitic effects due to packaging. This equation is reasonably expressive for the case where the multiplication factor is larger than $1/k_{eff}$. In the case where the multiplication factor is less than $1/k_{eff}$, the RC time-constant and carrier transit-time then determine the achievable bandwidth. Because the avalanche buildup time increases as the multiplication increases, the cut-off frequency of the APD decreases with increasing multiplication factor and a gain-bandwidth product can become evident [17, 18].

Avalanche Multiplication Noise

The average current flowing through the diode contributes the mean-square shot-noise current, which is $\langle i_s^2 \rangle = 2qIB$ in a bandwidth B. A carrier generated in the multiplication region is multiplied by a factor of $M(x)$, then the noise current at position x could be multiplied by a factor of $M(x)$. A small element of dx is considered and the variation of the spectral noise density for the unit bandwidth $d\phi$ is written as [26, 27]

$$d\phi = 2q \cdot dJ(x) \cdot M^2(x)$$

$$= 2q\{\alpha(x)J_e(x) + \beta(x)J_h(x) + qG(x)\}M^2(x)\,dx \qquad (9.14)$$

$\alpha(x)J_e(x)$ and $\beta(x)J_h(x)$ represent the electron–hole pair generation rate by the electron current J_e and hole current J_h at the position x, respectively. Integration of Eq. (9.14) gives the total noise spectral density ϕ. Consider a typical boundary condition so that photo-carrier generation in the multiplication region could be neglected, such as a thin multiplication region with $G(x) = 0$ and multiplication initiation by electron and hole currents injections at $x = 0$ and $x = L_m$, respectively. Assuming a constant ionization coefficient ratio in the multiplication region

$$\phi = 2q\, J_{h0} M_h F_h(M_h) + 2q\, J_{e0} M_e F_e(M_e) \qquad (9.15)$$

where $F_h(M_h)$ and $F_e(M_e)$ are the excess noise factors for pure hole injection and pure electron injection, respectively, and $J_e(0) = J_{e0}$, $J_h(L_m) = J_{h0}$, $M(0) = M_e$, $M(L_m) = M_h$. M_h and M_e are the multiplication by hole injection and electron injection, respectively, and $J_{h0} + J_{e0} = J_0$. The multiplication factor of the device is, then,

$$M = \frac{J_{h0} M_h + J_{e0} M_e}{J_{h0} + J_{e0}} \qquad (9.16)$$

k_{eff} is the reduced effective ionization coefficient ratio $(\alpha/\beta)_{eff}$, which is different from the α/β as a function of electric field. The excess noise factors are

$$F_h(M_h) = M_h \left\{ 1 - (1 - k_{eff}) \cdot \left[1 - \frac{1}{M_h} \right]^2 \right\}, \quad \text{and} \qquad (9.17a)$$

$$F_e(M_e) = M_e \left\{ 1 - \left(1 - \frac{1}{k_{eff}} \right) \cdot \left[1 - \frac{1}{M_e} \right]^2 \right\} \qquad (9.17b)$$

It is inevitable that the shot-noise is multiplied by M^2 times and extra noise is added by the multiplication process in APD operation. However, designing better noise performance by selecting materials with a smaller ionization coefficient ratio $k_{eff}(=\alpha/\beta$ or $\beta/\alpha)$ and making proper carrier injection profile is possible. Excess noise factor can be calculated from Eq. (9.15) to Eq. (9.17) for various k_{eff} values as shown in Fig. 9.5. Excess noise factor F depends strongly on the ionization coefficient ratio k_{eff}. When the hole injection current is dominant, a smaller k_{eff} value gives lower noise performance. When the ionization coefficient of holes is larger than that of electrons, it is obvious that the pure hole injection condition gives the low noise performance. $k_{eff} = \alpha/\beta = 0.4$ corresponds to the ionization coefficient ratio of InP [14, 15], which is one of the most important materials now composing long-wavelength APDs.

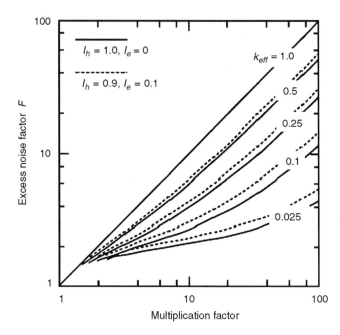

Fig. 9.5 Excess noise factors calculated from Eqs. (9.16) and (9.17). For various k_{eff} when $I_{e0} = 0.1$ and $I_{h0} = 0.9$ and $I_{e0} = 0$ and $I_{h0} = 1.0$.

Dark Current

The dark current I_d is another important characteristic of an APD, and it can be functionally expressed by the sum of the multiplied dark current I_{dm} and the unmultiplied dark current I_{d0} as

$$I_d = M_0 I_{dm} + I_{d0} \qquad (9.18)$$

The origin of the dark current can be classified into three components under a low reverse bias voltage. They are the surface leakage current that flows along the interface between the semiconductor and the dielectric passivation film $I_{d,surface}$, the generation-recombination (g-r) current in the depletion region under the reverse bias $I_{d,dep}$, and the diffusion current of minority carriers from outside the depleted region $I_{d,diff}$. The expressions connecting to the temperature dependence (activation energies) are given using the intrinsic carrier concentration $n_i \propto \exp(-E_g/2k_BT)$ as follows [28]:

$$I_{d,surface} = \frac{1}{2} q S n_i A_s \propto \exp\left(-\frac{E_g}{2k_BT}\right) \qquad (9.19a)$$

$$I_{d,dep} = \frac{1}{2}qW_d\left(\frac{n_i}{\tau_{eff}}\right)A_j \propto \exp\left(-\frac{E_g}{2k_BT}\right) \quad \text{and} \quad (9.19b)$$

$$I_{d,diff} = q\frac{n_i^2}{L_hN_D}A_j \propto \exp\left(-\frac{E_g}{k_BT}\right) \quad (9.19c)$$

where S is the surface recombination velocity, W_d is the depletion region width, τ_{eff} is the effective carrier lifetime in the depletion region, L_h is the diffusion length of holes, and N_D is the carrier concentration of the n-type substrate. A_s and A_j are the areas of the surface depletion region and the p-n junction of the APD. Tunneling dark current is the fourth component, which should be carefully taken into account for direct-transition narrow bandgap material such as InGaAs under a high reverse bias near the breakdown voltage [29,30]. Measuring temperature dependence and reverse bias voltage dependence of dark current can be a useful method to examine the source of the dark current.

9.2.3. SENSITIVITY OF APD RECEIVER

The processes from the conversion of the incoming optical signal to the output of the electrical signal to the external circuit have been reviewed in the previous subsections. The functional block diagram of Fig. 9.1 can be modified to describe an APD receiver in some detail with the addition of signal and noise sources onto the schematic circuit diagram. As the preamplifier design is not the subject of this chapter, we will only look into analytical expressions for the APD receiver used in the digital lightwave transmission system application and describe the advantage of APD over the p-i-n photodiode receiver.

Signal and Noise Expressions of APD Receiver

The equivalent circuit of an APD with the signal and noise current sources and electrical circuit elements can be combined with the equivalent input noise current of the following amplifiers as shown in Fig. 9.6. The noise sources in the equivalent circuit have the dimension of $pA/\sqrt{Hz}$ and will contain contributions from i_{shot} for the shot-noise associated with the signal photocurrent I_s, i_{dm} for the multiplied dark current I_{dm}, i_{d0} for unmultiplied dark current I_{d0}, and $i_{circuit}$ for the receiver electronics. The single equivalent noise current source $i_{circuit}$ is placed at the input of a noise-free amplifier with a transimpedance gain of $Zt(\omega)$. Considering the noise

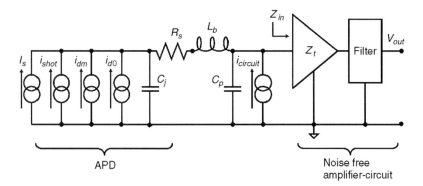

Fig. 9.6   Equivalent circuit of modified APD with a following amplifier circuit showing signal and noise sources and equivalent electronic components.

equivalent bandwidth of the receiver, B_{eq}, that is determined by the combination with filter to attain either the maximum signal-to-noise ratio, the minimum intersymbol-interference, or the maximum jitter tolerance, the total mean noise current from each noise current source in units of amps is [8, 31, 32]

$$I_{n\text{-}shot} = \sqrt{\langle i_{shot}^2 \rangle B_{eq}} = \sqrt{2q\, I_s M^2 F(M) B_{eq}} \qquad (9.20a)$$

$$I_{n\text{-}dm} = \sqrt{\langle i_{dm}^2 \rangle B_{eq}} = \sqrt{2q\, I_{dm} M^2 F(M) B_{eq}} \qquad (9.20b)$$

$$I_{n\text{-}d0} = \sqrt{\langle i_{d0}^2 \rangle B_{eq}} = \sqrt{2q\, I_{d0} B_{eq}} \qquad (9.20c)$$

and

$$I_{n\text{-}circuit} = \sqrt{\langle i_{circuit}^2 \rangle B_{eq}} \qquad (9.20d)$$

The optical signal is assumed to be the 50% duty cycle NRZ, and $P_{in\text{-}one}$ and $P_{in\text{-}zero}$ are the one-level and the zero-level optical input power, respectively. The signal currents observed at the input of the preamplifier for the one-level and the zero-level are

$$I_{s\text{-}one} = P_{in\text{-}one} R_{APD} M = (q\eta/h\nu) P_{in\text{-}one} M \qquad (9.21a)$$

$$I_{s\text{-}zero} = P_{in\text{-}zero} R_{APD} M = (q\eta/h\nu) P_{in\text{-}zero} M \qquad (9.21b)$$

where M is the average multiplication of the APD, R_{APD} is the responsivity of APD, and $R_{APD} = q\eta/h\nu$ using quantum efficiency of APD η and photon energy $h\nu$.

Using the Gaussian approximation in combination with the signal and noise currents of an APD, the probability densities for a one-level and a zero-level in the APD receiver are

$$P_{one}(z) = \frac{1}{\sqrt{2\pi}\,\sigma_{one}} \exp\left[-\frac{(z-\mu_{one})^2}{2\sigma_{one}^2}\right] \quad \text{and} \quad (9.22a)$$

$$P_{zero}(z) = \frac{1}{\sqrt{2\pi}\,\sigma_{zero}} \exp\left[-\frac{(z-\mu_{zero})^2}{2\sigma_{zero}^2}\right] \quad (9.22b)$$

where μ_{one} and μ_{zero} are the mean current at the one-level and at the zero-level, respectively, and σ_{one} and σ_{zero} are the total noise current at the one-level and at the zero-level, respectively. When optical excess noise (relative intensity noise from laser transmitter and waveform distortion caused by dispersion of lightwave channel) and optical background noise can be neglected, they are obtained from

$$\mu_{one} = M(P_{in\text{-}one}R_{APD} + I_{dm}) + I_{d0} \quad (9.23a)$$

$$\mu_{zero} = M(P_{in\text{-}zero}R_{APD} + I_{dm}) + I_{d0} \quad (9.23b)$$

$$\sigma_{one}^2 = 2qM^2F(M)P_{in\text{-}one}R_{APD}B_{eq} + 2q(I_{dm}M^2F(M) + I_{d0})B_{eq}$$
$$+ i_{circuit}^2 B_{eq} \quad (9.23c)$$

and

$$\sigma_{zero}^2 = 2qM^2F(M)P_{in\text{-}zero}R_{APD}B_{eq} + 2q(I_{dm}M^2F(M) + I_{d0})B_{eq}$$
$$+ i_{circuit}^2 B_{eq} \quad (9.23d)$$

B_{eq} is equivalent to Personick integrals that are often described as BI_1 and BI_2, where $I_1 = I_2 = 0.56$ for NRZ signal when the receiver implements a 100% raised-cosine filter function.

Q-Factor

One of the best ways to measure the data transmission quality of any digital transmission system is to determine the bit-error-rate (BER) characteristics. BER is greatly influenced by the optical signal-to-noise ratio (OSNR) of the system, which can't be easily tested because of the many aspects that contribute to noise in an optical transmission system, especially in the case of long-distance DWDM systems using many optical amplifiers. OSNR is the ratio between the received signal and the additive noise of the optical line. Many optical networks need to operate at low error rate, usually on the order of 1×10^{-12} to 1×10^{-15}. Use of Q-factor has become the new quality measure to approximate the BER in a system. Q-factor can be a quantitative parameter of quality of an eye as well as eye aperture. It is usually derived from making measurements of BER as a function of threshold level near the upper and lower rails of eyes at the input of the demodulator circuit to determine the σ_{one} and σ_{zero} in Eq. (9.23). The measured one-level is $\mu_{one}Zt$ and the measured variance is $\sigma_{one}Zt$. The mathematical expression of Q-factor is then obtained and the minimum BER can also be predicted by the Q-factor as follows:

$$Q = \frac{|\mu_{one} - \mu_{zero}|}{\sigma_{one} + \sigma_{zero}} \tag{9.24}$$

$$\mathrm{BER} = \frac{1}{\sqrt{2\pi}\,Q} \exp(-Q^2/2) \tag{9.25}$$

Because the variances of two Gaussian distribution in Eq. (9.22) are not equal, the optimum threshold is not positioned halfway between $\mu_{one}Zt$ and $\mu_{zero}Zt$. The threshold is set approximately at

$$Vth_{opt} = \frac{\sigma_{zero}\mu_{one} + \sigma_{one}\mu_{zero}}{\sigma_{one} + \sigma_{zero}} Zt \tag{9.26}$$

The calculation assumes that the signal levels, μ_{one} and μ_{zero}, are constant irrespective of the bit pattern (in the actual transmitter/receiver pairs, the signal levels could show any bit-pattern dependence if there is a considerable intersymbol-interference, etc.).

Receiver Sensitivity

For the case of $P_{in\text{-}zero} = 0$ (at a high extinction ratio, for example) and reasonable level of input signal power and multiplication factor, the multiplied signal shot-noise is the dominant factor during the one-level and the multiplied dark current becomes the dominant factor during the zero-level. Equation (9.24) can be written as

$$Q = \frac{P_{in\text{-}one} R_{APD} M}{\sqrt{2q\,P_{in\text{-}one} M^2 F(M) R_{APD} B_{eq}} + \sqrt{2q\,I_{dm} M^2 F(M) B_{eq}} + I_{circuit}^2 B_{eq}} \tag{9.27}$$

Solving this equation means the required optical power to the APD for the specified Q-factor can be obtained when the threshold of the receiver is

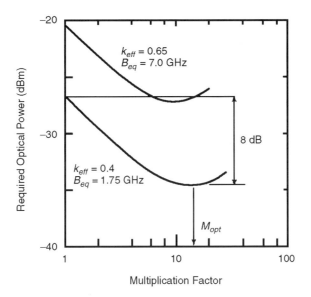

Fig. 9.7 Calculated receiver sensitivity for BER $= 10^{-12}$ versus multiplication factor. The data with $k_{eff} = 0.65$ represent the case of a 10 Gbps InP/InGaAs APD receiver and the data with $k_{eff} = 0.4$ represent the case of a 2.5 Gbps receiver.

assumed to be set near the optimum [32, 33].

$$P_{in\text{-}one} = \frac{2Q}{R_{APD}} \left[q\,F(M)\,Q\,B_{eq} + \sqrt{2q\,I_{dm}\,F(M)\,B_{eq} + \frac{I_{circuit}^2}{M^2}} \right] \qquad (9.28)$$

When a system needs the 10^{-12} error rate to specify the sensitivity of the APD receiver, $Q = 7$ is necessary. A calculated result is plotted in Fig. 9.7, where the parameters of the APD and circuit are assumed to resemble practical models ($k_{eff} = 0.4$ for InP/InGaAs APD). From the figure the advantage of the APD receiver can be obviously determined against the p-i-n photodiode receiver. From the calculation for 2.5 Gbps system applications ($B_{eq} = 1.75$ GHz), some 8 dB better sensitivity than PIN-PD receiver is confirmed, and about 8 dB improvement of sensitivity is also attainable for 10 Gbps applications ($B_{eq} = 7$ GHz). The effect of the circuit noise current is reduced by the multiplication factor of the APD, improving the sensitivity of the APD receiver. Note that the frequency variation of the multiplication factor is not considered in the discussion, which might influence when the gain-bandwidth product is considerably smaller than the bit rate of interest.

In higher bit-rate lightwave transmission systems, the minimum achieved sensitivity of the APD receiver becomes worse. This is due to widening of the receiver noise bandwidth and the ultimate achievable APD bandwidth at necessary gain. The first factor is a common subject for any wide-bandwidth receiver and the second factor determines the limitation of the APD receiver application. The largest GB product reported for InP/InGaAs APD so far is around 100 GHz, which implies that the highest bit-rate applicable to this APD and APD receiver would be 10 Gbps (OC-192). The superiority of APD receiver sensitivity then reduces beyond 10 Gbps.

9.3. Germanium Avalanche Photodiodes

The first Germanium avalanche photodiode (Ge-APD) was reported in 1966 by Melchior [19], and studies for putting them into practical use were activated after the minimum loss window of silica optical fiber was shifted from the 0.8-μm window to the 1.3-μm window in 1976 [4, 34–36]. In the early 1980s Ge-APDs achieved satisfactory results, and throughout the 1980s they were the most practical devices in the optical front-end of 1.3-μm wavelength lightwave transmission systems. In this section, we will briefly touch on structure, fabrication, and performance of Ge-APDs.

9.3.1. STRUCTURE AND FABRICATION

Ge has a large absorption coefficient at 1.3 μm (7.5×10^3 cm^{-1}) and large ionization coefficients ($\alpha, \beta > 10^4$ cm^{-1}) at a relatively low electric field of around 2×10^5 V/cm. Ge-APDs generally show high responsivity and low breakdown voltage around 30 volts. Of several structures that have been developed, the p$^+$-n structure Ge-APD was the only structure introduced into commercial application in volume [4, 36]. The cross section of the p$^+$-n structure Ge-APD is shown in Fig. 9.8 with illustrations of the electric field and the carrier absorption profiles in the device. The structure was fabricated from the n-type substrate with carrier concentration of 8×10^{15} cm^{-3} (resistivity of about 0.25 Ω-cm). The guard ring is formed by Beryllium (Be) ion implantation and annealing, which makes about 5-μm-deep graded junction by Be impurity profile tail. The thin p$^+$-region of about 0.2 μm in depth is formed by Indium (In) or Boron (B) ion implantation to produce an abrupt p$^+$-n junction. Realization of the thin p$^+$-region was extremely important to achieving both a high quantum efficiency and a small excess

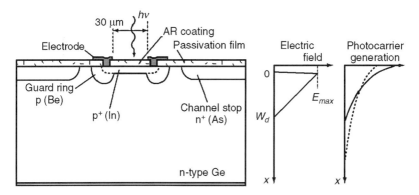

Fig. 9.8 Cross section of the p^+-n structure Ge-APD and schematic illustration of the electric field profile and photoabsorption profile in the device for $\lambda = 1.3$ μm (solid line) and $\lambda = 1.55$ μm (dotted line) lights.

noise. Photo-generated electrons near the surface of the p^+-region are easy to recombine because of a large surface recombination velocity of minority carriers, and they don't contribute to the quantum efficiency. Some of the photo-generated electrons are injected into the multiplication region. It results in a mixed injection scheme. The ionization coefficient of holes is larger than that of electrons in Ge, as shown in Fig. 9.3. The wider the p^+-region width is, the more the electrons are injected and the worse the excess noise factor becomes. Employment of ion implantation technique made high controllability of both the shallow and deep junction profiles possible so that the productivity was much improved from using the general thermal diffusion methods. Another feature of this structure is the channel stop that avoids the surface inversion layer extending farther away from the junction and increasing surface leakage current. Arsenic (As) was used for the ion implantation species to make the n^+-channel stop layer. SiO_2 by thermal chemical-vapor-deposition (CVD) method was used for surface passivation and SiN_x by plasma-assisted CVD was used for the antireflection (AR) film.

9.3.2.   DEVICE CHARACTERISTICS

Typical characteristics of dark current and multiplication factor versus reverse bias voltage for the p^+-n structure Ge-APD with the photosensitive area diameter of 30 μm are shown in Fig. 9.9. The breakdown voltage V_B is about 30 volts and the dark current at 90% of the breakdown voltage is about 50 nA. The multiplication was measured under 1.3 μm wavelength

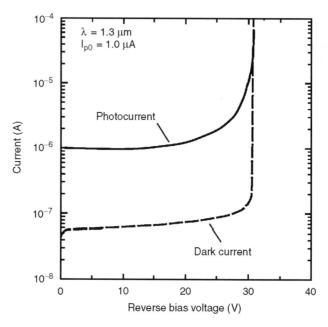

Fig. 9.9 Typical characteristics of dark current and photocurrent multiplication versus reverse bias voltage of p$^+$-n structure Ge-APD with 30 μm photosensitive area diameter.

input light. The multiplication factor M is defined as the ratio of the measured external photocurrent at increased reverse bias voltage to that at a lower bias voltage (1 volt is usually defined for Ge-APD). The maximum multiplication factor more than 100 was achieved at the primary photocurrent I_{p0} of 1 μA. The hole diffusion length in n-type Ge is much larger (about 50 μm) than the width of the depletion region (about 3 μm), making the dark current component by the carrier diffusion outside from the depletion region considerably larger and less reverse bias dependent. Then, increment of the dark current with the reverse bias voltage is all considered as the multiplied dark current component. Setting the two reference reverse bias voltages, V_1 and V_2, the multiplied dark current is [36]

$$I_{dm} = \frac{I_d(V_2) - I_d(V_1)}{M(V_2) - M(V_1)} \qquad (9.29)$$

The dark current and the multiplied dark current are severely dependent on the junction temperature T_j because of exponential dependence on the inverse of the junction temperature as $\exp(-E_g/k_B T_j)$, where k_B is the Boltzmann constant and $E_g = 0.67$ eV for Ge. Temperature dependence of I_d at 90% of V_B and I_{dm} is shown in Fig. 9.10 and components of

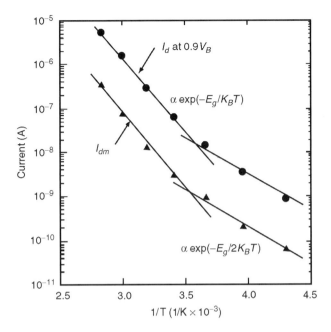

Fig. 9.10   Plot of dark current and multiplied dark current versus reciprocal temperature for p^+-n structure Ge-APD with 30 μm photosensitive area diameter.

dark current can be determined by referring to Eq. (9.19). Above 15°C the dark current varies depending on $\exp(-E_g/k_BT_j)$, which means the carrier diffusion from the undepleted region is dominant. The dependence on $\exp(-E_g/2k_BT_j)$ at lower junction temperature shows that the generation-recombination current and/or the surface leakage current is the dominant component of the dark current. The activation energy of the multiplied dark current is also 0.67 eV. Ge-APD receivers then suffer from sensitivity degradation at higher operating temperatures. For 1.3 μm wavelength light, the gain-bandwidth product of the p^+-n structure Ge-APDs is about 30 GHz and the maximum bandwidth is about 3.0 to 3.5 GHz for the 30 μm photosensitive area diameter devices.

For 1.55 μm wavelength application, the p^+-n structure Ge-APD can't support sufficient bandwidth. The small absorption coefficient of about 1×10^3 cm^{-1} causes undesirable photo-carrier generation outside the depletion region. The source of speed degradation is the slow diffusion of those carriers. The response speed depends on the time it takes for carriers to diffuse into the high electric field regions and then be collected. As such, the 3-dB cut-off frequency is only about 10 MHz, while it is more than 3 GHz at 1.3 μm wavelength light. The excess noise factor measured at

1.55 μm light, on the other hand, is as low as 6.2 at $M = 10$ because the carrier injection becomes almost the pure hole injection scheme.

The reach-through structure Ge-APD with p^+-n-$\nu(n^-)$ impurity concentration profile was studied and a small excess noise factor of 6.1 at $M = 10$ and 1.55 μm wavelength was reported by Mikawa [37]. The 3-dB cut-off frequency of this APD was more than 1 GHz. This improved Ge-APD did not, however, appear on the market in volume. Since InP/InGaAs APDs have become commercially available, applications of Ge-APDs have been transferred to and replaced by InP/InGaAs APDs.

9.4. InP/InGaAs Avalanche Photodiodes

Several types of InP/InGaAs APDs having satisfactory light-receiving performance for high-speed and long-distance lightwave transmission systems have been reported in the early stages of development [38–47]. Mesa-structure devices were reported to demonstrate high-frequency characteristics [35, 36], however, planar-type APDs with high-performance guard ring structure were recognized to be absolutely necessary for practical use in regard to electrical stability, long durability, and better productivity. The practical APD structure that has been widely deployed in long-wavelength optical transmission systems is the SAM (separated absorption and multiplication region) APD [38, 39]. In this section, we will begin from the explanation of the basic InP/InGaAs APD structure, then we will review several InP/InGaAs SAM structure APDs with different guard ring structures. Performance of InP/InGaAs APDs for 10 Gbps system applications will be then overviewed, and performance of receivers with hybrid integration of InP/InGaAs APDs and pre-amplifier-ICs will be reported. Reliability of planar structure InP/InGaAs APDs is discussed in the last part of this section.

9.4.1. SEPARATED ABSORPTION AND MULTIPLICATION STRUCTURE

Basic Structure

$In_{0.53}Ga_{0.47}As$ ($E_g = 0.75$ eV) is the narrowest bandgap mixed alloy material lattice-matched to InP ($E_g = 1.35$ eV). It has a large absorption coefficient for the whole 1-μm-wavelength range (absorption coefficient both at 1.3-μm and 1.55-μm is 1.2×10^4 cm^{-1} and 8×10^3 cm^{-1}, respectively). The narrow bandgap InGaAs light absorption layer can't support an electric field high enough to achieve avalanche breakdown without excessive tunneling leakage current [29, 30]. Avalanche breakdown is charged with the wide bandgap InP layer. The structure is called a Separated Absorption and Multiplication (SAM) structure. A schematic of SAM structure is explained in Fig. 9.11. As shown in Fig. 9.11(a), on an InP substrate, InGaAs light absorption layer, thin InGaAsP intermediate

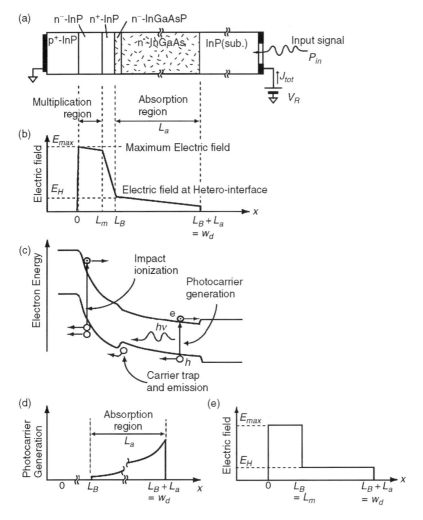

Fig. 9.11 Basic operation of InP/InGaAs SAM structure avalanche photodiode: (a) schematic structure, (b) electric field distribution, (c) schematic band diagram, (d) photocarrier generation distribution for back-illumination configuration, and (e) reduced electric field model for simplification.

layer, and InP layer are prepared by a certain epitaxial growth technique. A p-n junction is formed inside the InP layer. The n-InP, n-InGaAsP, and n-InGaAs layers are designed to be fully depleted under reverse biased operation of the p-n junction. Details of structures and fabrication process will be discussed later. Impact ionization coefficient of holes is larger than that of electrons in InP as shown in Fig. 9.3, so hole injection into the avalanche region (hole-initiated avalanche multiplication) is necessary for proper high performance. InP is transparent to 1-μm wavelength light and in the InGaAs layer the incident light is absorbed and photo-generated carriers are created as shown in Fig. 9.11(d). Carriers are transported over InP/InGaAsP/InGaAs hetero-interfaces into the InP layer making a pure hole-injection scheme. A typical electric field profile and a schematic band diagram at a reverse-biased operation condition are shown in Fig. 9.11(b) and (c), respectively. In the early stage of development, these APDs were slow due to hole trapping by a potential well of valence band discontinuity between InGaAs and InP. This problem was solved by inserting thin InGaAsP layers with intermediate bandgap energies between the InGaAs and the InP or by continuously grading the composition between InGaAs and InP [48–51]. A fundamental difficulty with SAM structure is the need to control simultaneously the maximum electric field strength at the p-n junction, E_{max}, and the field strength at the upper hetero-interface of the InGaAs layer, E_H. E_{max} must be designed to achieve a required multiplication gain, and E_H must be in the appropriate range. It is low enough for suppressing tunneling leakage current in the InGaAs layer and high enough for avoiding response speed degradation either by undepleting the InGaAs layer or by trapping of holes at the InP/InGaAsP/InGaAs hetero-interfaces.

Guard Ring Structure and Device Fabrication

Effective and reliable guard ring structures are strongly required for planar InP/InGaAs APDs. Difficulties in realizing the insertion of guard rings into InP/InGaAs SAM structure lie in the impurity control to achieve an ideally linear graded and deep-positioned junction that has sufficiently higher breakdown voltage than the one-sided abrupt p^+-n junction for the multiplication region. The guard ring is expected to be more useful with the low-high impurity profile of the InP layer, where a low impurity concentration n^--InP layer helps to suppress the edge breakdown at the periphery of the guard ring junction. Two other design issues combined with the guard ring structure are the electric field design in the n^--InGaAs layer beneath the deep guard ring junction to suppress the tunneling dark current, and the depletion layer width in the n^--InGaAs layer beneath the multiplication p^+-n junction to avoid undesired diffusion component. The first realization of effective guard ring structure was reported by Shirai *et al.* using a Be ion implanted junction and a low-high carrier profile in the InP layer [40], and so far almost all guard ring structures have followed this scheme.

Figure 9.12 shows several different planar InP/InGaAs APD structures with effective guard rings. The structures shown in Fig. 9.12(a) and (b) contain almost the ideal guard ring scheme. These structures were named

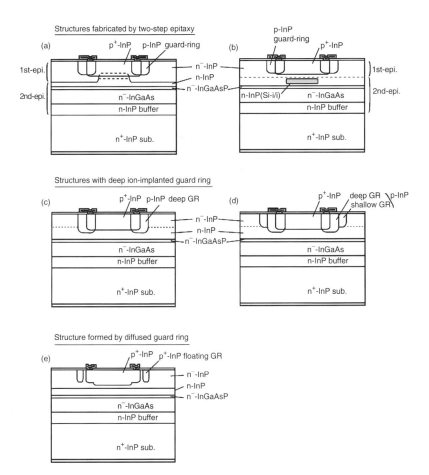

Fig. 9.12 Various guard ring structures employed for InP/InGaAs avalanche photodiodes: (a) and (b) structures made by two-step epitaxy methods, (c) and (d) structures made by Be ion implantation technique, (e) structure formed by multi-diffusion process.

"buried-structure APDs" [45–47]. The two-step LPE (liquid phase epitaxy) technique enables a completely embedded n^--InP multiplication layer and n-InP field-control layer into low carrier concentration n^--InP layers. The periphery of the multiplication p-n junction (abrupt junction) is covered by the guard ring p-n junction formed by Be ion implantation (graded junction). The combined effect of difference in carrier concentration and difference in p-n junction profile brings large breakdown voltage tolerance of the guard ring region (consisting of n^--InP, n^--InGaAsP, n^--InGaAs layers) to the active region (consisting of $n^-/n/n^-$-InP, n^--InGaAsP, n^--InGaAs layers). The fabrication procedure of the structure in Fig. 9.12(a) is as follows: In the first LPE growth, n^--InGaAs light absorption layer, n^--InGaAsP thin intermediate bandgap layer, and n/n^--InP layers are grown on a (111)A faced InP substrate. The n-InP layer is chemically etched into mesa shape to form the embedded field control layer. At the second growth the n^--InP layer is overgrown onto the mesa after a slight melt-back procedure. A proper growth condition gives a flat surface of the epitaxial wafer. Cadmium (Cd) selective thermal diffusion and beryllium (Be) ion implantation are used to make the multiplication and guard ring p-n junctions, respectively. A silicon-nitride (SiN_x) film deposited by a plasma-enhanced chemical-vapor-deposition (P-CVD) is used for a diffusion mask and the film is retained on the InP surface for junction passivation. An antireflection coating film of SiN_x is also formed by P-CVD. The structure of Fig. 9.12(b) employed Silicon ion implantation to form n-InP field-control layer to increase controllability of the amount of the charge [47]. Even those APDs with the two-step epitaxy include re-growth interface in or near the high electric field region. It has been shown that no excess leakage in dark current or any microplasma induced by local defects around the interface was observed. Large attainable multiplication factor and a quite uniform gain distribution within the active area were also reported. The structure of Fig. 9.12(a) was employed for the first mass-produced InP/InGaAs APD.

The structure shown in Fig. 9.12(d) is called the preferential lateral extended guard ring (PLEG) structure [7]. It is characterized by the deep guard ring region and by the shallow guard ring region. The junction front of the deep guard ring is located in the n-InP layer (it is deeper than the p^+-n junction of the multiplication region) in order to prevent the p^+-n junction edge breakdown. The junction front of the shallow guard ring is located around the n^-/n-InP interface to reduce the curvature of the deep guard ring and then suppress the deep guard ring edge breakdown in the n-InP. This dual guard ring increases the breakdown voltage of the guard ring sufficiently. The guard ring is formed by two-step selective Be ion implantation under different conditions and subsequent annealing. The first implantation for the deep guard ring is carried out with 5×10^{13} cm^{-2} dose at 110 keV energy followed by the second implantation at 3×10^{13} cm^{-2} dose and 60 keV energy for the shallow guard ring. The annealing is at 700°C for 20 to 30 minutes within evacuated and sealed quartz ampoules.

These sophisticated guard ring structures are useful when the breakdown voltage of the p$^+$-n junction is somehow higher than about 80 volts. In these APDs the breakdown voltage of the guard ring had to be more than 100 volts. Pursuing higher response speed with a wider 3-dB bandwidth and a larger GB product, a higher carrier concentration field control layer and a narrower multiplication region have been introduced. The breakdown voltages of APDs used for 2.5 Gbps system applications are now in the range of 50 volts and those for 10 Gbps applications are as low as around 30 volts. Single Be implanted guard ring structure as shown in Fig. 9.12(c) might be applicable for such a low breakdown voltage APD.

The structure in Fig. 9.12(e) uses a shaped junction profile and floating guard ring to control edge breakdown [25]. This design employs a widening of the multiplication region width at the junction periphery to reduce the electric field strength there. This shaping of the junction profile can be made using the multiple diffusion method. The p-n junction is formed by closed ampoule Zn diffusion through a patterned SiN$_x$ passivation film. The floating guard ring lowers the surface field intensity by spatially spreading surface equipotential lines and it is combined with the shaping junction to reinforce the guard ring effectiveness. In this structure, narrowing the multiplication region width reduces breakdown voltage due to reduction of the area for electric field profile and requires a higher maximum electric field to reach breakdown. To balance these two competing factors, there is an optimum width that brings the largest tolerance between the breakdown voltages for the center of the junction (multiplication region) and for the periphery of the junction. This type of guard ring structure would be useful for devices with lower breakdown voltage such as for 10 Gbps system application devices.

Multiplication Characterisitcs of SAM APD

A distinctive feature of SAM structure appears in a plot of photocurrent I_p versus reverse bias voltage V_R at 1.3 μm or 1.55 μm wavelength incident light, where kinks or knees exist on the I_p–V_R curve at a low reverse bias

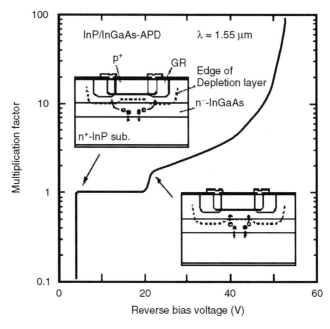

Fig. 9.13 Multiplication versus reverse bias voltage characteristics of InP/InGaAs-SAM-APD with 30 μm photo-sensitive area diameter.

voltage range. The I_p-V_R curve in Fig. 9.13 is for a typical InP/InGaAs APD with a planar guard ring structure. Two knees at lower reverse bias voltage correspond to the depletion layer of the p-n junction reaching through into the InGaAs layer from InP layer beneath the guard ring p-n junction and beneath the multiplication p^+-n junction, respectively. The insertions in the figure show the position of the depletion region edge and flow of photo-generated carriers at these knees. There is a valence band discontinuity about 0.36 eV between InP and InGaAs, which would form a potential well for holes. Photo-generated holes would be trapped in this well and they would be emitted into InP with the help of either an electric field or thermal energy. As the guard ring p-n junction is formed deeper than the p^+-n junction of the multiplication region, the first knee results from the guard ring p-n junction. Letting e_e and e_r represent the emission and recombination probabilities of holes from and/or at the well, respectively, then the kinks can be qualitatively explained. When reverse bias is low and the depletion region does not reaches into the InGaAs layer, $e_r \gg e_e$ and the holes are not transported into the InP layer nor collected by electrode. With the reverse bias condition for the depletion layer reaching into InGaAs layer, $e_r \ll e_e$ and photo-generated carriers go through the depletion layer and are collected by electrode. Estimation of the multiplication factor using the photocurrent at the first plateau appearing after the first knee as the reference for $M = 1$ has been often used. Because the photocurrent level at this plateau could depend on e_e and e_r through the hetero-interface quality such as abruptness of the barrier, it is unlikely that it corresponds to $M = 1$. The multiplication factor of InP/InGaAs APD could be estimated more quantitatively by a measurement of multiplication noise versus photocurrent. The noise power, ϕ, can be expressed as

$$\phi = 2q I_{p0} M^{2+x} R_L B = 2q I_p M^{1+x} R_L B \qquad (9.30)$$

Taking the logarithm of both the sides gives

$$\log(\phi) = (1 + x) \log(I_p) + \text{ const.} \qquad (9.31)$$

where M is the multiplication factor, $I_p = I_{p0} M$ is the total photocurrent, R_L is the load resistance (typically 50 Ω), and B is the noise bandwidth of the measurement system. The noise power in dB could be a linear function of $\log(I_p)$. The solid line in Fig. 9.14 is a plot of measured noise power against $\log(I_p)$ and the dotted line is for the ideal shot-noise. The intersection of the dotted line and the solid line in the figure gives the photocurrent for $M = 1$. The reverse bias voltage corresponds to this point, usually locate between the first and the second knees, but sometimes (depending on the device structure or fabrication condition) out of the first plateau. The multiplication factor at the edge of the second plateau after the second knee is usually less than $M = 2$, in which case the reverse bias voltage for $M = 2$ or a higher multiplication factor should be preferable as the reference point of the multiplication factor.

Assuming the double-path absorption in the InGaAs layer with a partial reflection of unabsorbed light by the front-side p$^+$ contact metal, the generation rate of photo-carrier in the InGaAs layer is

$$G(x) = a_T \exp[-a_T(w_d - x)] + R_B \exp(-a_T L_T) \cdot a_T \exp[-a_T(x - L_B)] \qquad (9.32)$$

where a_T is the absorption coefficient of InGaAs for the incoming light, R_B is the reflection rate at the front-side p$^+$ electrode, and w_d, L_B, and L_T are

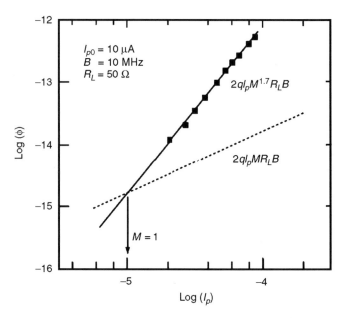

Fig. 9.14 Dependence of multiplication noise on photocurrent for InP/InGaAs APD. Closed squares are measured noise power at $I_{p0} = 10\ \mu A$, 10 MHz bandwidth, and 50 Ω load resistance.

as given in Fig. 9.11. When the InGaAs absorption layer is fully depleted through its whole thickness L_T, integrating Eq. (9.32) gives the primary photocurrent J_0 at input light power P_{in} and the quantum efficiency η_{APD} as

$$\eta_{APD} = (1 - R_F) \cdot [1 - \exp(-a_T L_T)] \cdot [1 + R_B \exp(-a_T L_T)] \quad (9.33a)$$

$$J_0 = q\frac{P_{in}}{h\nu}\eta_{APD} \quad (9.33b)$$

where q is the electronic charge, h is Planck's constant, ν is the photon frequency, and R_F is the reflection rate of incident light on the photodiode surface (mainly determined by the reflection rate at the AR coating film). Transparency of InP to long-wavelength light makes the double-path design possible, and in this case the APD usually employs the back-illumination configuration. This configuration also helps to shorten the total depletion region width and to reduce the total junction area, but it requires somewhat troublesome assembly compared to the surface illumination configuration.

9.4.2.   DESIGN OF LOWER NOISE AND HIGHER SPEED DEVICE

It has been assumed that all avalanche multiplication occurs in the InP region near the maximum electric field location until Kobayashi pointed out that both the excess noise factor and the gain-bandwidth products of InP/InGaAs SAM structure APDs depend strongly on the electric field at the hetero-interface [46]. When the electric field at the hetero-interface is high, impact ionization initiated by electrons occurs in the InGaAs absorption region. Holes generated in the InGaAs travel back to the InP multiplication region and initiate avalanche multiplication. Electrons generated in the InP multiplication region drift to the InGaAs. This process takes longer transit time from the InP avalanche region to the InGaAs and back again into the InP than the transit time across the InP avalanche region. The multiplication noise is also increased by this feedback process.

Multiplication and Noise Expressions of SAM APD

Let α_B, β_B, α_T, and β_T represent the electron and the hole ionization coefficients in InP and InGaAs, respectively. From Eq. (9.3), the position dependent multiplication at x in the InP and In GaAs depletion region is [52]

$$M_B(x) = \frac{1}{1 - P_B - Q_B P_T} \cdot \exp\left\{ -\int_0^x (\alpha_B - \beta_B) \, dx \right\} \quad (0 \leq x < L_B)$$

$$\tag{9.34a}$$

$$M_T(x) = \frac{Q_B}{1 - P_B - Q_B P_T} \cdot \exp\left\{ -\int_{L_B}^x (\alpha_T - \beta_T) \, dx \right\}$$

$$(L_B \leq x \leq w_d = L_B + L_T) \quad (9.34b)$$

where

$$P_B \equiv \int_0^{L_B} \alpha_B \exp\left\{ -\int_0^x (\alpha_B - \beta_B) dx' \right\} dx$$

$$P_T \equiv \int_{L_B}^{w_d} \alpha_T \exp\left\{ -\int_{L_B}^x (\alpha_T - \beta_T) dx' \right\} dx, \quad \text{and}$$

$$Q_B \equiv \exp\left\{ -\int_0^{L_B} (\alpha_B - \beta_B) dx \right\}$$

The excess noise factor for the SAM structure APD can be written when the depletion region is fully extending through the InGaAs absorption layer as

$$F(M) = \frac{2q \int_0^{w_d} G(x) M_T^2(x)\, dx}{J_0 M}$$
$$+ 2 \left\{ \left[\int_0^{L_B} \beta_B M_B^2(x)\, dx + \int_{L_B}^{w_d} \beta_T M_T^2(x)\, dx \right] - M_T^2(w_d) \right\}$$

(9.35)

where J_0 and M can be known from Eqs. (9.3), (9.4), and (9.34). Simplification can be made assuming the constant electric field in the InP layer and in the InGaAs layer as shown in Fig. 9.11(e). Calculation using the ionization coefficients in Fig. 9.3 shows that the hetero-interface electric field E_H of more than 1.4×10^5 V/cm at the breakdown voltage severely deteriorates the excess noise of InP/InGaAs SAM structure APD due to increase of randomness from considerable feedback by the multiplication in InGaAs. This critical electric field strength at the hetero-interface is lower than that for suppressing the tunneling dark current in the InGaAs layer. However, as discussed later in this section, the gain-bandwidth product is more sensitive to the electric field at the hetero-interface.

Hole Trapping at the Hetero-interface

The SAM structure of the InP/(InGaAsP)/InGaAs material system contains valence band discontinuity at the hetero-interface and it leads to the trapping of the primary holes before they are injected into the InP layer. The trapped holes are emitted with an exponential decay time that could be long enough to degrade the response speed of the device. The emission rate e_h is known as the function of the hetero-interface electric field E_H as [48]

$$e_h = C_0 \exp(-E_H / k_B T)$$

(9.36)

where k_B is the Boltzman's constant and C_0 is the constant over a limited temperature range. It was found that this effect gave rise to a long decay tail in the pulse response of InP/InGaAs APD and the observed bandwidth was typically less than 100 MHz at the beginning of development [48, 49]. There is some uncertainty with the emission process for the holes, but it is clear that the emission rate is very sensitive to changes in the barrier height and interface width. A general model incorporates a tunneling through the barrier at lower temperature and a thermionic-field emission at higher temperature [53]. Hot holes that are accelerated by the field in the InGaAs layer may not be in equilibrium with the lattice, and this results in enhanced emission rate by lowering the effective barrier height than that predicted by conventional thermionic emission models.

Several solutions have been proposed to eliminate the problem of hole trapping. Each proposal employs introduction of a transition layer between the InP multiplication layer and the InGaAs absorption layer [50, 51]. The optimum solution could be the use of continuously graded bandgap energy between InP and InGaAs, but there are technological difficulties in preparing the actual structure. A complete compositionally graded quaternary InGaAsP layer is considered the most effective in diminishing hole trapping delay at the hetero-interface. Kuwatsuka [57] first demonstrated a high-speed InP/InGaAs-SAM-APD with a compositionally graded quaternary InGaAsP layer between the InP multiplication region and the InGaAs absorption layer. The graded layer of 0.2 μm in thickness was grown by metal-organic vapor phase epitaxy (MO-VPE) by gradually changing the gas flow rate of TMI, TMG, AsH$_3$, and PH$_3$. The 3-dB cutoff frequency was measured at the fixed hetero-interface electric field of 1×10^4 V/cm in the temperature range between $-100°$C to $+50°$C. The cutoff frequency of the sample was unchanged in the measured temperature range, which means that hole trapping at the hetero-interface can be diminished by an engineering of the intermediate layer.

Gain-Bandwidth Product of SAM-APD

An approximate expression for the frequency response of the InP/InGaAs SAM structure APD can be modified from Eq. (9.13) of the conventional structure APD, and given by

$$F_{APD}(\omega) = F_{tr}(\omega) \cdot F_h(\omega) \cdot \frac{M(\omega)}{M_0} \cdot F_{RC}(\omega)$$

$$= \frac{\sin(\omega\tau_{tr}/2)}{\omega\tau_{tr}/2} \cdot \frac{1}{\sqrt{1 + (\omega/e_h)^2}} \frac{1}{\sqrt{1 + (\omega M_0 \tau_m)^2}} \cdot \frac{1}{\sqrt{1 + (\omega\tau_{RC})^2}}$$

$$(9.37)$$

where e_h is the emission rate for holes trapped at the hetero-interfaces, τ_m is the effective transit time through the avalanche multiplication region, τ_{RC} is the RC-time constant. At high multiplication factor the frequency response approaches a constant gain-bandwidth limit determined by τ_m. The 3-dB cutoff frequency at the lower multiplication region rolls off due to a combination of three other time constants, τ_{tr}, τ_h, and τ_{RC}. An example of cutoff frequency roll-off will be described in a later section.

The frequency response can be directly calculated using the time dependent equation for the current transportation of Eq. (9.2). A simplified electric field model in Fig. 9.11(e) assumes the constant electric field in the InP layer and in the InGaAs layer and neglects the thickness of the InP field control layer and the InGaAsP intermediate layer because they are much thinner than the InGaAs layer. Two pairs of transport equations can be set for InP and InGaAs from Eq. (9.2) and solved with suitable boundary conditions (details are not described here) [54].

The calculated gain-bandwidth product as a function of the multiplication region length L_m and the charge amount in the field control layer N_s is shown in Fig. 9.15. The GB product decreases when the multiplication region length increases because of the longer effective transit time associated with longer L_m. On the contrary, narrower multiplication region

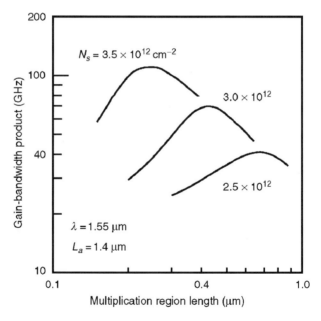

Fig. 9.15   Calculated gain-bandwidth product using the simplified electric field model of Fig. 9.11(e) as a function of the multiplication region length L_m and the charge amount of field-control layer N_s.

length increases the maximum electric field E_{max} and thus the electric field strength in the InGaAs layer, and degradation of the GB product results from occurrence of the avalanche multiplication in the InGaAs layer. The optimum L_m exists to achieve the maximum GB product at a given charge amount of the field control layer N_s. Larger GB product can be achievable by putting larger charge amount into the field control layer. The maximum GB product of about 100 GHz can be expected with the device parameters of $L_m = 0.23$ μm and $N_s = 3.5 \times 10^{12}$ cm^{-2}. The hetero-interface electric field E_H at the breakdown for this device parameter is about 1.0–1.1 $\times$ 10^5 V/cm. When the device has the parameters for the hetero-interface electric field at the breakdown of about 1.4 $\times$ 10^5 V/cm, which is the critical hetero-interface electric field strength for the excess noise factor degradation, the GB product falls about 60% from the maximum value. Tolerances of the charge amount in the field control layer and of the multiplication region length are very small if we want to fabricate a large GB product device with a reasonable reproductivity. Then, an extremely high controllability is absolutely necessary on the thickness and the carrier concentrations of the epitaxial layers and the diffusion depth of impurity species to form the p$^+$-InP layer.

9.4.3.   InP/InGaAs APDS FOR 10 GBPS SYSTEMS

In this decade, many kinds of InP/InGaAs APDs have been developed for application to 10 Gbps lightwave transmission systems [21–25]. Figure 9.16(a) shows the cross section of one of this kind of high-speed InP/InGaAs APDs reported [21], where the electric field profile has been carefully designed to suppress harmful avalanche multiplication in the InGaAs absorption layer [46, 54] and simultaneously to avoid hole pile-up at the hetero-interface. High cutoff frequency for wide dynamic range of the multiplication factors with a large gain-bandwidth product is obtained. This APD adopts, for the first time, a flip-chip bonding configuration. Flip-chip bonding using a back-illuminated structure could be the key technique due to minimization of the parasitic reactance and the capacitance in the surplus bonding pad area. The junction capacitance is reduced by shrinkage of the p-n junction area. The fiber-to-chip coupling tolerance with the small active area device can be managed by introducing a monolithically integrated micro-lens on the back-side of the InP substrate. A large fiber alignment tolerance of 40 μm was achieved with the micro-lens as shown in Fig. 9.16(b) despite the active p$^+$-n junction area (the avalanche multiplication area)

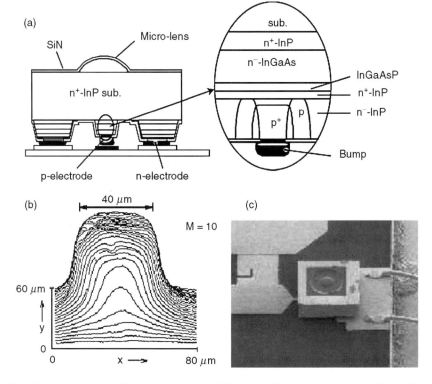

Fig. 9.16 Structure of flip-chip structure InP/InGsAs APD: (a) cross section of the APD, (b) photosensitivity of the APD when light is irradiated through the micro-lens, and (c) photograph of APD assembly on the evaluation substrate.

of about 15 µm in diameter. This wide photosensitive tolerance facilitates optical assembly for the receiver module. Input light passing through the micro-lens is focused on the active p-n junction area and reflected by the end facet mirror of p-type contact electrode. The quantum efficiency can also be increased by the double-path scheme in the InGaAs absorption layer. The InGaAs layer can be made narrower than conventional surface-illuminated structures to attain equivalent quantum efficiencies, and it contributes to a shorter carrier transit time so that the 3-dB bandwidth could be wider.

Figure 9.16(c) is the photograph of the flip-chip APD mounted on a evaluation substrate. This APD structure consists of n^+-InP buffer layer ($>5 \times 10^{17}$ cm^{-3}), n^--InGaAs absorption layer (1.4 µm thick and $1–2 \times 10^{15}$ cm^{-3} of carrier concentration), three thin n^--InGaAsP intermediate layers (0.05 µm thick each, $0.5–1 \times 10^{15}$ cm^{-3}, and bandgap

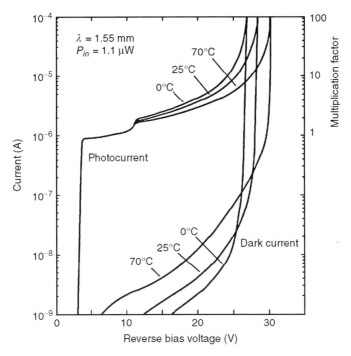

Fig. 9.17 Photocurrent multiplication and dark current characteristics of flip-chip structure InP/InGaAs APD measured at 0, 25, and 70°C.

energies are 0.80 eV, 0.95 eV, and 1.10 eV), n^+-InP field control layer (about
0.5 μm thick and $>5 \times 10^{17}$ cm^{-3}), and n^--InP top layer (1–2×10^{15} cm^{-3}).
The layers were grown by metal-organic vapor phase epitaxy (MO-VPE).
The thickness of the InP multiplication region defined by the depth of the
p^+-InP layer, which is made by selective Cd diffusion into the top n^--InP,
was about 0.2 μm. A Be ion implanted single guard ring structure was
employed for highly reliable planar p-n junction. An isolation mesa was
fabricated between the active p-n junction area and the n-type electrode.
The surface was passivated with SiN$_x$ film [21, 22].

Figure 9.17 shows photocurrent versus reverse bias voltage measured at
$\lambda = 1.55$ μm and 1.1 μW input light power and dark current characteristics
at three different ambient temperatures of 0, 25, and 70°C. The breakdown
voltage at 25°C is 28 volts and the temperature coefficient of the breakdown voltage is 0.08 V/°C. This APD does not have a flat plateau between
the first and the second kinks, which are for the reach-through of the depletion region beneath the guard ring junction and for the multiplication

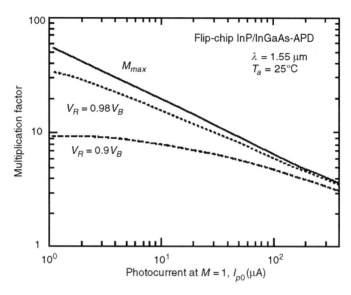

Fig. 9.18   Multiplication factor as a function of optical input power and reverse bias voltage.

region, respectively, as described previously. A photocurrent at the unity multiplication factor could be obtained by the method of Fig. 9.14 of this chapter or from calculation by integrating Eq. (9.4) in the whole depletion region with the ionization coefficient of both InP and InGaAs. The latter method includes fitting parameters such as the multiplication region thickness L_m. The dark current at 90% of the breakdown voltage is about 20 nA and the multiplied dark current is only about 2 nA. These data indicate that the flip-chip bonding structure does not influence the dark current characteristics. Photocurrent multiplication characteristics against input optical power are plotted in Fig. 9.18. The maximum attainable multiplication factor M_{max} at 1 μA of the primary photocurrent I_{p0} is more than 50. M_{max} decreases with increasing I_{p0} and it has a dependence of $\propto 1/\sqrt{I_{p0}}$ as expressed by Eq. (9.7).

The capacitance of the APD is as low as about 100 fF at 90% of the breakdown voltage measured with a die form. The APD die was flip-chip mounted on the top of a ceramic carrier with a 50 Ω impedance strip line and a 50 Ω load resistor used for performance evaluation. The 3-dB bandwidth as a function of the multiplication factor is shown in Fig. 9.19. The maximum bandwidth ceiling of about 10 GHz was obtained at the multiplication factor of 4. The cut-off frequency determined by the RC time constant is

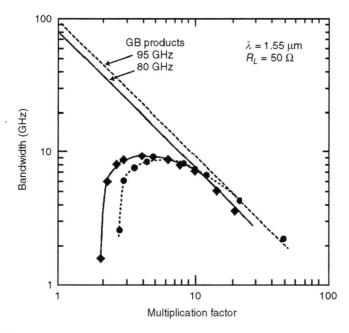

Fig. 9.19   3-dB bandwidth versus multiplication factors of the flip-chip InP/InGaAs APD.

calculated to be more than 20 GHz, so a combined effect of the carrier transit time across the depletion layer and the avalanche buildup time is expected to determine the bandwidth ceiling. In the low multiplication factor region, the turn-on of 3-dB cut-off frequency appears at around $M = 2.5$, and at $M = 3$ the bandwidth comes close to the maximum bandwidth. The gain-bandwidth product of 80 GHz has been obtained, which would be sufficient for 10 Gb/s systems applications. A larger GB product around 95 GHz can be achieved with a higher multiplication factor for the bandwidth turn-on point, such as $M = 3.5$. In this case, the maximum bandwidth ceiling becomes lower. Because APDs are the limiting element of receiver bandwidth in a 10 Gbps system application, it should be preferable to have a wider bandwidth within a wider range of multiplication factor for achieving a larger dynamic range operation. The excess noise factor of the APD designed for 10 Gbps applications is shown in Fig. 9.20. The measured k_{eff} is about 0.65 and this value is larger than the InP/InGaAs APDs designed for 2.5 Gbps applications (which were not described in this paper in detail).

Because the required wavelength range has been extended into the L-band region ($\lambda = 1.58$ μm to 1.62 μm) in the DWDM systems with

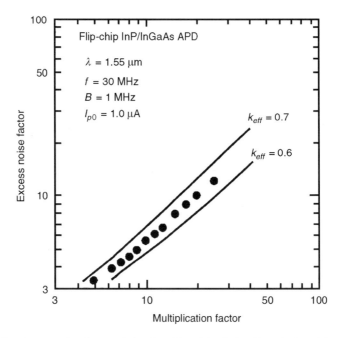

Fig. 9.20   Excess noise factor of the flip-chip InP/InGaAs APD designed for 10 Gbps application.

more than 100 wavelength channels, the wavelength dependent quantum efficiency is one of the interesting parameters for practical application. An optimization of device structure results in the performance shown in Fig. 9.21. Responsivity (in units of A/W) is plotted in place of the quantum efficiency, because it is more suitable to have direct understanding for the receiver design. At 1.62 μm wavelength, the responsivity is more than 0.8 A/W even at a low temperature of 0°C, and it is 0.9 A/W at 25°C. The response variation between 1.55 μm to 1.62 μm wavelength range is about 20% (1.0 dB). Typical characteristics of InP/InGaAs APDs designed for 2.5 Gbps system applications and for 10 Gbps applications are summarized in Table 9.1 [55, 56].

9.4.4.   INTEGRATED APD/PREAMPLIFIER RECEIVERS

High-performance optical front-end receivers are required to expand the application of high-speed lightwave transmission systems. Large-bandwidth, low-noise performance and high yield productivity are important for such

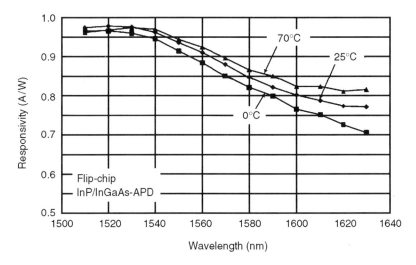

Fig. 9.21   Wavelength dependence of responsivity of the flip-chip InP/InGaAs APD designed for 10 Gbps system application at various temperatures.

applications of receivers. InP/InGaAs APDs with gain-bandwidth product of 80 to 100 GHz have been in practical use. Electronic amplifier elements made of advanced transistor technologies such as high-electron mobility transistors (HEMTs) and hetero-junction bipolar transistors (HBTs) using AlGaAs/GaAs, AlInAs/GaAs, and InGaP/GaAs systems have also shown significant improvements in cutoff frequency. Both bandwidth and noise of receivers are largely dependent on the circuit parameters and parasitic characteristics at the interconnection of the device elements. Hybrid flip-chip integration of APD and pre-amplifier is an attractive prospect to fabricate the elements at the closest distance so that the introduction of any parasitic reactance could be minimized. This approach has been studied and demonstrated so far using p-i-n photodiodes [57, 58] but has not grown to bear any practical applications in the case of APD receivers, primarily due to difficulties in achieving high manufacturing yield of such sub-assemblies. To migrate such an experimental effort to a productive product realization, many design constraints would have to be considered.

Figure 9.22(a) shows the photograph of a compact 10 Gbps APD/pre-amplifier module built by integrating the above-mentioned high-speed APDs with InGaP/GaAs hetero-junction bipolar transistor (HBT) pre-amplifier-ICs [55, 59, 60]. The schematic structure of the module is given in Fig. 9.22(b). The module has a single-mode fiber pigtail as the

Table 9.1   Characteristics of InP/InGaAs APD Designed for 2.5 Gbps and
10 Gbps System Applications. The Filp-Chip Structure with the
Back-Illumination Configuration is Adapted for 10 Gbps Apds, While the
Surface-Illumination Configuration is Employed for 2.5 Gbps APD

Parameter	Symbol	[2.5 Gbps Application] Characteristics			[10 Gbps Application] Characteristics			Units
		Min.	Typ.	Max.	Min.	Typ.	Max.	
Breakdown voltage at 25°C	V_B	45	50	60	25	28	33	V
Temperature coefficient of V_B	Γ	—	0.12	—	—	0.08	—	V/°C
Dark current at 0.9 V_B								
25°C	I_d	—	10	50	—	100	200	nA
70°C		—	100	—	—	500	—	
Multiplied dark current								
25°C	I_{dm}	—	1	5	—	10	20	nA
70°C		—	10	—	—	50	—	
Responsivity at λ = 1.55 μm	R	0.9	1.0	—	0.7	0.8	—	A/W
3-dB cut-off frequency at λ = 1.55 μm								
$M = 5$	f_c	—	4.0	—	—	9.0	—	GHz
$M = 10$		2.5	3.0	—	7.0	7.5	—	
Gain-bandwidth product	GB	—	40	—	—	80	—	GHz
Excess noise factor at λ = 1.55 μm & $M = 10$	F	—	5	5.6	—	7	8	—
Capacitance	C_t	—	0.25	0.3	—	0.08	0.12	pF

optical interface. The electrical interface for high-frequency signals is composed of co-planar leads through a ceramic feedthrough. An aspherical lens is used to achieve an efficient coupling into the small active area APD chip (coupling efficiency > 95%). The APD is flip-chip mounted on a ceramic carrier, whose use is necessary to assure quality of the APD such as rejection of infant failure (by the burn-in screening) and selection of required

(a)   (b)

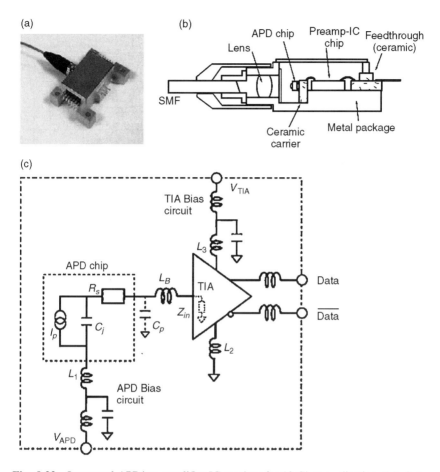

Fig. 9.22 Integrated APD/pre-amplifier-IC receiver for 10 Gbps application: (a) photograph of a compact-size receiver with coplanar high-speed signal output and single-mode fiber pigtail, (b) schematic internal structure of the receiver packaging, and (c) equivalent circuit of the receiver.

performances. Conventional wiring connection between the ceramic carrier and the preamplifier-IC chip is used. An electrical model of the module, including the equivalent circuit model of the APD chip, the interconnection between the APD and the pre-amplifier-IC, and other connections to the preamplifier-IC chip, is shown in Fig. 9.22(c). Parasitic elements at the interconnection between the APD and the preamplifier-IC influence the performance of the module such as the bandwidth. The effect of the ground inductance (L_1, L_2, and L_3) has also been taken into account in the design

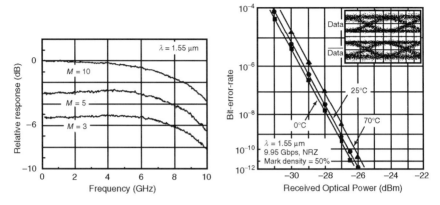

Fig. 9.23   Bandwidth and sensitivity of the APD receiver for 10 Gbps application. Inset of the BER chart is the output waveform from the receiver.

of the receiver because they often have influence on the stability of the receiver module.

Bandwidth characteristic of this module as a function of the multiplication factor is shown in Fig. 9.23(a). The receiver bandwidth is wider than 8 GHz at the APD multiplication factor from 3 to 10, which is sufficient for 10 Gbps systems applications. This results in a wide dynamic range of the receiver. The minimum sensitivity of -26 dBm is achieved at a bit-error-rate of 10^{-12} (NRZ with 50% mark density) and 1.55 μm wavelength as shown in Fig. 9.23(b). The inset is the output electrical waveform from the receiver when the optical input power is -25 dBm. A good eye opening and a symmetric differential output are confirmed. The sensitivity is some 8 dB better than the receiver module with integrated PIN-PD and the pre-amplifier-IC (with the same IC design). Sensitivity variation with the case temperature is less than 1 dB from 0 to 70°C. The APD receiver modules are now in volume manufacture. They have been found very useful to realize the cost-effective lightwave transmission system without fiber pre-amplifier, especially in short and intermediate reach applications such as metropolitan photonic network systems.

9.4.5.   RELIABILITY STUDIES OF InP/InGaAs SAM APDS

Even high reliability of InP/InGaAs APDs has already been established, due to relatively high operation voltages of InP/InGaAs-APDs in the early stage of development. Confirmation of their long-term stability had been

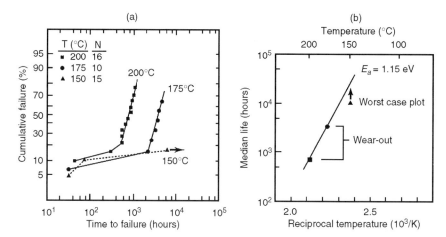

Fig. 9.24   Long-term life test data performed for the APDs with the structure shown in Fig. 9.12(a): (a) time-to-failure vs. aging time, and (b) plot of median lives.

the most important item for practical applications in long wavelength light-wave transmission systems throughout the 1980s. The first definite summary of reliability studies was reported in 1989 [6, 61], where these APDs with planar structure guard ring were concluded to meet even with the submarine grade reliability requirements.

Time-to-failure versus percent of cumulative failures for aging tests at three elevated temperature levels are plotted in Fig. 9.24(a) in a log-normal chart. During the elevated temperature aging tests, sample devices were biased at their breakdown voltage. A weak population was observed at an early stage of the test and was followed by a main population. At 200°C and 175°C tests, the devices that showed gradual increase of dark currents made up the main population, and they showed relatively good obedience to the log-normal distribution with a standard deviation of about 0.5. The main populations at these temperatures were consequently considered to be a wear-out failure. Median lives at 200°C and at 175°C were estimated to be 750 hours and 3500 hours. At 150°C, although a few devices failed at an early stage, others showed good stability within the aging test period of 6000 hours. Median lives from the aging tests are plotted against the reciprocal temperature as shown in Fig. 9.24(b). From an interpolation of the median lives obtained from 200°C and 175°C tests, the activation energy of wear-out failure was estimated to be 1.15 eV. The extrapolated median life to the actual operating temperature, for example at 50°C, exceeded 1×10^8 hours.

For all early failure devices, fluctuations of dark currents were distinctively observed at the initial measurement. Degree of the vibration was only a few nano-Amps at 90% of the breakdown voltages and increased at higher reverse bias voltage. The behavior was considered to be correlated with a microplasma that initially existed and localized at the periphery of the guard ring junction. Chin [62] reported the failure mechanism initiated by a microplasma in a planar PIN photodiode. The same degradation mechanism can be presumed for the early failure of InP/InGaAs-APDs. A higher electric field in APD due to a larger reverse bias voltage strongly triggered the degradation and the failure was observed in the early stage of the aging test. A hard burn-in screening at an elevated temperature such as $200°C$ and at breakdown condition was always employed for the planar InP/InGaAs APDs to eliminate those early failures.

The failure mode of the main population could be classified into a surface failure because the only degradation mode observed was an increase of dark current. The possible mechanism of the increase of the dark current was an accelerated degradation of the interface between InP and SiN_x at the periphery of the guard ring. Hot holes injection into the SiN_x film was likely the cause of deteriolation of the interface.

9.5.   Studies of Novel APDs

Along with the conventional APDs mentioned in the previous section, novel structure APDs have been studied for the enhanced ionization coefficient ratio for very high-speed and high-sensitivity performances. They are the superlattice (multiple quantum well structure) APDs, thin multiplication region APDs, and Si/InGaAs wafer fusion APDs. Also, from the DWDM system requirement, APDs with improved responsivity over 1.6 μm L-band wavelength range have been examined. In this section, those new APDs based on the new physics and technology are reviewed.

9.5.1.   IMPROVEMENT OF L-BAND RESPONSE

Rapid increase of aggregate throughput of information in recent years demands APDs successfully operating in the L-band DWDM systems over 1.6 μm wavelength. However, the responsivity of the current APDs for long wavelength tends to degrade, because the absorbing layer consists of the InP lattice matched $In_{0.53}Ga_{0.47}As$ material having lower sensitivities at 1.6 μm wavelength region. New APDs having a strain-compensated

InGaAs multiple quantum well (MQW) absorbing layer have been studied as the way to expand and improve the L-band responsivity.

An APD with an $In_{0.83}Ga_{0.17}As$ absorption layer and a thin $In_{0.52}Al_{0.48}As$ multiplication layer has been reported [63]. By introducing a high In molar fraction of $x = 0.83$ compared with the InP lattice matched value of $x = 0.53$, a high cutoff wavelength of 2.0 μm has been reported, although obtained response speed was still limited by hole trapping at the MQW hetero-interfaces. As demonstrated by this study, increasing the In molar fraction x of $In_xGa_{1-x}As$ is effective to create a higher absorption coefficient at the L-band wavelength because of narrowing the bandgap energy. There is a limitation of layer thickness according to the Matthews-Blakeslee calculation of the critical thickness to prevent misfit dislocation formation [64], and this results in inadequate quantum efficiency. Very recently, a study of the InGaAs strain-compensated MQW, where the strain of In-rich compressive InGaAs wells is compensated by the opposite tensile strain of Ga-rich InGaAs barriers, has been reported to optimize structural parameters of the InGaAs MQW absorbing layer [65]. The schematic band diagram of the MQW absorption layer is shown in Fig. 9.25(a). A strain compensation and a zero net strain condition can be achieved by adjusting the barrier strain and thickness to meet the equation as

$$e_w \cdot l_w + e_b \cdot l_b = 0 \qquad (9.38)$$

where e_w and e_b are the strain of well and barrier, l_w and l_b are the thickness of well and barrier, respectively. Thicker MQWs beyond the Matthews' critical thickness can be grown under this condition without misfit dislocations. Optimization on the structural parameters (e_w, e_b, l_w, and l_b) must be done in terms of epitaxial quality, which could be observed from the surface morphology of the epitaxial layer and the flatness of the MQWs interfaces. Figure 9.25(b) is the absorption coefficient at $-40°C$ for the optimized strain-compensated MQW absorption layer. In shorter wavelengths than 1550 nm, lattice-matched InGaAs has a larger absorption coefficient than the MQW. This is because the substantial absorption width of the MQW layer including the barrier layers is thinner than that of the lattice-matched homogeneous InGaAs. In longer wavelengths than 1550 nm, enhancement of the absorption coefficient by the MWQ is clearly demonstrated. A PIN photodiode utilizing this optimized structure has also revealed a responsivity as high as 0.46 A/W at a 1.62 um wavelength and even at a low operating temperature of $-40°C$ with a low dark current of 0.1 nA.

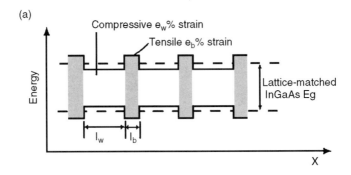

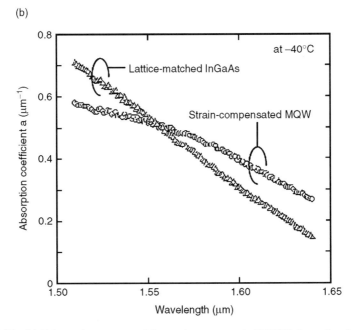

Fig. 9.25   (a) Schematic structure of the strain-compensated MQW absorption layer. (b) Wavelength dependence of the absorption coefficients for the strain-compensated MQW layer and the homogeneous InGaAs layer measured at −40°C.

The MQW absorption layer can directly replace the InGaAs aborption layer for the APD structure shown in Fig. 9.16(a). APDs employing this MQW have, for the first time, achieved excellent high-speed characteristics such as a gain bandwidth product of 70 GHz and a wide bandwidth of 9 GHz, eliminating the effect of the hole pile-up at the hetero-interfaces in the MQW absorption layer. These results confirm usefulness of this InGaAs strain-compensated MQW structure to realize L-band APDs operating up to a 1.62 um wavelength regime.

9.5.2. SUPERLATTICE APDS

The superlattice APDs of enhanced impact ionization coefficient ratio using the band offset of superlattice (multiple quantum well) structure have been studied since their first proposal by Chen [66] and implementation by Capasso [67]. In 1989, Kagawa [68] reported that InGaAs-InAlAs superlattice showed enhancement of electron ionization coefficient and suppression of hole ionization coefficient compared to InGaAs bulk material, which accelerated the development of superlattice APDs with InAlAs material systems on InP substrate. The main subjects in developing superlattice APDs in the initial stages have been the suppression of tunneling dark current from narrow bandgap wells, the achievement of higher gain-bandwidth products, and the trying out of several material systems [69–71]. The dark current, especially when it results in a large multiplied dark current, is the source of the degradation in receiver sensitivity as shown in Eq. (9.28). Some promising performances with mesa-structure InAlGaAs-InAlAs superlattice APD with back-illumination configuration were reported in 1996 [72, 73]. To realize the SAM structure, a thin p^+-InP field control layer has been introduced between the superlattice multiplication region and the InGaAs absorption region, and the light absorption layer is a lightly Be doped p^--InGaAs to make the electron injection into the superlattice avalanche multiplication region. Because the electron ionization coefficient is larger than that of holes in the superlattice structure, avalanche multiplication initiated by electron injection can minimize the excess shot-noise and enhance the gain-bandwidth products. Using the larger bandgap material for the well (InAlGaAs instead of InGaAs), low dark current less than 1 μA was obtained at a multiplication factor of 10 and a multiplied dark current of around 10 to 20 nA was reported. The maximum 3-dB cutoff frequency around 15 GHz was achieved, which is limited by the RC time constant, and the gain-bandwidth product was derived at about 120 GHz. The p^+-InP field control layer has been shown to play an important role of influence on the gain-bandwidth products. A thinner InP field control layer is preferable to suppress holes causing avalanche multiplication, or an InAlAs layer is suitable because it has larger electron ionization coefficient than the holes. The demonstrated performance could be enough for 10 Gbps system applications. The first work on the planar structure

InAlGaAs-InAlAs superlattice APDs was reported by Watanabe [74]. This APD has such distinctive features as Ti implanted guard ring to prevent periphery breakdown and Zn diffused p^+-regions around the InGaAs absorption layer to realize planar structure. Characteristics of this superlattice APD have shown fairly large gain bandwidth product of 110 GHz and the maximum bandwidth of 15 GHz with the effective ionization coefficient ratio k_{eff} of 0.4. As for the effective ionization coefficient ratio k_{eff} of the superlattice APDs, a record value of 0.1–0.2 leading to very small excess noise factor of 3 at a multiplication of 10 has been reported in other material systems, InGaAsP-InAlAs [71], InGaAs-InAlAs [69] superlattice structures.

In order to obtain high sensitivity in the ultra high-speed applications, APDs having both wide bandwidth and large quantum efficiency are crucial. As is well known, however, there is a trade-off between bandwidth and quantum efficiency in the conventional surface-illuminated and back-illuminated structures. To overcome this limitation, edge-coupled waveguide structure is very attractive. Possible improvement up to 30 GHz could be realized by introducing flip-chip configuration to reduce parasitic capacitance.

9.5.3.   THIN MULTIPLICATION REGION APDS

It has been observed that the excess noise from McIntyre's theory does not provide a good fit to the multiplication curves and that it overestimates the multiplication noise and underestimates the gain-bandwidth products for devices with thin multiplication regions [75–77]. After an ionization event a carrier needs to travel a certain distance before it can gain sufficient energy from the electric field to have a certain ionization probability. The distance is referred to as the "dead-space" or the "dead length." It can be ignored if it is small compared to the thickness of the multiplication region. However, when the thickness is reduced to the point that it becomes comparable to a few dead lengths, it is no longer valid to assume a continuous ionization process. The influence of dead length results in decrease of randomness of the impact ionization location inside the thinner multiplication region [78, 79]. A reduction in the multiplication noise permits the APD to be operated at a higher multiplication factor, which results in improvement of signal-to-noise ratio and higher receiver sensitivity. An added benefit with a thin multiplication region is a higher gain bandwidth product as expressed from Eq. (9.10) to Eq. (9.12).

Several achievements on large gain-bandwidth products with thin multiplication region have been reported so far with various materials [80–82]. Lenox reported InAlAs/InGaAs APD with a thin 200 nm thick InAlAs multiplication region and achieved an excellent ionization coefficient ratio of 0.18 and a large GB product of 290 GHz at near 1.55 μm wavelength [80]. They adopted a resonant cavity configuration with a thin InGaAs absorption layer of 60 nm so that at the same time they attained a high quantum efficiency of 70% and a large unity-gain bandwidth of 24 GHz. Nakata *et al.* demonstrated that 50 to 100 nm thick InAlAs layers can be used in separated absorption and multiplication (SAM) structure APDs [81] and achieved a high GB product of 140 GHz, a low operation voltage of 18.8 volts, and a quantum efficiency of 47% with a 100 nm thick multiplication layer and back-illuminated mesa structure [82]. They recently reported an APD composed of a multimode waveguide structure and a thin InAlAs multiplication layer, which could have an added advantage for high quantum efficiency. The waveguide (6-μm wide mesa structure) consisted of upper and lower AlGaInAs guide layers (0.7-μm thick and $\lambda_g = 1.3$ μm) and InAlAs/InGaAs SAM layers (0.5-μm thick p⁻-InGaAs absorption layer, 0.02-μm thick p⁺-InAlAs charge control layer, and 0.1-μm thick i-InAlAs multiplication layer) as a core region. All layers were grown by gas source molecular beam epitaxy on a semi-insulating InP substrate. The quantum efficiency at 1.54 μm wavelength was measured to be about 74% using a hemispherically ended fiber coupling to a 30-μm-long waveguide device. The maximum bandwidth of 32 GHz with 25 Ω load and a GB product of 180 GHz were obtained.

9.5.4.   Si/InGaAs HETERO-INTERFACE APD

Another interesting study is an APD with the wafer fusion technology, where p-type implanted Si substrate, which serves partially as a multiplication region, is heterogeneously integrated with the InGaAs absorption layer [83]. The scheme can take advantage of the absorption properties of InGaAs and the multiplication properties of Si, which exhibits nearly ideal avalanche multiplication for electron-initiated process. The large electron-to-hole ionization coefficient ratio of Si (about 50:1 at an electric field of 2×10^5 V/cm) results in APDs with little noise and high gain-bandwidth products with relatively a thick multiplication layer. The lattice mismatch between Si and InGaAs has prevented integration through epitaxy, however, the process of wafer fusion has allowed for the combination of the

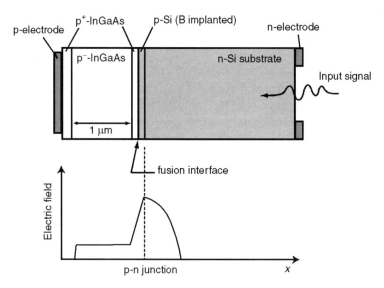

Fig. 9.26   (a) Schematic multiple layers structure of the Si/InGaAs hetero-interface APD fabricated by wafer fusion technology, and (b) the internal electric field distribution of the APD. (The figures are depicted after Ref. [84].)

two materials without threading dislocations or oxide barriers at the fused interface between Si and InGaAs. This is important for photodiode operation to achieve efficient transport of carriers across the hetero-interface with little loss or trapping of carriers.

Harkins [84] reported the fabrication of Si/InGaAs wafer fusion APD with a GB product of over 300 GHz. The device structure was a back-illumination structure through the Si substrate. Figure 9.26 is a schematic multi-layer structure of Si/InGaAs APD and its schematic internal electric field distribution. Illuminated light with 1.3–1.55 μm wavelength is absorbed in the InGaAs layer because of transparency of Si material in the near infrared, and photogenerated carriers are multiplied in the Si p-n junction. They prepared a 1.0-μm-thick undoped InGaAs layer and a thin Zn-doped p^+-InGaAs layer on InP substrate by MO-VPE and an n-type Si wafer with B implanted surface portion (providing p-doped layer after annealing). The top p^+-InGaAs layer was fused to the Si wafer by placing it in direct contact under pressure at around 650°C in a H_2 atmosphere. After fusion the InP substrate was selectively etched off from the InGaAs epitaxial layer. Their device was a circular mesa structure of 20 to 30 μm in diameter formed by etching through the InGaAs layer into the Si wafer. The series resistance and capacitance were less than 50 Ω and 0.1 pF, respectively. They measured 3-dB bandwidth of approximately 13 GHz at multiplication up to 20 and it decreased to 9 GHz at a multiplication of 35. The carrier transit time in the 1-μm-thick InGaAs absorption layer and the RC time constant limit the bandwidth for this sample. The highest reported GB-product of 315 GHz, to date, exceeding the superlattice APDs was successfully achieved due to the large ionization coefficient ratio of bulk Si material [84].

The novel APDs have exhibited excellent gain bandwidth products ranging 110 to 315 GHz, and a record effective ionization coefficient ratio of 0.1 corresponding to low multiplication noise around 3. These results indicate the great potential of the novel APDs for application to very high-speed systems over 10 Gbps. Especially, it is suggested that combination of an edge-coupled waveguide with superlattice or thin multiplication region APDs as well as Si/InGaAs wafer fusion type would be detector of choice for the coming ultra high-speed arena of 40 Gbps transmission systems.

9.6. Conclusions

Basic design and performance of avalanche photodiodes, current status of the InP/InGaAs APDs, and recent progress in the novel APDs have been reviewed in this chapter. Because the InP/InGaAs SAM structure APDs have been the most practical device for today's high-sensitivity and wide-bandwidth receiver in the lightwave transmission systems, some detailed explanation on the design and the performance of these APDs took up the major portion of the chapter.

It has been shown through the development of InP/InGaAs APDs that the suppression of avalanche multiplication in the light absorption InGaAs layer is the most critical development to improve the high-speed characteristics as well as the suppression of carrier trapping at the hetero-interface. Planar structure with effective guard ring has been indispensable for InP/InGaAs APDs to be put into practical applications, and the structures developed so far have shown satisfactory stability in productivity and reliability. The receiver modules with conventional InP/InGaAs APDs and wide-bandwidth pre-amplifier ICs have been confirmed very practical for application to 2.5 Gbps and 10 Gbps systems. Their advantage in system applications is their cost effectiveness and compact size without fiber pre-amplifier. This proves an attractive feature for lightwave transmission systems such as metropolitan and local area networks. Improvement of sensitivity in the L-band wavelength region also expands the potential of the InP/InGaAs Apds and the integrated receivers in the WDM systems. Reports on superlattice Apds, thin multiplication region Apds, and Si/InGaAs hetero-interface Apds (wafer fusion APD) have been summarized. Excellent performances of the novel Apds with very large gainbandwidth products and low multiplication noise have been demonstrated. Until these novel Apds become practical, much work remains, such as developing effective guard ring structure and confirming long-term reliability. The record GB-products of over 300 GHz achieved by the thin multiplication region APD and the Si/InGaAs hetero-interface APD indicate great potential for the next-generation high-speed systems applications.

References

1. H. Kuwahara, "Technology trend of photonic network system," FUJITSU, 51 (2000) 138–142.
2. H. Imai, "Evolution of devices for ultra-high speed optical transmission systems," FUJITSU, 51 (2000) 143–147.
3. T. Mikawa, M. Kobayashi, and T. Kaneda, "Avalanche photodiodes: present and future," Proc. of SPIE "Active and passive optical components for WDM communications," 4532 (2001) 139–145.
4. T. Mikawa, T. Kaneda, H. Nishimoto, M. Motegi, and H. Okushima, "Small-active-area germanium avalanche photodiode for single-mode fibre at 1.3 μm wavelength," Electron. Lett., 19 (1983) 452–453.
5. M. Kobayashi, T. Shirai, and T. Kaneda, "High reliability planar InGaAs avalanche photodiodes," Tech. Dig. IEDM'89 (Washington DC) (1989) 729–732.
6. J. N. Hollenhorst, D. T. Ekholm, J. M. Geary, V. D. Mattera, Jr., and R. Pawelek, "High frequency performance of planar InGaAs/InP avalanche photodiodes," Proc. of SPIE High Frequency Analog Communication, 995 (1988) 53–60.
7. K. Taguchi, T. Torikai, Y. Sugimoto, K. Makita, and H. Ishihara, "Planar-structure InP/InGaAsP/InGaAs avalanche photodiodes with preferential lateral extended guard ring for 1.0–1.6 μm wavelength optical communication," J. Lightwave Technol., LT-6 (1988) 1643–1655.
8. R. G. Smith and S. D. Personick, "Receiver design for optical communication systems," in Semiconductor Devices for Optical Communication, 2nd ed. (H. Kressel-Springer-Verlag, New York, 1982) 89–160.
9. G. E. Stillman and C. W. Wolfe, "Avalanche photodiodes," in Semiconductors and Semimetals, R. K. Willardson and A. C. Beer eds. (Academic Press, San Diego, CA, 1977) vol. 5, Chap. 5, 291–393.

10. S. M. Sze, "Photodetectors," in Physics of Semiconductor Devices, 2nd ed. (John Wiley & Sons, New York, 1981) Chap. 13, 743–789.

11. T. Kaneda, "Sillicon and Germanium avalanche photodiodes," in Semiconductors and Semimetals (Academic Press, San Diego, CA, 1985) Vol. 22, 247–328.

12. C. A. Lee, R. A. Logan, R. J. Batdrof, J. J. Kleimack, and W. Wiegmann, "Ionization rates of holes and electrons in sillicon," Phys. Rev., 134 (1964) A761–A773.

13. T. Mikawa, S. Kagawa, T. Kaneda, Y. Toyama, and O. Mikami, "Cristal orientation dependence of ionization rates in Germanium," Appl. Phys. Lett., 37 (1980) 387–389.

14. F. Osaka, T. Mikawa, and T. Kaneda, "Impact ionization coefficients of electrons and holes in (100)-oriented $Ga_{1-x}In_xAs_yP_{1-y}$," IEEE J. Quantum Electron., QE-21 (1985) 1326–1338.

15. K. Taguchi, T. Torikai, Y. Sugimoto, K. Makita, and H. Ishihara, "Temperature dependence of impact ionization coefficients in InP," J. Appl. Phys., 52 (1986) 476–481.

16. T. P. Pearsall, "Impact ionization rates for electrons and holes in $Ga_{0.47}In_{0.53}As$," Appl. Phys. Lett., 36 (1980) 218–220.

17. R. B. Emmons, "Avalanche-photodiode frequency response," J. Appl. Phys., 38 (1967) 3705–3714.

18. J. J. Chang, "Frequency response of PIN avalanche photodiodes," IEEE Trans. Electron Devices., ED-14 (1967) 139–145.

19. H. Melchior and W. T. Lynch, "Signal and noise response of high-speed Germanium avalanche photodiodes," IEEE Trans. on Electron Devices, ED-13 (1966) 829–838.

20. T. Kaneda and H. Takanashi, "Avalanche build up time of the Germanium avalanche photodiode," Japan. J. Appl. Phys., 12 (1973) 1091–1092.

21. T. Mikawa, H. Kuwatsuka, Y. Kito, T. Kumai, M. Makiuchi, S. Yamazaki, O. Wada, and T. Shirai, "Flip-chip InGaAs avalanche photodiode with ultra low capacitance and large gain-bandwidth product," Tech. Dig. OFC'91, San Diego, CA (1991) 186.

22. Y. Kito, H. Kuwatsuka, T. Kumai, M. Makiuchi, T. Uchida, O. Wada, and T. Mikawa, "High-speed flip-chip InP/InGaAs avalanche photodiodes with ultralow capacitance and large gain-bandwidth products," IEEE Photonic. Technol. Lett., 3 (1991) 1115–1116.

23. L. E. Tarof, J. Yu, R. Bruce, D. G. Knight, T. Baird, and B. Oosterbrink, "High-frequency performance of separate absorption, grading, charge, and multiplication InP/InGaAs avalanche photodiodes," IEEE Photonic. Technol. Lett., 5 (1993) 672–674.

24. W. R. Clark, A. Margittai, J. P. Noel, S. Jatar, H. Kim, E. Jamroz, G. Knight, and D. Thomas, "Reliable, high gain-bandwidth product InGaAs/InP avalanche

photodiodes for 10 Gb/s receivers," Tech. Dig. OFC/IOOC'99, San Diego, CA (1999) 96–98 (TuI).

25. M. A. Itzler, K. K. Loi, S. McCoy, and N. Codd, "High-performance, manufacturable avalanche photodiodes for 10 Gb/s optical receivers," Tech. Dig. OFC 2000, Baltimore, Maryland (2000) FG5-1–3.

26. R. J. McIntyre, "Multiplication noise in uniform avalanche diodes," IEEE Trans. on Electron Devices, ED-13 (1966) 174–168.

27. R. J. McIntyre, "The distribution of gains in uniformly multiplying avalanche photodiodes: Theory," IEEE Trans. Electron Devices, ED-19 (1972) 703–713.

28. C. J. Kircher, "Comparison of leakage currents in ion-implanted and diffused p-n junction," J. Appl. Phys., 46 (1975) 2167–2173.

29. S. R. Forrest, M. Didomenico, R. S. Smith, and H. J. Stocker, "Evidence for tunneling in reverse-biased III-V photodetector diodes," Appl. Phys. Lett., 36 (1980) 580–582.

30. H. Ando, H. Kanbe, M. Ito, and T. Kaneda, "Tunneling current in InGaAs and optimum design for InGaAs/InP avalanche photodiodes," Japan J. Appl. Phys., 19 (1980) L277–L280.

31. N. Z. Hakim, "Signal-to-noise ratio for lightwave systems using avalenche photodiodes," J. Lightwave Technol., LT-9 (1991) 318–320.

32. T. V. Moui, "Receiver design for high-speed optical-fiber systems," J. Lightwave Technol., LT-2 (1984) 243–257.

33. M. Brain and T. P. Lee, "Optical receivers for lightwave communication systems," J. Lightwave Technol., LT-3 (1985) 1281–1300.

34. H. Ando, H. Kanbe, T. Kimura, T. Yamaoka, and T. Kaneda, "Characteristics of Germanium avalanche photdiodes in the wavelength range of 1.0–1.6 μm," IEEE J. Quantum Electron., QE-14 (1978) 804–809.

35. T. Mikawa, S. Kagawa, and T. Kaneda, "Germanium avalanche photodiodes," Fujitsu Sci. Tech. J., 16 (1980) 95–118.

36. S. Kagawa, T. Kaneda, T. Mikawa, Y. Banba, Y. Toyama, and O. Mikami, "Fully ion-implanted p^+n germanium avalanche photodiodes," Appl. Phys. Lett., 38 (1981) 429–431.

37. T. Mikawa, S. Kagawa, and T. Kaneda, "Germanium reachthrough avalanche photodiodes for optical communication systems at 1.55-μm wavelength region," IEEE Trans. Electron Devices, ED-31 (1984) 971–976.

38. K. Nishida, K. Taguchi, and Y. Matsumoto, "InGaAsP heterojunction avalanche photodiodes with high avalanche gain," Appl. Phys. Lett., 35 (1979) 251–253.

39. H. Kanbe, N. Susa, H. Nakagome, and H. Ando, "InGaAs avalanche photodiode with InP p-n junction," Electron. Lett., 16 (1980) 155–156.

40. T. Shirai, S. Yamazaki, H. Kawata, K. Nakajima, and T. Kaneda, "A planar InP/InGaAsP heterostructure avalanche photodiode," IEEE Trans. Electron Devices, ED-29 (1982) 1404–1407.

41. T. Shirai, T. Mikawa, T. Kaneda, and A. Miyaychi, "InGaAs avalanche photodiode of 1 μm wavelength region," Electron. Lett., 19 (1983) 534–536.
42. J. C. Campbell, A. G. Dentai, W. S. Holden, and B. L. Kasper, "High-performance avalanche photodiode with separated absorption "grading" and multiplication regions," Electron. Lett., 19 (1983) 818–820.
43. J. C. Campbel, W. S. Holden, G. J. Qua, and A. G. Dentai, "Frequency response of InP/InGaAs avalanche photodiodes with separated absorption, grading, and multiplication regions," IEEE J. Quantum Electron., QE-21 (1985) 1743–1746.
44. Y. Sugimoto, T. Torikai, K. Makita, H. Ishihara, K. Minemura, and K. Taguchi, "High-speed planar-structure InP/InGaAsP/InGaAs avalanche photodiode grown by VPE," Electron. Lett., 20 (1984) 653–654.
45. M. Kobayashi, S. Yamazaki, and T. Kaneda, "Planar InP/InGaAsP/InGaAs buried-structure avalanche photodiode," Appl. Phys. Lett., 45 (1984) 759–761.
46. M. Kobayashi, H. Machida, T. Shirai, Y. Kishi, N. Takagi, and T. Kaneda, "Optimized GaInAs avalanche photodiode with low noise and large gain-bandwidth product," Proc. 6th Int. Conf. Integrated Opt. and Opt. Fiber Commun. Reno, ND (1987).
47. T. Shirai, M. Kobayashi, Y. Kishi, H. Machida, H. Nishi, and T. Kaneda, "A new InP/InGaAs reach-through avalanche photodiode," Proc. 19th Eur. Conf. Optical Commun. Helshinki (1987).
48. S. R. Forrest, O. K. Kim, and R. G. Smith, "Optical response time of In$_{0.53}$Ga$_{0.47}$As avalanche photodiode," Appl. Phys. Lett., 41 (1982) 95–98.
49. K. Yasuda, T. Mikawa, Y. Kishi, and T. Kaneda, "Multiplication-dependent frequency response of InP/InGaAs avalanche photodiode," Electron. Lett., 20 (1984) 373–374.
50. F. Capasso, H. M. Cox, A. L. Hutchinson, N. A. Olsson, and S. G. Hummel, "Psuedo-quaternary GaInAsP semiconductors: A new Ga$_{0.47}$In$_{0.53}$As/InP graded gap superlattice and its application to avalanche photodiodes," Appl. Phys. Lett., 45 (1984) 1193–1195.
51. H. Kuwatsuka, Y. Kito, T. Uchida, and T. Mikawa, "High-speed InP/InGaAs avalanche photodiodes with compositionally graded quaternary layer," Photonics. Technol. Lett., 3 (1991) 1113–1114.
52. F. Osaka and T. Kaneda, "Excess noise design of InP/GaInAsP/GaInAs avalanche photodiodes," IEEE J. Quantum Electron., QE-22 (1986) 471–478.
53. F. A. Padovani and R. Stratton, "Field and thermionic-field emission in Schottky barriers," Solid-State Electron., 9 (1966) 695–707.
54. R. S. Sussmann, R. M. Ash, A. J. Mosely, and R. C. Goodfellow, "Ultra-low-capacitance flip-chip-bonded GaInAs p-i-n photodetector for long-wavelength high-data-rate fibre-optic systems," Electron. Lett., 22 (1985) 593–595.

55. Y. Kito, H. Kuwatsuka, T. Uchida, and T. Mikawa, "High-speed Planar InP/InGaAs avalanche photodiodes with optimized electric field profile," Proc. Int. Conf. Solid State Devices & Mat., Yokohama, Japan (1991).
56. K. Satoh, I. Hanawa, and M. Kobayashi, "APD receiver module for 10 Gbps high-speed data communication," FUJITSU, 51 (2000) 152–155.
57. "2000 Lightwave Component & Modules Databook," Eds. Fujitsu Compound Semiconductor, Inc., San Jose, CA 2000.
58. Wada, M. Makiuchi, H. Hamaguchi, T. Kumai, and T. Mikawa, "High-performance, high-reliability InP/GaInAs p-i-n photodiodes and flip-chip integrated receivers for lightwave communications," J. Lightwave Technol., 9 (1991) 1200–1207.
59. Y. Yamashita, Y. Oikawa, T. Yamamoto, H. Ohnishi, T. Mikawa, and H. Hamano, "High performance 10-Gbit/s optical receiver using avalanche photodiode and hetero-junction bipolar transistor Ics," Tech. Dig. OFC'97, Dalas, TX (1997).
60. M. Kobayashi and A. K. Dutta, "Module/packaging technologies for optical components: current and future trends," Proc. of SPIE Active and passive optical components for WDM communications, 4532 (2001) 270–280.
61. M. Kobayashi and T. Kaneda, Semiconductor Device Reliability, A. Christou and B. A. Unger, eds., (Kluwer Academic Publishers, Dordrecht, Boston, and London, 1990).
62. A. K. Chin, F. S. Chen, and F. Ermanis, "Failure mode analysis of planar zinc-diffused In$_{0.53}$Ga$_{0.47}$As p-i-n photodiodes," J. Appl. Phys., 55 (1984) 1586–1606.
63. J. C. Dries, M. R. Gokhale, and S. R. Forrest, "A 2.0 µm cutoff wavelength separate absorption, charge, and multiplication layer avalanche photodiode using strain-compensated InGaAs quantum wells," Appl. Phys. Lett., 74 (1999) 2582–2583.
64. J. W. Matthews and A. E. Blakeslee, "Defects in epitaxial multilayer," J. Cryst. Growth, 27 (1974) 118–125.
65. T. Uchida, A. Yazaki, C. Anayama, A. Furuya, T. Shirai, and M. Kobayashi, "Improved responsivity at the L-band wavelength of a strain-compensated InGaAs multiple quantum well photodiode," Proc. of SPIE Active and Passive Optical Components for WDM Communications, 4532 (2001) 281–288.
66. R. Chin, N. Holonyak, Jr., G. E. Stillman, J. Y. Tang, and K. Hess, "Impact ionization in multilayered heterojunction structures," Electron. Lett., 16 (1980) 467–469.
67. F. Capasso, W. T. Tsang, A. L. Hutchinson, and G. F. Williams, "Enhancement of electron impact ionization in a superlattice: a new avalanche photodiode with large ionization rate ratio," Appl. Phys. Lett., 40 (1982) 38–40.

68. T. Kagawa, H. Asai, and Y. Kawamura, "An InGaAs/InAlAs superlattice avalanche photodiode with a gain bandwidth product of 90 GHz," IEEE Photonic. Technol. Lett., 3 (1991) 815–817.

69. T. Kagawa, Y. Kawamura, and H. Iwamura, "InGaAsP-InAlAs superlattice avalanche photodiode," IEEE J. Quantum Electron., QE-28 (1992) 1419–1423.

70. I. Watanabe, K. Makita, M. Tsuji, T. Torikai, and K. Taguchi, "Extremely low dark current InAlAs/InGaAlAs quaternary well superlattice APD," Proc. 4th Int. Conf. InP and Related Materials, Newport, RI, (1992) 246–249.

71. S. Hanatani, H. Nakamura, S. Tanaka, C. Notsu, H. Sano, and K. Ishida, "Superlattice avalanche photodiode with a gain-bandwidth larger than 100 GHz for very-high-speed systems," Tech. Dig. OFC/IOOC'93, San Jose, CA, (1993) 187–188.

72. I. Watanabe, M. Tsuji, K. Makita, and K. Taguchi, "Gain-bandwidth product analysis in InAlGaAs-InAlAs superlattice avalanche photodiodes," IEEE Photonic. Technol. Lett., 8 (1996) 269–271.

73. K. Makita, I. Watanabe, M. Tsuji, and K. Taguchi, "Dark current and breakdown analysis in In(Al)GaAs/InAlAs superlattice avalanche photodiodes," Japan J. Appl. Phys., 35 (1996) 3440–3444.

74. I. Watanabe, M. Tsuji, K. Makita, and K. Taguchi, "A new planar-structure InAlGaAs-InAlAs superlattice avalanche photodiode with a Ti-implanted guardring," IEEE Photonic. Technol. Lett., 8 (1996) 827–829.

75. J. N. Hollenhorst, "Ballistic avalanche photodiodes: ultralow noise avalanche diodes with neary equal ionization probabilities," Appl. Phys. Lett., 49 (1986) 516–518.

76. J. C. Campbell, S. Chandrasekhar, W. T. Tsang, G. J. Qua, and B. C. Johnson, "Multiplication noise of wide-bandwidth InP/InGaAsP/InGaAs avalanche photodiodes," J. Lightwave Technol., 7 (1989) 473–477.

77. C. Hu, K. A. Anselm, B. G. Streetman, and J. C. Campbell, "Noise characteristics of thin multiplication region GaAs avalanche photodiodes," Appl. Phys. Lett., 69 (1996) 3734–3736.

78. M. M. Hayat, W. L. Sargent, and B. E. A. Salhe, "Effect of dead space on gain and noise in Si and GaAs avalanche photodiodes," IEEE J. Quantum Electron., 28 (1992) 1360–1365.

79. A. Spinelli and A. L. Lacaita, "Mean gain of avalanche photodiodes in a dead space model," IEEE Trans. Electron Dev., 43 (1996) 23–30.

80. C. Lenox, H. Nie, P. Yuan, G. Kinsey, A. L. Holmes, Jr., B. G. Streetman, and J. C. Campbell, "Resonant cavity InGaAs-InAlAs avalanche photodiodes with gain-bandwidth product of 290 GHz," IEEE Photonic. Technol. Lett., 11 (1999) 1162–1164.

81. T. Nakata, T. Takeuchi, I. Watanabe, K. Makita, and T. Torikai, "10-Gb/s high sensitivity, low-voltage-operation avalanche photodiodes with thin InAlAs

multiplication layer and waveguide structure," Electron. Lett., 36 (2000) 2033–2034.

82. T. Nakata, I. Watanabe, K. Makita, and T. Torikai, "InAlAs avalanche photodiodes with very thin multiplication layer of 0.1 μm for high-speed and low voltage operation optical receiver," Electron. Lett., 36 (2000) 1807–1809.

83. W. Wu, A. R. Hawkins, and J. E. Bowers, "High-gain bandwidth product Si/InGaAs avalanche photodetectors," Tech. Dig. OFC'97, Dalas, TX, (1997) 35–36.

84. A. R. Hawkins, W. W. P. Abraham, K. Streubel, and J. E. Bowers, "High gain-bandwidth-product sillicon heterointerface photodetector," Appl. Phys. Lett., 70 (1997) 303–305.

Part 4 | Fabrication Technologies

Chapter 10 | Selective Growth Techniques and Their Application in WDM Device Fabrication

Tatsuya Sasaki
Koji Kudo

Photonic and Wireless Devices Research Laboratories,
System Devices and Fundamental Research, NEC Corporation,
2-9-1 Seiran, Ohtsu-shi, Shiga 520-0833, Japan

10.1. Introduction

Recent progress in dense wavelength-division multiplexing (DWDM) transmission systems has called for the development of advanced photonic devices with improved characteristics. The development of light sources has made it possible to achieve modulation at a rate exceeding 2.5 Gbps (OC-46) not only by combining a high-power distributed feedback laser diode (DFB-LD) with a $LiNbO_3$ external optical modulator, but also by compactly integrating an electroabsorption (EA) optical modulator with a DFB-LD. An increase in the transmission capacity of DWDM systems requires light sources that can be operated in precisely adjusted wavelength grids with high stability. Because the lasing wavelength of a DFB-LD is determined by the grating pitch and effective refractive index of the waveguide structure, a highly uniform waveguide structure as well as accurate grating pitch control are essential. Wavelength-tunable lasers, which can cover several or all WDM wavelength signals by an electronically or a thermally controlled scheme, will be key devices in future DWDM optical networks. Tunable lasers generally consist of multiple sections such as gain and tuning regions. Therefore, monolithic integration of various functional sections, each having active and passive waveguide layers, is important.

Device fabrication strongly affects the device performance and production yield, particularly for complicated integrated photonic devices. In the

443

WDM TECHNOLOGIES: ACTIVE
OPTICAL COMPONENTS
$35.00

conventional fabrication process of laser diodes, a double heterostructure (DH) including an active layer is first grown on a semiconductor substrate. Then, a mesa structure is formed by chemical etching, and both sides of the mesa structure are buried with current-blocking layers to form a buried heterostructure (BH). This fabrication process enables low threshold current and high output power operations of DFB-LDs that are suitable for WDM transmission applications. However, if a mesa structure is formed by wet chemical etching, accurate control of the active-layer width over the entire semiconductor wafer becomes difficult due to the fluctuation in the etching conditions, which results in poor controllability and reproducibility of lasing wavelengths for DFB-LDs. A dry etching technique can effectively improve the structure uniformity, but the deterioration of device performance due to the plasma-induced damage on the etched surfaces of the mesa structure is of great concern and should be carefully avoided.

To fabricate a monolithic integrated device such as a tunable light source, several waveguide structures for each functional section, including the active layer, the modulation layer, and the passive waveguide should be smoothly connected to each section of the device. Figure 10.1 shows different waveguide joint structures. A butt-joint waveguide (Fig. 10.1(a)) is commonly used in device fabrication [1]. This structure enables high optical coupling efficiency, but its relatively complicated fabrication process, including partial etching of the active layer and subsequent buried growth of a passive layer, should be optimized. Figure 10.1(b) shows an "evanescence wave-coupled" waveguide joint, which has an advantage of a simple fabrication process requiring only partial etching of the active layer in the passive section [2]. However, the coupling efficiency is not very high, due to a mismatch of the optical fields of the two sections. Figure 10.1(c) shows a new approach based on the disordering of a multiple quantum well (MQW) structure in the passive section that is attained by a cap annealing technique [3]. This technique is quite simple, and it enables high optical coupling efficiency. However, annealing at high temperatures may result in crystal deterioration, and this concern limits the flexibility of the waveguide design. Although several devices have been successfully fabricated by using the above-described techniques, the device fabrication process must still be significantly improved.

A selective area growth technique, in which a DH structure is grown on a mask-patterned planar substrate, is very attractive, especially for fabricating integrated photonic devices, because it enables bandgap energy control over the substrate (Fig. 10.1(d)). This bandgap energy control is

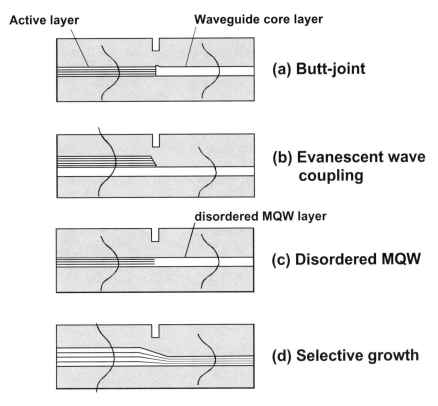

Active layer   **Waveguide core layer**

(a) Butt-joint

(b) Evanescent wave coupling

disordered MQW layer

(c) Disordered MQW

(d) Selective growth

Fig. 10.1   Waveguide joint structures: (a) butt-joint structure, (b) evanescent wave-coupled (taper) waveguide, (c) disordered MQW, (d) selectively grown structure.

mainly a result of diffusion in metal-organic vapor phase epitaxy (MOVPE, also called MOCVD, OMVPE, and OMCVD). By using this technique, all waveguide layers in monolithically integrated devices such as the active layer, the modulation layer, and the passive waveguide layer can be formed simultaneously. This significantly simplifies the fabrication process and improves the device yield. The technique also enables highly efficient optical coupling between each waveguide without reflection at the joint. Moreover, if a narrow stripe opening (which is typically 1 to 2 μm wide) is used for mask configuration, the layer grown on the narrow stripe opening can be directly used as an optical waveguide without any etching of the semiconductor layer.

Selective MOVPE has been extensively investigated with regard to its mechanism, and has been used in the fabrication of various photonic devices for optical communication systems. This technique is very attractive for WDM applications. In this chapter, we first describe several studies on selective area growth and then demonstrate various device applications mainly for WDM systems.

10.2. Selective MOVPE

10.2.1. PREVIOUS STUDIES

Selective growth, a process of growth on a mask-patterned semiconductor substrate, has been studied extensively [4]. Early studies of selective growth mainly focused on suppressed polycrystal formation on dielectric mask surfaces, on a characteristic crystalline shape of grown structures, and on buried growth of current-blocking layers into a mesa structure. Growth on an unmasked nonplanar substrate, in which the mask film used for substrate etching is removed prior to growth, also has been investigated. The resulting structures are shown in Fig. 10.2. Selective growth has been studied in various growth methods, such as liquid phase epitaxy (LPE), MOVPE, gas-source molecular beam epitaxy (GSMBE), metal-organic molecular beam

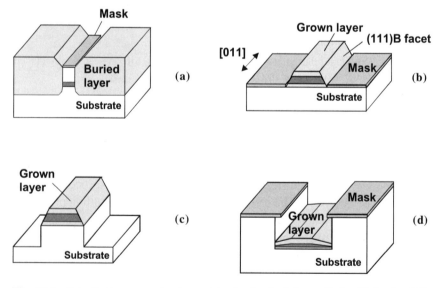

Fig. 10.2 Selectively grown structures: (a) selective buried growth for fabricating BH, (b) selective growth on a planer substrate, (c) a ridge growth on a mesa structure, (d) growth on an etched groove.

epitaxy (MOMBE), chemical beam epitaxy (CBE), and vapor phase epitaxy (VPE). The unique characteristics of a localized structure grown on an exposed semiconductor surface originate in their own growth kinetics, and have been investigated not only for the purpose of researching crystal growth but also for practical applications in photonic and electronic device fabrication.

In MOVPE, metal-organic species, which are a combination of group-III atoms and alkyls, diffuse in the stagnant layer and adsorb on the substrate surface. In the case of a mask-patterned substrate, that is a substrate used for selective growth, the density of species adsorbed on the mask surface should be relatively low to avoid nucleation; otherwise, polycrystals form on the mask surface especially if the species have high sticking coefficients [5, 6]. A low adsorption rate, a high migration rate, and a high desorption rate of the species on a mask surface can prevent mask-surface nucleation. Consequently, a low growth rate, low reactor pressure, and an adequately high substrate temperature are required to obtain high selectivity [7–9]. However, even under the optimized conditions, if the masked area is too large, polycrystals will still remain on the mask surface. One probable reason for this polycrystal formation is an insufficient migration length of group-III species on the mask surface. If the migration length is not greater than the mask width, the species will remain on the mask surface. Once polycrystals form, further nucleation occurs, and the size of the polycrystals increases. Based on this reasoning, the spacing between polycrystals, or the width between the mask edge and the nearest polycrystal, was believed to be related to the mean free path of migrated species [6, 10, 11].

Another study of selective MOVPE focused on low-index crystallographic planes as facets. As shown in Fig. 10.2(b), when the mask stripes on a (100)-oriented substrate are formed along the [011] direction, (111)B side facets usually form in the grown structure. This can be explained by a decrease in the number of surface dangling bonds on the (111)B plane compared to that on the (100) plane [12]. As the growth proceeds, the length of the (111)B facets increases, while the width of the (100) top surface decreases. In the case of GaAs, there is no growth on (111)B facets, and the growth does not proceed after the (100) top surface has disappeared. Thus a ridge structure with a triangular cross section is selectively formed on the window region between the mask stripes [7]. Note that the angle between the (111)B facets and the surface of the (100) substrate must be 54.7 degrees; therefore the mask-opening width determines the triangular shape. A similar GaAs/AlGaAs ridge structure can also be formed on a mesa structure with no mask stripes as shown in Fig. 10.2(c), and it can be used to fabricate laser diodes [13, 14] and optical waveguides [15].

InGaAs/InP ridge structures grown on mask-patterned substrates [16] and nonplaner substrates [17, 18] have also been described. As is the case with GaAs, InGaAs does not grow on (111)B side facets. However, a study on the growth rate of InGaAs layers on unmasked planar substrates with various surface orientations [9] shows that the InGaAs growth rate on a (111)B-oriented substrate is much lower than that on a (100)-oriented substrate, but it is not zero. In selective growth, the difference in the growth rates of InGaAs layers on these two planes causes a surface diffusion flux of group-III species from the (111)B facets to the adjacent (100) surface. In contrast, InP layers do grow slightly on (111)B facets. Therefore, in selective growth of an InGaAs/InP DH, the edge of the InGaAs thin layer is covered with the following InP cladding layer. This ridge-growth technique is used to form pn-junction diodes [19], modulator-integrated DFB lasers [20], optical modulators [21], and quantum dot structures [22]. Using this technique, narrow active waveguides of devices can be formed on mask-patterned and nonplanar substrates without conventional etching of semiconductor layers, as will be shown later. The width of the active layer is determined by the mask pattern.

When the mask stripes are patterned along the [0-11] direction, both (111)A and (111)B planes can be formed, and the lateral growth on the mask edge is relatively extensive [7]. Consequently, controlling ridge structures grown in the [0-11] direction is rather difficult.

10.2.2. GROWTH-RATE ENHANCEMENT

Studies of selective MOVPE have found that the growth rate is strongly affected by the mask configuration. The growth rate is locally enhanced in the region near the masked area by an extra supply of group-III species from the masked region as well as from the crystal facet region. This growth rate enhancement is significant when the adjacent mask area is large or the grown area is small. In the case of ridge structures, selectively grown between a pair of mask stripes (Fig. 10.2(b)), the grown layer thickness increases with an increase in the mask width or a reduction of the mask spacing [15, 16, 20, 23, 24]. Similar growth-rate variation has also been observed in the growth on nonplanar substrates [17, 25].

Figure 10.3 shows the process of selective MOVPE. In addition to regular vertical vapor phase diffusion of metal-organic group-III species such as

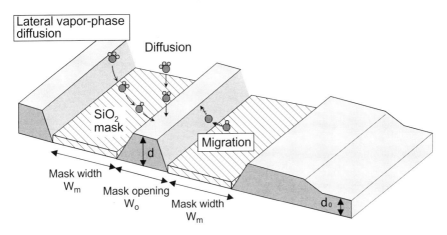

Fig. 10.3 Selective MOVPE process.

trimethylindium (TMIn), trimethylgallium (TMGa), and triethylgallium (TEGa), species are supplied from the masked region mainly as a result of the following two phenomena: lateral vapor phase diffusion and migration on the mask surface. Consequently, selectively grown layer thickness d is considerably larger than reference layer thickness d_0, which is measured far from the masked region. Growth rate enhancement d/d_0 increases with an increase in mask width W_m or a reduction in mask opening W_o. Though the mechanism of selective MOVPE is quite complicated and has not yet been fully understood, lateral vapor phase diffusion of group-III species has been recognized as the driving force of growth enhancement [9, 26, 27]. In the surface-exposed region, adsorbed species are incorporated into the surface atoms in a relatively short period of time, and the concentration of species on the surface is very low. In contrast, in the masked region, although most of the adsorbed species will evaporate or migrate, the concentration of species on the mask surface is not zero. This difference in the species concentration on the two surfaces results in a lateral concentration gradient in the gas phase. Consequently, lateral diffusion of group-III species occurs from the masked region to the growth region. Figure 10.4 shows two lateral thickness profiles of a selectively grown layer in the [0-11] direction, perpendicular to the mask stripes [27]. On the left side of the mask stripes, deeply etched grooves were formed on the substrate surface. If surface migration is the dominant cause of growth rate enhancement, the two profiles should be different because the layer thickness beyond the grooves should be affected by the grooves. However, both thickness profiles are exactly the same, and

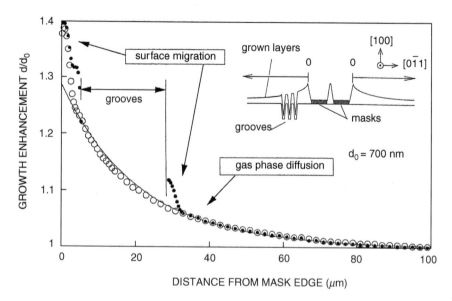

Fig. 10.4   Lateral growth enhancement profiles for selectively grown layers with a pair of 15-μm-wide mask stripes. Profiles of outer mask edges in the direction in which grooves are formed (●) and not formed (○) are shown. The solid line is an exponential fitting curve [27].

are not affected by the grooves. This result clearly shows that this thickness enhancement was not caused by the surface migration, but resulted from the vapor phase diffusion. Kayser [9] and Gibbon [26] described similar experiments. Note that extra growth can be observed in the limited regions near the edge of the mask and grooves, which is due to the species migrating from the mask surface and (111)B facets.

Because this lateral thickness enhancement profile is related to the diffusion profile of the species, we can control the profile by growth pressure [28–30]. Figure 10.5 shows lateral thickness enhancement profiles of InGaAsP layers grown under different pressures [27]. Under a high growth pressure, a reduced diffusion coefficient of the group-III species results in thickness enhancement near the mask edge, while the thickness enhancement in the region far from the mask decreases. Consequently, the growth rate enhancement of a selectively grown ridge structure increases with an increase in the growth pressure as shown in Fig. 10.6 [30]. The growth rate enhancement at an atmospheric pressure is four times that at 75 Torr.

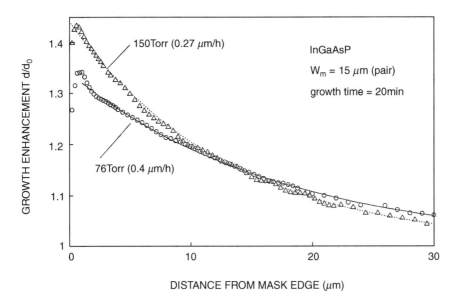

Fig. 10.5 Lateral growth enhancement profiles of InGaAsP layers with a pair of 15-μm-wide mask stripes under different growth pressures. Growth rates on unmasked substrates are shown [27].

This growth rate enhancement technique for selectively grown layers is very attractive for integrating an optical mode-field converter with an optical waveguide [31]. Spot-size converter-integrated laser diodes (SSC-LD) have already been developed [32, 33]. This technique is also used to control the Bragg wavelength in DFB/DBR-LD arrays by changing the effective refractive index of the waveguide structures [34–36].

10.2.3. COMPOSITION SHIFT

In the beginning of the 1990s, several studies reported that selective MOVPE enables not only growth-rate enhancement but also an alloy composition shift of the grown layers by changing the mask pattern [9, 37–39]. Figure 10.7 shows measured photoluminescence (PL) peak wavelength shifts of selectively grown InGaAsP ridge structures [40]. In InGaAs and InGaAsP layers, the PL wavelength increases with an increase in the mask stripe width, W_m. Direct measurements of the layer composition by using Auger electron spectroscopy and Raman spectroscopy showed an increase

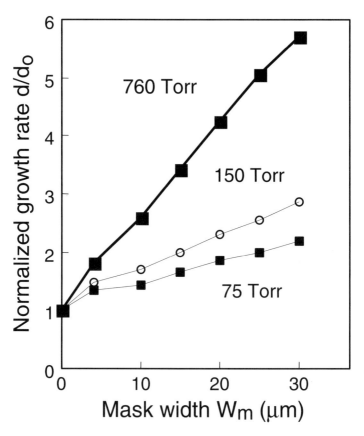

Fig. 10.6   Growth rate enhancement for InGaAsP layers ($\lambda_g = 1.5\,\mu m$) under different growth pressures [30].

in the indium content with an increase in the mask width. The group-III composition shifts in the figure are a result of accelerated gas phase diffusion and incorporation of indium-containing species as opposed to gallium-containing species [9, 39], or a result of the difference in the surface migration rates of these species [37, 38]. Due to a large V/III ratio in MOVPE, the transportation rates of As and P atoms to selectively grown layers seem comparable. Consequently, the shift in the group-V composition is rather small compared to the shift in the group-III composition, and is almost never observed. Caneau [41] reported that the indium-content enhancement in GaInAs and GaInP layers selectively grown by using TMGa as a gallium source was greater than that in the layers grown

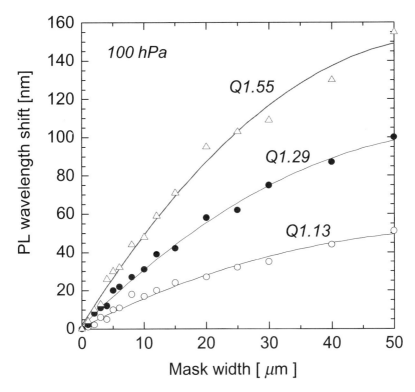

Fig. 10.7 Measured PL wavelength shift for different-composition InGaAsP layers as a function of mask width [40].

using TEGa. This phenomenon can be explained by the difference in the decomposition temperatures of the metal-organic species. The decomposition temperature of TMGa is higher than that of TEGa, and both TMGa and TEGa have a high decomposition temperature compared to that of TMIn. The decomposed group-III species appeared to have higher diffusion coefficients, which can account for their enhanced incorporation [42]. The difference in the indium-content shifts in GaInAs layers was also observed between regular TMIn and TMIn-di-isopropylamine-adduct [9], and TMIn and dimethylaminopropyl-dimethylindium [43], which can be explained by the difference in the decomposition temperatures of these sources. The indium-content enhancement of InGaAs increased under a low growth temperature of 400°C [44], which can also be attributed to a significant difference in the decomposition rates between TMIn and TEGa at low temperatures.

Caneau [45] also investigated growth-rate enhancement for binary materials based on the experimental results for GaInAs and GaInP layers. The growth enhancement was in the following order: R(InAs) > R(InP) > R(GaAs) > R(GaP), which coincides with the order of the decomposition rates of TMIn and TMGa in the presence of AsH$_3$ and PH$_3$, respectively. A higher decomposition rate of AsH$_3$ compared to that of PH$_3$ may enhance the decomposition of metal-organic species. Different composition shifts were also observed for InGaAsP layers grown with conventional group-V hydrides (AsH$_3$ and PH$_3$) and tertiarybuthyl-hydrides (TBA and TBP) [46]. These results indicate that indium-content enhancement for InGaAsP layer depends on the group-V composition and sources. On the other hand, a clear coincidence was observed in the growth enhancement of indium in InP and InGaAs layers with the same mask pattern [26, 41].

10.2.4. SELECTIVE MOVPE SIMULATION

Selective MOVPE has been numerically simulated to investigate the growth mechanism. Gibbon [26] developed a simple two-dimensional simulation model that can help understand the selective MOVPE process. The results of an earlier simulation [47] that investigated a two-dimensional concentration profile of group-III species in the gas phase, can be interpreted by using Laplace's equations. Figure 10.8 shows the simulation model developed by Gibbon. The boundary conditions in this model are essentially as follows;

1. The concentration of the group-III species at the top of the stagnant layer is constant.
2. The concentration profile is symmetric to the center of the mask pattern.
3. On the semiconductor surface, the flux of group-III species is equal to the adsorption rate of the species.
4. On the masked surface, the flux equals zero (no adsorption).

The third condition can be written in the form of the following equation:

$$D(\delta/\delta z) = kn,$$

where D is the diffusion constant of group-III species, n is the concentration of species, z is the growth direction, and k is the rate of adsorption of species on the semiconductor surface per unit of concentration in the

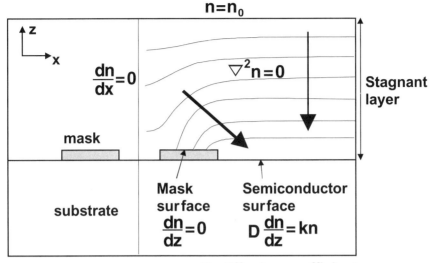

$$n = n_0$$

Fig. 10.8 Simulation model for selective MOVPE process.

gas phase. A useful fitting parameter is D/k, the ratio of the diffusion coefficient to the rate of adsorption on the semiconductor surface. In this simulation, the diffusion coefficient for each group-III species was kept constant and no decomposition process in the gas phase was taken into account. The measured thickness enhancement profiles plotted against the distance from the mask stripe were in good agreement with the simulated results. Consequently, Gibbon concluded that the difference in the growth enhancement between indium- and gallium-containing species results from the difference in adsorption rate k, while the diffusion coefficients of the species are almost the same. The estimated ratio of k_{Ga}/k_{In} was 0.14. These results show that the surface reaction of the species, including the breaking of alkyls from group-III atoms, should be taken into account in explaining the composition shift.

The simulated contour concentration plots of group-III species [4, 47] clearly show that the lateral concentration gradient causes an additional supply of species. Fujii [48] simulated a layer thickness profile along the stripe direction by considering the effects of surface migration and vapor phase diffusion of desorbed species, and found that the simulated and experimentally obtained profiles were in good agreement.

Simulation is a very effective way to estimate the layer thickness profile, because it takes a lot of time to determine the proper mask pattern by several growth experiments and characterizations. Although selective MOVPE is very complicated and its quantitative clarification by simulation is difficult, simulation is a powerful tool in the development of this selective growth technique.

10.2.5.   BANDGAP-ENERGY CONTROL

Based on the growth-rate enhancement and composition shift in selective MOVPE growth both on mask-patterned and nonplanar substrates, a longitudinal bandgap-energy control technique has been developed and its application to photonic devices has been demonstrated [15, 20, 24, 25]. Figure 10.9 shows a measured PL peak-wavelength profile for selectively grown InGaAsP/InGaAsP MQW layers [27]. By changing mask stripe width W_m from 4 to 10 µm, 50 nm wavelength shift was obtained. There was a transient area between two regions reflecting the gas phase

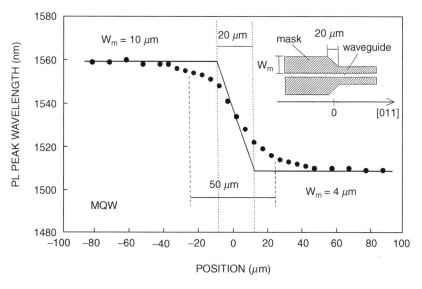

Fig. 10.9   PL peak wavelength profile for MQW structure along the stripe direction. The mask width was varied from 10 to 4 µm with a 20-µm-long transition region [27].

diffusion effects of group-III species. This technique is quite effective in fabricating photonic integrated devices because conventional etching/ regrowth processes used to form butt-joint waveguide structures are not needed. Electroabsorption modulator-integrated DFB-LDs are the most promising application [20, 24]. A similar bandgap-energy shift was also observed in a layer grown on an etched mesa and an etched groove [25, 49], but these techniques require additional etching to prepare the substrates, and their application in the fabrication of DFB-LDs is relatively difficult.

In a composition shift, the group-III content changes significantly while the group-V content remains almost unchanged. This means that a lattice mismatch occurs in the layers grown selectively under the growth conditions for a lattice-matched layer on an unmasked region. In selectively grown quantum wells, a compressive strain is induced. The strain is larger for InGaAsP layers with a lower bandgap energy. Therefore, for an InGaAsP/ InGaAsP MQW, the strain into the barriers is lower than that into the wells. As the mask width increases, both the compressive strain and the thickness of the InGaAsP wells increase, which results in a bandgap-energy decrease. Optimizing the mask pattern and growth conditions results in large bandgap-energy shifts [27, 30, 50, 51]. However, the largest shift of 253 meV in the preceding studies was obtained by using unstrained InP barrier layers. As will be explained later, a bandgap-energy shift of 354 meV was reported for an InGaAs/InAlAs MQW [52].

The bandgap-energy shift increases under high growth pressure, which is a result of suppressed lateral diffusion of group-III species over the grown region [35, 50, 53]. Figure 10.10 shows a PL peak wavelength shift for InGaAsP/InGaAsP MQW structures under different growth pressures [30]. In selective MOVPE under the atmospheric pressure, a 370-nm PL peak wavelength shift (related to a bandgap-energy shift of 240 meV) was observed when the mask stripe width was changed from 4 to 30 μm. Selective growth under the atmospheric pressure is also effective for obtaining a steep bandgap-energy shift along the waveguide due to a short diffusion length of group-III species.

Bandgap-energy control is the most attractive feature of selective MOVPE in device fabrication. However, to practically obtain a large bandgap-energy shift, the design of MQW structures and the growth conditions must be carefully controlled to prevent the degradation of crystal quality due to changes in thickness and induced strain.

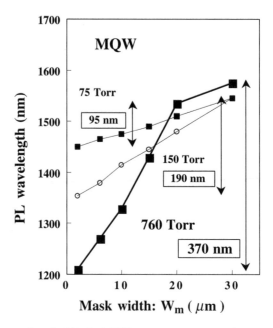

Fig. 10.10   PL wavelength shift for MQW structures grown under various growth pressures [30].

10.2.6. SELECTIVE GSMBE, MOMBE, AND CBE

In selective GSMBE, MOMBE, and CBE, a similar ridge structure is formed [19, 54]. However, no growth-rate enhancement such as that observed in MOVPE has been observed. The beam nature of the reactant supply transports the reactant to the mask surface without a gas-phase reaction. The species adsorbed on the mask surface evaporate and do not contribute to growth. Consequently, the growth rate and bandgap energy of the grown layers are independent of the mask pattern [39, 54]. When the growth conditions, especially the growth temperature, are optimized, no polycrystals remain even in the large masked area. These growth techniques enable growing a uniform structure with minor effect of mask structure variations [55]. In contrast, controlling the layer thickness and bandgap energy by the mask pattern is difficult. Surface migration of group-III species from grown (111)B facets to the (100) surface was also observed in selective MOMBE/CBE [56, 57], and the difference in the migration rates between the indium- and gallium-containing species resulted in indium content enhancement with a decrease in the grown layer width [19, 57].

10.3.   Narrow-Stripe Selective MOVPE

10.3.1.   FEATURES

The most interesting feature of selective MOVPE is its bandgap-energy control capability. By using this technique, photonic integrated devices and wavelength-tunable lasers can be fabricated by a simple process.

In one approach of selective MOVPE for device fabrication [24], a mask pattern with an enhanced thickness is formed in one region, and the other region is left unmasked, as shown in Fig. 10.11. The mask opening is usually larger than 5 μm, and only the center of a selectively grown structure is used for the active layer, which is typically 1.5 μm-wide. In this approach, only bandgap-energy and layer-thickness modulation on a substrate are used, and a conventional BH structure is formed by mesa etching and subsequent growth of buried layers (Fig. 10.11(c)).

In another approach, shown in Fig. 10.12, a narrow stripe opening (typically 1.5 μm wide) is used to grow waveguide layers [20, 27]. Mask stripes are formed in all regions (Fig. 10.12(a)). Changing the mask stripe width also enables bandgap-energy control. This approach, called "narrow-stripe

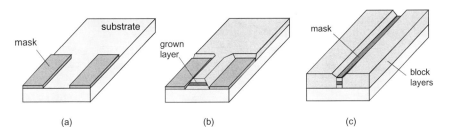

Fig. 10.11   Selective-area MOVPE for fabricating photonic integrated devices.

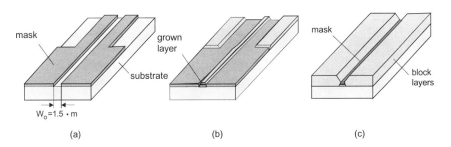

Fig. 10.12   Narrow-stripe selective MOVPE for fabricating photonic devices.

selective (NS) MOVPE," is effective for direct fabrication of active and passive optical waveguides, and it excludes mesa etching of semiconductor layers to form waveguides. The waveguide structure is determined by accurate control of the mask lithography and MOVPE growth, and the resulting structure is highly uniform. By using this technique, not only photonic integrated devices but also single-channel laser diodes with highly uniform device characteristics due to the uniform device structure have been fabricated [58]. A semiconductor optical amplifier (SOA) with a sub-micron-wide bulk active layer has also been successfully fabricated [59]. This approach is also effective for accurate control of the lasing wavelengths of DFB and DBR laser diodes [35, 60], because the effective refractive index of laser structures can be precisely controlled by the mask width.

Both "wide"- and "narrow"-stripe selective growth techniques are used to fabricate photonic devices for WDM systems. The main idea of device structure control in these techniques is the same. However, as will be explained in the following section, there are several differences between these growth techniques, mainly due to different growth mechanisms.

10.3.2. GROWTH MECHANISMS

In narrow-stripe selective MOVPE, the mask opening is typically as narrow as $1.5\,\mu m$, which means that the region close to the mask edge is used for a waveguide. As shown in Fig. 10.4, the layer grown near the mask edge is affected not only by the lateral gas-phase diffusion of group-III species, but also by the surface migration of the species from the mask surface. Both effects should be taken into account in the design of selectively grown layers. In contrast, when the mask opening is relatively wide, the migration effect is significantly small in the center of a grown layer. Sakata [61] investigated these processes in narrow-stripe selective MOVPE. Figure 10.13 shows a schematic drawing of the mask-width dependence on the growth-rate enhancement. Migration from the masked region (MMR) is saturated at several microns of the mask width because of a limited length of the surface migration. Lateral vapor phase diffusion (LVD), including the incorporation of once evaporated species from the mask surface, increases with an increase in mask width W_m, because the diffusion of the species is relatively long in the vapor phase. The actual growth-rate enhancement is related to both of these effects. The threshold mask width for the LVD effect is shown as W_{th}. Figure 10.14 shows experimentally obtained growth rates for 1.29- and 1.13-μm-wavelength composition InGaAsP and InP layers

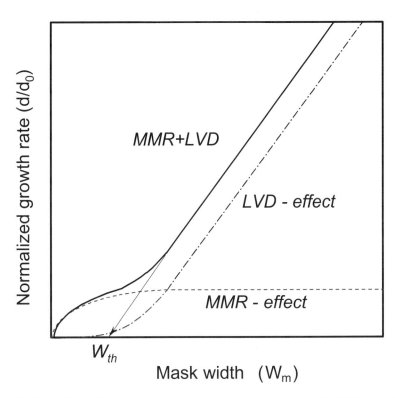

Fig. 10.13 (a) Dependence of growth-rate enhancement on mask width. (b) Calculation model for normalized growth rate (d/d_0) with MMR effect [61].

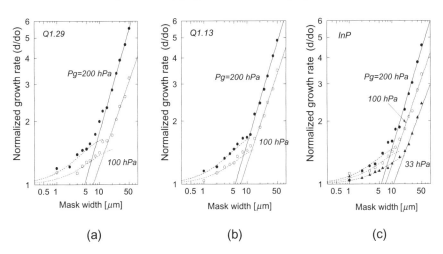

(a)   (b)   (c)

Fig. 10.14 Dependence of growth-rate enhancement on mask width as a parameter of growth pressure. Dashed lines show calculated results. (a) InGaAsP ($\lambda_g = 1.29\,\mu m$), (b) InGaAsP ($\lambda_g = 1.13\,\mu m$), and (c) InP [61].

grown under the pressures of 100 and 200 hPa, respectively. The growth temperature was 650°C. The growth rates were normalized by the growth rate on an unmasked substrate. On the logarithmic scale, the mask-width dependence can be divided into two regions. The W_{th} value is given by the intercept of the LVD dominant line. The W_{th} value is around 5 to 10 µm, and it is large for the layer grown under low pressure and for the layer grown with short-wavelength composition. Figure 10.15 shows the dependence of growth on the growth temperature for a 1.13-µm-wavelength composition InGaAsP layer. The W_{th} value increases at a high growth temperature.

The dependence of the threshold mask width on the growth conditions with respect to the LVD effect can be explained as follows [61]. When the mask width is smaller than W_{th}, the species that reach the mask surface immediately migrate to the growth region. In this case, the concentration of

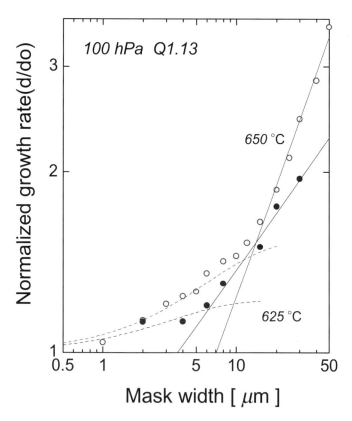

Fig. 10.15 Dependence of growth rate enhancement on mask width for InGaAsP layers ($\lambda_g = 1.13$ µm) as a parameter of growth temperature [61].

species on the surface of the mask is negligible as it is on the grown surface. Therefore, there is no concentration gradient between the narrow-masked region and the narrow-growth region, and there is no LVD effect. An increase in the W_{th} value under low growth pressure and high growth temperatures is due to enhanced migration of group-III species on the mask surface.

The MMR effect, which is dominant at a mask width smaller than W_{th}, is estimated by using the surface diffusion coefficient and the migration length [61]. The dashed lines in Figs. 10.14 and 10.15 represent the values calculated by assuming that the migration length is half the threshold mask width, W_{th}. The calculated results agree well with the experimental values, which shows the validity of estimating the migration length from the threshold mask width.

Figure 10.16 shows the dependence of the indium composition shift on the mask width for two 1.13-μm-wavelength composition InGaAsP layers

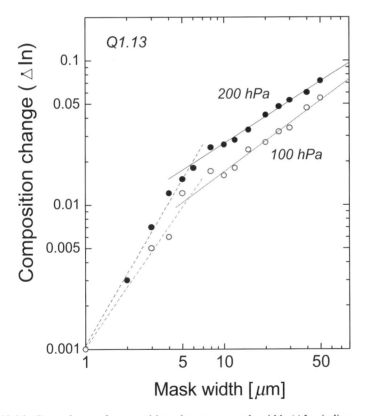

Fig. 10.16 Dependence of composition change on mask width (ΔIn: indium content increase) for InGaAsP layers ($\lambda_g = 1.13\,\mu$m) grown under different pressures [61].

grown under different growth pressures [61]. In both cases, the composition shift is composed of two lines, and the crossing points of these lines are close to the threshold mask width, W_{th} (see Fig. 10.14(b)). Therefore, the two lines are also a result of the MMR and LVD effects, respectively. The indium content increase due to the MMR effect is greater than that due to the LVD effect. This is consistent with the results of a previous study [51], in which relative composition shifts of layers grown selectively on a narrow ($W_o = 2\,\mu$m) and a wide ($W_o = 10\,\mu$m) mask opening were compared and evaluated at the same growth-rate enhancement. The study found that the relative composition shift in the narrow-stripe selective growth was greater due to the contribution of the MMR effect.

In conclusion, the migration of species from the mask surface is significant in narrow-stripe selective MOVPE, and this effect should be taken into account when estimating the growth-rate enhancement and composition shift.

10.3.3. SURFACE FLATNESS

Surface-migrated species sometimes generate significant edge growth on (100) surfaces [12, 27, 38, 42]. This migration is strongly related to (111)B facets. As the growth proceeds, the length of (111)B facets increases. Because the growth rate on a (111)B facet is very low, group-III species that have migrated from the mask surface to the facet quickly migrate toward the (100) surface. When the migration length of the species on the (100) plane is rather small compared to that on the (111)B facet due to non-optimized growth conditions, the migrated species will nucleate at the edge of the (100) surface. In narrow-stripe selective MOVPE, the width of the (100) top surface is generally comparable to the migration length, and the grown layer surface tends to be flat. Nevertheless, the growth conditions should be optimized to obtain a very flat surface. In the case of InP, the edge growth can be suppressed by enhancing growth on the (111)B facet at a high V/III ratio [9, 56, 62]. However, because the growth rate on (111)B facets is significantly low in the growth of InGaAs and InGaAsP layers, enhanced migration of species on the (100) surface is very important.

To enhance the migration length, the growth conditions such as the growth rate, temperature, and V/III ratio should be optimized. Figure 10.17 shows measured growth enhancement profiles for InGaAsP layers [27]. The profiles were measured outside the mask stripes in the direction perpendicular to that of the stripes, and the horizontal axis indicates the distance

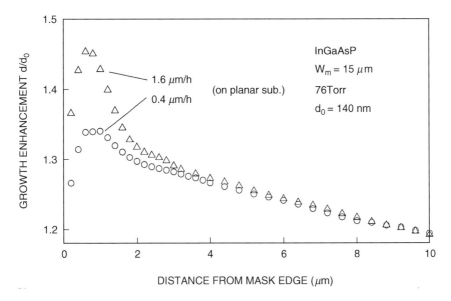

Fig. 10.17 Lateral growth-enhancement profiles of InGaAsP layers ($\lambda_g = 1.3\,\mu$m) with growth rates of 1.6 μm/h and 0.4 μm/h on an unmasked substrate [27].

from the mask edge. The growth rates on the unmasked region were 1.6 and 0.4 μm/h, respectively. The growth rate gradually increases toward the mask stripe, and it increases sharply near the mask edge. This growth enhancement over a wide area is due to the vapor phase diffusion of group-III species and is independent of the growth rate. In contrast, significant growth near the mask edge, resulting from the surface migration of group-III species, is suppressed at a lower growth rate. Thus, reducing the surface concentration of group-III species prevents edge growth.

The dependence of the surface flatness on the growth temperature and growth rate for selectively grown InP layers has been investigated by Sakata *et al.* [63]. Figure 10.18 shows a cross sectional structure of grown layer. A pair of SiO$_2$ mask stripes with stripe opening W_o of 1.5 μm was patterned on a (100) InP substrate. The narrow-stripe InP layer grown selectively between the mask stripes was almost flat and although the migration length could not be estimated, a flat (100) terrace formed outside of the mask next to the (111)B facet, as can be seen in the SEM image in Fig. 10.18. The (100) terrace near the mask edge formed as a result of the step-flow growth mode [30]. The species supplied from the (111)B facet migrated on the (100) terrace and were trapped at the steps and kinks. This step-flow

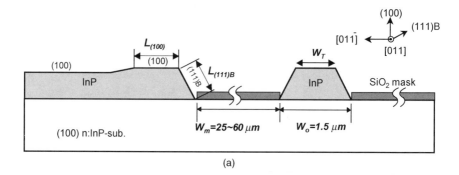

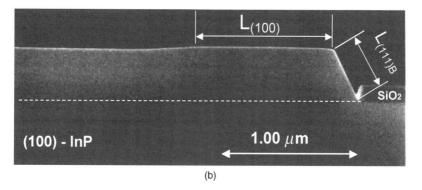

Fig. 10.18 (a) Cross section of a selectively grown structure. (b) SEM image of a region grown at the outer side of SiO₂ mask [63].

growth mode continued as long as the migration length was greater than the top surface width. Figure 10.19 shows the relationship between the length of the (100) terrace ($L_{(100)}$) and the length of the (111)B facet ($L_{(111)B}$) under different growth temperatures [63]. The formation of the (100) terrace was promoted by high growth temperatures and was a result of enhanced surface migration of indium-containing species on the (100) surface. A rapid decrease of the (100) terrace length at 580°C may be due to a lack of phosphorus pressure on the surface. The dashed lines in Fig. 10.19 show the estimated values of half the mesa-top width ($W_t/2$) for different mask spacing widths, W_o. At growth temperatures higher than 600°C, the length of the (100) terrace was greater than $W_t/2$, which means that the top of the InP mesa structure that formed on the mask opening narrower than 2.0 μm was relatively flat. Similarly, the dependence of the surface flatness on the growth rate is shown in Fig. 10.20. An enhanced (100) terrace width was observed at a lower growth rate.

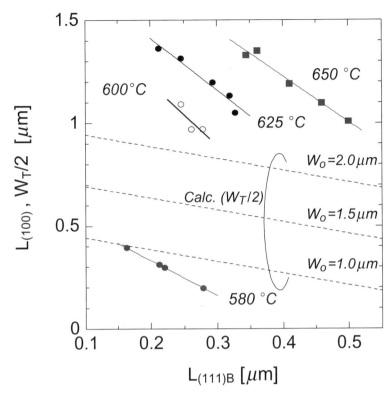

Fig. 10.19   Relationship between (100) facet length ($L_{(100)}$) and (111)B facet length ($L_{(111)B}$) as a parameter of growth temperature. Dashed lines show half of the calculated mesa-top width ($W_T/2$) [63].

To obtain a high-quality narrow-stripe MQW structure, the growth conditions should be optimized both for the well and barrier layers. The surface flatness of selectively grown narrow-stripe InGaAsP layers was investigated for the selective growth of InGaAsP/InGaAsP MQWs [30, 64]. Similar results were obtained at different growth pressures: 150 Torr and the atmospheric pressure. Figure 10.21 shows the results obtained under 150 Torr. Figure 10.21(a) compares the surface flatness of bulk and MQW layers grown under different temperatures and with different V/III ratios. The closed markers indicate a flat surface, and the open markers indicate a non-flat surface. The SEM cross sections of the surfaces are shown in Fig. 10.21(b). The growth rate on the unmasked substrate was 0.3 μm/h, which corresponds to a growth rate of about 0.7 μm/h on the narrow-stripe

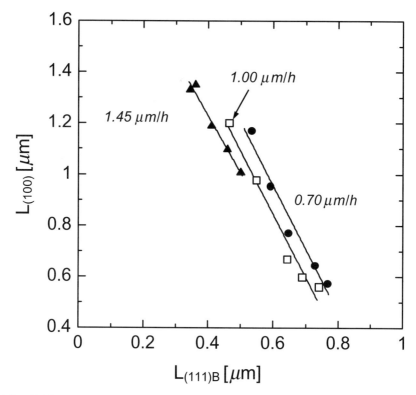

Fig. 10.20   Relationship between $L_{(100)}$ and $L_{(111)B}$ as a parameter of growth rate [63].

opening region. The results are divided into three groups. In the 1.26-μm-wavelength composition InGaAsP layer, used as a barrier layer in the 1.55-μm MQW, the surface was flat (sections II and III) under a high growth temperature, as it was in the results previously described for InP. Section I includes results obtained when the growth temperature was lower than the critical value. Here, the surface migration was insufficient and as a result, the cross section of the grown layer has a concave shape, as shown in Fig. 10.21(b). These results were observed for the InGaAsP layers with relatively high indium and phosphorus contents. In contrast, for the 1.46-μm-wavelength composition InGaAsP layer used as a well layer, higher growth temperatures and a higher V/III ratio (section III) caused deterioration of the surface flatness as shown in Fig. 10.21(b). This deterioration was not observed for InGaAs. The most plausible explanation for this degradation of flatness is step bunching [30].

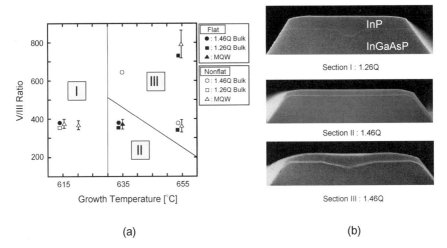

(a)   (b)

Fig. 10.21   (a) Flatness of layers selectively grown under different growth temperatures and with different V/III ratios. (b) Cross sectional SEM images of InGaAsP layers corresponding to section I, II, and III in Fig. 21(a) [64].

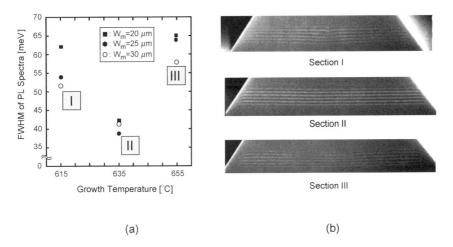

(a)   (b)

Fig. 10.22   (a) PL FWHM of selectively grown MQW layers as a function of growth temperature. (b) Corresponding SEM images [64].

Thus, flat surfaces can be obtained both for well and barrier layers under the conditions for section II results. Under these optimized conditions, fine InGaAsP/InGaAsP MQW structures have been obtained. Figure 10.22(a) shows measured photoluminescence FWHM for MQWs grown under different growth temperatures, while Fig. 10.22(b) shows cross sections of

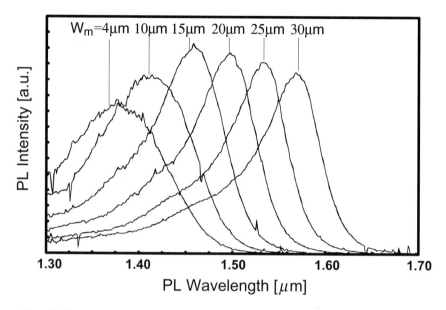

Fig. 10.23   PL spectra for selectively grown MQWs with different mask widths [64].

these structures [64]. In section II, flat MQW layers with narrow PL FWHM values were obtained, whereas in sections I and III, the MQW interfaces were not flat. Figure 10.23 shows PL spectra for simultaneously grown MQWs with different mask stripe widths, W_m [64]. A PL peak wavelength shift of 190 nm was obtained by changing the mask width from 4 to 30 μm. No specific degradation in the PL intensity was observed, because of the optimized growth conditions and design of the MQW structures.

In NS-MOVPE, crystal quality depends on the surface migration, because the enhanced strain is induced by surface migration [51]. By optimizing the growth conditions, we can obtain narrow-stripe MQWs whose optical quality is comparable to that of conventionally grown MQW. In NS-MOVPE, when the (111)B facet length increases, both the growth rate and the indium content in the grown layer change as the growth proceeds [27, 65]. Therefore, changing the growth time for each quantum well can effectively improve well-thickness uniformity and reduce the PL FWHM [66]. Figure 10.24 shows the dependence of PL FWHM on excitation Ar^+ laser power density. An Ar^+ laser beam was focused on an area, 1 μm in diameter, on the layer surface. A reference MQW was etched into the 1.5 μm-wide mesa to enable comparable excitation density. At a low excitation power density, the PL FWHM of the narrow-stripe MQW obtained

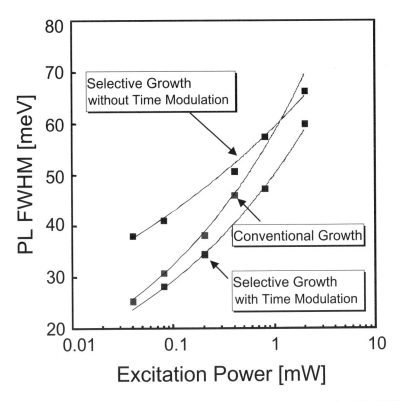

Fig. 10.24 PL FWHM as a function of excitation power for three samples: (a) a MQW structure grown entirely on a planar substrate followed by mesa etching, (b) a MQW structure grown selectively with constant growth time for each well, and (c) a MQW grown selectively with growth time modulation.

by using a time modulation scheme was comparable to that of the reference MQW. High crystal quality of the MQW obtained by NS-MOVPE was also evident in the device characteristics [58, 66].

10.3.4. SELECTIVE MOVPE GROWTH OF InAlAs AND InAlAs/InGaAs MQWs

Most studies on selective growth focus on InGaAsP materials, which are most commonly used in device fabrication. Recently, the number of studies on InAlGaAs materials has been increasing, because these materials enable a large conduction band offset especially when InAlAs layers are used for barriers. The device characteristics of laser diodes with an

InAlGaAs/InAlGaAs MQW active layer deteriorate less than those with an InGaAsP/InGaAsP MQW active layer at high temperatures. The use of EA modulators is also attractive because of their large extinction ratios. However, selective growth of aluminum-containing species is very difficult to achieve, mainly because of a large sticking coefficient of aluminum-containing species on mask surfaces. It is difficult to prevent polycrystal formation on a dielectric mask surface.

Selective growth of InAlAs and InAlGaAs layers can be achieved by choosing a proper mask material and optimizing the mask formation process, the wafer-surface cleaning process prior to growth, and the growth conditions [67, 68]. Figure 10.25 shows a SEM plane-view of a SiO_2 mask pattern after the growth of a 0.5-μm-thick InAlAs layer under different conditions [67]. The growth pressure was 70 Torr, and TMIn and trimethylaluminum (TMAl) were used as group-III sources. At a growth temperature of 700°C (Fig. 10.25(a)), polycrystals with a diameter smaller than

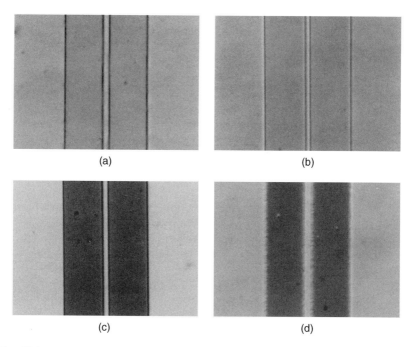

(a)   (b)

(c)   (d)

Fig. 10.25   Top view of selectively grown 0.5-μm-thick InAlAs layers: The growth temperature and growth rates were: (a) 700°C and 0.6 μm/h, (b) 650°C and 0.6 μm/h, (c) 650°C and 1.0 μm/h, (d) 650°C and 1.6 μm/h. The SiO_2 mask was 10 μm wide [67].

0.5 μm were deposited on the mask. No polycrystals remained on the mask at a growth temperature of 650°C (Fig. 10.25(b)). The selectivity deteriorated as the growth rate increased (Fig. 10.25(d)). Generally, InAlAs layers grown under low temperature have poor crystal quality due to the incorporation of impurity materials. However, the donor concentration in the layer grown under the conditions described in Fig. 10.25(b) was still around $3 \times 10^{15} \mathrm{cm}^{-3}$.

The growth-rate enhancement and the composition shift have been compared for InAlAs and InGaAs layers [52]. Figure 10.26 shows the measured

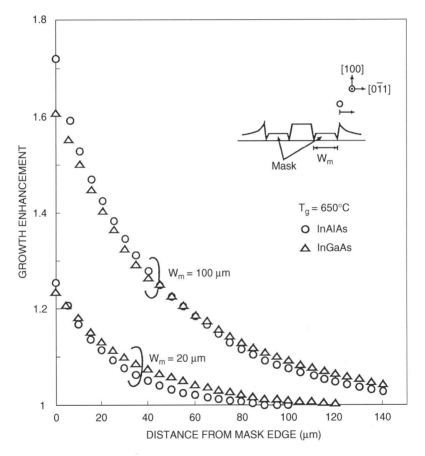

Fig. 10.26 Lateral growth-enhancement profiles of selectively grown InAlAs(○) and InGaAs(△) layers with 20- and 100-μm-wide mask stripes [52].

growth enhancement profiles versus the distance from the mask edge. The tendency observed in the layers with respect to growth enhancement was almost the same due to gas phase diffusion. In contrast, enhanced growth was observed in the InAlAs layer on the region near the mask edge, where migration of group-III species contributed to this enhanced growth. Figure 10.27 shows induced compressive strain versus mask width W_m,

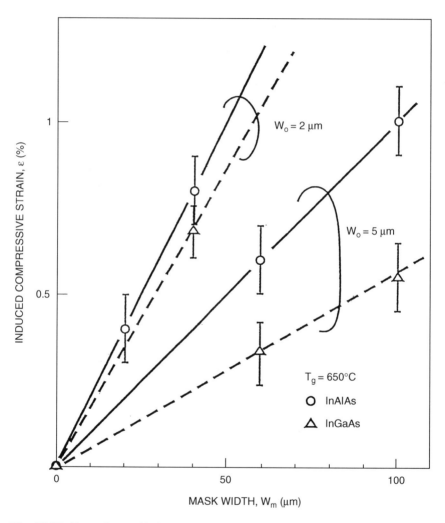

Fig. 10.27   Dependence of induced compressive strain on the mask width for selectively grown InAlAs($\bigcirc$) and InGaAs($\triangle$) layers [52].

for growths on narrow-stripe ($W_o = 2\,\mu m$) and wide-stripe ($W_o = 5\,\mu m$) spacings. The induced strain was larger in the InAlAs layer than in the InGaAs layer, which is a result of suppressed lateral gas-phase diffusion of aluminum-containing species as opposed to gallium-containing species. This can be explained by the fact that the decomposition temperature of TEGa is lower than that of TMAl. The difference in the induced strain between the InAlAs and InGaAs layers was small for the narrow-stripe selective growth, which indicates an increase in the incorporation of aluminum in the narrow mask opening due to surface migration.

After the investigation of bulk layers, a MQW consisting of InGaAs wells and InAlAs barriers was selectively grown. Figure 10.28 shows PL spectra of a 30-period MQW grown on a 5-μm-wide mask opening [52]. Both the well and barrier layers were designed to be lattice matched to an InP substrate with a mask width of around 20 μm. For the structure with a mask width of 80 μm, the PL peak wavelength was 1548 nm, while on the unmasked region, the PL peak wavelength was 1305 nm. The total bandgap energy shift was 149 meV. No polycrystals remained even on the $100 \times 200\,\mu m$ mask. Thus, narrow-stripe selective growth of Al-containing MQWs is quite attractive for obtaining BH-LDs with AlGaInAs MQW active layers.

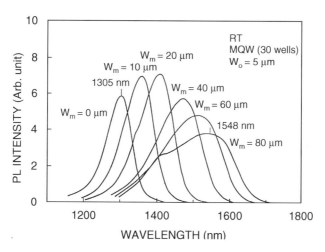

Fig. 10.28 PL spectra of an InGaAs/InAlAs MQW (30 periods) selectively grown on a 5-μm-wide mask spacing [52].

10.3.5.   MICROARRAY SELECTIVE GROWTH (MASE)

The multi-wavelength DFB/DBR laser array is a key device for WDM transmission systems. A promising approach here is the use of wavelength-selectable light sources (WSLs) that can handle several wavelength channels. In particular, array-based WSLs consisting of a DFB laser array, a waveguide combiner, and a semiconductor optical amplifier (SOA) are attractive for their simple tuning scheme. A detailed description of WSLs is given in Section 10.4.4, where a selective growth technique for fabricating arrayed DFB laser structures is also described.

Downsizing is one of the most important issues in designing photonic integrated devices. Techniques are needed that would enable obtaining a large number of devices from a processed wafer. To make the size of device structures smaller, a compact laser diode array structure with only 10-μm array spacing has been developed [69]. This "microarray" waveguide structure was obtained by using microarray selective epitaxy (MASE). Figure 10.29 shows a typical mask pattern used in MASE, where an eight-channel waveguide region is densely arrayed. As well as NS-MOVPE, grown layers are used directly for waveguide structures, and mask opening width W_o is typically 1.5 μm. In conventional NS-MOVPE, the stripe patterns are sufficiently separated from one another, and the effects of the mask stripes on the waveguide layers are isolated. In contrast, in MASE patterns, stripe

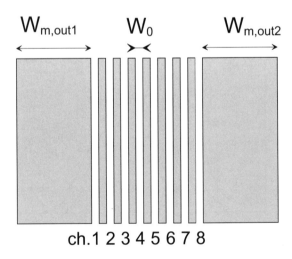

$$W_{m,out1} \qquad W_0 \qquad W_{m,out2}$$

ch.1 2 3 4 5 6 7 8

Fig. 10.29   Typical mask pattern for MASE (after [69]).

openings are densely arrayed with 10-μm spacing, and no surface-exposed region remains between the waveguide stripes. Therefore, each waveguide region is strongly affected by the number of neighboring mask patterns [69, 70]. This effect is called "mask interference." Figure 10.30 shows a mask pattern and its PL wavelength distribution for an eight-channel array waveguide [70]. When waveguide spacing L_s is relatively wide, as is the case in conventional NS-MOVPE, there is no interference between the waveguides, and the peak wavelength profile is essentially flat (Fig. 10.30(a)). However, when waveguide spacing L_s is close to inducing mask interference, the PL peak wavelengths shift to longer wavelengths and affect several mask stripes surrounding other waveguide channels (Fig. 10.30(b)). Because this effect affection is not constant over the waveguide array, the PL wavelength profile is no more flat. This mask interference depends on the growth pressure.

Figure 10.31 shows the dependence of the PL wavelength shift on arrayed waveguide spacing L_s under different growth pressures of 75, 150, and 740 Torr [70]. The center waveguide in an eight-channel arrayed waveguide structure (channels 3 and 4 in Fig. 10.30) was measured. As the waveguide spacing decreased, the PL wavelength increased sharply because of the effect of the neighboring mask stripes. The critical spacing at which mask interference occurred was 60 μm at 76 Torr, and it then decreased to 22 μm at 740 Torr. A short diffusion length at high growth pressure resulted in less occurrence of mask interference. Consequently, each waveguide in the array tends to be isolated with relatively narrow stripe spacing.

Figure 10.32 shows PL wavelength distribution for an eight-channel InGaAsP/InGaAsP MQW microarray grown under different pressures [70]. Only channels 1, 2, 3, and 4 are shown for simplicity. The mask pattern was symmetric, and the outermost mask width, shown as $W_{m,out}$ in Fig. 10.29, was varied. The arrayed MQW waveguides, grown under low pressure (75 and 150 Torr), had a relatively flat wavelength distribution over the eight-channel array, and the distribution could be controlled by the outer mask width. A mask width ranging from 20 to 60 μm resulted in an average wavelength shift of over 80 nm. Figure 10.33 shows PL wavelength distribution obtained for eight-channel MQW arrays grown under 150 Torr by using an asymmetric mask pattern. The PL wavelength profile could be controlled in the array by changing the outermost mask width asymmetry, and a wavelength shift of 90 nm was obtained for the mask widths of 20 and 70 μm.

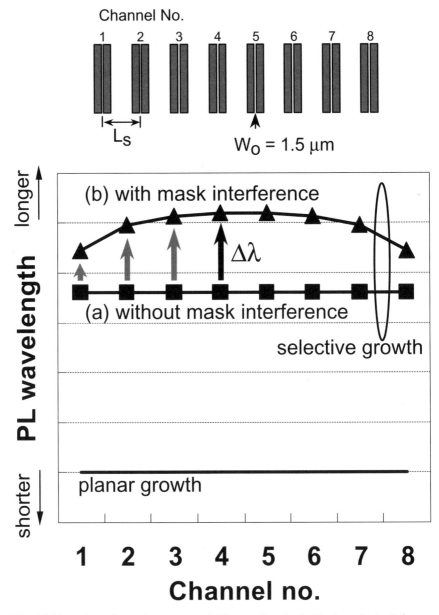

Fig. 10.30 Schematic mask pattern and PL wavelength distribution obtained for an 8-channel arrayed waveguide with and without mask interference [70].

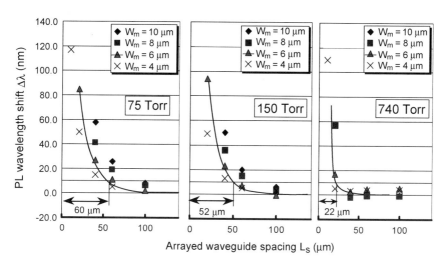

Fig. 10.31 Dependence of the shift in PL wavelength on arrayed waveguide spacing under growth pressures of 75, 150, and 740 Torr [70].

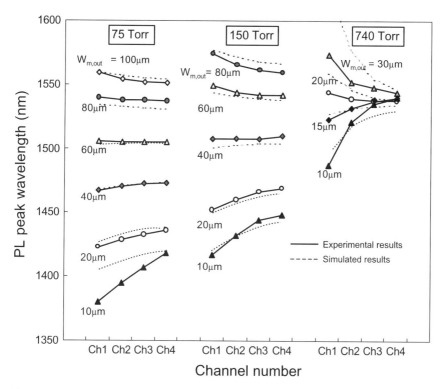

Fig. 10.32 PL wavelength distribution obtained for an 8-channel MASE pattern. Only channels 1 through 4 are shown [70].

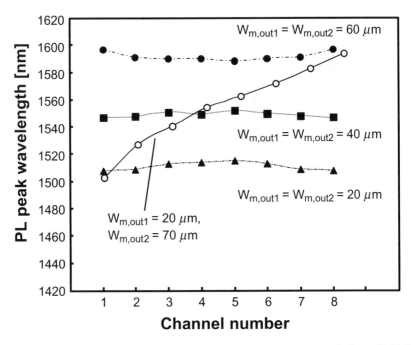

Fig. 10.33   PL wavelength distribution for MQW structures grown on an 8-channel MASE pattern under 150 Torr. The mask pattern is shown in Fig. 10.29 (after [69]).

The thickness variation between channels 1 and 8 was 30%, but a high crystal quality with flat interfaces was maintained. Thus, in low-pressure growth, the entire array waveguide structure can be effectively controlled by changing the outer mask width, because the outer mask width affects the entire array structure.

In contrast, as shown in Fig. 10.32, atmospheric pressure growth (740 Torr) resulted in a nonuniform PL wavelength profile in the eight-channel array, because the center waveguide is not significantly affected by the outermost mask width. In this case, inserting dummy waveguides between each waveguide and controlling their widths is effective in controlling the wavelength distribution. The dashed lines in Fig. 10.32 show calculated results for a model that takes the mask-interference effect into account [70]. In this model, the effect of neighboring mask stripes on the corresponding waveguide was introduced by using a mask interference constant that depends on the spacing between the mask stripes and on the growth pressure.

This MASE technique can be used with conventional NS-MOVPE, to fabricate wavelength-selectable light sources with a DFB-LD array, as will be shown in Section 10.4.4.

10.4. Application of Selective MOVPE in Fabricating WDM Light Sources

Due to its superior in-plane bandgap-energy controllability of semiconductor waveguides, selective MOVPE is now widely used for fabricating a variety of optical devices [71, 72]. This technique is quite effective, especially in the fabrication of photonic integrated circuits (PICs), because it can be used to form waveguides with different functions by using only single-step epitaxy and by simply changing the dielectric mask width on a wafer. An electroabsorption (EA) modulator-integrated DFB laser diode (DFB/MOD) is one example; this device will be described in Section 10.4.1. As was explained in Section 10.3, narrow-stripe selective MOVPE (NS-MOVPE) has an additional advantage; that is, it enables direct waveguide formability without semiconductor etching. By using this technique, a semiconductor etchingless optical device fabrication process has been developed that includes the formation of double heterostructure (DH) waveguides, current-blocking structures, and cladding structures. It is called "all selective MOVPE" (ASM). The ASM technique is used to fabricate many kinds of optical devices such as 1.3-μm-wavelength Fabry-Perot (FP) LDs, 1.48-μm-wavelength LDs for EDFA pumping, and optical spot-size converter (SSC)-integrated LDs. These ASM LDs are uniform and exhibit high device performance due to their semiconductor etchingless features. The ASM technique will be described in Section 10.4.2. By combining NS-MOVPE and electron-beam (EB) lithography, a simultaneous fabrication technique was developed to fabricate different wavelength light sources on a single wafer. This technique is a promising technique for cost-effective fabrication of WDM light sources with different wavelengths because it can reduce the number of wafers needed for device fabrication. This technique will be explained in Section 10.4.3, where the fabrication of different wavelength DFB/MODs on a single wafer will be described. Microarray selective epitaxy (MASE), described in Section 10.3.5, was used to fabricate wavelength-selectable light sources (WSLs) based on a DFB-LD array configuration. Although such WSLs have a relatively complicated and highly integrated structure consisting of an DFB-LD array, an optical combiner, a semiconductor optical amplifier (SOA), and an EA modulator, a simple process and a compact chip were obtained by using MASE due to its capability to densely integrate a DFB-LD array. The use of MASE in WSL fabrication will be explained in Section 10.4.4. Other applications of selective MOVPE, such as to fabricate SOAs and optical matrix switches, will be described in Section 10.4.5.

10.4.1. ELECTROABSORPTION (EA) MODULATOR-INTEGRATED DFB LDs (DFB/MODs)

DWDM systems require an external modulation scheme that guarantees low-chirp operation in order to prevent signal crosstalk between adjacent channels and to achieve long-distance, high-bit-rate transmission. Electroabsorption (EA) modulator-integrated DFB LDs (DFB/MODs) are light sources widely used for such systems not only because of their smaller wavelength-chirp characteristics compared to those of directly modulated LDs, but also because of their compactness, efficient optical coupling between an DFB LD and a modulator, and cost effectiveness.

The integration of a DFB LD and an EA modulator was first demonstrated by making full use of a semiconductor etching technique: the laser gain layers were locally etched away and modulator absorption layers with a higher bandgap wavelength were selectively regrown [73, 74]. The development of bandgap-energy-controlled selective MOVPE, described in Section 10.2, has enabled a simple fabrication process: proper masking during epitaxy produces a change in the thickness and composition of waveguide core layers along the length of the device. This enables forming a structure in which thinner MQW core layers in the modulator section have a higher bandgap wavelength compared to that in the laser section [20, 24, 75–77]. However, because the thickness of the MQW core layers in both the modulator and laser sections are directly related, individual control of each thickness is impossible in selective MOVPE. Thus designing proper MQW thickness combinations in both sections, and MQW detuning between the lasing and modulator-absorption-peak wavelengths, are important in achieving high-performance DFB/MODs based on selective MOVPE.

In the next section, the basic structure, design, fabrication process, and characteristics of 2.5-Gb/s DFB/MODs based on NS-MOVPE are described.

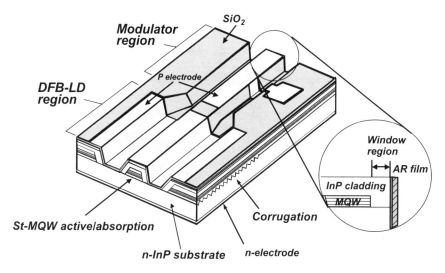

Fig. 10.34 Structure of a 2.5-Gb/s EA-modulator-integrated DFB LD (DFB/MOD) fabricated by NS-MOVPE.

Design and Structure

Figure 10.34 shows the structure of a DFB/MOD fabricated by using NS-MOVPE. The device has a DFB-LD section, an EA modulator section, and a window section in front of the modulator. A simple p-InP embedded layer was used for the buried heterostructure (BH). This device meets the 2.5-Gb/s application requirements. For applications requiring more than 10-Gb/s, a different BH such as Fe-doped semi-insulating, should be used to reduce the modulator capacitance [78]. To obtain a high single-mode yield, a $\lambda/4$ shift should be introduced into the center of the DFB cavity with an antireflection (AR) coating at both end facets. However, no $\lambda/4$ shift is used for devices with high-reflection (HR) films instead of AR films at DFB facet and this type of device is suitable for obtaining high-output-power characteristics.

To use DFB/MODs in long-haul optical transmission systems, the thickness of their selectively grown MQWs and the MQW detuning between the lasing and modulator absorption-peak wavelengths should be optimized in terms of the extinction ratio at a certain voltage, output power, and chirp characteristic. This optimization was done by Yamazaki *et al.* [79], who obtained low-drive voltage and high-power characteristics simultaneously for DFB/MODs by increasing the well thickness to enhance the quantum

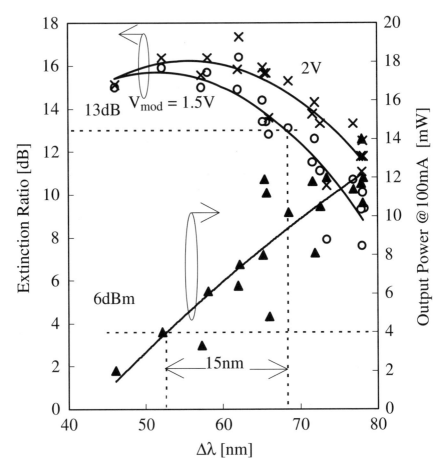

Fig. 10.35   Extinction ratio and output power characteristics as a function of detuning Δλ. Open circles and crosses show, respectively, extinction ratios at bias voltages of 1.5 and 2 V. Closed triangles show output power from the modulator facet at a bias voltage of 0 in the modulator and an injection current of 100 mA in the DFB LD [79].

confined Stark effect (QCSE) [80] and by increasing the electric field intensity by doping-profile optimization while maintaining a reasonably large fabrication tolerance. Figure 10.35 shows acceptable modulator detuning Δλ against the lasing wavelength, which simultaneously gives both an extinction ratio of more than 13 dB at 1.5 V and CW output power of more than +6 dBm at an injection current of 100 mA. This output power of +6 dBm means that it has sufficient output power to be coupled with single-mode fiber, even assuming a 3-dB coupling loss and 50% duty modulation.

In these experiments, the DFB/MODs had eight MQWs with wells 5.8-nm thick and an undoped layer 133.4-nm thick. The detuning tolerance was $\Delta\lambda = 15$ nm, which is high enough to enable the controllability and uniformity of currently used MOVPE. At a drive voltage of 2 V, the detuning tolerance was more than $\Delta\lambda = 20$ nm. Yamazaki et al. concluded that QCSE enhancement by increasing the well width and electric-field intensity through the use of a thin undoped layer can effectively reduce the modulator drive voltage without increasing the absorption loss and decreasing the fabrication tolerance [79]. Optimization of other parameters, such as modulator facet reflectivity R, electrical isolation r between the laser and the modulator, grating coupling factor κL, and differential gain dg/dN, is important for obtaining low-chirp characteristics to achieve long-haul transmission in DFB/MODs. This optimization was described in detail by Yamaguchi et al. [81], who experimentally derived the following optimum values: $R < 0.1\%, r > 2$ k$\Omega, \kappa L > 1.5$, and $dg/dN > 4 \times 10^{-12}$ m^3/s.

Fabrication Process

Figure 10.36 shows a typical fabrication process of a DFB/MOD by NS-MOVPE. First, a grating with pitch Λ determined by using effective-refractive index n_{eff} of the laser waveguide and target lasing wavelength λ_{DFB} ($\Lambda = \lambda_{DFB}/2n_{eff}$) is partially formed in the DFB section by holographic lithography or electron-beam (EB) lithography. Pairs of SiO$_2$ mask stripes for NS-MOVPE are formed in the [011] direction so that they sandwich a 1.5-μm-wide window stripe. The SiO$_2$ mask widths (W_m) for each section are determined so that 1) the active-layer gain-peak wavelengths coincide with the lasing wavelengths and 2) the absorption-layer bandgap wavelength follows the designed detuning, $\Delta\lambda$, from the lasing wavelength (for example, $\Delta\lambda = 60$ nm in Fig. 10.35.). Then, the active layer in the laser and the optical absorption layer in the modulator, both of which consist for example, of InGaAsP MQWs, are simultaneously grown by selective MOVPE, which results in no waveguide discontinuity. Next, SiO$_2$ masks for the buried heterostructure (BH) are re-formed in each section to obtain a ~7-μm-wide window stripe, and the MQW waveguide is covered with a p-InP embedded layer and a p$^+$-InGaAs contact layer also grown by selective MOVPE. A 25-μm-wide area in the contact layer at the joint of the DFB-LD and modulator sections is removed to enable high electrical isolation of $r > 8$ kΩ [81]. Finally, Ti/Au stripe electrodes with small bonding pad areas are formed on the p-side. An AR coating is applied to the

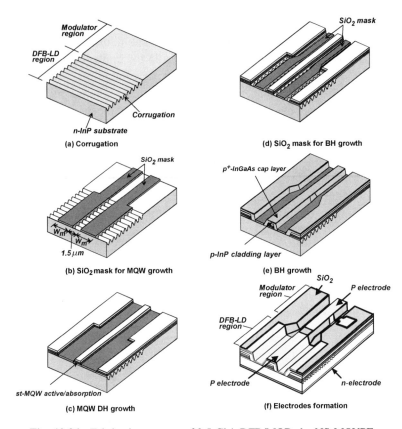

Fig. 10.36   Fabrication process of 2.5-Gb/s DFB/MODs by NS-MOVPE.

modulator window facet to reduce the effective reflectivity to less than 0.1%, which is done to prevent wavelength chirping caused by light reflection [81]. The choice of the coating film for the DFB facet between AR films and HR films depends on the device design described previously.

Characteristics

Figure 10.37(a) shows typical output power versus the injection current of a DFB/MOD fabricated by NS-MOVPE, while (b) shows the corresponding extinction ratio against the modulator-applied voltage. The DFB/MOD, fabricated under the optimized conditions, includes an InGaAsP well with eight MQWs and an InGaAsP barrier with a MQW bandgap wavelength of 1.55 μm in the DFB section (400 μm long) and that of 1.48 μm in the

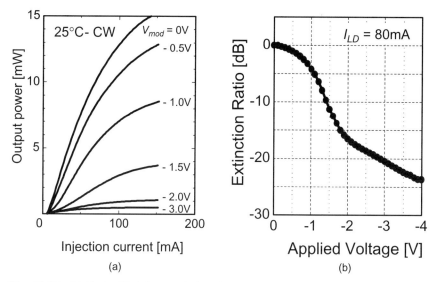

Fig. 10.37   (a) Typical L-I characteristics of DFB/MOD against applied modulator voltages and (b) typical extinction curves of DFB/MOD.

modulator section (200 μm long). The DFB/MOD showed output power of more than 10 mW CW from the modulator facet at an injection current of 100 mA and a rather high extinction ratio of more than 14 dB at −2 V.

Experiments with 2.5-Gb/s, 600-km transmission (120-km × 5 spans) were conducted by Ishizaka et al. [82], in which they used several 2.5-Gb/s DFB/MOD modules with a fiber pigtail (a 2.5-Gb/s DFB/MOD module is shown in Fig. 10.38). These devices had a κL value of around 2 and a negative laser detuning of around 10 nm for reducing the excess chirp [81]. Their measured 3-dB modulation bandwidth was 4 GHz. The 800-km transmission characteristics were also measured for several other DFB/MOD modules and for a discrete EA modulator used as a reference. The experimental setup used in these experiments is shown in Fig. 10.39(a). Conventional normal fiber with a chromatic dispersion of +17-ps/nm/km was used. Five Er-doped fiber amplifiers (EDFAs) were used as a booster and inline amplifiers, and the output power of each EDFA was kept constant at +8 dBm to eliminate the fiber nonlinear effect. Ishizaka et al. used an AC-coupled driver circuit with a modulation voltage of 2.0 V_{p-p}. Figure 10.39(b) compares the measured bit error rate of the DFB/MOD module with that of the discrete EA modulator. An optical isolator was inserted between the DFB LD and the modulator during the measurements

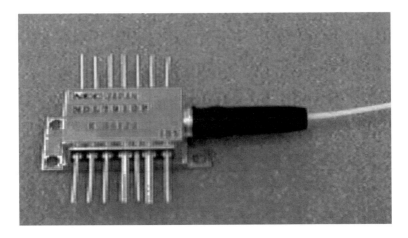

Fig. 10.38   2.5-Gb/s DFB/MOD module.

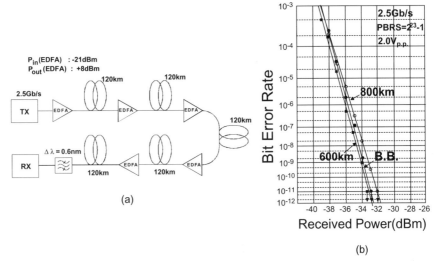

Fig. 10.39   (a) Experimental setup for 2.5-Gb/s, 600-km normal-fiber transmission. (b) Measured BER characteristics for DFB/MOD module and discrete EA modulator [82].

for the discrete EA modulator. Ishizaka *et al.* achieved a power penalty of less than 1 dB for DFB/MOD module even after 800-km transmission. The normal-fiber transmission characteristics of the DFB/MOD module were the same as those of the discrete EA modulator with an optical isolator.

10.4.2.   ALL-SELECTIVE MOVPE (ASM) TECHNIQUE

In the conventional BH-LD fabrication process, active-mesa stripes are formed by semiconductor etching followed by a regrowth process. However, such etching is detrimental to the LD performance, uniformity, and reproducibility. To overcome these problems, an all-selective MOVPE (ASM) technique for fabricating BH-LDs was developed by Sakata *et al.* by using NS-MOVPE [58]. In ASM, a MQW waveguide is formed directly on the mask-opening stripe by NS-MOVPE without any semiconductor etching. A current-blocking layer is also grown by selective MOVPE using a SiO_2 mask partially patterned on top of the DH mesa by a self-alignment process. ASM has many advantages: it renders semiconductor etching unnecessary, generates smooth mesa facets, and enables simultaneous growth of different wavelength composition layers simply by changing the mask width. Excellent performance has been reported for devices obtained by ASM including highly uniform 1.3-μm-wavelength LDs [58, 83, 84], thickness-tapered optical spot-size-converter (SSC)-integrated LDs [33, 85, 86, 87], high-power 1.48-μm-wavelength LDs for EDFA pumping [88, 89], and 10-Gb/s DFB/MODs with good transmission characteristics [78].

In the following section, the ASM technique is described in detail by introducing several ASM-based devices.

Fabrication Process

Figure 10.40 shows the basic fabrication process of ASM-based devices [84]. Typically, a 100-nm-thick SiO_2 mask is deposited on an n-InP wafer. A pair of SiO_2 mask stripes is patterned by a conventional photolithographic technique (Fig. 10.40(a)). Then, a DH mesa, consisting of an InGaAsP MQW active layer, is selectively grown by NS-MOVPE (Fig. 10.40(b)). Because this NS-MOVPE growth process induces changes in the InGaAsP composition and leads to growth-rate enhancement, the growth conditions are controlled to compensate for these changes. A SiO_2 mask is then formed only on top of the DH mesa using a self-alignment process (Fig. 10.40(c)). This self-alignment process is illustrated in Fig. 10.41 [84]. A SiO_2 film is deposited on the wafer by atmospheric-pressure thermal chemical vapor deposition (CVD) (Fig. 10.41(a)). The SiO_2 thickness, dt, at the top of the DH mesa is greater than thickness ds at the side facets. The SiO_2 film is then etched with a buffered HF solution without any mask, and the etching is stopped before $ds = 0$ (Fig. 10.41(b)). After this etching, SiO_2 remains only on the top and bottom of the ridge. The bottom SiO_2 is then removed by

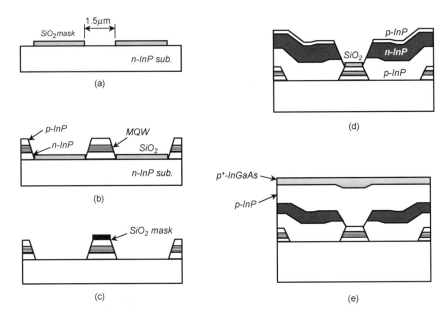

Fig. 10.40   Cross section of LD fabrication process. (a) SiO_2 pattern for NS-MOVPE. (b) Selective MOVPE of a strained InGaAsP MQW structure. (c) Self-aligned SiO_2 mask patterning. (d) Selective embedded re-growth of p/n-InP current-blocking layers. (e) MOVPE growth of p-InP cladding and p^+-InGaAs contact layers [84].

using SiO_2-side etching and a photoresist mask that is wider than the ridge (Fig. 10.41(c)). After forming a self-aligned SiO_2 stripe mask on the mesa top (Fig. 10.41(d)), a current-blocking layer consisting of a p-n-p-n thyristor is selectively grown using the SiO_2 mask (Fig. 10.40(d)). Then the SiO_2 mask is removed, and p-InP cladding and p^+-InGaAs cap layers are grown (Fig. 10.40(e)). Figure 10.42 shows a SEM cross sectional photograph of the current-blocking structure used in the 1.3-μm-wavelength ASM-DC-PBH LD described following [84]. Smooth (111)B facets are formed by the p-n-InP current-blocking layers around the ridge stripe without overgrowth on the SiO_2 mask (Fig. 10.42(a)). Generally, it is very difficult to obtain such a non-overgrown structure for embedded MOVPE-regrown layers around a mesa-etched ridge structure, but in ASM-BH, selectively grown ridge stripes have side walls with a perfect (111)B plane so that regrown layers are smoothly embedded around the ridge stripe with a change in the crystal plane from (111) to (100).

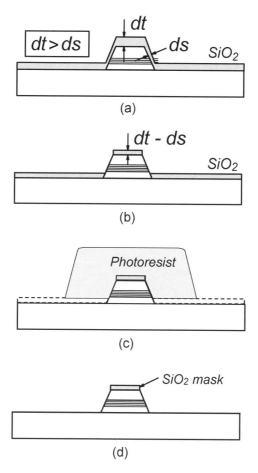

Fig. 10.41 Cross sectional self-aligned SiO_2 mask patterning. (a) SiO_2 CVD on a wafer. (b) SiO_2 etching over the whole area. (c) SiO_2 side etching. (d) Final structure after self-aligned process [84].

In this fabrication process, important device parameters, such as the active stripe width and the leakage-current path configuration, are precisely controlled by the epitaxial growth steps. This fabrication process enables precise control of the waveguide dimensions because mesa etching is no longer required. Therefore, light sources with uniform and reproducible characteristics can be obtained, which results in cost-effective optical devices. Moreover, smooth crystal planes are formed in selectively grown structures, which leads to low-propagation-loss waveguides.

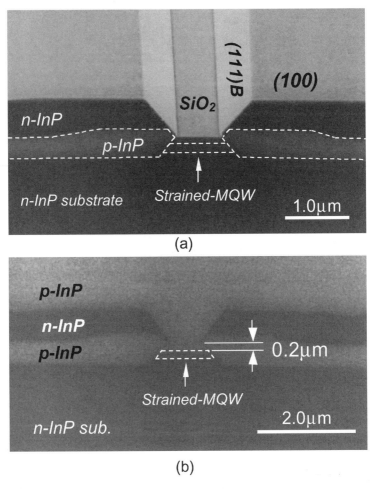

Fig. 10.42 Cross sectional SEM image of a 1.3-μm-wavelength ASM-DC-PBH LD (a) after second MOVPE step, (b) after third MOVPE step [84].

1.3-μm-Wavelength ASM-DC-PBH LDs

High-temperature and less-temperature-sensitive operations are very important for LDs used in optical access networks to reduce the module cost by using cooler-less packages. To meet these requirements, ASM was applied to 1.3-μm-wavelength LDs to obtain an excellent current-blocking structure for an optimized double-channel planer-buried-heterostructure (DC-PBH), which reduces the leakage current even under high-temperature conditions.

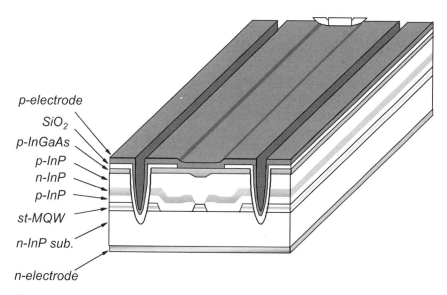

p-electrode
SiO₂
p-InGaAs
p-InP
n-InP
p-InP
st-MQW
n-InP sub.
n-electrode

Fig. 10.43 Schematic device structure of a 1.3-μm-wavelength ASM-DC-PBH LD [84].

Figure 10.43 shows a schematic structure of a 1.3-μm-wavelength ASM-DC-PBH LD reported by Sakata *et al.* [83, 84]. This device has a MQW active layer consisting of six 0.7% compressively strained InGaAsP wells (6 nm thick) and InGaAsP barriers (8 nm thick, $\lambda_g = 1.13$ μm), sandwiched by InGaAsP SCH layers (60 nm, $\lambda_g = 1.13$ μm). The current-blocking structure of this device is shown in Fig. 10.42. The structure has a theoretically optimized leakage-current path width of 0.2 μm [90], as shown in Fig. 10.42(b).

The fabricated devices had an internal quantum efficiency (η_i) of 100% at 25°C and that of 96.3% at 80°C. Such a high η_i for the ASM-DC-PBH LDs means that the leakage current is small even if the LDs operate under high-temperature conditions, due to the superior current-blocking characteristics of the optimized DC-PBH structure. Pulsed light-output-versus-current (L-I) characteristics for 35 consecutive LDs at 25 and 85°C are shown in Fig. 10.44 [84]. The cavity was 300 μm long, and the facet configuration had a 30 to 90% reflectivity. One can see the uniformity of both the threshold currents and slope efficiencies for the 35 consecutive LDs. For example, average operating current I_{op} at 15 mW was 36.1 mA with a standard deviation of ±0.85 mA at 25°C. This high uniformity was maintained even at a temperature of 85°C. Also average I_{op} at 15 mW was

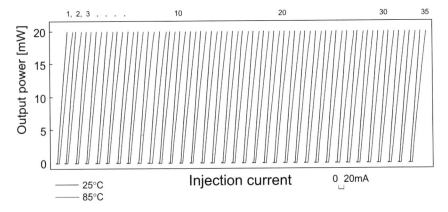

Fig. 10.44 Pulse *L-I* characteristics for 35 consecutive 1.3-μm-wavelength ASM-DC-PBH LDs at 25°C and 85°C [84].

51.6 mA with a standard deviation of ±1.25 mA. This excellent uniformity of the temperature characteristics can be attributed not only to the device structure, which has effective current-blocking layers and a high-quality MQW active layer, but also to the controllable laser fabrication process with no semiconductor etching.

1.48-μm-Wavelength ASM-DC-PBH LDs

The ASM technique is also used to obtain 1.48-μm-wavelength ASM-DC-PBH LDs for EDFA pumping [88, 89]. The schematic structure of such an LD is shown in Fig. 10.45. In this device, a DH mesa, including an MQW active layer and a recombination layer, is selectively grown by NS-MOVPE. To prevent leakage current and forward break-over, a DC-PBH structure is used. This structure has a recombination layer inserted below the current-blocking layer. Its 2.1-mm-long cavity and a front-facet reflectivity of 1% enable high output power without degrading the slope efficiency. An LD chip is mounted on a diamond heat sink in the junction-down configuration. The lasing and module performance is shown in Fig. 10.46 with the corresponding series resistance estimated from the I-V characteristics [89]. The solid line shows the output power from the LD chip, and the solid line with circles shows the output power from a butterfly laser module. The threshold current was 67 mA. The slope efficiency was 0.39 W/A. This high slope efficiency, even in a long-cavity structure, is due to the low front-facet reflectivity and low absorption loss in the cavity. The series

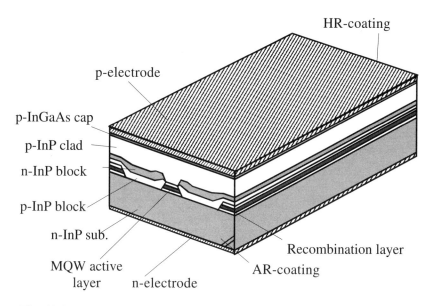

Fig. 10.45   Schematic structure of a 1.48-μm-wavelength ASM-DC-PBH LD [89].

resistance was constant over an injection-current of up to 2 A. This means that the leakage current through the current-blocking layer and forward break-over were prevented by the DC-PBH structure. The maximum output power was 504 mW. The maximum output power from the butterfly laser module was 337 mW.

Spot-Size Converter (SSC)-Integrated Laser Diodes (LDs)

Spot-size converter (SSC)-integrated LDs with low-current operation even under high temperatures [32, 33, 91] are needed for access network systems. These LDs are very cost effective, enable efficient fiber coupling without the use of a lens, and have a direct driving capability by an IC driver. Tohmori [91] described an SSC-LD consisting of a selectively grown thickness-tapered waveguide jointed with an active layer by butt-joint coupling. Selective MOVPE has also been used to fabricate both active and tapered waveguide layers simultaneously [32, 92, 93]. NS-MOVPE enables not only in-plane bandgap energy control but also direct waveguide formation, as previously described. Thus, NS-MOVPE with ASM is an attractive technique for fabricating SSC-LDs. Figure 10.47 shows a schematic structure of a 1.3-μm-wavelength SSC-LD [87]. The SSC-LD consists of an LD section and an SSC section with a thickness-tapered waveguide.

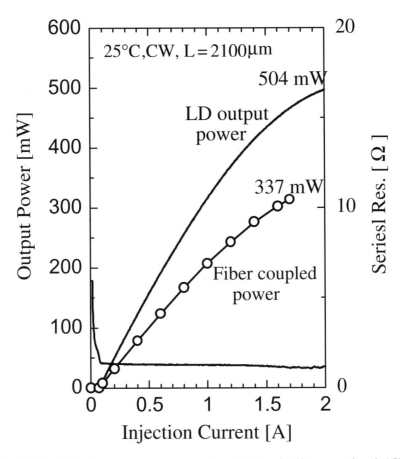

Fig. 10.46   CW light output versus current characteristics of 1.48-μm-wavelength ASM-DC-PBH-LD chip and module [89].

Both sections were simultaneously grown by changing the mask width in atmospheric-pressure NS-MOVPE. The mask width was fixed at 25 μm for the LD section and was varied from 25 to 5 μm for the SSC section. This mask-width combination was chosen to follow the optimized thickness profile at the SSC section for narrow-beam divergence operation to enable a large coupling tolerance between the LD chip and the optical waveguide, or optical fiber. Figure 10.48(a) shows typical lasing characteristics of the fabricated device at several temperatures [87]. The device had a highly re-flective coating with a 95% reflectivity on its rear facet, and was mounted on a diamond heat sink in the p-side-down configuration. The slope-efficiency

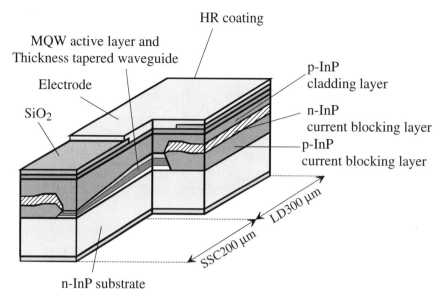

Fig. 10.47   Schematic structure of a 1.3-μm-wavelength SSC-ASM LD [87].

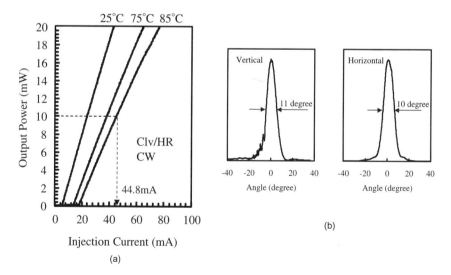

Fig. 10.48   (a) CW L-I characteristics of a 1.3-μm-wavelength SSC-ASM LD. (b) Far-field patterns of a 1.3-μm-wavelength SSC-ASM LD [87].

degradation at temperatures ranging from 25 to 85°C was only −1.6 dB. The operating current at 85°C was 44.8 mA. Figure 10.48(b) shows the measured far-field patterns [87]. The typical radiation angles were 11 and 10 degrees in the vertical and horizontal directions, respectively. Okunuki *et al.* obtained a single-mode fiber (SMF) coupling loss of 3.2 dB and a misalignment tolerance of ±1.8 μm for a 1-dB coupling loss increase. This high coupling tolerance, needed to maintain high coupling efficiency, enabled effective coupling between the LD chip and the optical waveguide by a passive alignment technique. This laser diode had a low threshold current, a low operating current, narrow beam divergence, and high misalignment tolerance, and it is thus suitable for use in low-cost optical modules for access networks.

10.4.3. SIMULTANEOUS FABRICATION OF DIFFERENT WAVELENGTHS LIGHT SOURCES

The rapid progress from 1997 in expanding the gain bandwidth of doped-fiber amplifiers has opened a new transmission window for WDM systems [94, 95], making it possible to greatly increase the system transmission capacity. For example, DWDM systems with narrow channel spacing of less than 100 GHz can potentially transmit more than 200 channels through single-mode fiber (SMF). This implies the possibility of developing WDM systems with a very wide band and a large capacity. In 2001, Fukuchi *et al.* achieved WDM transmission of over 10 Tb/s by using three gain bands, namely, the S, C, and L bands [96]. An important issue in the deployment of such DWDM systems is the need for inexpensive light sources, because many light sources with different wavelengths are needed. However, conventional techniques based on planar MOVPE for fabricating LDs require the use of many wafers to cover a wide wavelength range because the lasing wavelengths of LDs on one wafer are almost the same. This increases cost of WDM light sources. One way to reduce the cost of WDM light sources is to fabricate light sources with different wavelengths simultaneously on the same wafer. This was first demonstrated by Zah *et al.* [97], who fabricated different-wavelength DFB-LD arrays covering a wide wavelength range of 131 nm. However, the obtained characteristics were still primitive. This simultaneous fabrication technique was improved by Kudo *et al.* [60], who brought it up to the level of practical use by utilizing the in-plane bandgap energy controllability feature of NS-MOVPE for tuning the MQW gain-peak wavelengths to the designed lasing wavelengths on a wafer.

This simultaneous fabrication technique is flexible and can meet any customer's requirements in terms of the device wavelength and the number of devices.

In the following section, simultaneous fabrication of different-wavelength light sources is described in detail by demonstrating DFB/MODs fabricated on a single wafer that can emit at different wavelengths [98, 99]. This technique combines advanced electron-beam (EB) lithography to form different-pitch gratings with NS-MOVPE to grow MQW active and absorption layers.

Weighted-Dose Allocation Variable Pitch EB Lithography (WAVE)

The key techniques for simultaneous fabrication of different-wavelength light sources are (1) EB lithography for controlling the grating pitches of LDs to obtain the designed lasing wavelengths and (2) NS-MOVPE for controlling the MQW bandgap energy to trace the lasing wavelength maintaining the optimum detuning of LDs. The main features of NS-MOVPE have already been described in the preceding section; here, EB lithography is briefly described.

The wavelength grids of WDM systems have been standardized with uniform spacing of, for example, 100 GHz. The tolerable lasing wavelength error within a grid must be less than 10% of the wavelength, or less than 10 GHz (0.08 nm). The wavelength of a light source must therefore be controlled precisely over a wide range of wavelengths. A promising way to control the lasing wavelength of DFB-LDs is by using EB lithography and by accurately controlling the pitch of the gratings in the LDs. To set the lasing wavelength of each LD within the error of less than 10 GHz, the grating pitch must be controlled with a resolution higher than 0.01 nm. Conventional EB-lithography systems, however, have a fixed minimum drawing step, which in most cases is 1.25 nm. This means that grating pitches can be controlled in 1.25-nm steps. This is not fine enough for WDM applications, and the minimum drawing step must be reduced by more than two orders of magnitude. To achieve precise grating-pitch control, an advanced method of EB lithography was developed by Muroya *et al.*—weighted-dose allocation variable pitch EB lithography, called WAVE [100]. Figure 10.49 shows a schematic diagram of WAVE. In WAVE, an EB-dose profile for corrugation is composed of several weighted sub-dose profiles. The total dose profile is a superimposition of the sub-dose profiles approximated by Gaussian distribution. Therefore, the peak position of the total dose profile

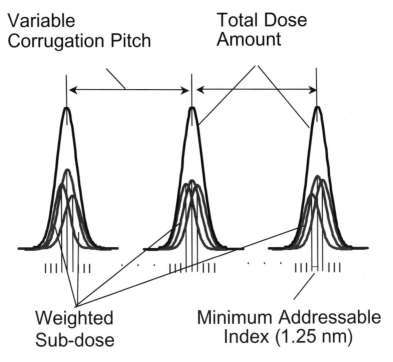

Fig. 10.49   Schematic diagram of weighted-dose-allocation variable-pitch EB lithography (WAVE) [100].

can be accurately controlled by a combination of both allocation and weighting of the multiple exposure for each minimum index. Compared to conventional methods based on deflection-amplitude control, WAVE is more suitable for EB lithography because it uses well-defined calibration of the EB system for different corrugation pitches. It enable easy control of corrugation pitches with a minimum resolution of less than 0.01 nm simply by using a software program. This resolution corresponds to 10 GHz and is sufficient for fabricating DWDM light sources.

Different-Wavelength 2.5-Gb/s-DFB/MODs

Figure 10.50 shows the configuration of a wafer ready for DH epitaxial growth of different-wavelength 2.5-Gb/s DFB/MODs reported by Kudo *et al.* [98], who designed 40-channel devices with lasing wavelengths ranging from 1523 to 1585 nm with 200-GHz spacing. Precise detuning control for each channel between the modulator-absorption-peak wavelength and

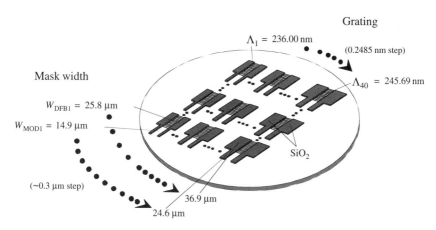

Fig. 10.50 Grating and SiO_2 mask design for simultaneous fabrication of different-wavelength DFB/MODs on a single wafer.

the lasing wavelength was achieved by combining accurately pitch controlled grating formation by WAVE and MQW bandgap-energy control by NS-MOVPE. The fabrication process of these devices is almost the same as that described in Section 10.4.1, except for the simultaneous formation of different-pitch gratings and the corresponding different-bandgap-energy MQWs on the same wafer. Pitches Λ of the simultaneously formed gratings varied from $\Lambda_1 = 236.00$ nm to $\Lambda_{40} = 245.69$ nm in 0.2485-nm steps. The NS-MOVPE mask width (W_m) for each channel was determined in order to trace the MQW detuning design, i.e., the active-layer gain peak wavelength coincided with the lasing wavelength, and the absorption-layer bandgap wavelength maintained an optimum detuning of about -70 nm relative to the lasing wavelength (see Section 10.4.1). In this case, $W_{MOD} = 14.9$ to 24.6 μm was used for the modulator sections, and $W_{DFB} = 25.8$ to 36.9 μm was used for the laser sections. The MQW strain increased from shorter wavelength MQWs to longer wavelength ones. The estimated well strain varied from $+0.55$ to $+1.21\%$, corresponding to the variation in W_m (14.9 to 36.9 μm). Yet, narrow photoluminescence (PL) spectra with the full-width at half maximum (FWDM) of less than 50 meV were maintained for all MQWs. This indicates that all the MQW layers, from the shortest bandgap wavelength layer (absorption layer for ch. 1) to the longest bandgap wavelength layer (active layer for ch. 40), were of high quality.

The complete DFB/MOD structure was almost the same as that shown in Fig. 10.34. The DFB was 400 μm long with a $\lambda/4$ shift located in the

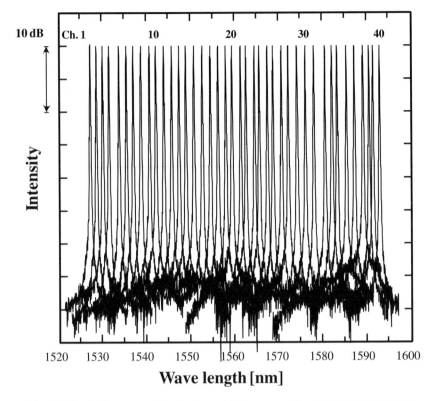

Fig. 10.51   Lasing spectra for 40-channel different-wavelength DFB/MODs [98].

center of the DFB cavity. The modulator was 250 μm long including a
25-μm-long window. Both end facets had an AR coating, and this resulted
in an effective modulator-facet reflectivity of less than 0.1%.

The lasing spectra of the forty channels are shown in Fig. 10.51 [98].
Stable single-mode lasing characteristics with a side-mode-suppression
ratio (SMSR) exceeding 35 dB were obtained for the devices with wavelengths ranging from 1527 to 1593 nm. Figure 10.52 shows the threshold-
current and extinction-ratio distribution. On average, a uniform, low threshold current of 9.9 mA was obtained. The extinction ratios were also
uniform and as large as 20 dB ±2 dB at −2 V for all the channels. The
output power at 100 mA was typically 4 mW. The modulation bandwidth
at −3 dB was 3.8 GHz. Experiments with 2.5-Gbit/s, 600-km transmission
were also conducted using normal fiber with a chromatic dispersion of
+17 ps/nm/km. Transmission with a power penalty of less than 1 dB was

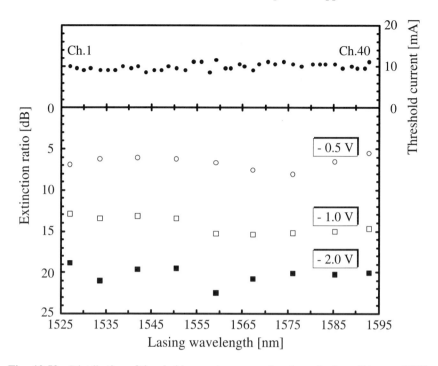

Fig. 10.52 Distribution of threshold currents measured under pulsed conditions at 25°C and DC extinction ratios for different-wavelength DFB/MODs [98].

achieved in these experiments. Such high performance of simultaneously fabricated DFB/MODs proves the effectiveness of the NS-MOVPE-based simultaneous-fabrication technique in DWDM device fabrication.

Different-Wavelength 10-Gb/s DFB/MODs

The simultaneous fabrication technique was further improved to enable the fabrication of different-wavelength 10-Gb/s-DFB/MODs on the same wafer [78, 101]. To achieve 10-Gb/s operability of DFB/MODs, ASM with semi-insulating BH growth was introduced. Kudo *et al.* also expanded the covered wavelength range to the full L band by using growth-time-modulation methods for NS-MOVPE [99]. Figure 10.53 shows a schematic structure of a 10-Gb/s DFB/MOD designed by Furushima *et al.* [78]. They fabricated 40-channel DFB/MODs with wavelengths covering 1560 to 1610 nm and obtained uniform L-I characteristics for each channel, as shown in Fig. 10.54. The devices had AR and HR coatings on their front and rear facets, which enabled high output power (over 10 mW at 100 mA)

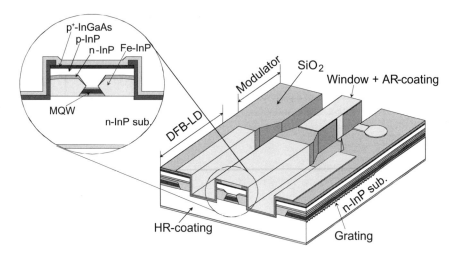

Fig. 10.53   Schematic structure of a 10-Gb/s DFB/MOD fabricated by ASM [78].

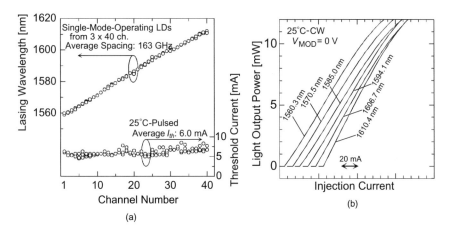

Fig. 10.54   Characteristics of simultaneously fabricated L-band 10-Gb/s DFB/MODs. (a) Lasing-wavelength and threshold-current distributions. (b) Representative CW L-I characteristics at 25°C [78].

even for the longest wavelength 1610-nm channel. The extinction ratios were over 15 dB for all the channels at a modulator bias of -2 V. The 3-dB bandwidth of the small-signal frequency response was typically 14 GHz. Furushima *et al.* also tested 10-Gb/s transmission by using standard SMF at $1.59\,\mu m$ ($\Delta \cong 20$ ps/nm/km). Figure 10.55(a) shows the experimental

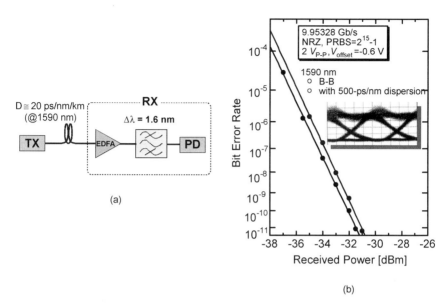

(a)

(b)

Fig. 10.55   10-Gb/s-transmission experiments for a DFB/MOD with a 1590-nm lasing wavelength. (a) Experimental setup. (b) Bit error rate (BER) characteristics [78].

setup, and Figure 10.55(b) shows the obtained BER characteristics and eye-opening pattern [78]. A 1.58-μm-band EDFA and a pin PD were used as a pre-amplifier and a detector, respectively. The receiver sensitivity at a BER of 10^{-10} was -32 dBm (back-to-back). They achieved error-free transmission with a power penalty of 0.5 dB over 25 km (500 ps/nm) and 2.0 dB over 50 km (1000 ps/nm). This dispersion tolerance is sufficient for both 10-Gb/s/ch DWDM inter-metro applications and 10-Gb/s/ch DWDM long-distance transmission systems that use a dispersion-management technique.

10.4.4. WAVELENGTH-SELECTABLE LIGHT SOURCES FABRICATED BY MASE

Wavelength-selectable light sources (WSLs) are key components of DWDM optical transmission systems because these devices can reduce the transmitter inventory costs and are cost-effective stand-by sources for restoring failed channels. They are also promising devices for use in re-configurable optical add/drop multiplexers (OADMs) and optical cross-connects (OXCs) of future photonic network systems. Many WSLs have been described and they can roughly be categorized into four types: 1) DFB-

LD-based [102–113], 2) DBR-LD-based [114–121], 3) VCSEL-with-MEMS-based [122], and 4) external-cavity LDs [123]. Among these types, WSLs based on DFB LDs have superior characteristics and stable and reliable wavelength properties. Tunability over a wide range of wavelengths can be achieved by using an array architecture. Wavelength tuning can then be implemented merely by selecting an appropriate DFB LD and using a thermoelectric cooler (TEC) to change the temperature. This type of WSL was first reported by Zah *et al.* in 1992 [102]. Since then, some progress was made in terms of device configuration, but the results were still poor due to a lack of sophisticated device fabrication techniques. DFB-LD-array-based WSLs with performance almost good enough for practical use were first reported in 2000 by Bouda *et al.* [106] and Yashiki *et al.* [110]. One such WSL was obtained by using an advanced NS-MOVPE technique, i.e., microarray selective epitaxy (MASE) developed by Kudo *et al.* [69].

In the next section, DFB-LD-array-based WSLs fabricated by using MASE are described. WSLs basically consist of a monolithically integrated $\lambda/4$-shift DFB-LD array, a multi-mode-interference (MMI) optical coupler, and a semiconductor optical amplifier (SOA). The performance of two types of WSLs is described. One type is a WSL with an EA modulator, called "EA-WSL," which was developed to obtain uniform 2.5-Gb/s modulation over the entire tuning range of $\sim$8 nm [110]. The other type is a WSL without an EA modulator, called "CW-WSL," which was developed for use in conjunction with external modulators [69] because it provides flexibility in making WDM transponders. A technique for simultaneously fabricating multi-range WSLs on a single wafer [109, 110, 111] is also described.

EA-Modulator-Integrated Wavelength-Selectable Microarray Light Sources

In an EA-modulator-integrated WSL (EA-WSL), wavelength tuning is implemented by selecting an appropriate DFB LD and changing its temperature by TEC. However, as described in Section 10.4.1, the extinction characteristics of EA modulators are strongly dependent on the detuning between the laser wavelengths and the modulator-absorption-edge wavelengths. Therefore, uniform modulation characteristics are difficult to obtain in such EA-WSLs because they have a wide tunable wavelength range and a large detuning variation. Yashiki *et al.* [110] reported one technique that overcomes this problem. They successfully developed an EA-WSL

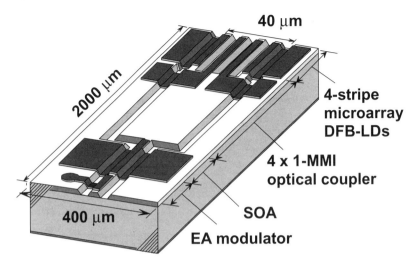

40 μm

2000 μm

400 μm

4-stripe microarray DFB-LDs

4 x 1-MMI optical coupler

SOA

EA modulator

Fig. 10.56 Schematic structure of EA-modulator-integrated microarray wavelength-selectable light sources (EA-WSL) [110].

with a wavelength-independent modulation capability.

Figure 10.56 shows the structure of their EA-WSL. They obtained a compact chip with an area of 0.8 mm² (0.4 mm wide and 2 mm long). This compactness was achieved by using MASE because MASE makes it possible to form densely arrayed (pitch $= 10\,\mu$m) multiple-quantum-well (MQW) waveguides without any complicated semiconductor etching and to individually control MQW bandgap wavelengths [69]. The DFB LDs, SOA, and EA modulator had a simple BH structure with a p-n InP homo-junction. The MMI optical coupler had a p-InP buried ridge waveguide geometry, and its length was optimized taking into account the covered wavelength range of the EA-WSL. Both end facets were AR-coated, and the output facet reflectivity, including the effect of the window structure, was less than 0.1%.

Yashiki *et al.* developed a simple device fabrication process, which requires only two selective MOVPE steps. First, four different-pitch gratings are partially formed in the DFB-LD-array section on an n-InP substrate by EB lithography with WAVE [100]. Then, SiO₂ masks with a MASE pattern in the DFB-LD-array section and an NS-MOVPE pattern in the SOA and MOD sections are formed on a wafer. Eight compressively strained MQW waveguides are selectively grown by using atmospheric-pressure MOVPE [70] so that the bandgap energy of each microarray MQW in the

DFB-LD-array section can trace the designed lasing wavelengths of each
LD within ±10 nm. MQW waveguides in the SOA and MOD sections are
simultaneously grown so that the typical lasing wavelengths were 60 nm
longer than the modulator-absorption-edge wavelengths and 20 nm shorter
than the SOA PL peak wavelengths. In the MMI section, a planar MQW
waveguide is grown. After this first selective MOVPE step, the SiO_2 masks
are patterned again, and a second selective MOVPE step is performed for
the p-InP and p^+-InGaAs over-cladding structure, in which a ridge structure for the MMI section is defined. Finally, electrodes are formed, and the
surface processing is completed. This process requires only two growth
steps, and it is very similar to the fabrication process described in Section
10.4.1 (Fig. 10.36). After that, the wafer is scribed into bars, and front and
rear facets are coated with AR films.

This technique enables overcoming the detuning variation problem between the tuned laser wavelengths and modulator-absorption-peak wavelengths by using an offset for each DFB-LD's operating temperature.
Figure 10.57 shows the temperature dependence of the lasing wavelengths
and modulator-absorption-edge wavelengths in the conventional and offset-
temperature techniques [110]. In the conventional technique, the lasing
wavelength of each LD is pre-assigned by a spacing of 2 nm so that an 8-nm-
wavelength range can be fully covered by controlling the temperature over
a range of ±10°C. However, this tuning scheme results in a 12-nm variation

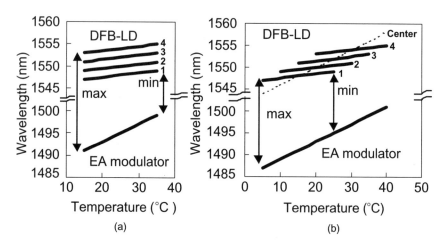

Fig. 10.57 Temperature dependence of DFB lasing wavelength and modulator-
absorption-peak wavelength. (a) Conventional method. (b) Offset-temperature method
[110].

in detuning between the laser and absorption wavelengths. This is because modulator-absorption-edge wavelengths have a temperature dependence of 0.4 nm/°C, which is four times greater than that of lasing wavelengths. In contrast, the offset-temperature technique reduces the detuning variation by 50%, to 6 nm. The operating temperature range of each LD shifts by 5°C so that the center lasing wavelengths of the LDs have the same slope as that of the modulator-absorption-edge wavelengths. This technique increases the tolerance in modulator bandgap control and results in uniform modulation.

The fabricated EA-WSL had a tunable wavelength range of ∼8 nm, stable single-mode operation with a SMSR of over 35 dB, and an optical signal-to-noise ratio (OSNR) of over 50 dB. Threshold current I_{th} was ∼6 mA at 25°C. Typical fiber-coupled power P_f was more than 5 mW at $I_{SOA} = I_{DFB} = 100$ mA. Figure 10.58 shows typical extinction characteristics of four different DFB outputs at center operating temperatures of 15°C for DFB1, 20°C for DFB2, 25°C for DFB3, and 30°C for DFB4 [110]. Uniform extinction curves were obtained as a function of the modulator bias voltage. The typical DC extinction ratio was 14 dB at −2 V. The 3-dB bandwidth of the small signal response was approximately 3.5 GHz. The RF extinction ratio at 2.5 Gb/s was 11.0 dB, and error-free transmission was obtained for all the DFB outputs at 2.5 Gb/s with a 600-km fiber span, with the power penalties ranging from 1.0 to 3.0 dB.

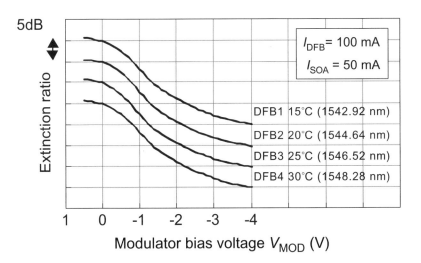

Fig. 10.58   Extinction characteristics of an EA modulator [110].

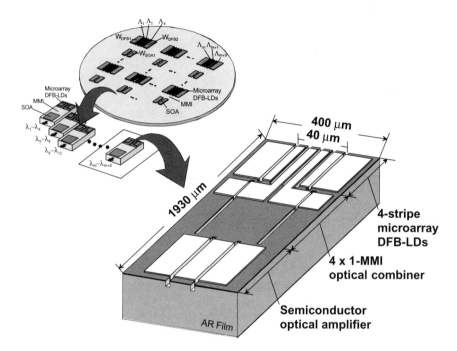

Fig. 10.59   Simultaneous fabrication of multi-range CW-WSLs on a single wafer and schematic structure of a fabricated ASM CW-WSL [111].

ASM-BH Wavelength-Selectable Microarray Light Sources

In this section, we describe simultaneous fabrication of multi-range CW-WSLs on a single wafer reported by Kudo *et al.* [111]. A conceptual drawing of this process and the device structure are shown in Fig. 10.59. The fundamental configuration of the CW-WSLs is the same as that of EA-WSLs except for the EA modulator. Five types of devices (C1~C5) were fabricated to cover the entire C band from 1530.334 to 1561.826 nm. The laser wavelength spacing was set to 2 nm so that a range of ~8 nm could be covered by one chip by varying the temperature between 15 and 40°C using TEC. All the devices had a compact chip of the same size (0.4×1.93 mm $= 0.77$ mm^2), an LD array spacing of 10 μm, and an LD length of 450 μm. The MMI lengths were optimized to minimize the insertion loss considering their wavelength ranges. The SOA lengths were determined so as to keep the total cavity length constant. The CW-WSLs also had antireflection films with 25-μm-long windows in both facets to obtain a reflectivity of less than 0.04%. The κL value of the DFB LDs was approximately 1.8.

The laser wavelength grids were designed to have an overlap of approximately 0.4 nm between the neighboring DFB LDs within the chip, and an overlap of approximately 2.5 nm between the CW-WSLs in the neighboring range. This was done to compensate for the wavelength fluctuation due to fabrication errors.

In the fabrication process, different-pitch gratings were formed by EB lithography, and corresponding SiO_2 masks for atmospheric-pressure MASE [70] were patterned onto a single wafer. A DH structure of compressively strained MQWs was grown by selective MOVPE so that the bandgap energy of each microarray MQW in the DFB-LD-array regions and that of each MQW in the SOA regions could trace the designed lasing wavelengths maintaining optimum detuning. To generate higher power, Kudo *et al.* introduced a pnp-InP current-blocking structure instead of a p-InP BH structure into both the DFB and SOA regions. A self-alignment process described in Section 10.4.2 was used to form SiO_2 masks on top of the DH mesa in the microarray DFB regions and in the SOA regions. This was followed by a second selective MOVPE step for the pnp-InP BH structure and a third selective MOVPE step for the p-InP and p^+-InGaAs over-cladding structure. After completing these steps, they used inductively coupled plasma (ICP) dry etching to electrically divide the DFB LDs and SOA. The MMI section had a ridge structure. Figure 10.60 shows a cross sectional SEM image of the DFB-LD-array region after ICP etching [111]. One can see a uniformly etched 5-μm-deep mesa structure. Finally, an electrode was formed to complete the device fabrication. The process thus consisted of four fabrication steps: three selective MOVPE steps and one dry-etching step.

The wavelength selectability of the fabricated CW-WSLs was examined by tuning their wavelengths. Figure 10.61 shows superimposed lasing spectra obtained for C1~C5 [111]. Each wavelength was tuned to 100-GHz-spaced ITU-T grids by choosing a DFB LD and varying the temperature between 15 and 40°C. Each device covered its eight pre-assigned channels, so that all forty channels, the entire C band of the 32-nm range, were covered by the devices. Stable single-mode oscillations with an SMSR greater than 45 dB were obtained for all tuning conditions. Kudo *et al.* obtained a uniform I_{th} of 8.3 ± 1.2 mA and a facet output power, P_o, of over 33 mW with $I_{DFB} = 75$ mA and $I_{SOA} = 200$ mA for C1 ~ C5. The total MMI insertion loss between the DFB LDs and SOA was estimated to be between 9 and 10 dB. Figure 10.62 shows the dependence of fiber-coupled power P_f on the SOA current at operating temperatures of 15 to 40°C at a fixed

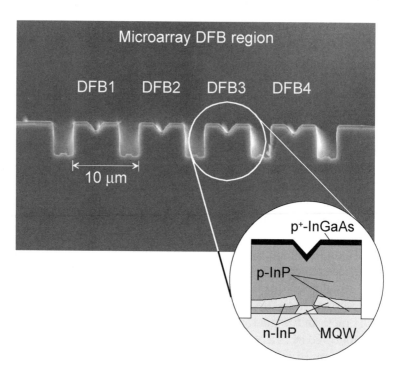

Fig. 10.60 Cross sectional SEM image of a microarray DFB region of an ASM CW-WSL after ICP mesa etching [111].

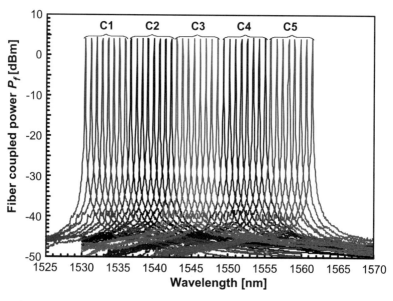

Fig. 10.61 Superposed lasing wavelength spectra tuned precisely to 100-GHz-spaced ITU-T grids obtained for fabricated ASM CW-WSLs [111].

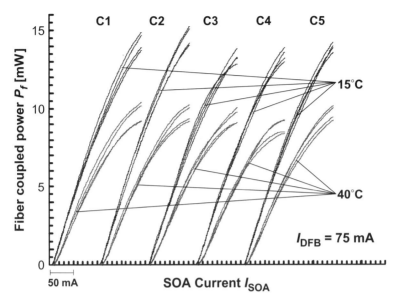

Fig. 10.62   Fiber-coupled power dependence on SOA current measured at 15 and 40°C for fabricated ASM CW WSLs [111].

I_{DFB} of 75 mA [111]. High values of P_f, between 8 and 15 mW, were obtained for C1 ~ C5.

10.4.5.   OTHER DEVICE APPLICATIONS

In addition to the optical devices described above, selective MOVPE is widely used for other devices. Some are based on in-plane waveguide-thickness and composition controllability of selective MOVPE; others take advantage of direct waveguide formability of NS-MOVPE. In the next section, we will describe some of these devices.

Tunable-Stair-Guide (TSG) DBR LDs

As described in Section 10.4.4, tunable DBR LDs have been extensively investigated for their use in WDM transmission systems and future optical cross-connection systems because they have fast wavelength tunability and can be integrated with other optical elements such as EA modulators. The only drawback of tunable DBR LDs is that two independent currents must be controlled to avoid mode jumping. There have been attempts to obtain single-current continuous-wavelength tuning in DBR lasers by adding a current divider to control the injection-current ratio in the phase-control (PC) and DBR regions [124]. However, continuous tuning was difficult to achieve and mode jumping still limited the tuning range because nonlinear I-V characteristics in the PC and DBR regions (typical diode I-V characteristics) made it difficult to keep the injection-current ratio constant. Continuously tunable DBR lasers with single-current injection without any special current control circuit were developed by Kudo *et al.* [116], who used the feature of MQW-waveguide-thickness and composition controllability of NS-MOVPE. Their idea about phase matching between the DBR and other regions resulted in a stair guide structure, in which the tuning layer thickness in the PC region is greater than that in the DBR region. They called the device "a tunable-stair-guide (TSG) DBR laser." A schematic structure of the TSG-DBR LD is shown in Fig. 10.63. The TSG structure is characterized by phase matching and it enables uniform tuning current injection

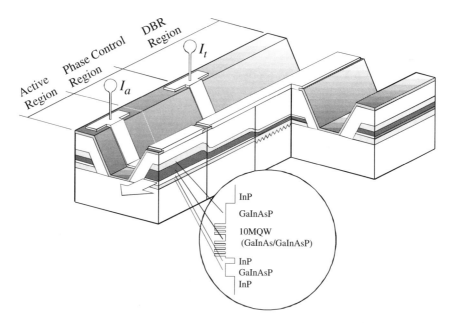

Fig. 10.63   Schematic structure of a tunable-stair-guide (TSG) DBR LD, where I_a and I_t are, respectively, injection current in the active region and tuning current in both PC and DBR regions [116].

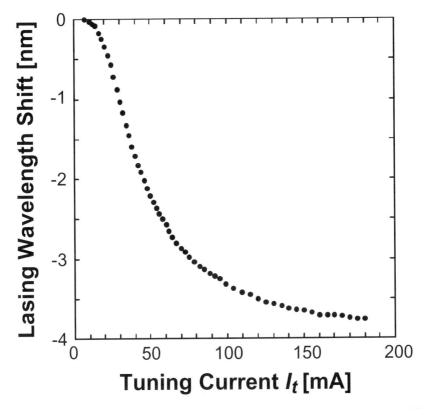

Fig. 10.64 Lasing wavelength tuning characteristics of an InGaAs/InGaAsP/InP TSG-DBR laser at $I_a = 100$ mA [116].

into the PC and DBR regions. The device design for phase matching is described in detail in Ref. [116]. This stair guide structure can be easily obtained by NS-MOVPE just by changing the SiO_2 mask width. The typical threshold currents for the fabricated TSG-DBR LDs were 15–30 mA at $25°C$. Figure 10.64 shows the measured lasing wavelength as a function of tuning current I_t at an active current of $I_a = 100$ mA. Continuous tuning of over 3.8 nm was achieved. They obtained a continuous tuning range of over 3 nm without mode jumping for I_a ranging from 50 to 130 mA. The output power was around 2 mW and the SMSR was over 30 dB. The maximum continuous tuning range for optimized TSG-DBR LDs is expected to be around 7 nm.

Semiconductor Optical Amplifiers (SOAs)

Semiconductor optical amplifiers (SOAs) are very attractive devices for various optical communications systems, such as parallel optical inter-connections, where high gain, low power consumption, and polarization insensitivity are needed. For polarization-insensitive operation, a narrow active waveguide of less than 1 μm is needed to obtain a square-shaped bulk waveguide structure. However, conventional wet etching cannot meet this requirement, because it is difficult to obtain uniform submicron-wide waveguides and the process is not reproducible. By using NS-MOVPE, submicron-wide active waveguides can be easily obtained [59, 125]. Figure 10.65 shows a schematic structure of an SOA and a SEM cross-section of a selectively grown bulk active layer. The 0.4-μm-wide, 0.3-μm-thick InGaAsP bulk active layer was selectively grown and was buried with a p-InP cladding layer. To suppress facet reflection, AR-coating films and window structures were used. A module gain of 10 dB, a saturation output power of 3 dBm, and an on/off ratio of 40 dB were obtained under a low injection current of 30 mA. The polarization dependence of the gain was less than 1 dB. Integration of SOA with a spot-size converter (SSC) is also effective for SOA arrays because it enables passive alignment of SOA arrays with silica waveguide arrays [126].

To obtain high output power and a high on/off ratio, an angled-facet waveguide with an offset between the input and output waveguides can also be used [127]. Figure 10.66 shows a schematic drawing of a bent waveguide

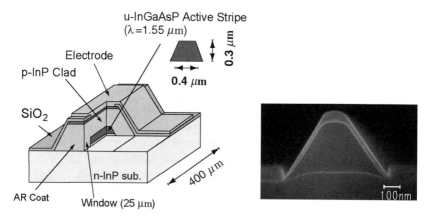

Fig. 10.65 Schematic structure of a SOA with a SEM cross section of a selectively grown narrow active layer.

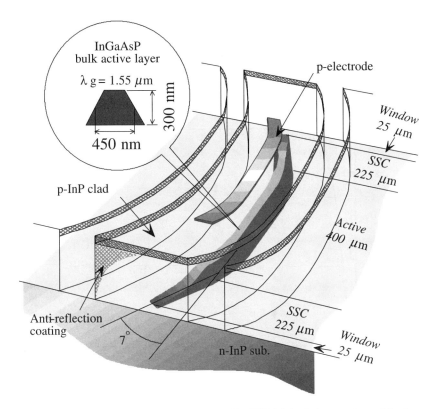

Fig. 10.66   Schematic view of a bent waveguide SSC SOA. Only one channel is shown [128].

SSC-SOA, which represents only one channel of an eight-channel SOA array reported by Hatakeyama *et al.* [128]. A quasi-square-shaped InGaAsP bulk active layer was selectively grown on a narrow (typically 0.7 μm wide) stripe opening between the mask stripes. The device had a bow-shaped active waveguide between the SSC waveguides. This configuration was used to reduce the facet reflectivity and obtain a high on/off ratio. A mask pattern with a sub-micron opening was easily made by using a step-and-repeat projection printer. This enabled obtaining SOA arrays with uniform characteristics. The InGaAsP active layer was 0.3 μm thick and 0.45 μm wide. The active region was 400 μm long, each SSC region was 225 μm long, and the window region was 25 μm long. These SOA arrays were also fabricated by two-step selective MOVPE. Figure 10.67 shows fiber-to-fiber gain characteristics of the eight-channel SOA array [128]. Hatakeyama *et al.*

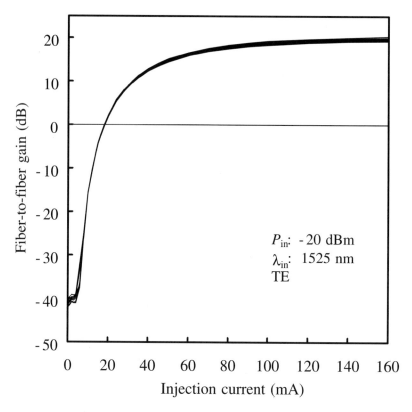

Fig. 10.67   Small signal gain characteristics for all eight channels of SSC SOA [128].

obtained operating currents of 18 and 32 mA for the fiber-to-fiber gains of 0 and 10 dB, respectively. The optical coupling efficiency and polarization dependence were, on average, −3.1 and 0.075 dB/facet. These results show that NS-MOVPE enables obtaining uniform sub-micron-wide active layers with a high gain.

Optical Matrix Switches

Among various kinds of optical switches, passive splitter/SOA gate switches [129] are particularly attractive because they are free from insertion loss, polarization insensitive, and have extremely low cross talk. Bandgap-energy-controlled selective MOVPE is the most suitable technique for fabricating such optical switch matrices [130, 131]. Figure 10.68 shows a 1 × 4 switch matrix for a 1.55-μm wavelength reported by

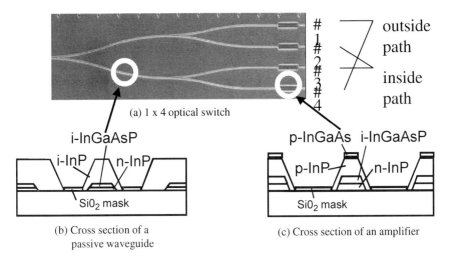

(a) 1 x 4 optical switch

(b) Cross section of a
passive waveguide

(c) Cross section of an amplifier

Fig. 10.68   A 1 × 4 passive splitter/SOA gate switch [131].

Hamamoto *et al.* [131]. It consists of three Y branches and four SOA gates. The chip was 5 × 1 mm, and the SOA was 500 μm long. Bulk InGaAsP layers were used as active and passive core layers, because of their superior polarization-insensitive characteristics when a square-shaped active waveguide is used. The InGaAsP core layer for the SOA gate was 0.4 μm thick and 1 μm wide. The core layer in the SOA-gate region was buried with a p-InP cladding layer and a p-InGaAs contact layer. In the passive waveguide region, the core layer was buried separately with an undoped InP cladding layer to reduce inter-valence-band absorption loss. Generally, the bandgap-energy shift for a bulk layer is smaller than that for an MQW layer, because the change in the layer thickness does not contribute to the bandgap-energy shift. Therefore, Hamamoto *et al.* used selective MOVPE under the atmospheric pressure to obtain a large bandgap-energy shift. Consequently, a 170-nm-wavelength shift was obtained. The core layer at the passive waveguide was 0.15 μm thick. They obtained a small propagation loss of 0.2 dB/mm for the passive waveguide due to a bandgap-energy difference of 170 nm between the core layer composition wavelength and input light wavelength. The measured gain characteristics are shown in Fig. 10.69 [131]. A 0-dB fiber-to-fiber gain was observed at an injection current of 58 mA. The estimated total loss was around 15 dB, and it consisted of a 6-dB fiber-coupling loss (3 dB/facet), a 6-dB splitting loss at the two Y branches, a 1-dB propagation loss from the measured loss of

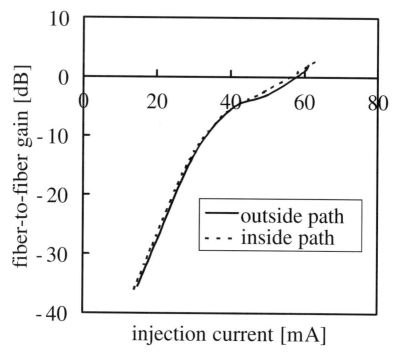

Fig. 10.69   Optical fiber-to-fiber gain of a 1 × 4 passive splitter/SOA gate switch. Solid line shows results for the outside path (port 1); dashed line shows results for the inside path (port 2) [131].

0.2 dB/mm, and a 2-dB excess loss. The total loss was small enough to be compensated for with a single SOA gate. A sufficiently high on/off ratio of more than 38 dB was obtained.

This integration technique was also used to fabricate a transceiver-integrated device [132], in which a 1.3-μm DFB LD, a PIN PD, and a monitor PD were integrated with a Y branch waveguide. Thus, NS-MOVPE can be effectively used to fabricate various devices from DFB LDs to integrated photonic devices.

10.5. Summary

Selective MOVPE has many unique characteristics such as in-plane layer-thickness control and bandgap-energy control, which is achieved simply by changing the mask pattern. This technique is very attractive for fabricating various photonic integrated devices used in WDM systems. Narrow-stripe selective MOVPE offers further advantages due to its direct waveguide formation capability, because no semiconductor mesa etching is needed to form active layers and optical waveguides. Consequently, highly uniform device characteristics can be obtained. This technique enables accurate control of Bragg lasing wavelengths for DFB-LDs, DFB/MODs, and wavelength-selectable light sources. This technique can be potentially used to fabricate various advanced photonic devices for future WDM networks.

ACKNOWLEDGMENTS

The authors thank K. Kobayashi, M. Ogawa, T, Suzuki, I. Mito, T. Uji, M. Fujiwara, M. Kitamura, K. Kobayashi, T. Torikai, K. Komatsu, and M. Yamaguchi for their continuous encouragement and helpful discussions. The authors also thank K. Mori, T. Morimoto, H. Yamazaki, S. Kitamura, K. Hamamoto, M. Ishizaka, T. Kato, M. Tsuji, Y. Sakata, Y. Muroya, T. Tamanuki, Y. Furushima, H. Hatakeyama, K. Yashiki, Y. Okunuki, and S. Sudo for their contribution to this work. The authors also thank N. Kida, T. Nakazaki, and K. Yasuda for their assistance in this work.

References

1. Y. Abe, K. Kishino, Y. Suematsu, and S. Arai, "GaInAsP/InP integrated laser with butt-jointed built-in distributed-bragg-reflection waveguide," Electron. Lett., 17 (1981) 945–947.
2. T. L. Koch and U. Koren, "Semiconductor photonic integrated circuits," IEEE J. Quantum Electron., 27 (1991) 641–653.
3. E. H. Li, Semiconductor Quantum Well Intermixing (Gordon and Breach Science Publishers, Amsterdam, 2000).
4. R. Bhat, "Current status of selective area epitaxy by OMCVD," J. Cryst. Growth, 120 (1992) 362–368.
5. R. Azoulay, N. Bouadama, J. C. Bouley, and L. Dugrand, "Selective MOCVD epitaxy for optoelectronic devices," J. Cryst. Growth, 55 (1981) 229–234.
6. Y. Takahashi, S. Sakai, and M. Umeno, "Selective MOCVD growth of GaAlAs on partly masked substrates and its application to optoelectronic devices," J. Cryst. Growth, 68 (1984) 206–213.
7. K. Kamon, S. Takagishi, and H. Mori, "Selective epitaxial growth of GaAs by low-pressure MOVPE," J. Cryst. Growth, 73 (1985) 73–76.
8. C. Tomiyama, A. Kuramata, S. Yamazaki, and K. Nakajima, "Deposition of InP particles on Si_3N_4 and SiO_2 films by low pressure MOCVD," J. Cryst. Growth, 84 (1987) 115–122.
9. O. Kayser, "Selective growth of InP/GaInAs in LP-MOVPE and MOMBE/ CBE," J. Cryst. Growth, 107 (1991) 989–998.

10. C. Blaauw, A. Szaplonczay, K. Fox, and B. Emmerstorfer, "MOCVD of InP and mass transport on structured InP substrates," J. Cryst. Growth, 77 (1986) 326–333.

11. K. Hiruma, T. Haga, and M. Miyazaki, "Surface migration and reaction mechanism during selective growth of GaAs and AlAs by metalorganic chemical vapor deposition," J. Cryst. Growth, 102 (1990) 717–724.

12. S. Ando and T. Fukui, "Facet growth of AlGaAs on GaAs with SiO_2 gratings by MOCVD and applications to quantum well wires," J. Cryst. Growth, 98 (1989) 646–652.

13. A. Yoshikawa, A. Yamamoto, M. Hirose, T. Sugino, G. Kano, and I. Teramoto, "A novel technology for formation of a narrow active layer in buried heterostructure lasers by single-step MOCVD," IEEE J. Quantum Electron., QE-23 (1987) 725–729.

14. K. M. Dzurko, E. P. Menu, C. A. Beyler, J. S. Osinski, and P. D. Dapkus, "Low-threshold quantum well lasers grown by metalorganic chemical vapor deposition on nonplanar substrates," IEEE J. Quantum Electron., 25 (1989) 1450–1457.

15. E. Colas, A. Shahar, B. D. Soole, W. J. Tomlinson, J. R., Hayes, C. Caneau, and R. Bhat, "Lateral and longitudinal patterning of semiconductor structures by crystal growth on nonplanar and dielectric-masked GaAs substrates: application to thickness-modulated waveguide structures," J. Cryst. Growth, 107 (1991) 226–230.

16. Y. D. Galeuchet, P. Roentgen, and V. Graf, "GaInAs/InP selective area metalorganic vapor phase epitaxy for one-step grown buried low-dimensional structures," J. Appl. Phys., 68 (1990) 560–568.

17. Y. D. Galeuchet, P. Roentgen, and V. Graf, "Buried GaInAs/InP layers grown on nonplanar substrates by one-step low-pressure metalorganic vapor phase epitaxy," Appl. Phys. Lett., 53 (1988) 2638–2640.

18. B. Garrett and E. J. Thrush, "Temporally resolved growth habit studies of InP/(InGa)As heterostructures grown by MOCVD on contoured substrate," J. Cryst. Growth, 97 (1989) 273–284.

19. Y. L. Wang, A. Feygenson, R. A. Hamm, D. Ritter, J. S. Weiner, and H. Temikin, "Optical and electrical properties of InP/InGaAs grown selectively on SiO_2-masked InP," Appl. Phys. Lett., 59 (1991) 443–445.

20. T. Kato, T. Sasaki, K. Komatsu, and I. Mito, "DFB-LD/modulator integrated light source by bandgap energy controlled selective MOVPE," Electron. Lett., 28 (1992) 153–154.

21. Y. Chen, J. E. Zucker, T. H. Chiu, J. L. Marshall, and K. L. Jones, "Quantum well electroabsorption modulators at 1.55µm using singlw-step selective area chemical beam epitaxial growth," Appl. Phys. Lett., 61 (1992) 10–12.

22. Y. D. Galeuchet, H. Rothuizen, and P. Roentgen, "*In situ* buried GaInAs/InP quantum dot arrays by selective area metalorganic vapor phase epitaxy," Appl. Phys. Lett., 58 (1991) 2423–2425.

23. K. Yamaguchi, K. Okamoto, and T. Imai, "Selective epitaxial growth of GaAs by metalorganic chemical vapor deposition," Japan J. Appl. Phys., 24 (1985) 1666–1671.

24. M. Aoki, H. Sano, M. Suzuki, M. Takahashi, K. Uomi, and A. Takai, "Novel structure MQW electroabsorption modulator/DFB-laser integrated device fabricated by selective area MOCVD growth," Electron. Lett., 27 (1991) 2138–2140.

25. L. Buydens, P. Demeester, M. Van Ackere, A. Ackaert, and P. Van Daele, "Thickness variations during MOVPE growth on patterned substrates," J. Electron. Mater., 19 (1990) 317–321.

26. M. Gibbon, J. P. Stagg, C. G. Cureton, E. J. Thrush, C. J. Jones, R. E. Mallard, R. E. Pritchard, N. Collis, and A. Chew, "Selective-area low-pressure MOCVD of GaInAsP and related materials on planar InP substrates," Semicond. Sci. Technol., 8 (1993) 998–1010.

27. T. Sasaki, M. Kitamura, and I. Mito, "Selective metalorganic vapor phase epitaxial growth of InGaAsP/InP layers with bandgap energy control in InGaAsP/InGaAsP multiple-quantum well structures," J. Cryst. Growth, 132 (1993) 435–443.

28. G. Coudenys, I. Moerman, W. Vanderbauwhede, and P. Van Daele, "Selective and shadow masked MOVPE growth of InP/InGaAs(P) heterostructures and quantum wells," J. Cryst. Growth, 124 (1992) 497–501.

29. T. Sasaki, M. Yamaguchi, and M. Kitamura, "Monolithically integrated multi-wavelength MQW-DBR laser diodes fabricated by selective metalorganic vapor phase epitaxy," J. Cryst. Growth, 145 (1994) 846–851.

30. K. Mori, H. Hatakeyama, K. Hamamoto, K. Komatsu, T. Sasaki, and T. Matsumoto, "Narrow-stripe selective growth of high-quality MQWs by atmospheric-pressure MOVPE," J. Cryst. Growth, 195 (1998) 466–473.

31. R. J. Deri, C. Caneau, E. Colas, L. M. Schiavone, N. C. Andreadakis, G. H. Song, and E. C. M. Pennings, "Integrated optic mode-size tapers by selective organometallic chemical vapor deposition of InGaAsP/InP," Appl. Phys. Lett., 61 (1992) 952–954.

32. H. Kobayashi, M. Ekawa, N. Okazaki, O. Aoki, S. Ogita, and H. Soda, "Tapered thickness MQW waveguide BH MQW lasers," IEEE Photonic. Technol. Lett., 8 (1994) 1080–1081.

33. H. Yamazaki, K. Kudo, T. Sasaki, J. Sasaki, Y. Furushima, Y., Sakata, M. Itoh, and M. Yamaguchi, "1.3μm spot-size-converter integrated laser diodes fabricated by narrow-stripe selective MOVPE," IEEE J. Sel. Topics Quantum Electron., 3 (1997) 1392–1398.

34. Y. Katoh, T. Kunii, Y. Matsui, H. Wada, T. Kamijoh, and Y. Kawai, "DBR laser array for WDM system," Electron. Lett., 29 (1993) 2195–2197.

35. T. Sasaki, M. Yamaguchi, and M. Kitamura, "10 wavelength MQW-DBR lasers fabricated by selective MOVPE growth," Electron. Lett., 30 (1994) 785–786.

36. M. Aoki, T. Taniwatari, M. Suzuki, and T. Tsutsui, "Detuning adjustment multiwavelength MQW-DFB laser array grown by effective index/quantum energy control selective area MOVPE," IEEE Photonic. Technol. Lett., 6 (1994) 789–791.

37. J. S. C. Chang, K. W. Carey, J. E. Turner, and L. A. Hodge, "Compositional non-uniformities in selective area growth of GaInAs on InP grown by OMVPE," J. Electron. Mater., 19 (1990) 345–348.

38. Y. D. Galeuchet and P. Roentgen, "selective area MOVPE of GaInAs/InP heterostructures on masked and nonplanar (100) and {111} substrates," J. Cryst. Growth, 107 (1991) 147–150.

39. J. Finders, J. Geurts, A. Kohl, M. Weyers, B. Opitz, O. Kayser, and P. Balk, "Composition of selectively grown $In_x Ga_{1-x}$ As structures from locally resorved Raman spectroscopy," J. Cryst. Growth, 107 (1991) 151–155.

40. Y. Sakata, Ph. D. Thesis, Hiroshima University (1999). In Japanese.

41. C. Caneau, R. Bhat, C. C. Chang, K. Kash, and M. A. Koza, "Selective organometallic vapor phase epitaxy of Ga and In compounds: a comparison of TMIn and TEGa versus TMIn and TMGa," J. Cryst. Growth, 132 (1993) 364–370.

42. M.-S. Kim, C. Caneau, E. Colas, and R. Bhat, "Selective area growth of InGaAsP by OMVPE," J. Cryst. Growth, 123 (1992) 69–74.

43. M. Eckel, D. Ottenwalder, F. Scholz, G. Frankowski, T. Wacker, and A. Hangleiter, "Improved composition homogeneity during selective area epitaxy of GaInAs using a novel In precursor," Appl. Phys. Lett., 64 (1994) 854–856.

44. M. Ida, K. Kurishima, and T. Kobayashi, "Highly selective InGaAs growth by metalorganic chemical vapor deposition with a high-speed rotating surceptor," J. Cryst. Growth, 145 (1994) 237–241.

45. C. Caneau, R. Bhat, M. R. Frei, C. C. Chang, R. J. Deri, and M. A. Koza, "Studies on the selective OMVPE of (Ga, In)/(As, P)," J. Cryst. Growth, 124 (1992) 243–248.

46. Y. Sakata, T. Nakamura, S. Ae, T. Terakado, Y. Inomoto, T. Torikai, and H. Hasumi, "Selective MOVPE growth of InGaAsP and InGaAs using TBA and TBP," J. Electron. Mater., 25 (1996) 401–406.

47. D. G. Coronell and K. F. Jensen, "Analysis of MOCVD of GaAs on patterned substrates," J. Cryst. Growth, 114 (1991) 581–592.

48. T. Fujii, M. Ekawa, and S. Yamazaki, "A theory for metalorganic vapor phase epitaxial selective growth on planar patterned substrates," J. Cryst. Growth, 146 (1995) 475–481.

49. F. Koyama, K.-Y. Liou, A. G. Dentai, G. Raybon, and C. A. Burrs, "GaInAs/GaInAsP strained quantum well monolithic electroabsorption modulator/amplifier by lateral bandgap control with nonplanar substrates," Electron. Lett., 29 (1993) 2104–2106.

50. C. H. Joyner, S. Chandrasekhar, J. W. Sulhoff, and A. G. Dentai, "Extremely large band gap shifts for MQW structures by selective epitaxy on SiO_2 masked substrates," IEEE Photonic. Technol. Lett., 4 (1992) 1006–1009.

51. M. Suzuki, M. Aoki, T. Tsuchiya, and T. Taniwatari, "1.24–1.66μm quantum energy tuning for simultaneously grown InGaAs/InP quantum wells by selective-area metalorganic vapor phase epitaxy," J. Cryst. Growth, 145 (1994) 249–255.

52. M. Tsuji, K. Makita, T. Takeuchi, and K. Taguchi, "Large bandgap energy control of InAlAs/InGaAs MQW selectively grown by LP-MOVPE," J. Cryst. Growth, 170 (1997) 669–673.

53. E. Colas, C. Caneau, M. Frei, E. M. Clausen, Jr., W. E. Quinn, and M. S. Kim, "*In situ* definition of semiconductor structures by selective area growth and etching," Appl. Phys. Lett., 59 (1991) 2019–2021.

54. H. Sugiura, R. Iga, T. Yamada, and T. Toriyama, "Selective area growth of InP and InGaAs layers on SiO_2-masked substrate by chemical beam epitaxy," Japan J. Appl. Phys., 30 (1991) L1089–L1091.

55. E. Veuhoff, "Exploitation of surface selective growth in metalorganic growth technologies for device applications," J. Cryst. Growth, 195 (1998) 444–458.

56. O. Kayser, R. Westphalen, B. Opitz, and P. Balk, "Control of selective area growth of InP," J. Cryst. Growth, 112 (1991) 111–122.

57. H. Sugiura, T. Nishida, R. Iga, T. Yamada, and T. Tamamura, "Facet growth of InP/InGaAs layers on SiO_2-masked InP by chemical beam epitaxy," J. Cryst. Growth, 121 (1992) 579–586.

58. Y. Sakata, P. Delansay, Y. Inomoto, M. Yamaguchi, T. Murakami, and H. Hasumi, "All selective MOVPE grown BH-LDs fabricated by the novel self-alignment process," IEEE Photonic. Technol. Lett., 8 (1996) 179–181.

59. S. Kitamura, K. Komatsu, and M. Kitamura, "Polarization-insensitive semiconductor optical amplifier array grown by selective MOVPE," IEEE Photonic. Technol. Lett., 6 (1994) 173–175.

60. K. Kudo, H. Yamazaki, T. Sasaki, and M. Yamaguchi, "Wide-wavelength range detuning-adjusted DFB-LD's of different wavelengths fabricated on a wafer," IEEE Photonic. Technol. Lett., 9 (1997) 1313–1315.

61. Y. Sakata, Y. Inomoto, and K. Komatsu, "Surface migration effect and lateral vapor-phase diffusion effect for InGaAsP/InP narrow-stripe selective metalorganic vapor-phase epitaxy," J. Cryst. Growth, 208 (2000) 130–136.

62. E. Veuhoff, H. Heinecke, J. Rieger, H. Baumeister, R. Scimpe, and S. Prohl, "Improvements in selective area growth of InP by metalorganic vapor phase epitaxy," IEEE Fourth International Conference on Indium Phosphide and Related Materials, WB4 (1992) 210–213.

63. Y. Sakata and K. Komatsu, "Migration effect on semiconductor surface for narrow-stripe selective MOVPE," J. Electron. Mater., 29 (2000) 37–41.

64. K. Kudo, T. Sasaki, and M. Yamaguchi, "Migration-controlled narrow-stripe selective MOVPE for high-quality InGaAsP/InGaAsP MQWs," J. Cryst. Growth, 170 (1997) 634–638.
65. K. Kudo, Y. Muroya, T. Nakazaki, Y. Inomoto, M. Ishizaka, and M. Yamaguchi, "Different-wavelength modulator-integrated DFB-LDs for 1.58µm band WDM systems," Elecron. Lett., 34 (1998) 1–2.
66. K. Kudo, T. Sasaki, and M. Yamaguchi, "Evaluation on MQW layers within directly formed waveguide by selective-area MOVPE growth," Eighth Int. Conf. on metalorganic vapor phase epitaxy (8th ICMOVPE), OJ. 4, Cardiff, UK (1996).
67. M. Tsuji, K. Makita, T. Takeuchi, and K. Taguchi, "Selective growth of InAlAs by low pressure metalorganic vapor phase epitaxy," J. Cryst. Growth, 162 (1996) 25–30.
68. T. Takeuchi, M. Tsuji, K. Makita, and K. Taguchi, "InAlGaAs selective MOVPE growth with bandgap energy shift," J. Electron. Mater., 25 (1996) 375–378.
69. K. Kudo, Y. Furushima, T. Nakazaki, and M. Yamaguchi, "Densely arrayed eight-wavelength semiconductor lasers fabricated by microarray selective epitaxy," IEEE J. Sel. Topics in Quantum Electron., 5 (1999) 428–434.
70. S. Sudo, Y. Yokoyama, T. Nakazaki, K. Mori, K. Kudo, M. Yamaguchi, and T. Sasaki, "Growth pressure dependence of neighboring mask interference in densely arrayed narrow-stripe selective MOVPE for integrated photonic devices," J. Cryst. Growth, 221 (2000) 189–195.
71. K. Komatsu, "Photonic integrated circuits using bandgap energy controlled selective MOVPE technique," 22nd European Conf. On Optical Communications (ECOC'96), ThC. 2.1, Oslo, Norway (1996).
72. T. Sasaki, M. Yamaguchi, K. Komatsu, and I. Mito, "In-plane bandgap energy controlled selective MOVPE and its applications to photonic integrated circuits," IEICE Trans. Electron., E80-C (1997) 654–663.
73. H. Soda, M. Furutsu, K. Sato, M. Matsuda, and H. Ishikawa, "5 Gb/s modulation characteristics of optical intensity modulator monolithically integrated with DFB laser," Electron. Lett., 25 (1989) 334–335.
74. H. Tanaka, M. Suzuki, M. Usami, H. Taga, S. Yamamoto, and Y. Matsushima, "5-Gb/s performance of integrated light source consisting of λ/4-shifted DFB laser and EA modulator with SI-InP BH structure," IEEE J. Lightwave Technol., 8 (1990) 1357–1362.
75. T. Kato, T. Sasaki, N. Kida, K. Komatsu, and I. Mito, "Novel MQW DFB laser diode/modulator integrated light source using bandgap energy control epitaxial growth technique," 17th European conf. on Optical Communication/ 8th Int. Conf. on Optical Communication, WeB7-1, Paris, France (1991).
76. M. Aoki, M. Suzuki, H. Sano, T. Kawano, T. Ido, T. Taniwatari, K. Uomi, and A. Takai, "InGaAs/InGaAsP MQW electroabsorption modulator

integrated with a DFB laser fabricated by band-gap energy control selective area MOCVD," IEEE J. Quantum Electron., 29 (1993) 2088–2096.

77. J. E. Johnson, T. Tanbun-Ek, Y. K. Chen, D. A. Fishman, R. A. Logan, P. A. Morton, S. N. G. Chu, A. Tate, A. M. Sergent, P. F. Sciortino, Jr., and K. W. Wechet, "Low chirp integrated EA-modulator/DFB laser grown by selective-area MOVPE," IEEE 14th Int. Semiconductor Laser Conf., M4.7, Hawaii (1994) 41–42.

78. Y. Furushima, K. Kudo, Y. Muroya, Y. Sakata, Y. Inomoto, K. Fukuchi, M. Ishizaka, and M. Yamaguchi, (1999). "1560 nm to 1610 nm wavelength EA-modulator integrated DFB-LDs for extended-band 2.5-Gbit/s/ch and 10 Gbit/s/ch WDM systems," Optical Fiber Communication Conference, (OFC'99), WH2, San Diego, CA.

79. H. Yamazaki, Y. Sakata, M. Yamaguchi, Y. Inomoto, and K. Komatsu, "Low drive voltage (1.5 Vpp) and high-power DFB-LD/modulator integrated light sources using bandgap energy controlled selective MOVPE," Electron. Lett., 32 (1996) 109–111.

80. G. Bastard, E. Mendez, L. Chang, and L. Esaki, "Variational calculations on a quantum well in an electric field," Phys. Rev., B28 (1983) 3241.

81. M. Yamaguchi, T. Kato, T. Sasaki, K. Komatsu, and M. Kitamura, "Requirements for modulator-integrated DFB LD's for penalty-free 2.5-Gb/s transmission," IEEE J. Lightwave Technol., 13 (1995) 1948–1954.

82. M. Ishizaka, M. Yamaguchi, Y. Sakata, Y. Inomoto, J. Shimizu, and K. Komatsu, "Modulator integrated DFB lasers with more than 600-km transmission capability at 2.5 Gb/s," IEEE Photonic. Technol. Lett., 9 (1997) 1406–1408.

83. Y. Sakata, Y. Inomoto, D. Saito, K. Komatsu, and H. Hasumi, "Low-threshold and high uniformity for novel 1.3μm strained MQW-DC-PBH-LDs fabricated by all selective MOVPE technique," IEEE Photonic. Technol. Lett., 9 (1997) 291–293.

84. Y. Sakata, T. Hosoda, Y. Sasaki, S. Kitamura, M. Yamamoto, Y. Inomoto, and K. Komatsu, "All selective MOVPE grown 1.3-μm-strained multi-quantum well buried-hetero laser diodes," IEEE J. Quantum Electron., 35 (1999) 368–376.

85. Y. Furushima, Y. Sakata, T. Sasaki, H. Yamazaki, K. Kudo, Y. Inomoto, and T. Sasaki, "1.3μm spot-size converter integrated ASM-BH LDs with low operating current and high coupling efficiency," Electron. Lett., 34 (1998) 767–768.

86. Y. Furushima, H. Yamazaki, K. Kudo, Y. Sakata, Y. Okunuki, Y. Sasaki, and T. Sasaki, "Improved high-temperature and high-power characteristics of 1.3-μm sopt-size converter integrated all-selective metalorganic vapor phase epitaxy grown planar buried heterostructure laser diodes by newly introduced

multiple-stripe recombination layers," Japan J. Appl. Phys., 38, Part 1 (1999) 1234–1238.

87. Y. Okunuki, H. Yamazaki, Y. Sasaki, T. Koui, Y. Sakata, K. Komatsu, A. Goto, and N. Kimura, "Spot-size-converter integrated ASM-LD's with 44.8 mA record low operation current at 85°C," Optical Fiber Communication Conference, (OFC'2000), WM26, Baltimore, MD (2000).

88. Y. Sasaki, T. Hosoda, Y. Sakata, K. Komatsu, and H. Hasumi, "1.48µm-wavelength ASM-DC-PBH LD's with extremely uniform lasing characteristics," IEEE Photonic. Technol. Lett., 11 (1999) 1211–1213.

89. H. Yamazaki, T. Hosoda, Y. Sasaki, and K. Komatsu, "Over half-watt output power 1.48-mm wavelength EDFA pumping ASM LD's," Optical Fiber Communication Conference, (OFC'2000), ThK4, Baltimore, MD (2000).

90. T. Ishida, H. Yamada, T. Torikai, and T. Uji, "Leakage current analysis in DC-PBH laser diodes using an optical device simulator (in Japanese)," Tech. Rep. IEICE, OQE92-165 (1993) 85 (in Japanese).

91. Y. Tohmori, Y. Suzaki, H. Fukano, M. Okamoto, Y. Sakai, O. Mitomi, S. Matsumoto, M. Yamamoto, M. Fukada, M. Wada, Y. Itaya, and T. Sugie, "Spot-size converted 1.3µm laser with butt-jointed selectively grown vertically tapered waveguide," Electron. Lett., 31 (1995) 1069–1070.

92. A. Kasukawa, N. Iwai, N. Yamanaka, Y. Nakahira, and N. Yokouchi, "Estimation of waveguide loss of 1.3µm integrated tapered MQW fabricated by selective area MOCVD," Electron. Lett., 32 (1996) 1294–1296.

93. T. Takiguchi, T. Itagaki, M. Takemi, A. Takemoto, Y. Miyazaki, K. Shibata, Y. Hisa, K. Goto, Y. Mihashi, S. Takamiya, and M. Aiga, "Selective-area MOCVD growth for 1.3µm laser diodes with a monolithically integrated waveguide lens," J. Cryst. Growth, 170 (1997) 705–709.

94. A. Mori, Y. Ohishi, M. Yamada, H. Ono, Y. Nishida, K. Oikawa, and S. Sudo, "1.5µm broadband amplification by Tellurite-based EDFAs," Optical Fiber Communication Conference (OFC'97), PD1, Dallas, TX (1997).

95. M. Fukushima, Y. Tashiro, and H. Ogoshi, "Flat gain Erbium-doped fiber amplifier in 1570 nm–1600 nm region for dense WDM transmission systems," Optical Fiber Communication Conference (OFC'97), PD3, Dallas, TX (1997).

96. K. Fukuchi, T. Kasamatsu, M. Morie, R. Ohhira, T. Ito, K. Sekiya, D. Ogasahara, and T. Ono, "10.92-Tb/s (273 × 40-Gb/s) triple-band/ultra-dense WDM optical-repeatered transmission experiment," Optical Fiber Communication Conference (OFC'2001), PD24, Anaheim, CA (2001).

97. C. E. Zah, P. S. D. Lin, F. Favire, B. Pathak, R. Bhat, C. Caneau, A. S. Gozsz, N. C. Andreadakis, M. A. Koza, T. P. Lee, T. C. Wu, and K. Y. Lau, "1.5µm compressive-strained multiquantum-well 20-wavelength distributed-feedback laser arrays," Electron. Lett., 28 (1992) 824–825.

98. K. Kudo, M. Ishizaka, T. Sasaki, H. Yamazaki, and M. Yamaguchi, "1.52–1.59-μm range different-wavelength modulator-integrated DFB-LD's fabricated on a single wafer," IEEE Photonic. Technol. Lett., 10 (1998) 929–931.

99. K. Kudo, Y. Muroya, T. Nakazaki, Y. Inomoto, M. Ishizaka, and M. Yamaguchi, "Different-wavelengths modulator-integrated DFB-LDs for 1.58-μm band WDM systems," Electron. Lett., 34 (1998) 1946–1947.

100. Y. Muroya, T. Nakamura, H. Yamada, and T. Torikai, "Precise wavelength control for DFB laser diodes by novel corrugation delineation method," IEEE Photonic. Technol. Lett., 9 (1997) 288–290.

101. Y. Furushima, K. Kudo, Y. Muroya, Y. Sakata, M. Ushirozawa, T. Sasaki, K. Yashiki, Y. Yokoyama, Y. Inomoto, M. Ishizaka, J. Shimizu, T. Kosugi, H. Tanaka, M. Yamaguchi, and K. Komatsu, "In-wafer wavelength distribution controlled L-band DFB/MODs for 10-Gb/s/ch long-haul DWDM systems," 26th European Conference on Optical Communications (ECOC'2000), Munich, Germany, 1 (2000) 3.3.4.

102. C. E. Zah, F. J. Favire, B. Pathak, R. Bhat, C. Caneau, P. S. D. Lin, A. S. Gozdz, N. C. Andreadakis, M. A. Koza, and T. P. Lee, "Monolithic integration of multiwavelength compressive-strained multiquantum-well distributed-feedback laser array with star coupler and optical amplifiers," Electron. Lett., 28 (1992) 2361–2362.

103. M. G. Young, U. Koren, B. I. Miller, M. Chien, T. L. Koch, D. M. Tennant, K. Feder, K. Dreyer, and G. Raybon, "Six wavelength laser array with integrated amplifier and modulator," Electron. Lett., 31 (1995) 1835–1836.

104. J. Hong, H. Kim, F. Shepherd, C. Rogers, B. Baulcomb, and S. Clements, "Matrix-grating strongly gain-coupled (MG-SGC) DFB lasers with 34-nm continuous wavelength tuning range," IEEE Photonic. Technol. Lett., 11 (1999) 515–517.

105. L. J. P. Ketelsen, J. E. Johnson, D. J. Muehlner, M. Dautartas, J. V. Gates, M. A. Cappuzzo, J. M. Geary, J. A. Grenko, J. M. Vandenberg, S. K. Sputz, M. W. Focht, E. J. Laskowski, L. T. Gomez, K. G. Glogovsky, C. L. Reynolds, S. N. G. Chu, W. A. Gault, M. S. Hybertsen, and J. L. Zilko, "Electroabsorption modulated 1.55μm wavelength selectable DFB array using hybrid integration," Optical Fiber Communication Conference (OFC'99), PD40, San Diego, CA (1999).

106. M. Bouda, M. Matsuda, K. Morito, S. Hara, T. Watanabe, T. Fujii, and Y. Kotaki, "Compact high-power wavelength selectable lasers for WDM applications," Optical Fiber Communication Conference (OFC'2000), TuL1, Baltimore, MD (2000).

107. K. Kudo, K. Yashiki, T. Sasaki, Y. Yokoyama, K. Hamamoto, T. Morimoto, and M. Yamaguchi, "1.55-μm Wavelength-selectable microarray DFB-LD's with a monolithically integrated MMI combiner, SOA, and EA-modulator," IEEE Photonic. Technol. Lett., 12 (2000) 242–244.

108. B. Pezeshki, A. Mathur, S. Zou, H.-S. Jeon, V. Agrawal, and R. L. Lang, "12 nm tunable WDM sources using an integrated laser array," Electron. Lett., 36 (2000) 788–789.

109. K. Kudo, T. Morimoto, K. Yashiki, T. Sasaki, Y. Yokoyama, K. Hamamoto, and M. Yamaguchi, "Wavelength-selectable microarray light sources of multiple ranges simultaneously fabricated on single wafer," Electron. Lett., 36 (2000) 745–747.

110. K. Yashiki, K. Kudo, T. Morimoto, S. Sudo, Y. Muroya, T. Tamanuki, M. Ishizaka, T. Hosoda, Y. Furushima, H. Hatakeyama, Y. Yokoyama, K. Mori, T. Sasaki, and M. Yamaguchi, "Multi-range wavelength-selectable microarray light sources with detuning adjustable EA-modulator," 26th European Conference on Optical Communication, (ECOC'2000), 6.3.2, Munich, Germany (2000).

111. K. Kudo, K. Yashiki, T. Morimoto, Y. Hisanaga, S. Sudo, Y. Muroya, T. Tamanuki, H. Hatakeyama, K. Mori, and T. Sasaki, "Wavelength-selectable microarray light sources simultaneously fabricated on a wafer covering the entire C-band," Optical Fiber Communication Conference, (OFC'2001), TuB4, Anaheim, CA (2001).

112. H. Oohashi, Y. Shibata, H. Ishii, Y. Kawaguchi, Y. Kondo, Y. Yoshikuni, and Y. Tohmori, "46.9-nm wavelength-selectable arrayed DFB lasers with integrated MMI coupler and SOA," Indium Phosphide and Related Materials (IPRM'2001), FB1-2, Nara, Japan (2001).

113. D. M. Adams, C. Gamache, R. Finlay, M. Cyr, K. M. Burt, J. Evans, E. Jamroz, S. Wallace, I. Woods, L. Doran, P. Ayliffe, D. Goodchild, and C. Rogers, "Module-packaged tunable laser and wavelength locker delivering 40 mW of fiber-coupled power on 34 channels," Electron. Lett., 37 (2001) 691–693.

114. S. Murata, I. Mito, and K. Kobayashi, "Over 720GHz (5.8nm) frequency tuning by a 1.5μm DBR laser with phase and Bragg wavelength control regions," Electron. Lett., 23 (1987) 403–405.

115. M. G. Young, U. Koren, B. I. Miller, M. A. Newkirk, M. Chien, M. Zirngibl, C. Dragone, B. Tell, H. M. Presby, and G. Raybon, "A 16 × 1 wavelength division multiplexer with integrated distributed Bragg reflector lasers and electroabsorption modulators," IEEE Photonic. Technol. Lett., 5 (1993) 908–910.

116. K. Kudo, P. Delansay, M. Yamaguchi, and M. Kitamura, "Tunable stair-guide (TSG) DBR lasers for single current continuous wavelength tuning," Electron. Lett., 31 (1995) 1843.

117. S. Lee, M. E. Heimbuch, D. Cohen, L. A. Coldren, and S. Denbaars, "Integration of semiconductor laser amplifiers with sampled grating tunable lasers for WDM applications," IEEE J. Sel. Topics in Quantum Electron., 3 (1997) 615–627.

118. F. Delorme, G. Alibert, C. Ougier, S. Slempkes, and H. Nakajima, "Sampled-grating DBR lasers with 101 wavelengths over 44 nm and optimized power variation for WDM applications," Electron. Lett., 34 (1998) 279–281.

119. A. Talneau, M. Allovon, N. Bouadma, S. Slempkes, A. Ougazzaden, and H. Nakajima, "Agile and fast switching monolithically integrated four wavelength selectable source at 1.55μm," IEEE Photonic. Technol. Lett., 11 (1999) 12–14.

120. L. J. P. Ketelsen, J. E. Johnson, D. A., Ackerman, L. Zhang, K. K. Kamath, M. S. Hybertsen, K. G. Glogovsky, M. W. Focht, W. A. Asous, C. L. Reynolds, C. W. Ebert, M. Park, C. W. Lentz, R. L. Hartman, and T. L. Koch, "2.5 Gb/s transmission over 680 km using a fully stabilized 20 channel DBR laser with monolithically integrated semiconductor optical amplifier, photodetector and electroabsorption modulator," Optical Fiber Communication Conference (OFC'2000), PD14, Baltimore, MD (2000).

121. F. Delorme, "1.5-μm tunable DBR lasers for WDM multiplex spare function," IEEE Photonic. Technol. Lett., 12 (2000) 621–623.

122. D. Vakhshoori, J.-H. Zhou, M. Jiang, M. Azimi, K. McCallion, C.-C. Lu, K. J. Knopp, J. Cai, P. D. Wang, P. Tayebati, H. Zhu, and P. Chen, "C-band tunable 6 mW vertical-cavity surface-emitting lasers," Optical Fiber Communication Conference (OFC'2000), PD13, Baltimore, MD (2000).

123. T. Day, "External-cavity tunable diode lasers for network deployment," Optical Fiber Communication Conference (OFC'2001), TuJ4, Anaheim, CA (2001).

124. O. Ishida, Y. Tada, and H. Ishii, "Tuning-current splitting network for three-section DBR lasers," Electron. Lett., 30 (1994) 241–242.

125. S. Kitamura, K. Komatsu, and M. Kitamura, "Very low power consumption semiconductor optical amplifier array," IEEE Photonic. Technol. Lett., 7 (1995) 147–149.

126. H. Hatakeyama, S. Kitamura, K. Hamamoto, M. Yamaguchi, and K. Komatsu, "Spot size converter integrated semiconductor optical amplifier gates with low power consumption," 2nd Optoelectronics Communications Conf. (OECC'97), 9C3-2, Seoul, Korea (1997).

127. S. Kitamura, H. Hatakeyama, T. Tamanuki, T. Sasaki, K. Komatsu, and M. Yamaguchi, "Angled-facet S-bend semiconductor optical amplifiers for high-gain and large extinction ratio," IEEE Photonic. Technol. Lett., 11 (1999) 788–790.

128. H. Hatakeyama, T. Tamanuki, K. Mori, T. Ae, T. Sasaki, and M. Yamaguchi, "Uniform and high performances of eight-channel bent waveguide SOA array for hybrid PICs," IEEE Photonic. Technol. Lett., 13 (2001) 418–420.

129. M. Janson, L. Lundgren, A.-C. Morner, M. Rask, B. Stoltz, M. Gustavsson, and L. Thylen, "Monolithically integrated 2×2 InGaAsP/InP laser amplifier gate switch arrays," Electron. Lett., 28 (1992) 776–778.

130. K. Hamamoto and K. Komatsu, "Insertion-loss-free 2×2 InGaAsP/InP optical switch fabricated using bandgap energy controlled selective MOVPE," Electron. Lett., 31 (1995) 1779–1780.

131. K. Hamamoto, T. Sasaki, T. Matsumoto, and K. Komatsu, "Insertion-loss-free 1×4 optical switch fabricated using bandgap-energy-controlled selective MOVPE," Electron. Lett., 32 (1996) 2265–2266.

132. T. Takeuchi, T. Sasaki, M. Hayashi, K. Hamamoto, K. Makita, K. Taguchi, and K. Komatsu, "A transceiver PIC for bidirectional optical communication fabricated by bandgap energy controlled selective MOVPE," IEEE Photonic. Technol. Lett., 8 (1996) 361–363.

Chapter 11 | Dry Etching Technology for Optical Devices

Stella W. Pang

Dept. of Electrical Engineering & Computer Science,
304, EECS Bldg., University of Michigan, 1301 Beal Ave.,
Ann Arbor, MI 48109-2122, USA

11.1. Introduction

In this chapter, dry etching technology used for the fabrication of high-performance optical devices is reviewed. Various plasma systems are used for dry etching. Among the different plasma system designs, high-density plasma systems with an inductively coupled plasma source provide better control and more flexibility because ion density and ion energy can be adjusted using separate power supplies. A number of dry etched optical devices are described, including high aspect ratio vertical mirrors in Si as optical switching arrays, dry etched mirrors in GaAs/AlGaAs for micro-cavities and triangular ring lasers, high reflectivity horizontal distributed Bragg reflector mirrors with air gaps between semiconductor layers, and photonic bandgap lasers. All these optical devices require precise control of etch depth, etch profile, and smooth morphology. In addition, the importance of developing low-damage dry etching technology is discussed and techniques to minimize surface damage are reviewed. With precisely controlled, low-damage dry etching technology, high-performance optical devices using dry etched components can be achieved.

WDM TECHNOLOGIES: ACTIVE
OPTICAL COMPONENTS
$35.00

11.2. Dry Etching Equipment

In dry etching, plasma systems are used to dissociate etch gases and generate reactive species. A typical system for dry etching consists of parallel plates with an rf power supply connected to the bottom electrode. The rf energy is capacitively coupled into the etch gases, resulting in accelerated electrons which collide with gases and excite the plasma. The reactive neutrals and ions generated in the plasma are used to etch away materials from a wafer placed on the bottom electrode. The ions are accelerated toward the wafer by an electric field that forms between the plasma and wafer due to the self-induced potential difference between plasma and electrode, which depends mainly on the rf power applied. The density of ions is also largely controlled by the rf power. However, in a parallel plate plasma system, ion density and ion energy cannot be controlled independently.

In addition to parallel plate plasma systems, high density plasma sources including helical resonator, helicon, electron cyclotron resonance (ECR), and inductively coupled plasma (ICP) sources have been used [1]. These systems provide denser plasmas with more efficient coupling of the applied power. Compared to conventional parallel plate systems with plasma density on the order of 10^9–10^{11} cm^{-3}, high density sources offer plasma density that is 10–100 times higher [2]. A denser plasma can provide higher etch rates. Because separate energy sources are used to generate the plasma and accelerate ions toward the wafer, ion density and ion energy could be controlled almost independently. Ion density has a large effect on the etch rate, while ion energy has a large effect on selectivity and plasma-induced damage. With separate power supplies for the plasma source and the stage, these high-density plasma systems allow greater control and more flexibility for dry etching compared to conventional parallel plate reactors.

A typical high-density plasma system with an ICP source is shown in Fig. 11.1. This ICP source consists of an rf coil as an antenna to inductively couple energy into the plasma, which acts as the secondary of an rf transformer. The inductive discharge is a surface process, where the secondary current is carried into the process gases, and it extends only a few skin depths into the gases. The rf current in the coil generates a magnetic flux that penetrates into the adjacent discharge. This time-varying magnetic flux induces a solenoidal rf electric field to accelerate the electrons that sustain the discharge. Large rf currents can be driven into the plasma in the inductive mode. Etch gases are injected into the system through a radial port in the top plate. The water-cooled ICP source is powered by an rf supply

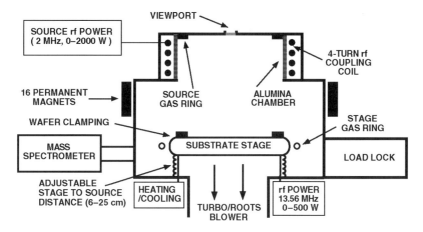

Fig. 11.1 Schematic of an inductively coupled plasma source with an rf powered stage.

at 2 MHz. The stage where the wafer is placed is powered by a supply at the rf frequency of 13.56 MHz. This power allows a self-induced dc bias ($|V_{dc}|$) to develop across the sheath or dark space between the plasma and the stage, which creates an electric field and accelerates the ions toward the wafer.

11.3. High Aspect Ratio Vertical Mirrors in Si

The fast-growing optical fiber communication network requires low cost, highly reliable optical devices [3]. For a typical moving fiber switch, the optical fiber physically moves between two positions to allow optical signal transmission from one channel to the other [4, 5]. Although this kind of switch is compact, power efficient, and has low insertion loss, it suffers from low switching speed and is difficult to integrate into a large array. An alternative is the moving light-beam switch [6, 7]. Optical beams are redirected into different stationary fibers by actuating reflective mirrors to achieve switching functions. This design has the drawbacks of large size and substantial power consumption due to mechanical moving compartments. It does, however, have the advantages of easier integration into a large matrix and faster switching speed.

The advances in micromachining technologies make it possible to precisely fabricate small movable mirrors. Hence, a moving beam switch array with micromachined mirrors can be realized with reduced size and power consumption. Vertical Si micromirrors have been demonstrated for optical

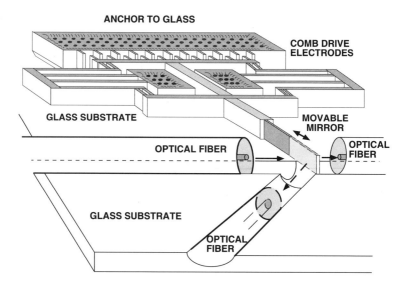

Fig. 11.2 Schematic of an optical switch with a movable mirror driven by electrostatic comb structures.

switching applications [8]. These high aspect ratio mirrors were fabricated by a deep etching and shallow diffusion process [9]. Tall Si micromirrors with comb drive were first dry etched, followed by a short B diffusion to fully convert the etched microstructures to p^{++} layer. Subsequently, an unmasked second dry etch, anodic bonding to glass substrates, and selective wet etch were used to release the microstructures. It is simpler to assemble and batch fabricate a large matrix using vertical mirrors as opposed to horizontal mirrors. The micromirrors were actuated by the electrostatic comb drives with minimal applied voltages.

Figure 11.2 shows the schematic of a micromachined mirror for optical switching applications. The device consists of a vertical micromirror driven by high aspect ratio electrostatic comb fingers. The mirror and comb are supported by either folded beams or serpentine (meander) springs to provide sufficient flexibility for large mirror displacement. The height of the mirror is >30 μm to accommodate the light beam diameter from a single-mode fiber ($\sim$10 μm) and to minimize optical loss due to diffraction. Because glass substrates can be recessed by dicing or etching to the required depth and width, fibers can be easily mounted onto the glass substrates. Depending on the mirror position, which is controlled by the applied voltage, input light can either be redirected to another stationary fiber through a 45° reflection or it can continue its path until it hits another micromirror.

11.3.1. CONTROLLING SIDEWALL SMOOTHNESS OF DRY ETCHED Si MICROMIRRORS

Smooth sidewall morphology and vertical profile are necessary to minimize light scattering and maintain collimated light beam when etched sidewall is used as a mirror surface. To maintain smooth and vertical sidewalls, it is important that the etch mask have smooth edges and high etch selectivity to prevent erosion during etching. Due to interference fringes along the photoresist sidewalls in conventional optical lithography, etch mask often has rippled edges. This causes rough sidewalls for the Si mirrors after etching. Baking the photoresist after exposure and development can reduce the photoresist edge roughness due to the rippled optical interference fringes. Because a large mask thickness is needed to prevent mask erosion for the dry etching of thick mirrors, a multiple layer resist scheme can be used to ensure high resolution and vertical profile in thick photoresist. For example, baking the top imaging photoresist at 160°C for 10 min can smooth out the interference fringes along the edges. Using atomic force microscopy to characterize sidewall roughness, the smoothness of dry etched Si mirrors improved from 30 to 19 nm after baking the photoresist mask [10, 11].

After Si micromirrors are formed by dry etching, the roughness along the sidewalls can be further reduced by thermally oxidizing the Si microstructures, followed by removal of SiO_2. Because Si oxidation rates are influenced by feature geometry and growth temperatures, roughness along Si sidewalls can be smoothed out with optimized oxidation conditions. The amount of available oxidizing species depends on surface geometry, so thermal oxidation rates are different for concave and convex surfaces [12, 13]. Typically, oxide grows faster on convex surfaces than on concave ones. Ripples along Si sidewalls can be considered to consist of continuous concave and convex surfaces. The different oxidation rates on convex vs. concave surfaces result in smoother morphology after the SiO_2 is removed. Figure 11.3(a) shows the atomic force micrograph of the dry etched Si micromirror sidewall with a roughness of 30 nm. The mirror was etched in a Cl_2 plasma with 100 W power to the plasma source and 100 W rf power to the stage at 3 mTorr for 2 h to a depth of 20 μm. After oxidation at 1175°C for 2 h followed by oxide removal, smoother sidewall topography was obtained, with a roughness of 5 nm as shown in Fig. 11.3(b).

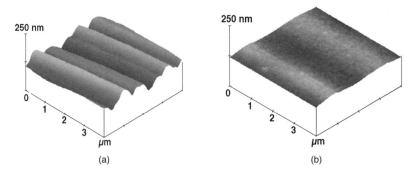

Fig. 11.3 Scanned etched Si mirror sidewalls using atomic force microscopy. (a) Si sidewall after dry etching and (b) smoother sidewall after oxidation at 1175°C for 2 h.

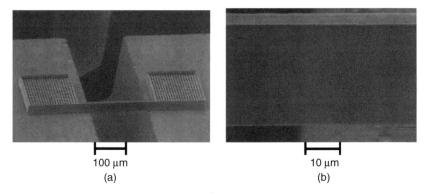

Fig. 11.4 Scanning electron micrographs of a 40-μm-thick, 2-μm-wide Si vertical micromirror bonded onto glass. Deep grooves in glass were formed by cutting 300 μm deep into the glass with a dicing saw.

This shows smooth Si sidewalls for vertical micromirrors can be obtained using smooth photoresist mask and thermal oxidation of Si.

11.3.2. MICROMACHINED VERTICAL Si MICROMIRRORS

Released Si micromirrors were fabricated by the deep etch–shallow diffusion process. The micromirrors were etched down to 40 μm deep and diffused with B for 6.5 h. After an unmasked second dry etching to remove the heavily doped layer near the bottom of the trenches, these micromirrors were bonded to Corning 7740 glass and released in ethylenediamine pyrocatechol for 4 h. Figure 11.4(a) shows a released, 40-μm-tall, 2-μm-wide vertical micromirror with both sides anchored on the glass substrate.

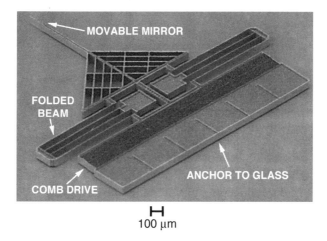

H
100 μm

Fig. 11.5   Scanning electron micrograph of a dry etched Si mirror that is 50 μm tall with 3-μm-wide folded suspension beams.

The deep grooves in glass, formed by cutting halfway through the glass wafer, were designed to increase the clearance to minimize light scattering from the glass substrate. A close-up view of the mirror surface is shown in Fig. 11.4(b). With the smooth sidewall and vertical profile, these micromirrors have high reflectivity >90%.

Figure 11.5 shows a laterally driven, dry etched Si micromirror fabricated by the deep etch–shallow diffusion process. This vertical micromirror is 50 μm tall and it can be moved by the electrostatic comb drives. The microresonator is 50 μm thick with 2-μm-wide interdigitated comb fingers, and the gaps between comb electrodes are 3 μm. The high aspect ratio Si dry etching technology allows large number of tall and narrow fingers with small gaps to be formed for large electrostatic force at low voltage to reduce power consumption for the comb drives. The movable Si micromirror is supported by 3-μm-wide, 800-μm-long folded suspension beams. The narrow suspension beams with large beam thickness increase the beam rigidity along the vertical direction while keeping the beams compliant along the lateral direction. This ensures that during the optical switching, stiffeners at the ends of these long folded beams are not in contact with the glass substrate due to gravity. These dry etched micromirrors in Si were used to form optical switching arrays. A mirror displacement of 34 μm and a resonant frequency of 987 Hz was achieved by a dc voltage of only 30 V. Typically, the mirror displacement increases with beam length and decreases with beam width.

11.4. Dry Etched Mirrors for Triangular Ring Lasers and Microcavities

Low damage dry etching technology can be used to fabricate optoelectronic devices including dry etched mirrors, triangular ring lasers, and microcavities [14–16]. Etched mirrors are necessary for optoelectronic circuit integration and for surface-emitting laser arrays and diode ring lasers. Reflectivity reported on wet etched or cleaved mirrors was only ~30%, and the surface morphology after wet etching was typically rough. Dry etching allows these mirrors to be fabricated with high reflectivities, vertical profiles, smooth sidewalls, and high packing density. In order to achieve high reflectivity on these mirrors, the surfaces of the etched sidewalls should be smooth because surface roughness can cause light scattering and reduce reflectivity. In addition, etch-induced defects have to be minimized because these defects can act as non-radiative recombination centers. A vertical etch profile is also crucial to ensure that all the light will be reflected with the same direction.

11.4.1. DRY ETCHED VERTICAL MIRRORS AND MICROCAVITIES

Dry etching conditions can be optimized to fabricate mirrors with high reflectivity. Dry etched mirrors for the $In_{0.20}Ga_{0.80}As/GaAs$-based waveguides were fabricated using a Cl_2/Ar plasma. In a low pressure plasma, the directionality of the etch species can enhance etch anisotropy so that vertical etch profile for the mirrors can be produced. Figure 11.6(a) shows an $In_{0.20}Ga_{0.80}As/GaAs$ mirror etched down to 2.8 μm deep. The mirrors were etched in a Cl_2/Ar plasma using 10% Cl_2 generated with 50 W source power and 70 W stage power at 0.5 mTorr. It can be seen that vertical profile and smooth sidewall morphology have been obtained. The etch rate was 0.11 μm/min. The selectivity of GaAs to the Ni etch mask was 42 under these etch conditions. These highly controllable etch profile and surface morphology are critical for high reflectivity mirrors. This dry etching technique has also been applied to the fabrication of microcavities in $In_{0.20}Ga_{0.80}As/GaAs$. Significant improvement in the threshold current density has been predicted for microcavity structures in which the width of the optical cavity is reduced to 0.3 μm [17]. Figure 11.6(b) shows 0.25-μm-wide microcavities etched to a depth of 2.8 μm with vertical profile and smooth surface. In addition, dry etching of $In_{0.90}Ga_{0.10}As_{0.22}P_{0.78}/InP$

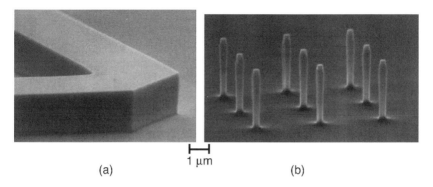

(a)   1 μm   (b)

Fig. 11.6 Scanning electron micrographs of dry etched $In_{0.20}Ga_{0.80}As/GaAs$ for (a) a total internal reflecting mirror and (b) 0.25-μm-wide microcavities with 2.8 μm etch depth, smooth sidewall, and vertical profile.

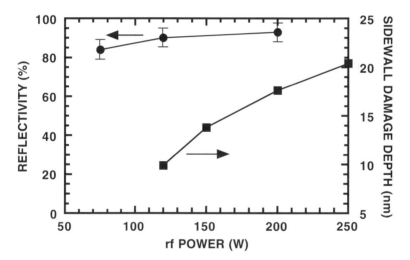

Fig. 11.7 Reflectivity (●) and sidewall damage depth (■) as a function of rf power. The samples were etched with rf power on the stage ranging from 70 to 200 W.

mirrors has also been achieved with vertical profile and smooth sidewalls. Fast etch rates in InP can be obtained using a Cl_2/Ar plasma at 25°C as long as the microwave power and the reactive species density are high.

The reflectivity of etched mirrors depends on the dry etch conditions. Figure 11.7 shows the changes in the mirror reflectivity as a function of rf power applied to the stage used for etching. The sidewall damage depths extracted from GaAs conducting wires etched at different rf power are also shown for comparison [18, 19]. The mirrors were etched with 10% Cl_2

in a Cl_2/Ar plasma generated at 0.5 mTorr with 50 W source power and rf power on the stage ranging from 70 to 200 W. In this rf power range, $|V_{dc}|$ increased from 120 to 380 V. Vertical profile and smooth sidewalls were achieved on all the mirrors etched within this rf power range. The etch depth was 2.8 μm. It can be seen that even though the sidewall damage depth increases substantially from 9.9 to 20.4 nm with rf power, there is no significant change in the mirror reflectivity. The reflectivity is 84% at 70 W rf power and 93% at 200 W rf power. Because this is within the range of measurement uncertainty, reflectivity is mostly independent of the rf power used during etching. These results suggest that the additional electrical degradation on the sidewalls induced at higher rf power does not cause any reduction in the mirror reflectivity. Similarly, varying the ion flux does not change the mirror reflectivity as long as the mirror sidewalls remain smooth and vertical.

However, the mirror profile affects the mirror reflectivity significantly. Changing the etch gases from 10 to 80% Cl_2 in a Cl_2/Ar plasma would change the etched $In_{0.20}Ga_{0.80}As$/GaAs mirrors from having vertical profile to slightly undercut profile with rougher sidewalls. The mirror reflectivity would be reduced from 90 to 47% due to the changes in the dry etched mirror sidewalls. The decrease in mirror reflectivity could be caused by light being scattered at different angles from the undercut mirror sidewalls or by surface roughness. These results show that the reflectivity of the mirrors is highly sensitive to the etch profile and sidewall roughness, but is not influenced significantly by the ion energy or ion flux used for etching. In order to produce a vertical profile with smooth sidewalls for the mirrors, dry etching techniques should be optimized to generate directional reactive species at low pressure to reduce scattering. High selectivity to the etch mask is also important to maintain smooth sidewalls and to avoid the formation of tapered profile due to mask erosion.

11.4.2. TRIANGULAR RING LASERS WITH DRY ETCHED MIRRORS

Triangular ring lasers with high-quality etched mirrors were fabricated. Figure 11.8 shows the scanning electron micrograph of a triangular ring laser dry etched with 120 W rf power on the stage. The exit mirror of the triangular ring laser was below the critical angle, while the other two mirrors were kept at above the critical angle for total internal reflection. The critical angle for this laser structure is 15.9°. The etched mirrors should have high

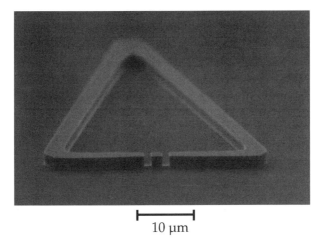

10 μm

Fig. 11.8 Scanning electron micrograph of an $In_{0.20}Ga_{0.80}As/GaAs$ triangular ring laser with dry etched mirrors. The laser was etched down to 2.8 μm deep using 120 W rf power on the stage.

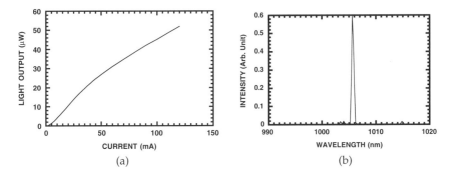

(a)   (b)

Fig. 11.9 (a) Light vs. current characteristics of the $In_{0.20}Ga_{0.80}As/GaAs$ triangular ring laser. (b) Emission spectrum of the triangular ring laser.

reflectance in order to minimize the threshold current for lasing. As can be seen from the micrograph, the etched triangular ring laser has smooth sidewalls and vertical profile. The entire structure was etched through to the bottom n-$Al_{0.30}Ga_{0.70}As$ cladding layer with a total etch depth of 2.8 μm.

Figure 11.9(a) shows the light vs. current characteristics of the triangular ring laser. The threshold current measured from these lasers was only 3.7 mA. The low threshold current obtained here shows that the etched mirrors are of very high quality. This is made possible by the high controllability of this dry etching technique using the Cl_2/Ar plasma.

Further improvement in the threshold current would be possible if the mirrors and the waveguides were etched separately in a two-step etch process. The waveguides only need to be etched through the top p-$Al_{0.30}Ga_{0.70}As$ cladding layer, but the mirrors have to be etched all the way to the bottom n-$Al_{0.30}Ga_{0.70}As$ cladding layer. A lower guiding loss is expected from waveguides fabricated with this two-step etch process because etch-induced defects in the active multiple quantum well layers can be minimized. As shown in Fig. 11.9(b), this ring laser has a single longitudinal mode with the emission peak at 1005.6 nm using current level 20× the threshold current.

11.5.   Nanostructures for Horizontal Distributed Bragg Reflector Mirrors

An important component of any laser system is the mirror. The mirrors define the cavity and allow light to leave the cavity. If the mirror reflectivity is too low, the gain medium must be pumped harder to reach threshold and higher current densities are required. A high threshold current increases the power dissipation, causing substrate heating and device failure. Initially, mirrors for ridge-waveguide lasers were made by cleaving the sample along crystal planes [20, 21]. The processing constraints due to cleaved mirrors are a major disadvantage of these lasers. Work has begun on etching the mirrors, which would allow for optoelectronic integration and novel semiconductor laser designs [22]. However, the reflectivity of etched mirrors is typically lower than cleaved mirrors. In a study on etched mirrors, it was found that the laser fabricated with cleaved mirrors had a lower threshold current than one with etched mirrors [23]. Using total internal reflection and optimized etch conditions, reflectivity of etched mirrors up to 93% has been reported [16].

Other alternatives have been investigated. For example, high reflectivity mirrors can be obtained by modifying the material structure such that a distributed Bragg reflector (DBR) mirror is formed. Distributed Bragg reflector mirrors fabricated using focused ion beam implantation [24] or etch and regrowth [25] rely on a very small change in the index of refraction, n, and several hundred periods are required to obtain high reflectivity. The passive mirror region can be quite long for these structures. However, one can obtain a high reflectivity mirror in a relatively small area by etching a periodic structure consisting of the semiconductor material and an air gap [26]. This structure will be referred to as a horizontal DBR mirror, as a counterpart to the vertical-cavity surface-emitting laser (VCSEL), which also uses DBR mirrors to obtain high reflectivity. The difference in refractive index between the semiconductor material and air is much larger than between different semiconductor layers, so fewer layers need to be used.

11.5.1. REQUIREMENTS FOR DISTRIBUTED BRAGG REFLECTOR MIRRORS

As an example, for an InP-air DBR and 1.5 μm wavelength light, $n_{InP} = 3.175$ and $n_{air} = 1$. The reflectivity for this DBR mirror is 96% with a two-period structure. Setting $n_{r1}d_1 = n_{r2}d_2 = \lambda/4$, requires that the thickness of the air layer be 388 nm and the thickness of the InP layer be 122 nm for $\lambda = 1.55$ μm. Typical ridge waveguide lasers are etched 1.5 to 3.0 μm deep [27]. For a 1.55-μm-tall waveguide, etching a 122-nm-wide InP stripe requires an aspect ratio of 13. Although this structure requires very good control in lithography and dry etching, the fabrication of this waveguide is feasible.

The two most important issues for the etching of the DBR mirrors are etch selectivity to the masking material and the etch profile. The etch should remove the upper cladding region, the active region, and part of the lower cladding layer. For a typical InP-based device structure [28], this requires an etch over 2.1 μm deep, while for a GaAs/AlGaAs waveguide laser an etch of 1.4 μm is sufficient [29]. The major part of the InP-based device in terms of layer thickness is the InP cladding layer. For the GaAs/AlGaAs laser, the cladding is AlGaAs, but the etch characteristics of GaAs are similar to those of AlGaAs.

The thickness of the etch mask is limited because high-resolution patterning is required for the DBR mirrors. For a 100-nm etch mask, a selectivity > 34 is required for InP. High-density plasma systems allow ion energy to be controlled by adjusting the rf power applied to the stage. Figure 11.10 shows the effect of increasing the rf power on the InP and Ni etch rates. The etch conditions were 250 W of microwave power to the ECR plasma source, Cl_2/Ar flow of 3/17 sccm, 1.5 mTorr chamber pressure, and stage temperature of 350°C. The $|V_{dc}|$ changed from 130 to 260 V as rf power was increased from 100 to 250 W. At rf power of 100 W, the InP etch rate is 879 nm/min and the Ni sputter rate is 6 nm/min. This yields a selectivity of 147. As the rf power increases, both etch rates increase. At 250 W rf power, the InP etch rate is 1.11 μm/min and the Ni sputter rate is 20 nm/min.

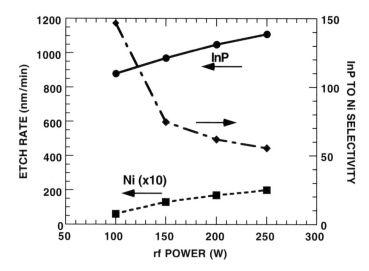

Fig. 11.10   Etch rates of InP and Ni increase with increasing rf power, and the selectivity of InP to Ni decreases.

However, this leads to a decrease in the selectivity of InP to Ni to 55. To etch the deep mirror structures, high selectivity must be maintained. At the low rf power of 100 W, where the selectivity was largest, the etched surface is somewhat rough due to inefficient removal of the $InCl_x$ etch products, leading to preferential etching and roughening of the surface. By increasing the ion energy, the physical component of the etch is increased. This tends to equalize the etch rates of the In and P compounds and a smooth surface is obtained, but the etch selectivity is reduced. Therefore, a tradeoff exists between how deep a mirror can be etched and how smooth the surface can be.

To obtain high selectivity of InP to the masking material, it was necessary to use high temperature to increase the $InCl_x$ volatility [30]. This can be done by heating the stage or by using a high-density plasma at low pressure. It has been shown previously that at higher temperatures the sidewall etches significantly [31]. The Cl_2 percentage in Ar was 50%, and there was 1 μm undercut in InP for a 4-μm-deep etch. The sidewall etch rate has been suppressed by reducing the Cl_2 percentage in Ar to 20%.

By using source power of 250 W with the stage at room temperature, the surface temperature increases due to the energy transferred to the wafer by the high ion density. An average etch rate of 785 nm/min is obtained with the stage at room temperature. The average etch rate given is for a 5 min etch. However, the etch rate for the first min is slower before the wafer is heated.

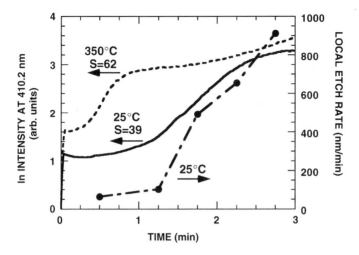

Fig. 11.11 In emission intensity at 410.2 nm increases sooner at higher stage temperature. Use of high stage temperature increases selectivity of InP to the Ni mask. The etch conditions were 250 W source power, 200 W rf power, Cl_2/Ar at 3/17 sccm, and 1.5 mTorr. S is the selectivity of InP to the Ni etch mask.

The results are shown in Fig. 11.11, where the In optical emission intensity at 410.2 nm increases with time for two different stage temperatures. At 25°C, the In emission intensity begins to increase substantially only after the first min of etching [31, 32].

Also shown in Fig. 11.11 is the local etch rate of InP at 25°C. This was obtained by etching InP samples for 1 to 3 min, in 30 s intervals, and calculating the etch rate for each 30 s interval. The local etch rate begins increasing after 1 min of etching, in agreement with the optical emission signal. At 350°C, the In emission intensity begins increasing after only 15 s of etching, signifying that the etch rate starts to increase shortly after the etching begins. By heating the stage to 350°C, a higher average etch rate of 1.05 μm/min is obtained. This also improves the average selectivity (S) of InP to Ni from 39 at 25°C to 62 at 350°C.

11.5.2. HORIZONTAL DISTRIBUTED BRAGG REFLECTOR MIRRORS IN InP

A λ/4 period InP mirror etched at 200 W rf power and 350°C is shown in Fig. 11.12. The gaps are etched 2.0 μm deep and the region outside the waveguide is etched 3.4 μm deep. The gap is etched through the active region into the lower cladding while the area outside the mirror stack is

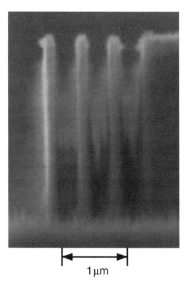

Fig. 11.12   Distributed Bragg reflector mirrors etched in InP with $\lambda/4$ period. The etch conditions were 250 W microwave power, 200 W rf power, Cl_2/Ar at 4/16 sccm, 1.0 mTorr pressure, and stage temperature of 350°C.

etched through the lower cladding. The reflectivity of the mirror should not be degraded because the gap is etched into the lower cladding. It can be seen that a slower etch rate and a rough surface occur near the waveguide and in between the mirror plates. This is related to the microloading effect, which is caused by the limitation of the transport of reactive species and etch products through high aspect ratio trenches.

One method to reduce the surface roughness is to increase the gap spacing. This allows more ions to enter the gap and assist in removal of etch products and redeposited materials. The mirrors can be designed as a $3\lambda/4$, $5\lambda/4$, or $7\lambda/4$ period grating structure. The reflectivity of the mirror will decrease slightly but the operation of the mirror is the same. Figure 11.13 shows how the microloading effect is reduced as the gap between mirrors is increased from 388 nm ($\lambda/4$) to 2.72 μm ($7\lambda/4$). At 1 mTorr, the normalized etch rate (obtained by dividing the etch rate of the gap by the etch rate obtained for an open area) increases from 53% to 100% as the gap spacing is increased. Additionally, Fig. 11.13 shows that the microloading effect can be reduced by decreasing the pressure. For the $3\lambda/4$ linewidth, at 0.7 mTorr the normalized etch rate is 96%, while at 1.3 mTorr the normalized etch rate is only 76%.

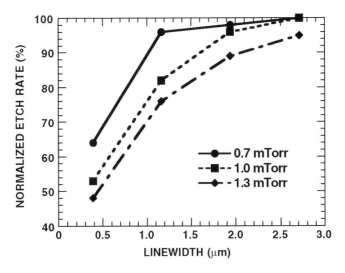

Fig. 11.13   Effects of linewidth and pressure on the etch rate of InP for various gap widths. The etch conditions were 250 W source power, 200 W rf power, Cl_2/Ar at 2/18 sccm, and 350°C.

11.5.3.   HORIZONTAL DISTRIBUTED BRAGG REFLECTOR MIRRORS IN GaAs

High selectivity and vertical profile could also be obtained for the GaAs-based DBR mirrors. The rf power was found to have a large effect on the selectivity of GaAs to the Ni etch mask. For 50 W rf power, the GaAs etch rate is 287 nm/min, the Ni sputter rate is 2.8 nm/min, and the selectivity is 102. At 200 W rf power, the selectivity decreases to 46, due to an increase in the Ni sputter rate to 8.6 nm/min. The $3\times$ increase in the Ni sputter rate is caused by the increasing ion energy with rf power. At 50 W rf power, $|V_{dc}|$ is only 40 V, while at 200 W rf power, $|V_{dc}|$ is 255 V. The etch conditions were 50 W microwave power, Cl_2/Ar flow at 4/16 sccm, 1 mTorr pressure, and 25°C stage temperature.

The Cl_2 percentage in Ar also has a large effect on the selectivity. In this case, the Ni sputter rate is essentially constant for varying Cl_2 percentage. However, the GaAs etch rate increases significantly. At 10% Cl_2 in Ar, the GaAs etch rate is only 217 nm/min and the selectivity is 46. At 30% Cl_2 in Ar, the GaAs etch rate is increased to 455 nm/min and the selectivity is increased to 105. The large increase in the GaAs etch rate is caused by the increased chlorine reactive species for GaAs etching; however, this causes

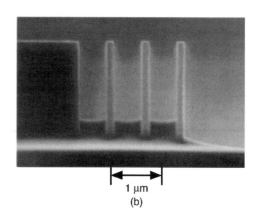

1 μm
(a)

1 μm
(b)

Fig. 11.14   Three period horizontal Bragg reflector mirror structures etched in GaAs
(a) with $\lambda/4$ grating and (b) with $3\lambda/4$ grating. The etch conditions were 50 W source
power, 100 W rf power, Cl_2/Ar at 4/16 sccm, and 1 mTorr.

the profile to become more undercut. For etching the DBR mirrors, the Cl_2
percentage was kept at 20% and the profile remained vertical.

Figure 11.14(a) shows a 1.5-μm-deep, 59-nm-wide, $\lambda/4$ GaAs mirror
etched with 20% Cl_2 and 100 W rf power. The etch depth of the 213-nm-
wide gap was 64% of the open area. Figure 11.14(b) shows $3\lambda/4$ GaAs
mirror that was 177 nm wide and had 638-nm-wide gaps. The etch depth
was 1.8 μm and the etch rate inside the gaps was 75% of the open area.
As in the case of InP mirror stacks, the gaps between mirrors would be
etched into the lower cladding and the region outside the waveguide would
be etched through the lower cladding to a depth of 2.2 μm. All the dry
etched DBR mirrors have vertical profile and smooth sidewalls to ensure
high reflectivity.

11.6.   Photonic Bandgap Lasers

Typical semiconductor lasers have broad spontaneous emission spectra that
only fractionally contribute to the stimulated emission. As a result, many
generated photons are lost to non-lasing modes. The photonic bandgap
(PBG) laser seeks to increase the spontaneous emission in the lasing mode

by suppressing emission in the non-lasing modes with etched holes in the laser structure to modify the photonic bandgap [33, 34]. Semiconductor lasers often have high thresholds due to the large linewidth of spontaneous emission spectra and the resulting low fraction of emission in the lasing mode. By reducing the linewidth of the spontaneous emission and designing the structure so that the emission coincides with the lasing mode, zero threshold lasers are possible.

11.6.1. DRY ETCHING TECHNOLOGY TO FORM PHOTONIC BANDGAP

These PBG lasers require very demanding dry etch capabilities. The sidewalls must be vertical with no undercutting. The aspect-ratio-dependent etching effects and etch-induced damage must be minimized. Previously, optically pumped PBG lasers have been fabricated but surface recombination from etch-induced damage has prevented demonstration of electrically pumped lasers [35, 36]. For emission at 0.94 μm, the etched holes should be arranged in a honeycomb pattern with diameters of 190 nm and a pitch of 300 nm as shown in Fig. 11.15 [37, 38]. The etched holes also have to be deep enough to penetrate the active region of the vertical-cavity surface-emitting laser, typically ∼1 μm below the surface. High-precision

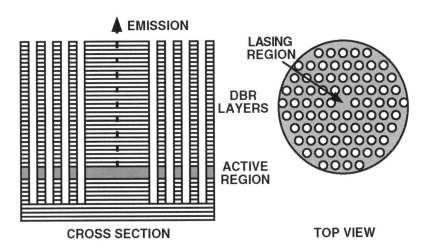

EMISSION

LASING REGION

DBR LAYERS

ACTIVE REGION

CROSS SECTION   TOP VIEW

Fig. 11.15   Illustration of the photonic bandgap structure showing the etched holes in the vertical-cavity surface-emitting laser.

dry etching techniques are needed to fabricate high efficiency, low threshold lasers.

The critical fabrication technology of the PBG laser is the etching of the PBG holes. For light emission at 940 nm, the hole dimensions were 200 nm diameter with 300 nm pitch. These PBG structures were defined by electron beam lithography and a 150-nm-thick Ni etch mask was lifted off. The device heterostructure was grown by metal-organic vapor phase epitaxy. It consists of an n^+ contact layer, an n-type lower 29 pair of GaAs/Al$_{0.2}$Ga$_{0.8}$As DBR mirrors, an undoped λ cavity region with two 7 nm pseudomorphic In$_{0.15}$Ga$_{0.85}$As wells in the middle and p-type AlGaAs and contact layers on the top. n- and p-type Al$_{0.96}$Ga$_{0.04}$As layers are also inserted on the respective sides of the cavity region for eventual lateral wet oxidation during device processing. The Al$_x$O$_y$ regions created by wet oxidation help to funnel the charge carriers more efficiently into the center of the PBG region. The etch conditions were 150 W source power, 30 W stage power, Cl$_2$ flowing at 5 sccm, and a chamber pressure of 0.12 mTorr. The $|V_{dc}|$ was 45 V and the etch depth in the holes was 1.57 μm. The deep etch goes through the entire cavity region and well into the bottom DBR to ensure a good overlap with the optical field. Top and cross section views of the etched PBG holes are shown in Fig. 11.16. In (a), the area in the middle without holes is protected from etching and serves as a vertical laser cavity and the emission point of the device. The small lateral dimensions and presence of the PBG holes reduce the number of modes supported by the device. Due to the high aspect ratio, the etch rate in the holes was limited

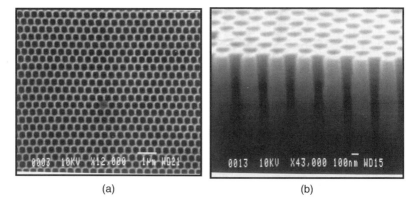

(a)   (b)

Fig. 11.16 Micrographs of (a) the etch mask prior to etching with the cavity region in the center and (b) a cross section of a photonic bandgap structure after dry etching. The holes are 200 nm in diameter and are etched to a depth of 1.57 μm.

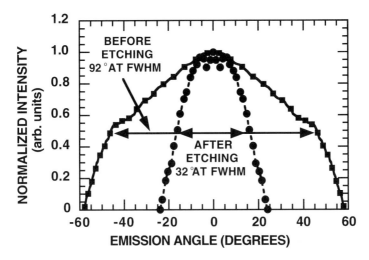

Fig. 11.17   Improved collimation of light emitted from etched photonic bandgap structure. Beam dispersion decreased from 92° to 32°, proving effectiveness of photonic bandgap structure at suppressing unwanted spontaneous emission modes.

by the etch species and etch product transport. This effect was reduced by the extreme low pressure used during etching.

11.6.2.   EMISSION CHARACTERISTICS OF PHOTONIC BANDGAP LASERS

Figure 11.17 shows the far field radiation patterns, with and without PBG formation. The devices were tested in the pulsed mode via probe contacts. The threshold current is 300 μA and the maximum measured output power is 14.4 μW. The peak emission in the spectral output at 931 nm has a linewidth of 0.8 nm. The samples were measured before and after etching of the PBG holes to determine the effectiveness of the structure at suppressing unwanted emission modes. Prior to etching, the emission was found to be Lambertian, as expected for light-emitting diodes [39], and the emission angle at full-width-half-maximum (FWHM) was 92°. After etching the PBG holes into the sample, the emission angle at FWHM was reduced to 32°, thus improving the directionality of the spontaneously emitted light and confirming that lasing originates from the defect mode. As the light is confined by the microcavity formed by etching, the number of supported modes in the structure was reduced and the associated collimation of emitted light was observed.

11.7.   Effects of Dry Etching on Optical Properties

Dry-etch-induced damage has been shown to cause changes in material that can degrade device characteristics [40–43]. The demands for integration of electronic and optoelectronic devices on the same wafer have made it important to understand how the etch-induced damage can affect the electrical and optical properties of the material. Characterization of etch-induced damage to the optical properties of AlGaAs/GaAs and InP/InGaAs have shown that defects can propagate deep into the material by a combination of ion channeling and diffusion. Etch-induced damage has been shown to affect the electrical properties of materials as well by changes in mobility, sidewall damage depth, contact resistance, and Schottky diode ideality factor. Previous work has also shown that the use of higher stage power during etching can affect the residual damage depth and distribution of the etch-induced defects [19, 44]. Furthermore, optical characteristics of near-surface quantum wells for epitaxial layers grown on InP substrates have been shown to be adversely affected by etch-induced damage to a greater extent than epitaxial layers grown on GaAs substrates [45]. Therefore, it is important to develop low-damage dry etching technology for electrical and optical devices.

11.7.1.   DECREASED PHOTOLUMINESCENCE
DUE TO DRY ETCHING

Changes in the photoluminescence (PL) intensity and conductance of quantum well structures can be used to study dry-etch-induced damage in semiconductors. For the nanostructures used in the study of dry-etch-induced damage, 160-μm-long conducting wires with lateral dimensions down to 120 nm and 500 μm $\times$ 500 μm gratings with lines and spaces of 110 and 240 nm, respectively, were used. These wires and gratings were defined by electron beam lithography and dry etched using the ICP plasma system to a depth of 400 nm. Typical etch conditions were 150 W source power and 50 W stage power with Cl_2 flowing at 6 sccm. The chamber pressure was 0.15 mTorr, the $|V_{dc}|$ was 80 V, and an etch rate of 250 nm/min was achieved. It has been shown that etching at low pressures in a Cl_2 plasma yields vertical profiles, smooth surfaces, and high etch rates for gratings and wires with small dimensions and high aspect ratio [46].

For the PL measurements, gratings were used instead of single wires to improve the intensity of the PL signal. The $In_{0.15}Ga_{0.85}As$ PL signals

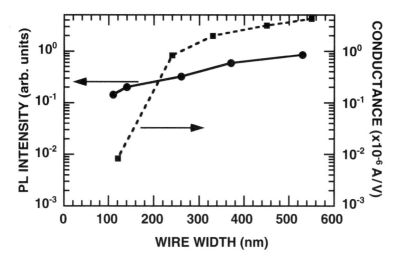

Fig. 11.18 Comparison of photoluminescence intensity and conductivity for wires of different widths showing different cutoff width for optical and electrical signals.

at 925 nm were normalized to the GaAs substrate signal at 820 nm and the intensities were all divided by a fill factor for each grating so that the active area was the same for all gratings. The PL signal, as a function of etch conditions, was found to be independent of the illuminating laser intensity. Figure 11.18 shows that both the PL intensity and conductance decreased with wire width. It can be seen that the conductance of the wires degrades more quickly for smaller wire width as compared to the PL intensity [47]. The wires became effectively non-conducting when the wire width decreased to 120 nm. The conductance decreased from 4.3×10^{-6} to $8.3 \times 10^{-9} \ \Omega^{-1}$ as the wire dimension was reduced from 550 to 120 nm. However, a measurable PL signal was still detected from gratings with linewidth of 110 nm. The PL intensity decreased by 83% as the linewidth was varied from 530 nm to 110 nm, and the gratings were still producing an optical signal at these sizes. The extracted cutoff width for the optical signal was 33 nm, while the cutoff width for the electrical signal was 136 nm. These results indicate a difference in how dry-etch-induced damage can affect the electrical and optical properties of an InGaAs quantum well as evidenced by the different cutoff widths of etched wires and gratings. While the etch-induced damage may render the material non-conducting, the quantum well could still be sufficiently intact to allow for radiative recombination of the carriers generated by the incident laser in the PL measurement to allow the optical signal to be detected.

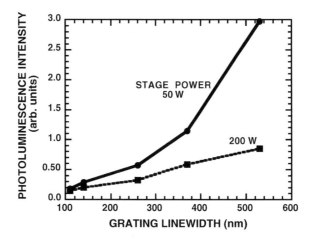

Changes in stage power during etching cause a degradation in PL intensity from etched gratings. The stage power was varied between 50 and 200 W.

The effect of the stage power used during etching on the optical characteristics of etched gratings is shown in Fig. 11.19. The rf power on the stage was varied from 50 to 200 W to control the ion energy bombarding the devices. The PL intensity from the gratings etched with 200 W stage power were significantly lower than those etched with 50 W. The PL intensity from the gratings with 530 nm linewidth decreased by 70% and the 110 nm lines decreased by 22% as the stage power was increased from 50 to 200 W. The $|V_{dc}|$ varied from 80 to 270 V in this rf power range. The degradations in the PL signal with increasing stage power were probably related to the increase in ion energy, which could increase the etch-induced damage along the sidewalls and reduce the luminescence from the quantum well [48].

11.7.2. DAMAGE REMOVAL BY PLASMA PASSIVATION

It is desirable to develop techniques to remove or minimize dry-etch-induced damage in devices. Low-energy chlorine species, generated in a high-density plasma system with no stage bias, have been shown to passivate the etched surface and sidewalls of GaAs and InGaAs structures [49, 50]. No etching was observed as the materials were protected by either native or plasma-grown oxides. To study the improvement on the optical properties using plasma passivation, gratings were defined by electron beam lithography and a Ti/Ni (30/50 nm) etch mask was lifted off. After dry

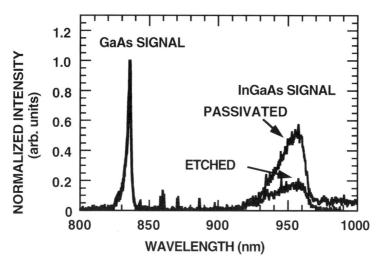

Fig. 11.20   Improved photoluminescence intensity from etched InGaAs quantum well gratings after 10 min Cl_2 passivation.

etching, the mask was stripped in buffered HF and a plasma oxide was grown for 1 min. The samples were then passivated in a low-energy plasma using 150 W source power, Cl_2 flowing at 20 sccm, and a chamber pressure of 2 mTorr. No stage power was used during passivation and the $|V_{dc}|$ was 0 V. Therefore, the reactive Cl species generated in the plasma will have very low energy.

As Fig. 11.20 shows, PL measurements were performed and the signal from the GaAs substrate was used to normalize the signals from the wells in different samples. The signal from the $In_{0.15}Ga_{0.85}As$ well showed an improvement of 2.7 times as compared to the etched sample. This indicates that the Cl_2 passivation techniques can be used to improve the optical signals from materials and can repair damage along the sidewalls of structures.

11.8.   Summary

Optical devices in Si or III-V semiconductors have been fabricated using dry etching technology. In order to achieve the desirable optical properties, the dry etched devices should have precisely controlled etch depth, etch profile, smooth morphology, and low surface defects. In many cases, these dry etched optical components consist of high aspect ratio ($>20:1$) structures with very small dimensions (<100 nm). The multiple requirements for high-performance optical devices make it very challenging to develop the needed dry etching technology.

High aspect ratio Si micromirrors with folded suspension beams for optical switching were fabricated using dry etching technology. These vertical Si mirrors are 50 μm tall and they have high reflectivity because the etched surface has smooth morphology. These micromachined Si mirrors can be actuated using the electrostatic comb drives. The high aspect ratio comb resonators are supported by the folded beams that are 50 μm thick, 3 μm wide, and 800 μm long. These optical switching arrays can be operated at low power, achieving a mirror displacement of 34 μm at 30 V.

High reflectivity micromirrors and microcavities in $In_{0.20}Ga_{0.80}As/GaAs$ have been dry etched using a Cl_2/Ar plasma. These mirrors and cavities have dimensions down to 0.25 μm and a depth of 2.8 μm. Mirror reflectivity up to 93% is obtained, independent of the ion energy or ion flux used during etching, as long as smooth sidewalls and vertical profile are maintained. These high-quality etched mirrors were used to form total internal reflection in the triangular ring lasers. The high reflectivity etched mirrors resulted in low threshold current of only 3.7 mA for the triangular ring lasers.

Nanostructures have been dry etched in GaAs and InP to form horizontal DBR mirrors. With alternating layers of semiconductor and air gap, reflectivity >95% can be obtained using only a two-period structure. For $\lambda/4$ DBR mirrors, mirror width should be 59 nm for GaAs and 122 nm for InP. To clear the top cladding layer and the active region of the DBR mirrors, etch depths up to 2 μm are required. Therefore, dry etching technology for very high aspect ratio DBR mirrors is needed. Horizontal DBR mirrors in GaAs and InP have been demonstrated. These high aspect ratio mirrors have vertical profile and smooth sidewalls for high reflectivity applications.

Photonic bandgap crystal has been fabricated using dry etching to form surface-emitting PBG lasers. By suppressing emission from the non-lasing modes, PBG lasers have the advantage of increased spontaneous emission into the lasing mode with very low threshold. For light emission at 940 nm, PBG crystal with 200 nm diameter holes and 300 nm pitch is needed. Dry etching technology for vertical holes with low defects in $GaAs/Al_{0.2}Ga_{0.8}As$ DBR mirrors and pseudomorphic $In_{0.15}Ga_{0.85}As$ wells has been developed. In order to get through the entire cavity region and into the bottom DBR mirror to ensure a good overlap with the optical field, an etch depth >1 μm is needed. These high aspect ratio PBG structures are dry etched in a Cl_2 plasma with pressure down to 0.1 mTorr to enhance the transport of the etch species and etch products through these deep, narrow trenches. The dry etched PBG lasers show low threshold current and improved collimation of the emitted light.

Most importantly, for the optical devices to be functional and provide the desired performance, low-damage dry etching technology should be used to minimize surface defects. To avoid dry-etching-induced damage, low-energy ions and uniform etching are needed, while impurities from the gases, etch mask, and plasma system should be avoided. In addition, low-energy Cl species have been shown to passivate surface defects and improve optical device performance. With optimized dry etch conditions, high-performance optical devices can be obtained.

ACKNOWLEDGMENTS

The results and materials presented in this chapter are based on the research and development carried out at the University of Michigan. The author would like to thank the collaborators, E. W. Berg, K. K. Ko, W. H. Juan, and S. Thomas III, for their significant contributions to the work reviewed here.

References

1. M. A. Lieberman and A. J. Lichtenberg, Principles of Plasma Discharges and Materials Processing (John Wiley and Sons, New York, 1994).
2. D. L. Book, NRL Plasma Formulary (Naval Research Laboratory, Washington, DC, 1987).
3. R. L. Carroll, "Optical architecture and interface lightguide unit for fiber-to-the-home feature of the AT&T SLC series 5 carrier system," IEEE J. Lightwave Technol., 7 (1989) 1727.
4. S. Nagaoka and Y. Suzuki, "Compact optomechanical switches and their applications in optical communication and testing systems," Proc. IEEE Micro Electro Mechanical Systems (MEMS'97), Nagoya, Japan (Feb. 1997) 366–371.
5. E. Ollier, P. Labeye, and P. Mottier, "A micro-opto-mechanical switch for optical fiber networks," Proc. SPIE, 2401 (1995) 116.
6. M. F. Dautartas, A. M. Benzoni, Y. C. Chen, G. E. Blonder, B. H. Johnson, C. R. Paola, E. Rice, and Y.-H. Wong, "A silicon-based moving-mirror optical switch," IEEE J. Lightwave Technol., 10 (1992) 1078.
7. H. Toshiyoshi and H. Fujita, "Electrostatic micro torsion mirrors for an optical switch matrix," IEEE J. Microelectromech. Syst., 5 (1996) 231.

8. W. H. Juan and S. W. Pang, "High aspect ratio Si vertical micromirror arrays for optical switching," IEEE J. Microelectromech. Syst., 7 (1998) 207.

9. W. H. Juan and S. W. Pang, "Released Si microstructures fabricated by deep etching and shallow diffusion," IEEE J. Microelectromech. Syst., 5 (1996) 18.

10. W. H. Juan and S. W. Pang, "Controlling sidewall smoothness for micromachined Si mirrors and lenses," J. Vac. Sci. Technol. B, 14 (1996) 4080.

11. W. H. Juan and S. W. Pang, "High reflectivity micromirrors fabricated by coating of high aspect ratio Si sidewalls," J. Vac. Sci. Technol. B, 15 (1997) 2661.

12. R. B. Marcus and T. T. Sheng, "The oxidation of shaped silicon surfaces," J. Electrochem. Soc., 129 (1982) 1278.

13. D. B. Kao, J. P. McVittie, W. D. Nix, and K. C. Saraswat, "Two-dimensional thermal oxidation of silicon-II. Modeling stress effects in wet oxides," IEEE Trans. Electron. Dev., 35 (1988) 25.

14. T. Yuasa, T. Yamada, K. Asakawa, S. Sugata, M. Ishii, and M. Uchida, "Short cavity GaAs/AlGaAs multiquantum well lasers by dry etching," Appl. Phys. Lett., 49 (1986) 1007.

15. P. Buchmann, H. P. Dietrich, G. Sasso, and P. Vettiger, "Chemically assisted ion beam etching process for high quality laser mirrors," Microelectron. Engin., 9 (1989) 485.

16. K. K. Ko, K. Kamath, O. Zia, E. Berg, S. W. Pang, and P. Bhattacharya, "Fabrication of dry etched mirrors for $In_{0.2}Ga_{0.8}As/GaAs$ multiple quantum well waveguides using an electron cyclotron resonance source," J. Vac. Sci. Technol. B, 13 (1995) 2709.

17. F. M. Matinaga, A. Karisson, S. Machida, Y. Yamamoto, T. Suzuki, Y. Kadota, and M. Ikeda, "Low-threshold operation of hemispherical microcavity single-quantum-well lasers at 4 K," Appl. Phys. Lett., 62 (1993) 443.

18. K. K. Ko and S. W. Pang, "Surface damage on GaAs etched using a multipolar electron cyclotron resonance source," J. Electrochem. Soc., 141 (1994) 255.

19. K. K. Ko, S. W. Pang, T. Brock, M. W. Cole, and L. M. Casas, "Evaluation of surface damage on GaAs and GaInAs etched with an electron cyclotron resonance source," J. Vac. Sci. Technol. B, 12 (1994) 3382.

20. C. Thirstrup, S. W. Pang, O. Albrektsen, and J. Hanberg, "Effects of reactive ion etching on optical and electro-optical properties of GaInAs/InP based strip-loaded waveguides," J. Vac. Sci. Technol. B, 11 (1993) 1214.

21. M. R. Krames, N. Holonyak, Jr., J. E. Epler, and H. P. Schweizer, "Buried-oxide ridge-waveguide in InAlAs-InP-InGaAsP ($\lambda \sim 1.3\mu m$) quantum well heterostructure laser diodes," Appl. Phys. Lett., 64 (1994) 2821.

22. J. J. Liang and J. M. Ballantyne, "Self-aligned dry-etching process for waveguide diode ring lasers," J. Vac. Sci. Technol. B, 12 (1994) 2929.

23. N. C. Frateschi, M. Y. Jow, P. D. Dapkus, and A. F. J. Levi, "InGaAs/GaAs quantum well lasers with dry-etched mirror passivated by vacuum atomic layer epitaxy," Appl. Phys. Lett., 65 (1994) 1748.

24. A. J. Steckl, P. Chen, H. E. Jackson, A. G. Choo, X. Cao, J. T. Boyd, and M. Kumar, "Review of focused ion beam implantation mixing for the fabrication of GaAs-based optoelectronic devices," J. Vac. Sci. Technol. B, 13 (1995) 2570.

25. M. Takahashi, Y. Michitsuji, M. Yoshimura, Y. Yamazoe, H. Nishizawa, and T. Sugimoto, "Narrow spectral linewidth 1.5μm GaInAsP/InP distributed bragg reflector (DBR) lasers," IEEE J. Quantum. Electron., 25 (1989) 1280.

26. S. Thomas, III and S. W. Pang, "Dry etching of horizontal distributed Bragg reflector mirrors for waveguide lasers," J. Vac. Sci. Technol. B, 14 (1996) 4119.

27. M. R. Krames, N. Holonyak, Jr., J. E. Epler, and H. P. Schweizer, "Buried-oxide ridge-waveguide in InAlAs-InP-InGaAsP ($\lambda \sim 1.3$μm) quantum well heterostructure laser diodes," Appl. Phys. Lett., 64 (1994) 2821.

28. A. L. Gutierrez-Aitken, H. Yoon, and P. Bhattacharya, "High-speed InP-based strained MQW ridge waveguide laser," Proc. Inter. Conf. on InP and Rel. Mat. (1995) 476.

29. D. Hofstetter, H. P. Zappe, and J. E. Epler, "Ridge waveguide DBR laser with nonabsorbing grating and transparent integrated waveguide," Electron. Lett., 31 (1995) 980.

30. N. L. DeMeo, J. P. Donnelly, F. J. O'Donnell, M. W. Geis, and K. J. O'Connor, "Low power ion-beam-assisted etching of indium phosphide," Nuc. Instr. Meth. Phys. Res. B, 7 (1985) 814.

31. S. Thomas III, E. W. Berg, and S. W. Pang, "*In situ* fiber optic thermometry of wafer surface etched with an electron cyclotron resonance source," J. Vac. Sci. Technol. B, 14 (1996) 1807.

32. S. Thomas III, H. H. Chen, C. K. Hanish, J. W. Grizzle, and S. W. Pang, "Minimized response time of optical emission and mass spectrometric signals for optimized endpoint detection," J. Vac. Sci. Technol. B, 14 (1996) 2531.

33. E. Yablonovitch, "Photonic band-gap structures," J. Opt. Soc. Am. B, 10 (1993) 283.

34. J. D. Joannopoulos, R. B. Meade, and J. N. Winn, Photonic Crystals (Princeton University Press, New York, 1995).

35. O. Painter, R. K. Lee, A. Scherer, A. Yariv, J. D. O'Brien, P. D. Dapkus, and I. Kim, "Two-dimensional photonic band-gap defect mode laser," Science, 284 (1999) 1819.

36. B. G. Levi, "Lasing demonstrated in tiny cavities made with photonic crystals," Physics Today, 52 (1999) 20.

37. W. D. Zhou, J. Sabarinathan, B. Kochman, E. W. Berg, O. Qasaimeh, S. W. Pang, and P. Bhattacharya, "Electrically injected single-defect photonic

bandgap surface-emitting laser at room temperature," Electron. Lett., 36 (2000) 1541.

38. W. D. Zhou, J. Sabarinathan, P. Bhattacharya, B. Kochman, E. W. Berg, P. C. Yu, and S. W. Pang, "Characteristics of a photonic bandgap single defect microcavity electroluminescent device," IEEE J. Quantum Electron., 37 (2001) 1153.

39. P. Bhattacharya, Semiconductor Optoelectronic Devices (Prentice Hall, New Jersey, 1994), 267.

40. S. W. Pang, "Surface damage on GaAs induced by reactive ion etching and sputter etching," J. Electrochem. Soc., 133 (1986) 784.

41. E. L. Hu, C. H. Chen, and D. L. Green, "Low-energy ion damage in semiconductors: a progress report," J. Vac. Sci. Technol. B, 14 (1996) 3632.

42. S. Murad, M. Rahman, N. Johnson, S. Thoms, S. P. Beaumont, and C. D. W. Wilkinson, "Dry etching damage in III-V semiconductors," J. Vac. Sci. Technol. B, 14 (1996) 3658.

43. E. W. Berg and S. W. Pang, "Time dependence of etch induced damage generated by an electron cyclotron resonance source," J. Vac. Sci. Technol. B, 15 (1997) 2643.

44. S. Thomas and S. W. Pang, "Dependence of contact resistivity and Schottky diode characteristics on dry etching induced damage of GaInAs," J. Vac. Sci. Technol. B, 12 (1994) 2941.

45. C. H. Chen, D. G. Yu, E. L. Hu, and P. M. Petroff, "Photoluminescence studies on radiation enhanced diffusion of dry-etch damage in GaAs and InP materials," J. Vac. Sci. Technol. B, 14 (1996) 3684.

46. E. W. Berg and S. W. Pang, "Low pressure etching of nanostructures and via holes using an inductively coupled plasma system," J. Electrochem. Soc., 146 (1999) 775.

47. E. W. Berg and S. W. Pang, "Electrical and optical characteristics of etch induced damage in InGaAs," J. Vac. Sci. Technol. B, 16 (1998) 3359.

48. M. Rahman, N. P. Johnson, M. A. Foad, A. R. Long, M. C. Holland, and C. D. W. Wilkinson, "Model for conductance in dry-etch damaged n-GaAs structures," Appl. Phys. Lett., 61 (1992) 2335.

49. K. K. Ko and S. W. Pang, "Plasma passivation of etch-induced surface damage on GaAs," J. Vac. Sci. Technol. B, 13 (1995) 2376.

50. E. W. Berg and S. W. Pang, "Cl_2 passivation of etch induced damage in GaAs and InGaAs with an inductively coupled plasma source," J. Vac. Sci. Technol. B, 17 (1999) 2745.

Part 5 | Optical Packaging Technologies

Chapter 12 | Optical Packaging/Module Technologies: Design Methodologies

Achyut K. Dutta

Achyut K. Dutta

*Fujitsu Compound Semiconductors Inc., 2355 Zanker Road,
San Jose, CA 95131, USA*

Masahiro Kobayashi

*Fujitsu Quantum Devices, Kokubo Kogyo Danchi, Showa-cho,
Nakakoma-gun, Yamanashi-ken 409-3883, Japan*

12.1. Introduction

Today's fiber optics components being used in WDM networks are expensive, with most of the cost coming from their packaging/assembly technologies. To achieve higher market penetration and greater volumes, costs must come down further. Technology developments are underway to lower the cost associated with the packaging and assembly.

Packaging/assembly technologies are also one of the keys to assuring good performance and reliable field of application for the optical components. These packaging technologies for optical components are varied depending on their area of application. For example, packaging technologies required for subscriber networks such as fiber-to-the home (FTTH) or fiber-to-the curbs (FTTC) are different from those being used in long-haul transmission systems. The components in the subscriber networks need to be inexpensive, highly stable, and exhibit low loss. These qualities depend not only on the chip fabrication and design technologies, but also on the packaging technology. Lack of any of these would make it difficult to realize high performance and cost-effective optical components.

12.1.1. BACKGROUND

Fiber optics components research has been going on since the early 1970s, following the discovery of the glass fiber [1–3]. However, practical implementation of optical components, especially light-emitting diode (LED)

565

based components [4], started in the late 1970s. After confirming reliable operation of the laser diode (LD) in mid 1980, implementation of the LD gradually focused on use in the optical networks. Optical packaging technologies have been changed since its implementation. If we look back at the pre-LD packaging technologies, the components packaging at the initial stage was mainly based on the metallic, and alignment technique was mainly active alignment. The uses of optical components in real systems were also limited at that time. However, with increasing demands of the optical components for WDM network applications, the trends in packaging diverted from expensive to inexpensive by replacing the active alignment and metallic package with passive alignment and flat-type (plastic) package. This is especially true for the components required for a subscriber network, where cost-effective components are necessary. So far, small changes are noticeable in the optical packaging, especially used in long-haul application.

Packaging is much more complex for photonics than for electronic ICs. Packaging-related issues are fiber-chip coupling, which leads to the use of tapers and lenses, the influences of reflections and the accompanying use of isolators and temperature sensitivity, requiring thermoelectric coolers (TEC) and thermistors. It is important to note that these packaging issues form, on one hand, the major economic incentive for integration, but on the other, a technical obstacle against integration. Therefore, it is highly desirable to design the packaging that is simple and cost effective.

This chapter reviews the design methodologies required for optical package design for future photonic components. Each design criteria will be explained with its practical implementation in packages as the examples. The next chapter will focus on more recent optical packaging trends with the examples.

12.2. Package Types

Optical packages are mainly divided into two types: coaxial type package and butterfly type package. The former is generally uncooled and the latter is cooled or uncooled based on the application area. Figure 12.1 shows examples of the two types of packages frequently used in telecommunication industries. Each package type could be with or without fiber attached, and the fiber could be multimode or single-mode type fiber. As multimode fiber is not used for WDM systems, optical package with multimode fiber will not be considered in this chapter. The area of application of the optical

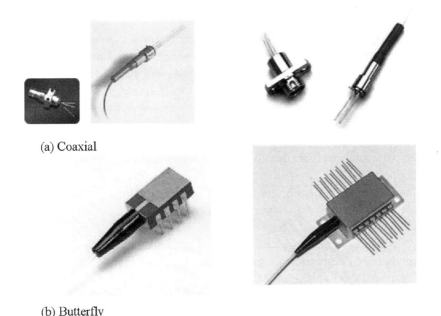

(a) Coaxial

(b) Butterfly

Fig. 12.1 Pictures of different types packages used in telecommunication applications: (a) coaxial package, (b) butterfly package.

components, such as laser diode or detector, dictates which type of package is used. For example, the optical components for subscriber networks are generally packaged without cooler, so in this case either coaxial or butterfly packages could be used. On the other hand, in long-haul transmission systems where environmental conditions are severe on component performance, a butterfly type package with cooler is used. The package types shown in Fig. 12.1 are the standard packages, and their footprints are almost the same for single a functional optical module. For more than one functional optical device, however, hybrid or monolithic integration schemes are used and package sizes vary from industry to industry.

12.3. Package Classifications

Based on how the enclosure of the packages (coaxial and butterfly) is assembled and how their fibers are connected, optical packages are also classified as hermetic and nonhermetic, based on their permeability to moisture exposure.

12.3.1. HERMETIC VERSUS NON-HERMETIC PACKAGES

The package might be assembled hermetically for a number of reasons, including protection of the LD or photodetector (PD) materials, reliability requirements, and the use of environment. For low-cost applications in benign environments using the silicon or even InGaAs/InP detectors or LD, the device will be sealed in plastic packages with flat or lensed windows to input the light. This type of package will permit normal handling and assembly operations and keep away the dust and dirt (which will degrade performance) from the surface of the detectors (or LD). If the optical components will be used in dusty applications, the window needs to be cleaned without damaging the device. For higher reliability applications where there is a concern about organic contaminations affecting the long-term device reliability or where harsh environments can cause condensation and/or freezing of moisture on the internal surfaces, hermetic packages are normally used. Hermetic case materials are typically metals and ceramics, and do not allow permeation of moisture through them.

Traditionally, optical components for long-haul terrestrial and submarine applications are hermetically packaged to prevent chemical degradation of the chip and the interconnection due to the ionically contaminated moisture and to assure long-term reliability. On the other hands, the optical components for short-haul transmission applications are low-cost and reduced size. These are typically encapsulated by a plastic package, and permeable to moisture, which is called non-hermetic packaging. This trends are likely to change in the future by development of the chip itself, especially the fabrication technology along with the passivation technology. Recent studies have shown that high reliability could be achievable with even nonhermetic packaging due to the improvement in chip quality.

12.4. Design Methodologies/Approaches

For designing the optical package for the specific communication application, a design methodology is required. This reduces the chances of any design fault happening in the packaging. Figure 12.2 shows the design methodology for the optical component packaging. Design methodology, as shown in Fig. 12.2, combines with system requirements—packaging technology and components/layout options—to obtain the various package designs. Each of these options must be assessed with respect to optical, electrical, thermal, and mechanical designs.

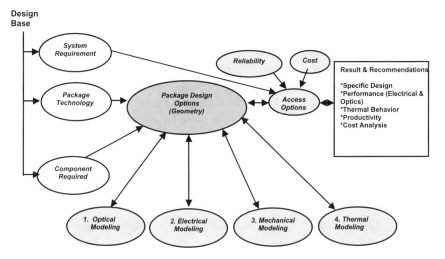

Fig. 12.2   Design methodology for optical component packaging.

First, optical design must consider the package geometries, and the type of alignments, active or passive, available in the assembly process. In optical design, achieving maximum coupling to/from the single fiber provides the greatest chance for using less optics. Especially in assembly where passive-alignment technique is used, internal component designs are required. This technology takes the much simpler approach of forming the optical coupling by positioning components in predefined locations [5–10]. In order to minimize optical reflection occurring due to transmitting of light through different refractive index media, index-matching of materials is also considered before launching to the optical fiber [6]. In long-haul system application, where the back reflection would cause optical performance degradation, an isolator is used for reducing the back reflection.

Second, electrical design is required, especially for the high-speed optical components. Because of the short rise time and fall time requirement, a transmission line model is required for internal electrical interconnection along with appropriate driving, receiving, and loading circuitry design to minimize electrical reflection. The relationship between materials parameters and electrical performance has to be established using a combination of analysis and experimental verification. The system-level simulation would be helpful to estimate system performance in the package design stage, and to allow necessary design changes to be made, if required.

Third, after handling the optical and electrical issues, mechanical designs, comprising thermal design, material selection, and package design (geometry), are to be carried out based on the area of applications. Thermal analyses using computer-aided design (CAD) can address both component and module levels heat dissipation, helping to estimate temperature characteristics of the module. Direct heat sinking from the optical chips using high thermal conductivity material is mandatory. Thermoelectric cooler (TEC) is used in some packages for keeping the temperature fixed for optical performance stability. As different types of materials are used in optical packaging, careful material selection is also mandatory for avoiding any possible stress formation. This selection is not only required to focus on the higher dielectric constant for high-frequency packaging, but also on the higher thermal conductivities and lower difference in thermal expansion between the materials used for avoiding the stress creation under temperature changes. This is essential, as the formation of thermal step either in manufacturing or in operation would not only degrade the optical performance by changes of the original optical alignment, but would degrade the long-run reliability substantially. Surface-mounted packages using the plastic package would face more thermal stress [11] if the packaging material were not selected properly. Extra effort on the chip development side is also sometimes required, if the chip is encapsulated by plastic package.

Finally, design of assembly steps for components placement and attachment inside the optical package must consider vibration, shock, thermal cycling, and acceleration. Thermal design can provide feedback in the design stage about the effects of the large thermal excursion. In many applications, optical components will be subjected to adverse environmental conditions such as high humidity, contamination, and temperature. For this reason, some optical components are required to be packaged in hermetic moisture-free enclosures. Selection of all components inside the optical module must consider whether the package is hermetic or non-hermetic.

12.4.1. OPTICAL

Optical Design Issues

In optical module design, the first and foremost detail of the design is a good coupling scheme, considering the space available inside the package and the allowance required in manufacturing. Optical coupling in optoelectronics devices is as important as electrical connection. Optical performance is mainly dependent on the coupling scheme design. The source of the light can be either another fiber or a light source such as semiconductor LD or LED. Regardless of what the source of the light might be, the recipe for efficient coupling is to match the light field from the source with the mode field of the fiber; the coupling scheme design plays the key role in optimizing optical module performance. Alignment precision of micron order or higher is required for optical coupling because of the small optical beam sizes in the LD. In coupling system design, the incident light angle should be smaller than the total reflection angle that occurs at the interface between the core and cladding layer in the optical fiber or waveguide. The light incident at the angle larger than the total reflection angle will leak out through the cladding layer and be lost. It is essential to estimate the coupling efficiency and to find a way to optimize it for the specific optical package. As shown in Fig. 12.3, if P_1 is the optical power output from the LD, and P_2 is the power received in the optical fiber, the coupling efficiency

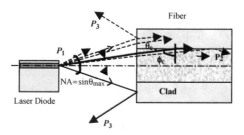

(a) *Without lens*

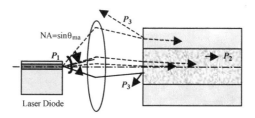

(b) *With lens*

Fig. 12.3 Laser diode to fiber coupling.

can be represented by

$$\eta = -\log 10(P_2/P_1) \quad \text{(in +ve dB)} \tag{12.1}$$

If the light passes from one medium to another medium of a different refractive index, some portion of the light P_3 will be reflected back at the interface, and the reflectance can be expressed by

$$R = -\log 10(P_3/P_1) \quad \text{(in +ve dB)} \tag{12.2}$$

Based on the coupling system design, a coupling efficiency η and reflectance R of close to 1 could be obtained. Generally, a lens is used between the optical source and the fiber, to enhance the coupling efficiency as shown in Fig. 12.3(b). This could be a single lens or multiple lenses and/or optics for optimizing the optical coupling. As in the optical module, different materials having separate refractive indices are used, and single or multiple reflections may occur, which causes noise in the transmitter and spurious signals at the receiver, possibly worse for the optical amplifier. These reflections can limit the data rate of the transmission due to the inherent noise caused by excess light in the system. For digital systems, reflectance as low as -30 dB are required, and most components can meet this requirement. However, in the analog systems where amplitude modulation is required, reflectance as low as -55 dB is desirable. The reflection can be minimized by using index matching materials, optical coatings, angled interfaces (to deflect the light from the transmission path), and optical isolators. The use of an isolator in the optical module rotates the polarization of light passing through it so that impinging reflected light is further rotated and blocked.

Coupling of Optical Light into the Optical Fiber

Mode Field Matching

For efficient transferring of optical energy from the LD source to the fiber, the respective mode profiles should be overlapped as much as possible. To understand what parameters are important for efficient coupling of light into an optical fiber, it is essential to have a more quantitative definition. It can be shown that for linearly polarized modes LP LP_{lm}, coupling efficiency can be represented by

$$\eta = \left| (n_2/Z_0) \int_A E_i \cdot E_{lm} * dA \right|^2 \qquad (12.3)$$

where n_2 is the refractive index of the cladding, Z_0 is the free-space impedance, and E_i and E_{lm} are the electric field amplitudes for the incoming light and for the light propagation in mode LP_{lm} in the fiber, respectively. The fields in Eq. (12.3) are just the transverse fields, because in the LP mode approximation the longitudinal component is neglected. In the optical module/package, the single-mode fiber (allowing only one LP mode to propagate) or multimode fiber (allowing several LP modes to propagate) is used to transmit or to receive the light. In the step-index fiber, the core with higher refractive index is surrounded by the clad with lower refractive index. The index differences of the most important single-mode fiber (e.g., standard SMF-28) for the optical communication are about 0.3 to 0.8%. The core and the clad diameters are 8 μm (2a) and 125 μm, respectively. If the optical signal in the fiber is assumed to be a Gaussian distribution, then the optical signal $P(r)$ variation in the fiber can be expressed by

$$P(r) = P(0) \exp[-2(a/w_g)]^2 \qquad (12.4)$$

where $2w_g$ is the mode field diameter of the fundamental mode, considering the step index fiber, a is the core diameter. The mode field diameter $2w_g$ is the point where the intensity is 13.5% of the center intensity, and it is dependent on the wavelength of the guided wave. It is greater than the core diameter. For example, mode field diameter is 6% larger than the core diameter for the 9 μm core, at an operating wavelength of 1.3 μm. For a wavelength of 1.5 μm, however, MFD is 20% larger than the core diameter. The larger the MFD is, the more power lies in the cladding layer. This power is loosely bound and more susceptible to bending and being radiated away. In the multimode fiber, where core diameter varies from 50 to 200 μm, the mode distribution is not Gaussian shaped, as in Fig. 12.4. The discussion on multimode fiber is out of context, however, as mostly single-mode fiber is in packaging of optical components.

Losses Due to Different Mismatches

In the optical package, the optical axis alignment is the prime requirement for achieving optimum coupling efficiency. In the assembly, it is difficult to achieve submicron ranges of alignment due to instrument and

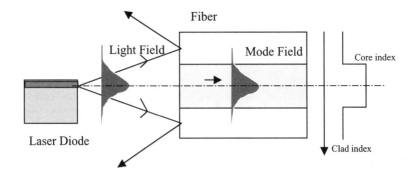

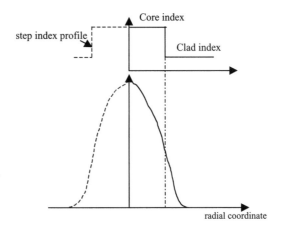

Fig. 12.4   Coupling of optical mode.

optics tolerances. It is necessary to design the coupling system considering all tolerances, from component level to the instrument itself. There are a few cases in which the mismatches (MFD, offsets, and misalignment) cause the reduction of the coupling efficiency, and these mismatches are mainly due to the MFD, axial, and angular misalignment, as shown in Fig. 12.5. If there are multiple mismatches happening in the coupling system, the losses are increased dramatically.

The following relations, as shown in Fig. 12.5, can express the mismatches;

Case 1: MFD mismatches: This is caused mainly by spot-size mismatches in the coherent system, and the coupling efficiency

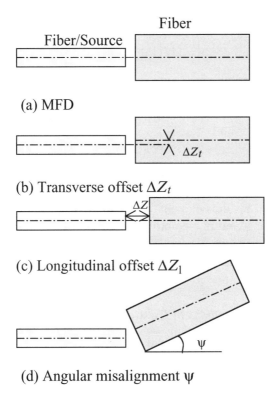

(a) MFD

(b) Transverse offset ΔZ_t

(c) Longitudinal offset ΔZ_1

(d) Angular misalignment ψ

Fig. 12.5 Possible mismatches causing the optical losses.

can be expressed by

$$\eta = \left[2S_1 S_2 / \left(S_1^2 + S_2^2\right)\right]^2 \qquad (12.5)$$

where S_1 and S_2 are the field size of the laser diode and the fiber itself, respectively.

Case 2: Matched MFD and Transverse Offset ΔZ_t: In the optical packaging, this results if the height of the components is different as compared with the designed ones. The η can be expressed as

$$\eta = \exp[-(\Delta Z_t / S_2)^2] \qquad (12.6)$$

Case 3: Matched MFD and Longitudinal Offset ΔZ_1: In the optical packaging, this results from the axial displacement due to the instruments and/or due to component thickness variation presence.

The η can be expressed as

$$\eta = [1/\{1 + (\Delta Z_1/2Z_0)^2\}] \tag{12.7}$$

where Z_0 is the Rayleigh range.

Case 4: Matched MFD but Angular Misalignment Ψ: The η can be expressed by

$$\eta = \exp[-(\Psi/\Psi_o)^2] \tag{12.8}$$

From the set of equations it can be seen that losses due to the longitudinal offsets are more critical than any other mismatches. The preceding relationship is valid for the coherent source to the single-mode fiber, and the details could be found in Nemoto et al.'s report [12]. On the other hand, losses due to mismatches happening in the multimode fiber are complicated for the incoherent sources, and that context is not the part of this chapter.

Different Coupling Methods in Packaging

In optical package, proper optical coupling system would improve the optical performance of the module. The fiber could act as either a receiver (from LD) or transmitter (to PD) in the optical module. As such fiber in most cases is single-mode type fiber having a core diameter of 8 to 9 μm, the coupling scheme is to be designed to minimize coupling loss as much as possible. Typically, in most cases, single or multiple lenses are used between the fiber and the chip to enhance the coupling efficiency. In some cases, however, tapered or lensed fiber is used for increasing coupling by compensating the mode-field diameter mismatching [13–16]. Figure 12.6 shows typical coupling schemes used in optical packaging. Increasing the number of the discrete components in coupling would not only complicate the alignment system, but also increase the packaging cost. The following section describes some of the main optics technologies related to the packaging of the optical devices.

Different Packages

Optics in Packaging

As mentioned in the previous section and shown in Fig. 12.6 (case of laser-diode to fiber), the coupling scheme must be designed properly prior to proceeding to the next design stage. After the coupling scheme design, the

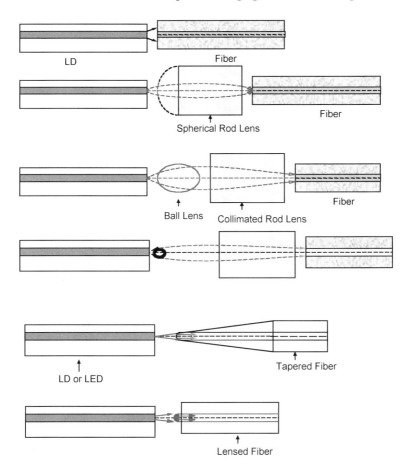

Fig. 12.6   Different coupling methods for single-mode fiber-to-LD.

next step in the optical design is to consider the way of implementing the designed scheme inside the package. Based on the package geometry, and also the alignment technique available, there are several ways to implement the optics inside the package. One of the optics scheme, frequently used in a typical LD package, is shown in Fig. 12.7. A discrete lens is used to focus the beam into the fiber. In some packages, the isolator (not shown in Fig.12.7) is used for reducing the optical reflection. Tapered fiber or lensed fiber is also generally used for focusing the beam into the fiber. Reducing the radius of tapered fiber lens can increase the coupling efficiency, but with a trade-off in optical alignment tolerance. Sometimes, LD with spot-size converter is also used in packaging for easy coupling. Tapered waveguide LD having a narrower far field pattern (FFP) is used for this purpose.

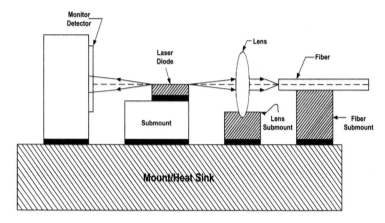

Fig. 12.7 Schematic showing the laser diode packaging considering the optics.

Very recently, FFP of <9 degrees has been achieved in well-designed tapered LD [17]. Figure 12.8 shows the cross sectional views and the far field pattern of tapered LD in comparison with conventional LD. The coupling efficiency, along with the alignment tolerance with single-mode fiber, can be enhanced using the tapered LD [17–20], as shown in Fig. 12.8. Usually, alignment associated cost can also be reduced using such LD. This LD as a transmitter source is targeted for the customer premises link application, where cost issue is the main factor.

Another way to enhance the coupling efficiency of the LD-to-SMF fiber is to use the index matching resin between the conventional LD and the fiber interface. This would reduce the reflection of the different medias, while the light beam passes from LD to SMF. Normally, silicon bench is used as the platform for keeping both fiber and LD. Figure 12.9 shows the schematic as well as the picture showing the LD to fiber, and their coupling characteristics. Using index-matching resin having a refractive index of 1.38 can enhance both the tolerance and coupling losses [6]. This coupling scheme also reduces the packaging cost, and is used in low-end applications (customer premises).

Coaxial Package: Uncooled Package

Coaxial package (hereafter called coax) is generally the package type in which cooler is not used. Optical components used in data or subscriber networks are usually encapsulated by coax package. These package types are connectorized or non-connectorized based on the application area. Connectorized packages are currently used in applications that employ

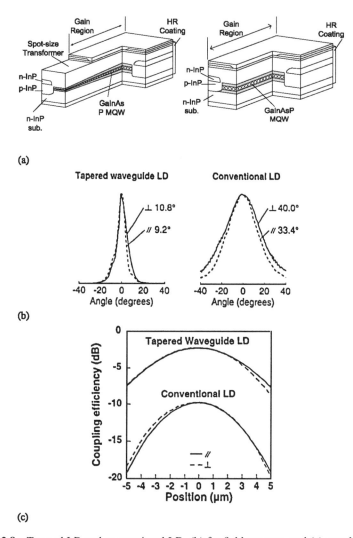

Fig. 12.8 Tapered LD and conventional LD, (b) far field patterns, and (c) coupling efficiencies variations.

multimode fiber such as data links, or single-mode fiber links operating in the lower bit rate and with a modest dynamics range. For high-performance links where high reliability, maximum dynamic range, and stability are required, pigtailed type packages are required. A multitude of designs are based on the package to be mounted on to the IC board. Some are shown in Fig. 12.1. Internal structure of the CAN package is simple, where the LD or PD is mounted on the submount carrier, and fiber is attached using

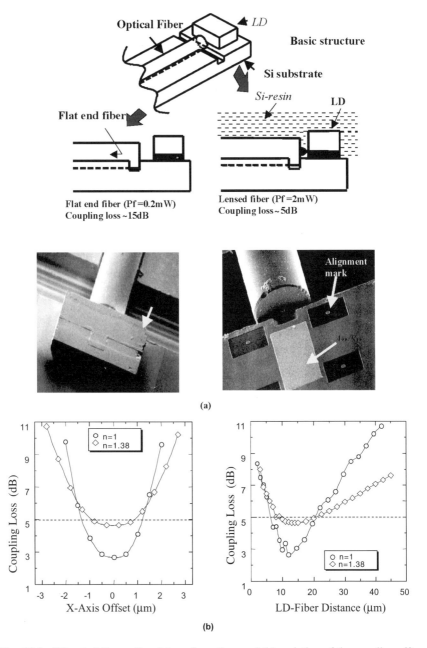

Fig. 12.9 Fiber to LD coupling (a) configuration, and (b) variation of the coupling efficiency with and without index matching gel.

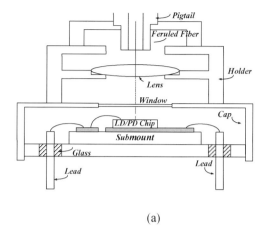

(a)

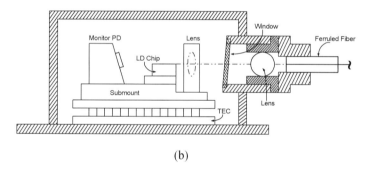

(b)

Fig. 12.10 Schematic representing the cross sectional views of (a) coaxial, and (b) butterfly packages.

the epoxy or Tack welding using YAG laser. The schematic representing internal structure of the typical coax package is shown in Fig. 12.10(a). Generally, based on the assembly, a lens is used for focusing the light into the fiber or onto the PD. This could be a hemispherical lens, discretely implemented inside the module, or the fiber with the lens at the tip, as mentioned earlier.

Butterfly Package: Cooled or Uncooled

This package has been named butterfly because of its external geometry. The 14 pins (7 pins each side) butterfly package has been in telecommunication industries quite a while from its first origin. The packages for the

optical components could be with or without cooler, based again on the
application. The optical components, especially the transmitter source,
being used in long-haul communication systems come with the cooler for
better stability of the source wavelength and long-run reliability. On the
other hand, the components used in the data or subscriber networks, especially comparatively short-haul applications, are being packaged without
cooler. Figure 12.10(b) shows a schematic diagram representing the internal structure of the butterfly package. The optics of this package usually
comprise two lenses. The lens inside the package is usually for collimating
the beam and the outside lens is for focusing purposes. Tapered or lensed
fiber is also used instead of discrete lenses.

12.4.2. ELECTRICAL

Design Aspects/Considerations

The demand for high-speed optical components in the WDM communication system pushes the requirement of the electrical design side-by-side
with other considerations (optical and mechanical) in package designing.
As the electrical signal is necessary as an input for any optical devices,
all components (inside the package) carrying electrical signal are to be
designed for achieving good electrical performance. The main issue to
be considered in the design is the electrical needs of the circuits inside
the package. All connected signals should be considered to be propagated through the well impedance matched transmission line. Figure 12.11
shows the schematics comparing the interconnections in the optical package with and without considering the electrical design. Discontinuities
in electrical signal line due to conventional wire bonding as shown in
Fig. 12.11 cause the reflection, which degrades the signal waveform reaching the optical device. Proper electrical design considering all discontinuities is necessary, without which total module performance (both optical and electrical) cannot be improved. The electrical characteristics that
are important for the design of the high-speed package include the fundamental transmission line properties of the interconnects as well as the
dynamic characteristics of the signals propagated to the optical devices.
The fundamental transmission line properties include the line resistance
(R), capacitance (C), inductance (L), and dielectric conductance (G), all
per unit length, and the characteristic impedance (Z_0), and the propagation constant (δ). These characteristics are functions of the cross sectional

Conventional
Interconnection: <2.5 Gbps

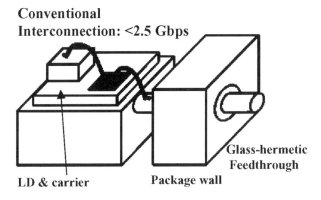

Glass-hermetic
Feedthrough

LD & carrier   Package wall

Controlled impedance
Interconnection: >2.5 Gbps

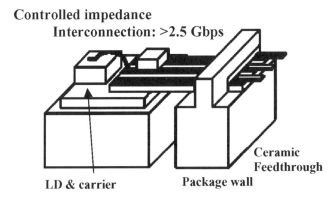

Ceramic
Feedthrough

LD & carrier   Package wall

Fig. 12.11 Schematic representation comparing the conventional wire bonding and controlled impedance matching in the optical packaging.

geometry of the interconnect, the electrical properties of the driver and the optical device, and other interconnect structures such as wire bond, vias, and stubs affect the dynamic behavior of the signals reaching the optical devices. Important dynamic characteristics include the propagation delay, the rise and fall time of the receiving signals, attenuation, noise due to the signal reflections, and cross talk between the adjacent lines. The interconnection between the transmission lines could be accomplished by wire or ribbon bonding or other flip-chip bonding. At multi-GHz frequencies, interconnect lengths become a significant fraction of the wavelength of the highest frequency harmonics, and therefore the interconnects must be designed as transmission lines with proper concern for impedance, cross talk, and attenuation [21, 22]. Impedance mismatch must be minimized to reduce the reflections and prevent ringing, which can cause false decision (switching) in the receiver signal. Significant attenuation and rise time degradation can be caused by losses in the transmission line. The transmission line loss is the sum of the conductor loss and dielectric loss, both of which are dependent on the frequency. Some issues to be considered in electrical design are explained in the following sections.

Interconnections: Signal Lines

In optical packages, interconnections play a key role for overall module performance (both electrical and optical). Interconnection structure and interconnection performance issues are to be considered seriously in electrical design. Figure 12.12 shows a simplified view of an interconnection for the case of a modulator-integrated laser diode (MILD) package. In this case, assume that signal is fed through from the electronics driver module. An actual interconnection inside the optical package is, in fact, a much more complex object than the simple straight interconnection, as shown in Fig. 12.12. Common variations include bends in lines, vias between interconnection layers, connectors between levels of packaging, etc. Figure 12.13 illustrates the levels of packaging common in optical packages: the various packaging levels generally defined as *the first level*, corresponding

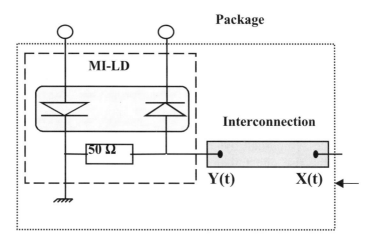

Fig. 12.12 Idealized interconnection connecting the modulator integrated laser diode (MI-LD) to feedthrough connection.

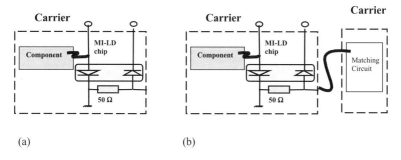

(a)   (b)

Fig. 12.13   Interconnection and packaging levels: a) interconnection between components on the same carrier, and (b) interconnection between circuits, on different carriers.

to components and carrier (substrates); and the second level, corresponding to subassemblies.

An ideal interconnection would in all cases deliver the near-end signal $X(t)$ from the driver module (for example), undistorted, to the far-end optical component, i.e., $Y(t) = X(t - \tau_d)$, where τ_d is the propagation delay across the interconnection. Surprisingly, it is true to some extent that this idealized and simplified view of the interconnection is commonly used in the optical packaging. Only recently it is understood that such a view is no longer valid in the case of high bit rate (over gigabit rate) optical component packaging. As interconnection performance has become an increasingly critical issue in high-speed optical devices, delay/performance optimized interconnection layout has become a priority in the electrical design.

The ideal interconnection is a simple delay line, propagating the signal unchanged at some propagation velocity. A real interconnection is more properly viewed as the circuit function, and the interconnection medium can be assumed as the linear medium (i.e., the electrical parameters of the interconnection are independent of the signal level). With respect of thinking about the linear medium, it is appropriate to consider it as the linear analog circuit. If the signal is considered as the electromagnetic wave propagating down the interconnection, then the Fourier analyses based on the frequency component of the waveform are convenient, because the interconnection parameter is always frequency dependent. If instead, the interconnection parameter appears as the localized lumped element, then the analysis is considered as the Laplace transform, instead of frequency dependent. These thoughts are mainly dependent on the operating wavelengths and whether they are comparable with the interconnection length.

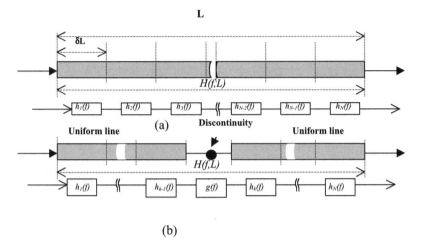

Fig. 12.14   Long line as a series of shorter segments: (a) N number of uniform lines, and (b) two uniform lines connected by the discontinuity having transfer function $g(f)$.

Let's consider the interconnection (see Fig. 12.14), as the Fourier transform. In this case the signal transfer function associated with the line is $H(f, L)$, where L, length, is the the parameter of the transfer function. If L is long, then the line itself could be considered as the number of section having the length of δL, and the transfer function $h(f, \delta L)$ is written as

$$H(f, L) = \prod_{1}^{N_L} h(f, \delta L) = [h(f, \delta L)]^{N_L} \qquad (12.9)$$

where N_L is the number of the segments having each segment length δL. If the segments are of different types having the transfer function of $h_i(f, \delta L_i)$ at ith section of the segment, then the overall transfer function of the interconnection is merely

$$H(f, L) = \prod_{i=1}^{N_L} h_i(f, \delta L_i) \qquad (12.10)$$

where again N_L is the number of the segments having each segment length δL. This provides a convenient means of concatenating a sequence of interconnection structures (straight lines, bends, vias, etc.), and obtaining the response of overall connections. The casual effects of multiple reflections are also to be considered for each section. If $H(f, L)$ is the overall interconnection transfer function, the far end signal $Y(t)$, as shown in Fig. 12.12, is obtained from the near-end signal $X(t)$, by decomposing its near-end signal into its Fourier components $a_x(f)$, propagating through individual frequency component across the line to obtain the far-end frequency component $a_y(f) = H(f, L) \cdot a_x(f)$. Then far-end signal $Y(t)$ could be reconstructed from its frequency components $a_y(f)$. The signal distortion in this propagating wave model are the consequences of the changes (both amplitude and phase) of the frequency components as the signal propagates down the line. Thereby, after propagating different types of lines, if any, the frequency components $a_x(f)$ becomes distorted, and leads to $a_x(f) \neq a_y(f)$. In particular, $H(f, L)$ can be represented by general form as, $H(f, L) = A_t(L) \exp(N\theta)$, where A_t is the total attenuation and θ is the phase-shift of the signal over a distance of δL (total length $L = N \cdot \delta$). In this consideration, the line length is larger than the wavelength of the signal, and the signal is treated as the wave propagating across the line. However, if the line length is very small as compared with the signal wavelength, and the signal is occupied in fraction of time period over the space. Voltage across the line is essentially constant, and the far end signal can be estimated using the Laplace transform.

An active or passive optical device (requiring the electrical signal) is a localized and invariant structure. The electrical lumped model is necessary to estimate the interconnection behavior. The other localized elements in interconnection structures are the vias, bend, wire bonds, connectors, etc., and their lumped model can also help to estimate the overall interconnection behavior. In packaging, however, the line behavior could be obtained from its scattering parameters (s-parameters). The overall line performance can be estimated calculating each segment's (of different types) s-parameters.

Discontinuities

In the preceding section, we assumed the simple model for a continuous, uniform interconnection line. However, interconnection lines between the optical device and the signal feedthrough connector encounter discontinuities as shown in Fig. 12.14(b), due to bends, vias between the metallization levels, and interpackage level connection using the wire/ribbon bonding and/or flip-chip bonding. These discontinuities can significantly impact the performance of the interconnection line, the effect depending to some extent on the frequency of the propagating signal. At low–frequency signal, the lumped circuit model remains valid, and the discontinuities lead to changes in the net capacitance, inductance, and resistance. The RLC model of such discontinuities must consider the nonuniformity of the charge distribution and the current distribution at the discontinuities.

For example, a right-angle bend in the line will exhibit the concentration of the field lines at the sharp edges of the bend, impacting the calculation of the capacitance of the bend. In addition, the current distribution will not be uniform as it flows around the bend, affecting the inductance and the resistance of the bend. At higher frequency signal, corresponding to long interconnections, as the signal propagates through the uniform transmission line with characteristic impedance Z_0 and encounters discontinuities, the discontinuities appear as the new transmission line with the impedance Z_{dis} or as lumped circuit elements acting as the imperfect termination. In both cases, signal reflection occurs, and it mixes with incident signal. The resultant electrical signal launching into the optical components is distorted, and degrades the optical performance accordingly. In order to avoid or reduce signal reflection, the proper interconnection design with impedance matching considering both load (in this case optical device) and source (e.g., electronics driver) is required.

Carrier Materials

In high-speed package design, the materials of the signal carrier must be selected first for their electrical characteristics and then for their mechanical properties. That is not to say that mechanical performance is secondary in importance to electrical performance, rather that the electrical properties must be met before mechanical properties are considered. For example, if a dielectric material has adequate mechanical properties and stability against all anticipated temperatures, but has a dielectric loss factor in excess of what can be tolerated by the design under consideration, that material cannot be used and another one must be chosen that has adequate mechanical properties and meets electrical requirements as well. Consideration is also to be given to the environment to which the package is to be exposed (in the case of the nonhermetic type package). A wide variety of carrier materials with good electrical characteristics is available. These materials range from inorganic, such as aluminum, aluminum nitride, and beryllia, to organics, such as polymide, polysulfone, and FR4 epoxy-glass. The properties of these materials are to be studied to ascertain the compatibility with the anticipated requirements. Usually, in packaging of an optical device of speed >2.5 Gb/s, mostly aluminum, and aluminum nitride are used as the carrier materials because of their high electrical characteristics and also good mechanical characteristics. Detailed characteristics will be explained in the Section 12.4.3 (mechanical design).

Bonding/Attachment

The chip carrier inside the optical package is generally required to connect with another carrier to launch the electrical signal to the optical devices (e.g., LD). For an electrical connection, wire and ribbon bonds, and flip-chips bonds are frequently used. Ribbon bonding is generally used in the package where inductance is to be minimized or where electrical current may be greater, and wire bond cannot be handled. Flip-chip bonding is used to connect the chip on the interconnection substrate, especially where the high-speed signal is to be carried by minimizing interconnection inductance, and also where many chips have to be connected having a pitch that cannot be connected by wire/ribbon bonding. Flip-chip bonding involves inverting the chips that have bumps or balls of compliant metal such as solder adhered to its bonding pads and attaching the chip on an interconnecting substrate.

Electrical Design in Packaging: An Example

The previous section has discussed the importance of the electrical design and the way of estimating the electrical characteristics considering each component inside the package. We will show an example of electrical design in high-speed optical device package where the electrical signal path mainly comprises the connector, transmission line, and optical device such as MILD, as shown in Fig. 12.15. The signal launching to the optical device from connector is via the microstrip transmission line (MSL) or coplanar transmission line (CPL). Single or multiple MSL (or CPL) carriers may be required for launching the electrical signal to the optical component. Interconnection between two carriers for high-speed electrical signal path is usually done using the multiple wire or ribbon bondings. The discontinuity of the transmission line, experienced from the feedthrough section (connector to the MSL or CPL) and interconnection section (one MSL carrier to another MSL carrier), must be understood, and interconnection design must be done properly. As mentioned earlier, these discontinuities change the line impedance and cause the reflection of a part of energy to the incident signal. Perfect impedance matching circuit (network) is required to reduce the reflection loss, considering the optical component's impedance, interconnection elements, and also other parasitic effects. The matching circuit for controlling the impedance may require it be designed on the MSL or CPL carriers. Prior to designing the passive components or the interconnection, and also estimating the total module-level electrical

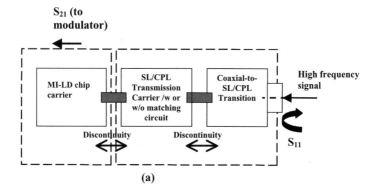

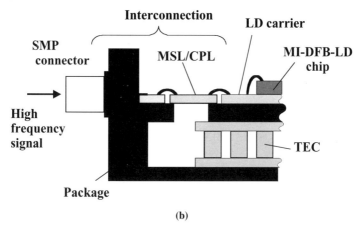

Fig. 12.15 Schematic representing the electrical signal path inside the package (a), and schematic showing the interconnection including the feedthrough path inside the module (b).

characteristics, the most important issue is to derive the equivalent circuit (lumped model) of the optical device, which could be the laser diode, the modulator, or the photodiode.

Optical Device Modeling: Modulator Integrated Laser Diode (MI-LD)

Deriving the equivalent circuit of the optical device is very important for designing the components such as MSL or CPL carrier, matching circuit, and related interconnection. This is also necessary to estimate the module performance in the design stage. In deriving the exact equivalent circuit (lumped model) of an optical device, it is required to know its approximate or measured frequency response characteristics, mainly S_{11} parameters. Generally, the electrical parameters in an equivalent circuit

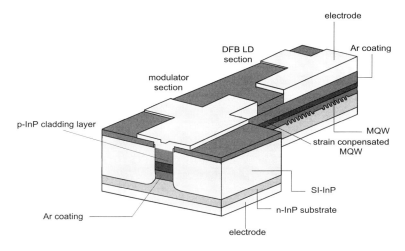

electrode

DFB LD
section

Ar coating

modulator
section

p-InP cladding layer

MQW
strain conpensated
MQW

SI-InP

n-InP substrate

Ar coating

electrode

Fig. 12.16   Schematic diagram showing the typical MI-LD chip structure.

can be estimated from the frequency characteristic following the method of extraction.

Frequency Response Characteristics: MI-LD.   We consider here the modulator integrated laser diode (MI-LD), as shown in Fig. 12.16. Both LD and modulator have a p-n structure with a common n-electrode and a separate p-electrode, electrically isolated from each other [23–25]. The working principle of the modulator is based on the Frantz-Keldysh effect, i.e., the effect of a strong electrical field on light absorption in semiconductors at the fundamental absorption edge [26]. The absorption edge or the bandgap of the semiconductor shifts to the lower energy with the increase of the electric field. In MI-LD, the modulator material is a semiconductor with its band edge very close to the LD photon energy. The absorption edge of the modulator can then be modulated slightly higher (light passing) or lower (light absorption) than the LD photon energy by the input electrical signal. The modulator is a high impedance component and, on the other hand, the LD is a low impedance component. As both LD and modulator are generally fabricated monolithically, and they are electrically isolated from each other, the modulation characteristic is mainly determined by frequency response of the modulator. Figure 12.17 shows the frequency response (S_{21}) characteristic (under operation) of a typical 10G MI-LD chip terminated with 50 Ω thin-film resistor on the carrier.

A few factors influence MI-LD frequency response. The response of change in refractive index and absorption coefficient is very high under

E/O (S21)

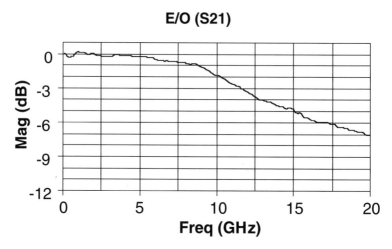

Fig. 12.17   Frequency response (S_{21}) of typical MI-LD chip on carrier under operation.

modulation. The frequency response of the change in absorption coefficient is, for example, around 100 GHz in III-V compound semiconductor intensity modulators using the electroabsorption effect. The response is determined by the plasma oscillation frequency. In most cases the frequency response is not determined by its physical effects, but is limited by its physical structural dimension. The scale factor limiting its frequency response is mainly due to the RC time constant and lightwave transit time through the modulator. The impedance of the modulator gradually decreases as the modulation frequency increases and thus the intensity of the modulated electric field decreases. The 3-dB bandwidth can be defined by the frequency at which the voltage decreases by $2^{-1/2}$ from the value at low frequencies or under dc, and approximated by $f_{3\text{-dB/CR}} \sim 1/\pi R_L C_m$, where R_L is the load resistance and C_m is the modulator capacitance. Another factor limiting the frequency response is the transit time, which is $\tau_m = n_m L_m / C_a$, where n_m and L_m are the refractive index and length of the modulator, and C_a is the velocity of light in air. The 3-dB bandwidth limited by transit time can be expressed [27] by $f_{3\text{-dB/transit}} \sim (1.4/\pi)\tau_m$. For $n = 3.2$, and 200 μm modulator length, the bandwidth limited by transit time is estimated to be 208 GHz. Mainly, the 3-dB bandwidth is limited by RC time constant, mainly due to its physical dimension. This 3-dB bandwidth of the module is affected/reduced if internal interconnection is not optimized properly and/or other parasitic components are included inside the module.

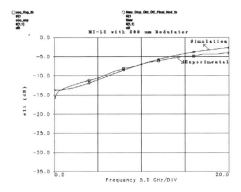

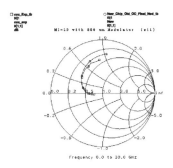

Fig. 12.18 Measured and fitted small signal frequency responses of the MI-LD in (a) Cartesian, and (b) Smith chart. Chip carrier having 50 Ω coplanar lines is used for the measurement. The MI-LD is terminated with 50 Ω thin-film-resistor.

Equivalent Circuit: MI-LD. The exact electrical parameters induced in MI-LD are generally estimated from its measured S_{11} parameter. Figure 12.18 shows the S_{11} of the MI-LD (with chip carrier) under operation. The equivalent circuit of the modulator section of the MI-LD is similar to that of the pin photodiode. After fitting with the measured S_{11}, as shown in Fig. 12.18, the equivalent circuit of the MI-LD can be extracted from the measured chip-on-carrier. Figure 12.19 shows the MI-LD chip equivalent circuit, comprising the modulator junction capacitance C_A, and series resistance R_{MI}, and the bonding pad inductance L_B and bonding capacitance C_B. The bias-dependent current source $(I_D - I_P)$ is also included, representing the current induced due to the light absorption inside the modulator. The laser part mainly consists of junction resistance R_{LD}, junction

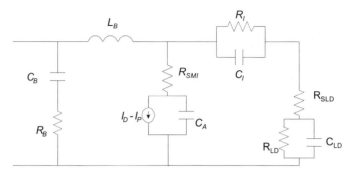

Fig. 12.19   Approximated large signal equivalent circuit of MI-LD.

capacitance C_{LD}, and series resistance R_{SLD}. The modulator and LD part are separated electrically by RC parallel network.

Module Performance: Controlled versus Uncontrolled Impedance.
Once the lumped parameter of the optical device is estimated, then the equivalent circuit of the passive component to be used inside the module—for example, thermoelectric cooler, feedthrough section carrying electrical signal—is generally estimated from its behavioral characteristics (measured S_{11}) and/or physical dimension (simulated S_{11}). The electromagnetic models of all key passive interconnect structures are helpful for very high-speed module design. Module-level electrical response could be derived from all key components equivalent circuits (including optical device), and lumped model (or electromagnetic models of passive interconnect structures). As discussed in an earlier section, significant signal reflection is expected, if the signal line impedance is not properly controlled. This signal reflection causes degradation in the signal waveform. Figure 12.20 shows module-level s-parameters, and the waveform with and without using the controlled impedance in signal line. From module design stage, module performances could be estimated considering all effects of passive components and various interconnect structures. The key passive interconnect structures could be optimized for the best module performance.

12.4.3. MECHANICAL

This section describes detail mechanical issues required for packaging of the optical components. All materials as well as thermal issues to be considered in packaging will also be included in this section.

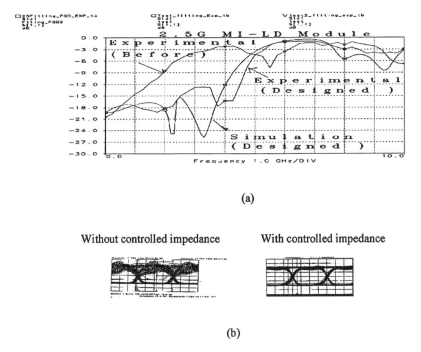

(a)

Without controlled impedance   With controlled impedance

(b)

Fig. 12.20 (a) Small signal frequency response of 2.5 Gb/s MI-LD module including the total effects of the connectors, MSL carrier with or without matching circuit and chip (MI-LD) on carrier, and (b) Eye patterns (back-to-back) under $I_{LD} = 60$ mA and $V_{mod} = -0.5$ V: without and with the controlled impedance. Simulation (designed) in (a) is the simulated S_{11} characteristic of the equivalent circuit of a 10G MI-LD chip-on-carrier terminated with 50 Ω resister. Experimental (before) and experimental (after) in (a) are without and with controlled impedances, respectively.

Material Types and Properties

Prior to designing the optical components and their packages, it must be known where the system will be used, and the different temperature environments. The optical components, as well as the materials used in the optical modules, experience different temperature conditions from manufacturing to application. These are steady-state temperature, temperature gradients, and rates of temperature changes, temperature cycles, and thermal shocks, through manufacturing, storage, and operation. Thermal properties enduring such life-cycle profiles include thermal conductivity, deflection temperature, glass transition temperature (typical for polymeric material), and the coefficient of thermal expansion (CTE).

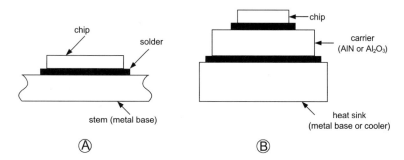

Fig. 12.21   Basic configuration of the optoelectronics device packaging.

Chip Materials

In packaging of optical components, different types of materials from the solder material to package enclosure are used. For designing highly reliable optical components, material selection is of foremost importance in package design. Figure 12.21 shows the basic configuration of packaging, requiring the die (chip), attachments process, the carrier (or the heat sink). The chip or die in the optoelectronics module is generally bonded on the stem (metal base, acting as the heat sink) or on the carrier with or without cooler.

The semiconductor materials used for optical device fabrication must be direct bandgap, and they are fixed for the optical device (chips) used for the communication. Electrons and holes can recombine directly across the bandgap without needing third particle to conserve the momentum. Only a direct bandgap material has the radiative recombination sufficiently high to produce the adequate level of optical emission. These semiconductor materials are III-V material, made from the compounds of a III element (such as Al, Ga, or In) and a group of V elements (such as P, As, or Sb). Various ternary and quaternary combinations of binary compounds of these materials are also direct bandgap materials and suitable for optical sources and optical detection.

For operation of 0.8 μm thru 0.9 μm wavelength spectrum the principle material used is the ternary alloy of $Ga_{1-x}Al_xAs$. The content x determines the wavelength of the peak-emitted radiation. The substrate used for this optical device is GaAs, because of having good lattice matches of the ternary compound materials. On the other hand, at longer wavelengths (1.3 μm and 1.5 μm), the quaternary alloy of $In_xGa_{1-x}As_yP_{1-y}$ is the primary material candidate. By varying the mole fractions of x and y contents, the peak wavelength could be varied from 1.0 μm to 1.7 μm for the

Table 12.1   **Chip Materials and Their Mechanical Properties**

Material Types	Thermal Conductivity $W (m \cdot °K)$	Thermal Coefficient of Expansion $(\times 10^{-6}/°K)$
AlP	90	
GaAs	54	6.0
InP	67	4.56
Si	125	3.5
$Al_{0.5} Ga_{0.5}As$	11	5.4
$Al_{0.03} Ga_{0.97}As$	29	5.4
$Ga_{0.515} In_{0.485}P$	5	
$In_{0.47} Ga_{0.53}As$	66	6.4
$(Al_{0.5} Ga_{0.5})_{0.525} In_{0.475}P$	6	

emitter and the detector. The substrate used for these long wavelength optical components is Indium Phosphide (InP), which comes very close to a crystal lattice match to the InGaAsP, and helps to reduce the interfacial defects and to minimize the strains in the device as the temperature increases. Both substrate materials (GaAs and InP) have low thermal conductivity, and it is essential to choose the material that could match with this material for reliable package/module operation. Table 12.1 summarizes the mechanical properties of the chip materials used for optical devices fabrication.

The thermal and mechanical properties of the semiconductor crystal are critical for die attach design. These properties are directional dependent due to the isotropic nature of the semiconductor crystal. The material properties involved in the chip attaching design are modulus of the elasticity, Poisson ratio, and the coefficient of thermal expansion. Of them, the key property involved is the thermal expansion. Therefore, it is required to choose the packaging material so that it closely matches the thermal conductivity.

Attachment Materials

The chip or die bonding on the heat sink carrier is the first step of package assembly. The selection of heat sink (hereafter called carrier) and solder materials, as well as the bonding processing, would influence the temperature characteristics of the optoelectronics devices, especially laser diode and LED (light-emitting diodes). The bond strength of the attachment material is an essential property to ensure that the die and the carrier stay

598 Achyut K. Dutta and Masahiro Kobayashi

in place despite the stresses imposed during manufacturing, storage, and operation. The die must not be detached from the carrier during power or thermal shock or exposure to extreme temperature. The properties for the bonding material are tensile strength, shear strength, and fatigue endurance. Thermal conductivity is another property, as the attachment must conduct heat from the chip to the carrier.

There are a number of ways that chips can be bonded to the substrates. Generally, solder, conductive plastics, and filled glasses can be used for bonding. In optical components packaging, however, eutectics solders are mostly used for the chip attachment, which provide good thermal conduction, and adhere despite elevated temperatures. This process requires that the bonding surfaces first be plated with noble metals with a thickness in the fractions of a micron. Solder material is usually chosen that cannot create any mechanical stress due to the differences in thermal expansion, if any, between the chip and carrier.

Solder is a fusible alloy with a liquid temperature below 400°C. The common elements used for the solder alloys and their melting points are summarized in Table 12.2. Solder alloys used in optical device packaging are classified as hard solder and soft solder. Table 12.3 summarizes different solder alloys (and their properties), generally used in optical packaging. The melting point of hard solder is relatively higher, and this is usually Au:Sn or Au:Ge. Because of temperature tolerance level of the chip's epitaxial growth, the carrier-level solder for attaching optical device chips and other components inside the package is limited to eutectic or near eutectic solders. In some cases, tin/silver eutectic and soft solder are used. Gold-based

Table 12.2 **Summary of the Common Elements Used in Solder and Their Melting Points**

Material Types	Melting Point (°C)
Ag	961
Bi	271.5
Cd	321
In	157
Pb	328
Sb	630.5
Sn	232

Table 12.3   **Melting Points of Different Solder Alloys**

Solder Alloys	Melting Ranges (°C)	
	Liquidus	*Solidus*
80Au-12Ge	356	335
80Au-20Sn	280	250
10Au-90Sn	217	205
94Au-6Si	370	345
42Sn-58Bi	138	138
52Sn-48In	131	118
25Sn-75Pb	266	183
30Sn-70Pb	255	183
40Sn-60Pb	238	183
50Sn-50Pb	216	183
60Sn-40Pb	190	183
63Sn-37Pb	193	183

eutectics such as Au:Sn, and Au:Ge are frequently used for chip attachment, because they have higher flow stresses (onset of plastic flow) and therefore offer excellent fatigue and creep resistance. The main disadvantages are primarily due to their lack of plastic flow, which leads to high stresses in the semiconductor, if the thermal expansion mismatches between the chip and the carrier. On the other hand, the soft solders have relatively lower melting point and their compositions contain Bi or In. These solders help to reduce the mechanical stress from the bonds to the chip. Soft solder such as In-Pb-Ag, Bi-Sn, and Pb-Sn are commonly used in optical packaging. This solder alloy choice is also suitable for attaching other components inside the package.

Reliabilility in Attachment

Process and Material dependent. The reliability of the optoelectronics devices is dependent on the type of solder alloy used for the chip attachment, and its processes. Reduction of stress between the chip carrier and chip, and also reduction of oxidation formation during reflowing, has to be considered for a long reliable optoelectronic package. Stress reduction in

Indium-bonded GaAs is possible by choosing the Si or Silicon Carbide (SiC) heat spreader instead of the Diamond. The thermal and electrical resistance of the indium-bonded device is found to be decreased with time, as the gold can diffuse into the Indium solder material and form the inter-metallic layer of InAu, which is thermally resistant and brittle. Solder joint embrittlement also formed in the tin-lead-based solder. Avoiding indium-based material in die bonding is better for long reliable optoelectronics packaging. More stable bonds are formed using Au:Ge and Au:Sn eutectic alloys for optical device (chip) bonding.

The whisker growth may be expected in lasers bonded on a diamond heat sink with Sn or Pb-Sn solders. In laser case, the whisker formation increases slightly to the laser threshold, but in long run case, the risk of failure is involved because of the possibility of an electrical short between the p- and n-side regions. This whisker formation is generally seen for the laser bonded with PbSn, In, or Au-Sn (Sn 90%) solders, and the current density promotes the growth of whisker due to the electromigration mechanism [28–31] . This formation is also enhanced by internal strain temperature, environment, and also depends on materials used in the solder, and the type of metalization used on the heat sink or packaged stem. Whisker formation can be eliminated using hard solder containing the Au-Sn (80% Au) or Au:Ge solder alloys [32].

Mismatching thermal expansion between the carrier and the die also creates undesired stress, and could be minimized using the soft, low melting temperature solder. However, soft solders are subject to the slow, continuous deformation known as creep, which reduces module reliability [33]. Fatigue crack is also formed if the tin without silver exists in the alloys. Using soft solder, especially In-based alloy, in LD attachment causes degradation in bonding in long-run operation, and causes thermal resistance to increase [32]. This degradation results from the reaction of the In solder and the neighboring Au layer [28]. The use of soft solder for LD bonding must be avoided in LD packaging.

Void and Oxide Formation in Chip/Carrier/Heat Sink Attachment.   Air has a very poor thermal conductivity (0.026 W/m-°C) [34], and must be eliminated from the thermal path from the semiconductor junction (p-n junction) to the heat sink. Every package type, consisting of many materials, is neither perfectly flat nor perfectly smooth. The portion of heat sink that attaches with the carrier or chip is also not perfectly flat nor smooth. At the microscopic level, the interface consists of the point-to-point

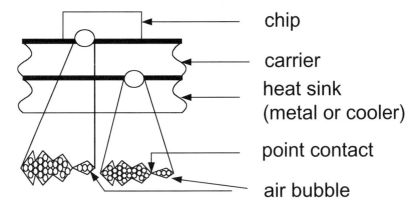

Fig. 12.22 Microscopic view of air bubble, accumulated in the interface.

contacts surrounding an air bubble, as shown in Fig. 12.22. To minimize the thermal resistance caused by the non-smooth surfaces and the resulting air pockets, several techniques are available. One technique is to apply mechanical pressure to smooth out the interfaces. This works to some extent if soft solders are used. However, the pressure required to produce an acceptable thermal condition might exceed the material strength and produce degradation or failure. Solder can be used in two different ways to fill the air gaps in the heat sink attachment interface. (i) First, solder is reflowed between the device and the heat sink. Usually the solder used for this purpose is Pb:Au having a thermal conductivity of 50 W/m°C. This provides several orders of magnitude improvement over the gap. (ii) Second, solder can be used to fill gaps, and in this case solder (perform) in precut form is to be used between the interface. By applying mechanical pressure, the solder having high lead content compresses easily and fills the gaps.

Formation of void and/or surface oxide inside/outside of the solder bonding would not make reliable bonding. As the molten solder solidifies during the cooling to form solder joins, the cooling process, such as cooling rate, has a direct bearing on the resulting solder joints as to their microstructure and void developments. The presence of oxide appearing on the carrier surface or chip surface also interferes with soldering during the bonding process. The void and oxide formation not only increases the device contact resistance, but also increases the thermal resistance, which can cause a thermal dissipation breakdown. A void free, crack free, and good wetted contact is necessary in optical packaging.

This oxide formation should be minimized or prevented by using the capping layer of gold. Wettable layer using the liquid fluxes, traditionally, could minimize the oxide formation. Flux is a chemical solution that contains a reducing agent to convert metal oxides to metals and is generally used just prior to solder joint formation. The most common fluxes are the organic acids; nearly all fluxes used for this purpose use this same chemistry. After the soldering, the flux must be removed or deactivated to prevent corrosion later on. The presence of acid in contact with metal can cause a host of problems later. Corrosiveness of liquid fluxes and residue left by fluxes can reduce reliability and complicate the cleaning process. Oxide formation can also be fully alleviated or minimized if the bondings are performed inside an oxygen-free atmosphere. A slow heating or in situ vacuum bake prior to the solder reflow or brazing assists in removing the contamination from the assembly.

Carrier Materials

For attaching the chip, ceramic-type materials are frequently used as the carrier for optoelectronics packaging. Ceramics offer two important advantages for chip bonding. The first and the most important feature is that the CTE is low enough that there is only a small thermomechanical mismatch between the chip material and the ceramic carriers. The second feature is high-temperature stability as well as superior thermal conductivity. Ceramic can handle 300°C or higher, a valuable feature for the chip carrier. From a frequency point of view, ceramic materials have also low attenuation in higher frequency, which makes them advantageous for use in high-frequency optoelectronics packaging. The other advantage is that transmission line based on micro-strip line or coplanar line design having desirable impedance can also be made on the ceramics for reducing electrical reflection.

Only a few ceramic materials have been used extensively in optoelectronics packaging. These are aluminum oxide (Al_2O_3), aluminum nitride (AlN), beryllium oxide (BeO), and various formulations of silicon carbide. Other ceramic materials also used in the packaging are boron nitride (BN), covering hexagonal, and cubic. They are not frequently used yet. Other than ceramic materials, Si and SiC materials are also used as the chip carrier, which have their thermal expansion closer to the chip materials, and higher thermal conductivity. Copper and Copper-Tungsten (Cu:W) are also used as carriers because of their high thermal conductivity and TCE close to

Table 12.4 **Materials Frequently Used as the Carrier**

Material Types	TCE @ 25°C (ppm/°C)	Thermal Conductivity @ 25°C (W/m°C)	Thermal Conductivity @ 100°C (W/m°C)
Alumina (961)	6.2	25.1	14.6
AIN	4.6	170	140
Copper	16.5	397	388
GaAs	6.0	58.0	
Kovar^{tm}	5.5	16.5	17.6
LTCC (Dupont 951)	5.9	3.0	
Molybdenum	4.8	137	134
Silicon	2.6	147	98.8
Tungsten Copper (85/15)	7.0	230	

chip material. Table 12.4 summarizes the carrier materials types frequently used in the optoelectronics packages.

Aluminum Oxide (Al$_2$O$_3$).   Aluminum oxide (hereafter called alumina) is the most frequently used ceramic in optoelectronics packaging. The alumina is among the most hard and heat resistant ceramic, and could resist temperature up to 2000°C [35]. It also has excellent resistance to chemical heat. Its properties vary based on the percentage of alumina contents. For high-speed interconnection in optoelectronics and electronics packaging, alumina containing 90% or more alumina is used. The most common alumina available for packaging is 85, 90, 94, 96, and 99.9. Table 12.5 summarizes the thermal and mechanical properties, and Table 12.6 summarizes the electrical properties of the alumina ceramic, containing different contents of alumina percentage. The 85% grade alumina is good grade alumina in use in much electronics packaging. They have the good wear resistance. The 92% grade alumina has more wear resistance and is being used in moderate frequency packaging [35, 36]. 96% grade and above alumina is used in many optoelectronics packaging. These are to be fired at comparatively lower temperature. They have smooth surface, and high mechanical strength, and excellent electrical properties. The 99.5% alumina is usually

Table 12.5   **Thermal and Mechanical Properties of Alumina (A_2O_3)**

Property	Units	A_2O_3, Wt%-as Indicated		
		85	94	99.9
Density	g/cm^3	3.41	3.62	3.96
Thermal Expansion (25–200°C)	W/°C	5.3	6.3	6.5
Thermal Conductivity @25°C	W/m-°C	15	18	39
Tensile Strength @25°C	10^3 lb/in^2	22	28	45
Compressive Strength @25°C	10^3 lb/in^2	280	305	550
Modulus of Elasticity	10^6 lb/in^2	32	41	56
Shear Modulus	10^6 lb/in^2	14	17	23
Hardness	R45N	73	78	90

Table 12.6   **Electrical Properties of Alumina (A_2O_3)**

Al_2O_3 Contents (Wt%)	Dielectric Constant			Loss Tangent			Resistivity (Ω-cm)	
	1 KHz	1 MHz	100 MHz	1 KHz	1 MHz	100 MHz	25°C	1000°C
Al_2O_3 (85 Wt%)	8.2	8.2	8.2	0.0014	0.0009	0.0009	$>10^{14}$	
Al_2O_3 (94 Wt%)	8.9	8.9	8.9	0.002	0.001	0.005	$>10^{14}$	5.0×10^5
Al_2O_3 (99 Wt%)	9.8	9.7		0.002	.0002		$>10^{14}$	1.1×10^7

used for high-frequency applications because their high purity improves the dielectric loss characteristics.

Aluminum Nitride (AlN).   Aluminum nitride (AlN) has thermal conductivity five times that of alumina at room temperature and above 200°C higher than that of berryllia. Tables 12.7 and 12.8 summarize the various

Table 12.7   **Thermal and Mechanical Properties of AlN**

Property	Units	Value
Density	g/cm^3	3.27
Thermal Expansion (TCE)	10^{-6}/K	4.4
Thermal Conductivity (25°C)	W/m-K	170
Compressive Strength	Mpa	2000
Modulus of Elasticity	Gpa	300
Poison's Ratio		0.23
Hardness	Knoop 100 g	1200
Melting Point	°C	2232

Table 12.8   **Electrical Properties of AlN**

Property	Unit	Value
Dielectric Constant		
1 MHz		8.9
10 GHz		9.0
Loss Tangent		
1 MHz		0.0004
10 GHz		0.0004
Resistivity	Ω-cm	
25°C		$>10^{12}$
150°C		2×10^8

properties of AlN. As depicted, mechanical strength is comparable with alumina. Its thermal expansion of coefficient is 4 ppm/C, better matched with Si, GaAs, and InP as compared with alumina, which makes it advantageous to use in high-power optical devices packaging. Thermal conductivity of AlN can vary widely from its maximum value, reported to be 320 W/m-K, and is typically in the range of 170 to 200 W/m-K [37–39]. These properties are process dependent variations. This AlN is suitable for blank substrate for high-power and high-frequency applications. Most optoelectrics packaging, especially for telecommunications applications, use AlN as the chip carrier. As AlN's thermal expansion of coefficient is closer to Cu:W, it is also used with Cu:W as the chip carrier for efficient heat dissipating in the optoelectronics packaging [40, 41].

Other Carrier Materials.   Other ceramic materials, which are also used for optoelectronics packaging, are boron nitride, silicon carbide, and low-temperature cofired ceramic. Table 12.9 summarizes their properties. Beryllium oxide or beryllia has many characteristics comparable with alumina, which makes it advantageous to use in many types of high-power density packaging from electronics to optoelectronics. Pure beryllia exhibits a thermal conductivity higher than that of all metals except silver, gold,

Table 12.9   **Properties of BeO, SiC, and LTCC**

Property	Units	BeO 96.1 BeO	BeO 99.5 BeO	SiC	Low-Temperature Cofired Ceramics DUPONT 951	NEC	NGK	IBM
Density	g/cm^3	2.85	2.90	3.21	3.1			2.62
Thermal Expansion of Coefficient (TEC)	10^{-6}/K	6.3	6.4	3.8	5.8	5.0	5.1	3.0
Thermal Conductivity	W/m-K	159	281	70–260	3.0		3	<4
Flexural Strength	Mpa			450	320	280	380	210
Dielectric Constant								
1 MHz			6.7	40–79	7.8	7.1	7.0	5.0
10 GHz			6.8					
Loss Tangent								
1 MHz		0.000	0.001	0.050	0.0015	0.002	0.001	
10 GHz		7 0.003	0.003			0.005		
Volume Resistivity 25°C	Ω-cm	>10^{17}	>10^{17}	>10^{13}	>10^{12}	>10^{14}		

copper, and high-purity aluminum. Beryllia has excellent dielectric constant, outstanding resistance to wetting and corrosion by many metals and chemicals, and mechanical properties comparable to high-grade alumina [42]. Despite its favorable properties, beryllia use is limited, due to its toxicity for approximately 1% of the populace when submicron size particles are ingested into the lungs.

Silicon carbide is also attractive for use in packaging, due to its two main advantages; high thermal conductivity and thermal expansion coefficient closer to silicon than any other available ceramic. Because of its high thermal conductivity, this is also popular for use as the carrier for the multi-chip-module (MCM).

Low-temperature co-fired ceramics (LTCC) are very famous for their ability to stack multilayered interconnection configurations for high-density packaging, and also to print conductor on the smooth surface. These are cost-effective, and could be used in the comparatively low-speed optoelectronics packaging.

Bonding

Signals produced in the chip or required for the chip need to be connected to the interconnects. After the chip attachment, the second step in packaging hierarchy is to connect the chip pad to the carrier electrode. There are three techniques for chip-to-package interconnection. These are (a) wire/ribbon bonding, (b) flip-chip bonding, and (c) tap-assisted bonding.

Wire or Ribbon Bonding.   Usually, in optical packaging the bond pad of the chips is electrically interconnected to the package by wire or ribbon bonding. Figure 12.23 shows the basic package configuration with wire bonding. Gold wires of 25 μm diameter are used for bonding. Either a thermosonic or thermocompression ball-wedge process using gold wire or an ultrasonic wedge-to-wedge process using gold wire/ribbon can accomplish the wire bonding. In high-speed application, however, sometimes ribbon is preferable to wire for reducing the line impedance, which may help to minimize the discontinuities in the line at higher frequency operation. The wire/ribbon bonding process could create a defect inside the crystal if special precaution is not taken. In particular, the substrate for optical devices such as GaAs and InP is not as hard or as fractured as silicon, and is more susceptible to crystal damage during the wire bond. Hardness and fracture toughness is lower for GaAs and InP than for Si, which makes them much

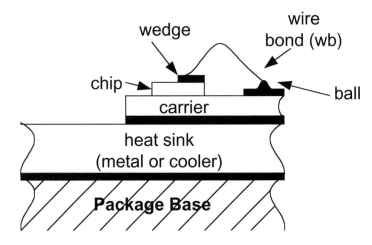

Fig. 12.23   Package configuration with wire bond, as the second step.

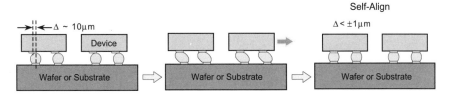

Fig. 12.24   Flip-chip process showing the self-alignment technique.

more subject to cracking. The fracture properties depend on the substrate's defect density, and damage to substrate such as GaAs is shown to occur at stress as low as 15 to 20 Mpa. Thermosonic ball bonding instead of ultrasonic wedge bonding is recommended for wiring to avoid fracture and damage to the chip. The minimum ultrasonic energy should be used during wire bonding, and the temperature should be increased to compensate the wire bonding.

Flip-Chip Bonding (FCB).   In flip-chip bonding, as shown in Fig. 12.24, direct attachment and electrical interconnection of the chip to package (carrier) is done without using any additional pads. This helps to increase interconnection density inside the package. In this process, the chip terminals are placed facedown to the footprints of the package carrier with the help of electrically conducting bumps. FCB is used in optical packaging/assembly where a passive alignment technique is utilized [5–10]. In FCB, positioning accuracy of <1 μm in both vertical and horizontal directions could be achieved using a self-alignment technique [43, 44]. The bump materials are mainly eutectic solder materials, such as Sn:Pb, Au:Sn, etc., dependent on the reflow temperature that both substrate and chip can withstand. The short solder bump interconnections between the chip and carrier would help to reduce interconnection inductance lower even than TAB. The lower inductance and higher interconnection density capability helps to enhance the electrical performance. In optical packages where higher input/output (I/O) connections are required, such as arrays of vertical-cavity surface-emitting laser (VCSEL) [45, 46], FCB with self-alignment technique is generally used. In high-speed (especially >10 Gb/s) optoelectronics packaging [47], FCB is also used for enhancing module electrical performance. Especially in optical packaging, the selection of the FCB depends on the type of devices to be bonded, and effective designing of the device (including solder material and also process), is absolutely necessary to achieve high reliability. As the chip is placed facedown for this process, especially for the case of laser diode or other electronics IC bonding, proper heat dissipation is difficult through the small area solder bumps, which may cause poor reliability. Heat dissipation for such high-power device bonding using FCB may require under-fill, which is frequently used in electronics packages [48–50].

Tape-Automated Bonding (TAB).   TAB is the bonding process where innerconnects attach to the outer pad leads using the automated tape bonding instead of using the slow wire bonding. This bonding has the advantages of lower inductance, controlled lead geometry, higher bond strength, and higher impedance controlled capability, helping to have better electrical performance as compared with wire/ribbon bonding. This process is fast and can be used in low-cost packaging. In optical packaging, this TAB bonding is not used so far. However, in lowering the assembly cost, this TAB bonding has a greater chance of use in future optoelectronics packaging.

Packaging/Enclosure Materials

Figure 12.25 shows the basic configuration of the optoelectronics packaging used for a butterfly type package. The types of materials used for enclosure also influence the module performance, especially its mechanical and thermal characteristic. It is desirable to use material having good

Table 12.10   **Thermal Properties of the Materials Used for Package Case or Enclosure**

Material Types	TCE @25°C (ppm/°C)	Thermal Conductivity @25°C (W/m°C)
Copper	16.5	397
Kovar[tm]	5.5	16.5
Molybdenum	4.8	137
Silica (SiO$_2$)	0.6	12
Stainless steel	16	173
Tungsten Copper (85/15)	7.0	230

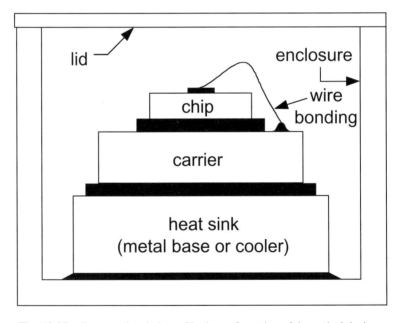

Fig. 12.25   Cross sectional view of basic configuration of the optical devices.

mechanical and thermal characteristics. Table 12.10 summarizes the materials frequently used for optoelectronics packaging and their mechanical and thermal properties. Based on the applications, the type of materials used for enclosure for optoelectronics packaging is varied. For example, package case made from KOVAR with a copper-tungsten (Cu:W) base is usually used in high-end optical component packaging (14-pin butterfly package is a good example). This is due to its high thermal conductivity, and also low thermal expansion as compared with other materials [40], helping to dissipate the heat instantaneously [41]. Besides, this material is also hard compared with KOVAR and stainless steel, which helps to prevent the enclosure from any unintentional mechanical stress, which could be intrinsic (in-module) or extrinsic (off-module) stress. In data or subscriber applications, the stainless steel (e.g., coax package) and plastic package [51, 52] cases are used due to their low cost and friendliness for large-scale manufacturing.

12.4.4. THERMAL

Thermal Delima in Packaging (Why Thermal Management?)

Increasing the temperature in the optical module would not only change the electrical performance but also affect the optical performance of the module such as power, wavelength, etc. This happens due to the stress build-up with changes of the temperature, which may change the optical alignment, and subsequently change the module performance. For LD made from InP and GaAs materials, the wavelength changes are 0.1 nm/°C. It is required to keep the temperature fixed inside the module despite the changes of the operating condition. For this, the package design has to be done keeping in mind that heat must be dissipated instantaneously from the module. To alleviate the temperature dependences, efficient thermal management in package design is essential. The following sections describe the thermal design, including thermal load estimation, and cooler selection.

Heat Transfer Model

As mentioned earlier, in optoelectronics packaging, different types of materials (varying from chip to enclosure) are generally used. The characteristics of all materials must be considered together for the module's optimum thermal, optical, and electrical performance. Let us consider a case where an optical chip (active device) is attached on the carrier, and the carrier is attached on the heat-sink, as shown in Fig. 12.26. Let us consider that k_1 and k_2 are the thermal conductivity of the chip and carrier, assuming that those are not varied over the length.

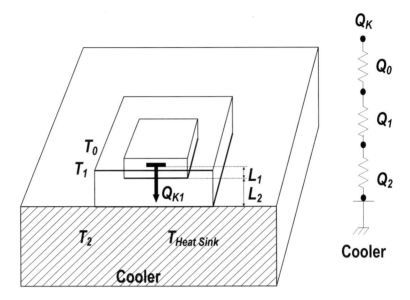

Fig. 12.26   Basic heat flow model.

Changes in temperatures can be written as

$$T_{\Delta 1} = T_0 - T_1 = Q_{k1}(L_1/k_1 A_1) \tag{12.11}$$

$$T_{\Delta 2} = T_1 - T_2 = Q_{k2}(L_2/k_2 A_2) \tag{12.12}$$

where $T_{\Delta 1}$ and $T_{\Delta 2}$ are the temperature difference along the length L_1 and L_2, respectively; T_1 and T_2 are the temperature in the interfaces of the chip and carrier and the carrier and the heat sink, respectively. Thermal resistances θ_1 and θ_2 can be defined as

$$\theta_1 = L_1/k_1 A_1 \tag{12.13}$$

and

$$\theta_2 = L_2/k_2 A_2 \tag{12.14}$$

assuming that the active device, considered as the heat source, is producing the heat of Q_{k1}, and dissipating into the carrier. The temperature at the interfaces could be calculated as

$$T_0 = T_1 + \theta_1 Q_{k1} = T_1 + T_{\Delta 1} \tag{12.15}$$

and

$$T_1 = T_2 + \theta_2 Q_{k2} = T_2 + T_{\Delta 2} \qquad (12.16)$$

So that Eqs. (12.15) and (12.16) give

$$T_0 - T_2 = \theta_1 Q_{k1} + \theta_2 Q_{k2}$$
$$T_{heat\ source} - T_{heat\ sink} = \theta_1 Q_{k1} + \theta_2 Q_{k2} \qquad (12.17)$$

If we neglect any heat dissipation due to convection and radiation, then the heat dissipation process is done by the heat conduction. In this case, the Q_{k1} and Q_{k2} are the same, and equal to $Q_k(Q_{k1} = Q_{k2})$, heat generated by the heat source.

If there are j number of interfaces between the heat source and heat sink, then Eq. (12.17) can be written as

$$T_{heat\ source} - T_{heat\ sink} = Q_k \sum \theta_j \qquad (12.18)$$

and

$$T_{j,j-1} = T_{heat\ sink} + Q_k \sum \theta_{j,j-heat\ sink} \qquad (12.19)$$

where

$$T_{j,j-1} = \text{temperature at the interface of } j \text{ and } j - 1, \text{ and}$$
$$\sum \theta_{j,j-heat\ sink} = \text{the sum of the thermal resistances from the interface of layers } j \text{ and } j - 1 \text{ to the heat sink.}$$

The temperature at the particular interface can be calculated by using Eq. (12.19) when several materials are attached in series, such as the chip is attached on the carrier, which is again attached on the submount. The equivalent thermal resistance is equivalent to the sum of the thermal resistance in series, and could be written as for n, number of the layers, as

$$\theta_{Equivalent} = \theta_1 + \theta_2 + \theta_3 + \cdots + \theta_n \qquad (12.20)$$

Heat Spreading Model for Thermal Design

In optoelectronics packaging, the active device (heat source) is generally attached to the carrier before it is attached on the heat sink. The thickness of the carrier in the optoelectronics packaging is generally decided based on the thermal coupling system, and usually the carrier is thick. As shown in Eqs. (12.13) and (12.14), the thermal resistance is inversely proportional to the area of the carrier, and in order to dissipate the heat instantaneously,

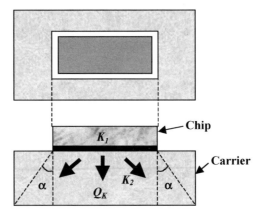

Fig. 12.27   Basic heat spreading configuration.

not only the thermal conductivity, but also the area, must be large. As the thermal conductivity is needed to be kept high, and also as the carrier is thick, the heat propagated through the carrier is not perfectly propagated $90°$ toward the heat sink (z direction). The heat is also propagated toward the heat sink with a spreading angle of α, as shown in Fig. 12.27. This spreading angle is dependent on the thermal conductivity of the chip and carrier itself. Usually, in packaging, the carrier of higher thermal conductivity, as compared with the chip's, is used, and in this case, the heat is spread laterally due to the random nature of the heat, and the spreading angle of greater than $0°$. On the other hand, if the carrier thermal conductivity is lower than the chip, the spreading angle is close to $90°$.

The spreading angle can be expressed as

$$\alpha = \tan^{-1}(K_1/K_2) \tag{12.21}$$

where K_1 and K_2 are the thermal conductivities of the chip and the underlying carrier, respectively. Eq. (12.21) indicates that when $K_1 = K_2$, the spreading angle becomes $45°$, and in the case of high K_1/K_2, the spreading angle becomes close to $90°$. In the case of the slab where the heat spread in trapezoidal and rectangular slab as shown in Fig. 12.28, the equivalent thermal resistance is the summation of the trapezoidal and rectangular heat path, and can be written as

$$\theta_{Equivalent} = \theta_1 + (\theta_{21} + \theta_{22}) \tag{12.22}$$

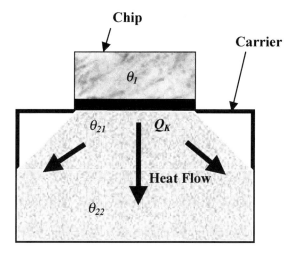

Fig. 12.28 Heat spreading approximation.

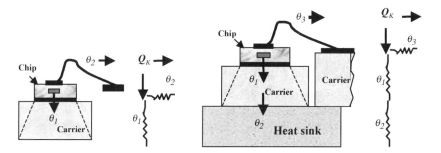

Fig. 12.29 Schematic showing parallel heat path in the packaging.

where θ_1 and θ_2 are the thermal resistances of the trapezoidal and the rectangular heat path. In the optoelectronics packaging, the heat is not always dissipated toward the heat sink. If the chip is bonded by the wire bond or ribbon bond, the heat dissipation in this case is multiple, and the heat dissipation is toward the heat sink and along the other heat path across the wire or ribbon bond, as shown in Fig. 12.29. In this case, the equivalent thermal resistances are the parallel of the two thermal resistances. For n number of thermal paths, equivalent thermal resistances can be expressed as

$$1/\theta_{Equivalent} = 1/\theta_1 + 1/\theta_2 + 1/\theta_3 + \cdots + 1/\theta_n \qquad (12.23)$$

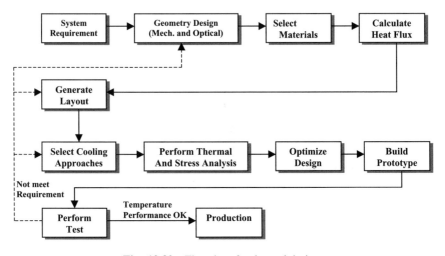

Fig. 12.30 Flowchart for thermal design.

Thermal Design

Thermal design is one of the important design considerations required to be done in optoelectronics packaging. The temperature characteristics of the module are dependent on how effectively the thermal design or thermal management is done. Before proceeding on to the thermal design, many factors are required to be known. Figure 12.30 shows the flowchart of the thermal design, essential in the optoelectronics package design. Before doing the thermal design, it is required to know (i) the system requirement, especially the worst condition of device operation; (ii) total heat dissipation for all active and passive devices; (iii) the worst condition of system operation; (iv) availability of cooling approaches and permissible for the system application. This follows the preliminary selecting of the material and design of the layout. These layouts are usually done after the optical, electrical, and mechanical designs. A thermal analysis is required in component and module level, which includes calculating the temperature in each point of the assembly, and the interface, and also calculating quantitatively the stress build up during worst operating condition. This is required not only for high reliable module design, but also to check whether the optical coupling system is changed due to stress build up. The total thermal results are required for analysis to check whether the module performance meets the system requirements. As the thermal design is an iterative process, the material selection along with process choices, may need to be taken into consideration for optimum thermal design. The prototype and test are performed subsequently to check whether system requirements are met at worst condition. If all performances meet the system requirements, the module is then considered for production.

While the speed requirement for the optoelectronics module continues to rise, it pushes the importance of the integration of more functional devices, including the electronics and the optical devices, inside the same module [53]. In this case, total power requirement for the integrated module goes up, and thereby the challenges increase for highly integrated packages. The net result is that the amount of energy that must be dissipated will continue to rise for the foreseeable future. For example, if the laser (transmitter source) will be integrated with the driver circuit inside the module, the net power dissipation is increased to 3 to 5 W, and this is only one electronics IC case. However, for the optoelectronics module with the multiplexer/ demultiplexer and other amplifier circuits, the total power dissipation is increased to close to 10 W, a case that will occur soon if it has not already done so. To assure the optimum module performance, thermal design considering each functional chip would be necessary. This is a hybrid type integration, and in all cases, the chips are used. The simple fact remains that the chip assembly approach taken must make allowance for this important issue. Employment of direct thermal paths away from the chip to a suitable heat exchange medium is vitally important. Failure to address this issue will likely result in significant reduction in product life due to thermal degradation of the chip. Direct chip attachment technologies, with the capability of rapid temperature transients, can help to solve the problem.

Heat Sink Selection

The heat dissipation from the active and passive devices to be packaged must be estimated based on the heat flux, and the cooling system has to be selected accordingly. For low heat flux module normally the conduction and/or natural convection is sufficient to meet the design requirement. This optoelectronics module is low cost, and lower size. If the heat flux increases, then the cooler is to be used inside the module for efficiently dissipating the heat from the active device to outside of the module. Mostly, peltier cooler is used in certain optoelectronics modules where stabilizing the temperature characteristics under worse operating condition is very important for optimum system performance, such as long-haul, metro applications.

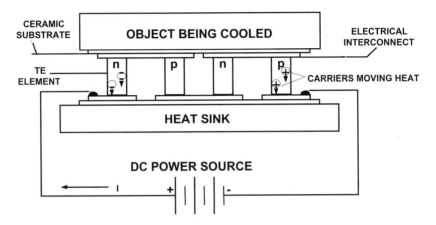

Fig. 12.31   Typical thermoelectric cooler.

Thermoelectric Cooler

Peltier effect is the basis of the thermoelectric cooler (TEC), which is a semiconductor-based electronic component that functions as a small heat pump. This is basically a semiconductor p-n junction, and is operated by low-voltage DC power source, and could be used for heating and cooling just by reversing the flow of current direction. Typical TEC consists of two parallel plate ceramic, separated by the p and n type of materials, as shown in Fig. 12.31. The elements are connected with each other electrically in series, and thermally in parallel. When a positive DC current is applied to the n-type thermo element, electrons pass from the p to n type thermo element and the cold side temperature decreases as heat is absorbed. This heat is transferred to the hot side of the cooler, where it is dissipated into the heat sink and surrounding environment. The heat absorption (cooling) is proportional to the current and the number of thermoelectric couples. The thermoelectric semiconductor material most often used in TEC is bismuth telluride (Bi-Te) that has been suitably doped to provide the individual blocks or elements having distinct n and p characteristics. These materials are most often fabricated by either directional crystallization from a melt or pressed powder metallurgical. Each manufacturing method has its own particular advantages, but directionally grown materials are most common. In addition to Bi-Te, other thermoelectric materials used for particular types of TEC are lead telluride (PbTe), silicon germanium (Si:Ge), and bismuth-antimony (Bi-Sb) alloys. TEC modules are made in a great variety of sizes, shapes, operating currents, operating voltages, and ranges of heat pumping capacity. In optoelectronics packaging, different types of packaging are based on the heat flux load.

Thermoelectric Cooler Selection.

Heat Load Estimation.   It is necessary to estimate the total heat load of the package in worst operating condition before proceeding to select the TEC. The heat load may consist of two types: active and passive, or a combination of the two. An active load is the heat dissipated by the device being cooled. It generally equals the input power to the device. Passive heat loads are parasitic in nature and may consist of radiation, convection, or conduction.

In optical packaging, the carrier is generally placed on the TEC for cooling. The bonding parts from the active device (e.g., LD) to the TEC are acting as the barriers of heat dissipation, as long as carriers, solder materials are not selected properly. This is dependent on the thermal resistance of the device, and the interface of device and carrier. The thermal resistance of the pn junction device is easily known from its junction voltage changes under the forward bias. The forward junction voltage is determined by the bandgap of the semiconductor, and linearly dependent on the junction temperature. Heat load for the active device is generally expressed by

$$Q_{active} = V^2/R = V_j I_F \qquad (12.24)$$

where Q_{active} is the active heat load (*Watt*), V_j (*Volt*) is the junction voltage, $R(\Omega)$ is the device junction resistance under operation, and I_F is current (A) flowing through the device.

If V_b is the applied bias voltage, R_c is the contact resistance, I_F is the forward current, and T_a is the ambient temperature, then the junction temperature can be written using Eq. (12.15) as

$$T_j = T_a + \alpha_{thermal} I_F (V_b - I_F R_s) = T_a + \alpha_{thermal} I_F V_j \quad (12.25)$$

where $\alpha_{thermal}$ is the junction thermal resistance, and V_j is the junction voltage for the active device. The $\alpha_{thermal}$ could be expressed using Eq. (12.14) as

$$\alpha_{thermal} = L/(K_{active} A), \qquad (12.26)$$

where A is the junction area, L is heat flow length, and K_{active} is the thermal conductivity.

In the active device such as a laser diode or light-emitting diode, if P_{out} is the power output, then the total heat dissipated is

$$Q_{active} = Q_{laser} = I_F V_j - P_{out} \qquad (12.27)$$

and the junction temperature in the laser diode can be written from Eq. (12.25) as

$$T_{j-laser} = T_a + \alpha_{thermal} Q_{laser} = T_a + \alpha_{thermal}(I_F V_j - P_{out}) \quad (12.28)$$

For example, a laser diode, if operated by 100 mA injection current, and having the junction voltage of 1.8 V, lases 10 mW of power. The active load is equal to 170 mW ($=180 - 10$) of heat dissipation. If there are other passive devices and other active devices such as the electronics integrated circuit (IC) inside the package, total heat load is the sum of heat produced by each device under the operating condition, and can be written as

$$Q_{total} = \sum Q_{i\,active} + \sum Q_{j\,passive} \qquad (12.29)$$

where

$\sum Q_{i\,active}$ = Total heat flow from i numbers of the active devices
$\sum Q_{j\,passive}$ = Total heat flow from j numbers of the passive devices

TEC's Heat Estimation.   In Eq. (12.29), the heat flow from/to the different path through the wire is neglected. Heat flow through the TEC to be selected could be written as

$$Q_{Tec} = \theta_{Tec}(T_{heat} - T_{cold}) \qquad (12.30)$$

where θ_{Tec} is the thermal resistance of the TEC, and the T_{heat} and T_{cold} are the temperatures at heat and cold side of the TEC. The TEC's heat dissipation Q_{Tec} in Eq. (12.30) is equal to total heat dissipated from all devices (mounted on TEC). From Eq. (12.13) or (12.14),

$$Q_{Tec} = K_{TEC}(A/L)(T_{heat} - T_{cold}) \qquad (12.31)$$

where K_{TEC} is the thermal conductivity of the TEC material, A is total contact area, and L is length of the heat flow. As the contact area for the TEC is equal to $A = 2NA_{each}$, where N and A_{each} are the total number of elements and area of p or n type junction, then heat flow from TEC could be written as

$$Q_{Tec} = 2N K_{TEC}(A_{each}/L)(T_{heat} - T_{cold}) \qquad (12.32)$$

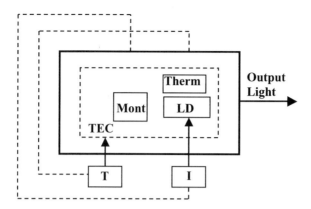

Fig. 12.32 Example of output circuitry frequently used to measure the temperature of the LD chip by using the NTC thermistor. That temperature is fed back to the TEC to control the temperature. The monitor detector is used to measure the light output of LD, and fed back to the LD to control the output power.

Thermistor

For keeping the temperature constant in an optical package where TEC is used, it is required to monitor the exact temperature of the package, especially the chip. Thermistors (thermally sensitive resistors), which are a solid-state device, are used for this purpose. They are mounted onto the chip carrier very close to the chip for exact temperature monitoring, and feedback the variation to the compensation circuit for cooling/heating as required by using the TEC. Figure 12.32 is the typical example of the compensation circuit frequently used for controlling the temperature and the light output of the LD package. The wavelength of LD is controlled using a similar configuration.

Thermistor is a kind of semiconductor and made from the solid solution or sintered of oxides or carbonates of Cu, Fe, Mn, Mo. Based on the temperature changes, thermistors are classified as negative temperature coefficient (NTC) and positive temperature coefficient (PTC) thermistors. A typical resistance versus temperature characteristic is shown in Fig. 12.33. The NTC has larger delta R vs. T than PTC. In an optical package, as the precision measurement is necessary, (although resistance changes nonlinear vs. temperature), NTC type of thermistor is mostly used. The resistance R of a thermistor at a temperature $T(K)$ can be related to the resistance R_{ref} at a reference temperature T_r by

$$R = R_{ref} \exp \{(B/T) - (B/T_r)\} \qquad (12.33)$$

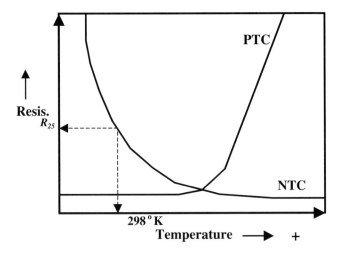

Fig. 12.33 Example of resistance versus temperature behavior of the thermistor.

where B is the characteristic temperature of the thermistor in K, expressed by

$$B = \{(T_r T)/(T_r - T)\} \, ln \, \{R/R_{ref}\} \tag{12.34}$$

The B value is determined by the thermistor material and is a measure of thermistor sensitivity to the changes in temperature. The higher B values give greater percentage changes in resistance.

Reliability in Optical Packaging Due to TEC and Thermistor

The characteristics of the optical package, especially LD, frequently used for WDM application is to tune the wavelength. In the DFB LD, it is difficult to predetermine the laser wavelength with the desired accuracy (~0.1 nm/C), and generally the temperature dependence of the DFB laser is used to tune the wavelength channel. The tuning requirement of the LD package implies that the package needs to control and also monitor the temperature accurately, and this are done using the TEC and thermistor, respectively as mentioned in the previous section. The stability of the TEC and thermistor is important, and influences the reliability of the modules.

In the LD package case, wavelength drift (changes in operational wavelength at constant operating condition), occurs in long-run operation, which is temperature dependent. This wavelength drift is mainly caused by (i) the LD chip's wavelength drift at constant temperature [54, 55], (ii) the aging of the thermistor [56], and (iii) also the bonding process especially for TEC and thermistor used in the package. The wavelength drift due to the chip is mainly dependent on the fabrication process, and the drift due to other factors is dependent on the package. A 1% change of initial $10\,k\Omega$ thermistor resistance results in a temperature increase of $0.25°C$, and the laser wavelength red shift of $0.02\,nm$. This dependency puts stringent requirements on the reliability of the thermistor to be used for the WDM LD module.

The soft solder used for bonding the thermoelectric material to the electrode mainly causes the instability of the TEC. Degradation similar to that of the LD bonded with the soft solder, mentioned earlier, is also expected in the TEC. The solder instability increases the electrical resistance and reduces the stability to control the temperature. On the other hand, the thermistor itself is quite stable, and the instability is added to the package if soft solder is used for bonding to the package. This instability could be alleviated if hard solder is used. The degradation of the cooler and thermistor both could sometimes lead to failure, even if the photonics device itself is not degraded.

12.5. Conclusions and Future Challenges in Optical Module Designing

The basic design methodologies as required for optoelectronics package designs are discussed in detail. Achieving high performance in the module requires not only the chip design, but also requires the package design, which includes optical, electrical, mechanical, and thermal designs. Without any of these, it is hard to improve the total performance of module.

Increasing demands of bandwidth in optical communication pushes the requirements of the high-speed integrated optoelectronics module. Future developments will see miniaturization and incorporation of both electrical and optical functional devices inside the single packages, reducing the space requirements and cutting system complexities. Packaging for hybrid or monolithic integration of electronics and photonics devices in the single package will be a trend in development. These push the consideration of design methodologies, described in this chapter. Efficient convergence of optical components and electronics components (communication IC) requires well-designed packaging technologies.

Development of advanced design tools is required for improving the optical module performance, by effectively estimating system level performance from module design level. As the needs for higher speed optical devices in the optical networks system get higher and higher, advanced design tools comprising microwave CAD, thermal CAD, and photonics simulation will help to develop the advanced packaging technologies for the optical device module.

References

1. F. P. Kapron, D. B. Keck, and R. D. Maurer, "Radiation losses in glass optical waveguides," Appl. Phys. Lett., 17 (1970) 423.
2. D. B. Keck, R. D. Maurer, and P. C. Schulz, "On the ultimate lower limit of attenuation in glass optical waveguides," Appl. Phys. Lett., 22 (1973) 307.
3. I. Hayashi, M. B. Panish, P. W. Foy, and S. Sumski, "Junction lasers which operate continuously at room temperature," Appl. Phys. Lett., 17 (1970) 109.
4. H. Kressel and M. Ettenberg, "A new edge-emitting (AlGa)As heterojunction LED for fiber-optic communications," Proc. IEEE, 63 (1975) 1360.
5. K. Kurata, K. Yamauchi, A. Kawakami, H. Tanaka, H. Honmou, and S. Ishikawa, "A surface mount type single mode laser module using passive alignment," VI International Workshop on Optical Access Networks, S3.4 (1994).
6. K. Kurata, K. Yamauchi, A. Kawakami, H. Tanaka, H. Honmou, and S. Ishikawa, "A surface mount type single mode laser module using passive alignment," Proc. ECTC'95, 765 (1995).
7. H. Kobayashi, M. Ekawa, N. Okazaki, O. Aoki, S. Ogita, and H. Soda, "A corner illuminated structure PIN photodiode suitable for planner lightwave circuit," Conference Proc. LEOS'96, 9th annual meeting, MA5 (1996).
8. G. Nakagawa, S. Sasaki, N. Yamamoto, K. Tanaka, K. Miura, and M. Yano, "High power and high sensitivity PLC module using a novel corner-illuminated PIN photodiode," Proc. ECTC'97, 607 (1997).
9. A. Goto, S. Nakamura, K. Kurata, M. Funabashi, T. Tanabe, K. Komatsu, O. Akiyama, N. Kitamura, T. Tamura, and S. Ishikawa, "Hybrid WDM transmitter/receiver module using alignment free assembly technique," Proc. ECTC'97, 620 (1997).
10. J. J. Koh, M. H. Choir, H. J. Song, and J. U. Bu, "On wafer process for mass production of hybridly integrated optical components using passive alignment on silicon motherboard," Proc. ECTC '99, 216 (1999).
11. M. Fukuda, F. Ichikawa, Y. Yamada, Y. Inoue, K. Sato, H. Sato, T. Sugie, H. Toba, and J. Yoshida, "Highly reliable plastic packaging for laser diode

and photodiode modules used for access network," Electron. Lett., 33 (1997) 2158.

12. S. Nemoto and T. Makimoto, "Analysis of splice loss in single mode fibers using a Gaussian field approximation," Optics and Quantum Electron., 11 (1979) 447.

13. M. Saruwatari and T. Sugie, "Efficient laser diode to single mode fiber coupling using two lenses in confocal condition," Electron Lett., 16 (1981) 955.

14. Y. Odagiri, M. Shikada, and K. Kobayashi, "High Efficiency laser to fiber coupling circuit using a combination of a cylindrical lens and Selfoc lens," Electron. Lett., 13 (1977) 395.

15. H. Kuwahara, M. Sasaki, N. Tokyo, M. Saruwatari, and K. Nakagawa, "Efficient and reflection insensitive coupling from semiconductor lasers into tapered hemispherical end single mode fiber," 6th ECOC Tech Digest, 191 (1980).

16. J. Yamada, Y. Murakami, J. Sasaki, and T. Kimura, "Characteristics of a hemispherical microlens for coupling between semiconductor lasers and single mode fiber," IEEE J. Quantum Electron. QE-16 (1980) 1067.

17. T. Yamamoto, H. Kobayashi, M. Egawa, T. Fujii, H. Soda, and M. Kobayashi, "High temperature operation of 1.3 μm narrow beam divergence tapered-thickness waveguide BH MQW lasers," Electron. Lett., 31 (1995) 2178.

18. T. Yamamoto, H. Kobayashi, T. Ishikawa, T. Takeuchi, T. Watanabe, T. Fujii, S. Ogita, and M. Kobayashi, "Low threshold current operation of 1.3 μm narrow beam divergence tapered-thickness waveguide lasers," Electron. Lett., 33 (1997) 55.

19. T. Yamamoto, H. Kobayashi, M. Ekawa, S. Ogita, T. Fujii, and M. Kobayashi, "156 Mbit/s, penalty-free 40 km transmission using a tapered-thickness waveguide BH MQW laser with external optical feedback," Electron. Lett., 33 (1997) 65.

20. S. Ishikawa, "A passive alignment technique using the Au/Sn/Au multiplayer solder for LD to fiber waveguide coupling," Tech. Dig. Series of the Integrated Photonics Research, 3 (1994) 364.

21. R. Sainati, S. L. Palmquist, and T. J. Moravec, "Packaging high-speed multigigahertz GaAs Digital Integrated Cricuits," Microwave System News and Com. Tech., 18, Part. 1:1, 94096, Part 2:2, 68 (1988).

22. K. G. Gupta, R. Garg, I. Bahl, and P. Bhartia, Microstrip Lines and Slotlines (Artech House, Inc., Boston, 1996), Chapter 1.

23. M. Suzuki, Y. Noda, H. Tanaka, S. Akiba, Y. Kushiro, and H. Issiki, "Monolithic integration of InGaAs/InP distributed feedback laser and electroabsorbtion modulator by vapor phase epitaxy," J. Lightwave Technol., 5 (1987) 1277.

24. H. Soda, M. Furutsu, K. Sato, N. Okazaki, S. Yamazaki, H. Nishimoto, and H. Ishikawa, "High power and High-speed semi insulating BH structure monolithic electroabsorbtion modulator/DFB laser light source," Electron. Lett., 26 (1990) 9.

25. T. Watanabe, Y. Saito, K. Sato, and H. Soda, "High Reliability modulator integrated distributed feedback lasers," Tech. Dig. Conf. Optical Fiber Comm. (OFC), WH9, 9 (1993).

26. S. Aytac and A. Schlachetzki, "Franz-Keldysh effect in InP," J. Opt. Commun., 6 (1985) 82.

27. A. Yariv, Quantum Electronics, 2nd ed. (John Wiley & Sons, New York, 1975).

28. H. Kressel, "Thermal resistance increase in In Solder," in Characterization of Epitaxial and Semiconductor Films (Elsevier, Amsterdam, 1976).

29. K. Mizuishi, Sawai, S. Todoroki, S. Tsuiji, M. Hirao, and M. Nakamura, "Reliability of InGaAs/InP buried heterostructure 1.3 μm lasers," IEEE J. Quantum Electron., QE-19 (1983) 1294.

30. M. Fukuda, O. Fujita, and G. Iwane, "Failure modes of InGaAs/InP lasers due to adhesives," IEEE, Trans. Comp. Hybrids Manufactur. Technol., CHMT-7 (1984) 202.

31. K. Mizuishi, "Some aspects of bonding solders deterioration observed in long-lived semiconductor lasers: Solder migration and Whisker growth," J. Appl. Phys., 55 (1984) 289.

32. K. Fujiwara, T. Fujiwara, K. Hori, and M. Takusagawa, "Aging characteristics of GaAlAs double heterostructure laser bonded with gold eutectics alloy solder," Appl. Phys. Lett., 34 (1979) 668.

33. O. Mitomi, T. Nozawa, and K. Kawano, "Effects of solder creep on optical component reliability," IEEE Trans. Comp. Hybrids, and Manufacturing Technol., CHMT-9 (1986) 265.

34. J. E. Sergent and A. Krum, Thermal Management Handbook (McGraw-Hill, New York, 1998).

35. D. M. Mattox, Electronic Materials and Processes Handbook, 2nd ed., C. A. Harper and R. M. Sampson, eds. (McGraw-Hill, New York, 1994).

36. M. M. Schwartz, Handbook of Structural Ceramics (McGraw-Hill, New York, 1992).

37. Y. Krakow, K. Utsumi, H. Takamizawa, T. Kamata, and S. Noguchi, "AlN substrates with high thermal conductivity," IEEE Trans. On Comp., Hybrids, and Man. Technol., CHMT 8:2 (1985) 247.

38. Y. Kurokawa, H. Hamaguchi, Y. Shimada, K. Utsumi, H. Takamizawa, T. Kamata, and S. Noguchi, "Development of highly thermal conductive ALN substrates by green sheet technology," IEEE Trans. On Comp., Hybrids, and Man. Technol., CHMT 9:2 (1986) 412.

39. E. S. Dettmer and H. K. Charles, "Fundamental characterization of aluminum nitride and silicon carbide for hybrid substrate applications," Int. J. Hybrid Microelectron, 10:2 (1987) 9.

40. K. Kohmoto, and M. Osada, US Patent # 5,481136, Semiconductor element mounting composite heat-sink base.

41. C. C. Lee and D. H. Chien, "Thermal and package design of high power laser diode," Proc. Semiconductor Thermal Measurement and Management Symp. Ninth Annual IEEE, 75 (1993).

42. R. E. Holmes, "Choices widen for Ceramic substrates," Electronic Packaging and Production, (Jul. 1991) 56.

43. M. J. Wale and C. Edge, "Self aligned flip chip assembly of photonics devices with electrical and optical connections," IEEE Trans. on Component, Hybrids, Man. Technol., 13:4 (1990) 780.

44. M. Itho, J. Sasaki, A. Uda, I. Yoneda, H. Honmou, and K. Fukushima, "Use of solder bumps in three dimensional passive aligned packaging od LD/PD arrays on Si optical benches," Proc. ECTC'96, 1 (1996).

45. H. Kosaka, A. K. Dutta, K. Kurihara, Y. Sugimoto, and K. Kasahara, "Giga-bit-rate optical signal transmission using vertical cavity surface emitting lasers with large core plastic cladding fibers," IEEE Photonic. Technol. Lett., 7 (1995) 926.

46. A. K. Dutta, H. Kosaka, K. Kurihara, Y. Sugimoto, and K. Kasahara, "High speed VCSEL of modulating bandwidth over 7.0 GHz and its application to 100 m PCF datalink," IEEE J. Lightwave Technol., 16 (1998) 870.

47. Y. Oikawa, H. Kuwatsuka, T. Yamamoto, T. Ihara, H. Hamao, and T. Minami, "Package technology for a 10-Gb/s photoreceiver module," IEEE J. Lightwave Technol., 12:2 (1994) 343.

48. S. Rzepka, M. A. Korhonrn, E. Meusel, and C.-Y. Li, "The effect of underfill and underfill delamination on the thermal stress in flip-chip solder joints," ASME J. Electron. Packag., 120 (1998) 342.

49. E. Madenci, S. Shkarayev, and R. Mahajan, "Potential failure sites in a flip-chip package with and without underfill," ASME J. Electron. Packag., 120 (1998) 336.

50. J. B. Nysather, P. Lundstrom, and J. Liu, "Measurements of solder bump lifetime as a function of underfill material properties," IEEE Trans. Comp. Packag. Manufact. Technol., 21 (1998) 281.

51. N. Takehashi and M. Horii, "A new premolded packaging technology for low cost E/O device applications," 47th ECTC'97, 16 (1997).

52. See also for example, J. W. Osenbach and T. L. Evanosky, "Temperature humidity bias behavior and acceleration model of InP planar pin photodiodes," J. Lightwave Technol., 14 (1996) 1865.

53. See for example, D. Yap, Y. K. Brown, R. H. Walden, T. P. E. Broekaert, K. R. Elliott, M. W. Yung, D. L. Persechini, W. W. Ng, and A. R. Kost, "Integrated Optoelectronic Circuits with InP-Based HBTs," SPIE Proc. 4290, (2001) 1–11 and also W. Zhou and M. H. Ervin, "Fabrication and Analysis of optoelectronic integrated circuits with one selective area growth," SPIE Proc., 4290 (2001) 21.

54. F. Delorme, G. Terol, H. de Bailliencourt, S. Grosmarie, and P. Devoldere, "Long-term wavelength stability of 1.55 um tunable distributed bragg reflector lasers," IEEE J. Selected Topics Quantum Electron., 5 (1999) 480.

55. F. Delorme, B. Pierre, H. de Bailliencourt, G. Terol, and P. Devoldere, "Wavelength aging analysis of DBR lasers using tuning section IM frequency response measurements," IEEE Photonic. Technol. Lett., 11 (1999) 310.

56. C. G. M. Vreeburg and C. M. Groeneveld, "Packaging trends for semiconductors lasers for DWDM applications," Proc. ECTC'99, 1 (1999).

Chapter 13 | Packaging Technologies for Optical Components: Integrated Module

Achyut K. Dutta

Fujitsu Compound Semiconductors Inc., 2355 Zanker Road,
San Jose, CA 95131, USA

Masahiro Kobayashi

Fujitsu Quantum Devices, Kokubo Kogyo Danchi, Showa-cho,
Nakakoma-gun, Yamanashi-ken 409-3883, Japan

13.1. Introduction

The basic design methodologies, such as optical, electrical, mechanical, and thermal designs, as required for different kinds of optical packages/modules used for different applications, have been described in the previous chapter. Appropriate examples related to the design criteria have also been cited. Those are the basic criteria to be considered for high-reliability and cost-effective optoelectronics package designs.

The demands for high-speed data transmission are increasing, and there are subsequent needs for an integrated optical module comprising more and more functional optical devices, and electronics devices, as well. The reasons are twofold; first to improve the total module performance, and second to make packaging doing integration cost effective. Technology developments in integrated packages are underway in the 10 Gb/s application area, and will soon be shifted toward the 40 Gb/s application area. This chapter is an extension of the previous chapter, to include the basic technologies, as required for the integrated module, and also examples of the state-of-the-art of integrated module technologies. Future road map/trends in the optical packaging area will also be included in this chapter.

WDM TECHNOLOGIES: ACTIVE
OPTICAL COMPONENTS
$35.00

13.2. Integrated Modules: Hybrid versus Monolithic

Highly functional optical modules are only possible if numbers of integrated photonics devices are available. Today's technology constraints make it difficult to achieve all types of optical components (active and passive) on a single substrate. To increase the functionality, it is necessary to use the hybridization technique to integrate a number of devices. Many technologies have already developed, focusing on hybrid integration, and some are simultaneously underway to realize monolithic integration on a single substrate. An important challenge for the packaging of integrated optics with multiple optical inputs or outputs is the attachment and alignment of multiple optical fibers and devices in the tiny package. This challenge is significantly less in the case of monolithic integration. For the hybrid-integrated package, the silica-on-silicon (Si) technologies (i.e., planar lightwave circuits:PLC technologies) are the most popular technologies, as conventional IC technology can be used. The idea behind silica-on-Si is similar to the use of the printed circuit board in the electronics industry. A network of silica-based optical waveguides can be fabricated using precision photolithography and planar patterning techniques on the Si substrate [1–9]. Alignment as well as attachment of the fibers and laser diode (LD) or photodiode (PD) can be done using the Si-V grooves, made by means of an isotropic wet etching [10–12]. For example, Si-V grooves can hold multiple fibers at extremely well-defined regular positions. Passive alignment, eliminating the complicated optical axis precise alignment, can be used. Because the Si substrate has good thermal conductivity and also closer thermal expansion of III-V semiconductors, thermal control can also be possible. But however attractive silica-on-Si technologies are as the integration platform, they are already close to their limits for monolithic integration. Silica does not possess attractive material properties for manufacturing active components such as lasers, modulators, switches, and detectors. An interesting alternative to the silica-on-Si platform is Lithium Niobate ($LiNbO_3$)-based integrated optics. Low loss waveguides can be made by titanium diffusion into $LiNbO_3$. Rare-earth ion doping of these waveguides creates a media suitable for optically pumped amplifiers and lasers. Unlike Silica, $LiNbO_3$ is an excellent electrooptical material; the same property that made $LiNbO_3$ modulators a big commercial successful monolithic integration of a DBR waveguide laser and Mach-Zehnder (MZ) intensity modulator as a part of transmitter module is reported. Despite all the benefits, $LiNbO_3$ has not been successful for integration so far. The main reasons are its higher optical loss (>0.1 dB/cm) than silica, and also the small size and expense of wafers.

In contrast to hybrid integration, monolithic integration using the indium phosphide (InP)-based compound semiconductors is the most advantageous integration method. This not only promises to put active and passive optical components interconnected by a network of waveguides in the compact assemblies, but also allows electronics and micromechanical components to be manufactured on the same chip. InP provides an excellent substrate for epitaxial growth of indium gallium aluminum arsenide (InGaAlAs) and indium gallium arsenide phosphide (InGaAsP), the two major materials for telecom lasers and detectors and the most promising candidates for ultra-high-speed electronics. InP also has good thermal conductivity and a high optical damage threshold and eliminates material compatibility problems typical of Si-based integration. Overall, InGaAlAsP/ InP monolithic integration has huge potential for delivering devices with novel functionality for all-optical networks. Since its introduction for monolithic integration, successful products include a photodetector with transimpedance amplifier module, commercially available for 40 Gb/s application [13–15]. The DFB laser diode with electroabsorption modulator [16–18] or MZ modulators [19–21] is another successful integrated module available and commercially implemented in systems. More work on monolithic integration is already underway to increase the functionality [22–29], and these devices are expected to be available commercially within a few years. Despite its many advantages in both performance and manufacturing costs, the monolithic integrated module is advancing for implementation very slowly. The main reasons are that these materials are lossy (>0.25 dB/cm) and brittle. Their optical properties are strongly nonlinear and temperature dependent, implying the necessary use of thermoelectric coolers (TEC). Also, the InP is very expensive, and the integration doesn't exceed over 3 inches. Besides the substrate itself, there are many technologies required for successfully making a highly integrated module (especially for hybrid integration). Following sections explain several technologies required for integrated electrooptics packaging.

13.3.   Technology Requirements

13.3.1.   FIBER ALIGNMENT/ATTACHMENT IN PACKAGE

In packaging, the outcoming or incoming optical signals can generally be handled using optical fiber. To do this, the optical fiber must be attached to the package in such a way that it can provide higher coupling efficiency and mechanical stability. Many technologies for making this attachment

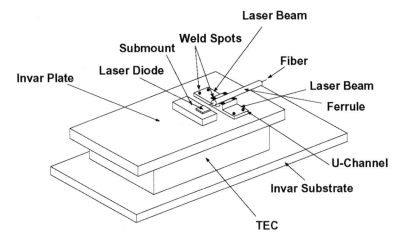

Fig. 13.1   Fiber attachment technique inside the optical module using YAG laser welding.

have developed, and these are mainly based on the type of the modules. These attachments can be accomplished using (a) YAG laser welding, (b) soldering, and (c) Silica-on-Si platform.

Fiber Attachment: YAG Laser Welding and Soldering

Modules frequently used in optical communications are usually hermetically encapsulated, and their fibers are usually attached with the package by soldering or laser welding [29–30]. Soldering often causes a large excess coupling loss due to movement caused by thermal shrinkage of the assembly components and the solder volume while the solder is cooling. Laser spot welding has the advantage of being able to strongly and quickly join high-activation energy materials without contact. Figure 13.1 is the schematic showing the pigtailed fiber and other components assembled inside the package using laser welding. Active alignment is generally used to achieve maximum coupling efficiency. In the laser weld process, reduction of optical coupling efficiency generally occurs due to post-weld shift, and recovery of the coupling efficiency is generally achieved using the laser hammering process. These technologies are well matured and widely used for the single optical device (LD or PD) and single optical fiber attachment.

Fiber Attachment: Using Silica-on-Silicon Platform

An important challenge for the packaging of integrated optics with multiple inputs and outputs is the attachment and alignment of multiple optical fibers. One technique that has evolved to address this challenge is the use of Si-V groove arrays to hold single or multiple fibers at extremely well-defined regular positions [11, 12]. A maximum of 96 fibers, with high precision defined by photolithography techniques, can be attached on the Si-substrate, and coupled with the waveguide arrays. The isotropic etching technique [17], frequently used in conventional IC technology, can also be used for making the V-grooves in Si. Figure 13.2 is an example of the fiber attachment configuration onto the silica-on-silicon platform. In the V-groove approach, the fiber is aligned by the two angled walls of the groove and is held either by a flat top or another V-groove Si platform, as

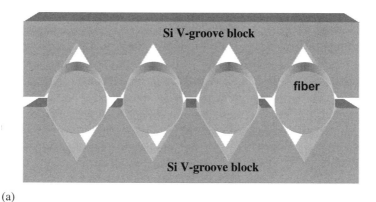

(a)

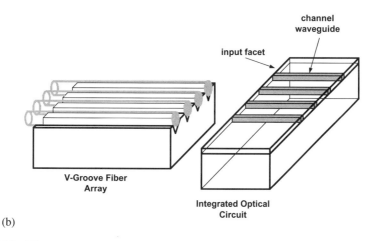

(b)

Fig. 13.2 Fiber attachment and coupling techniques using the PLC platform: (a) Fiber arrays attachments using the V-groove blocks, and (b) Coupling of fiber arrays to waveguide arrays.

shown in Fig. 13.2(a). This V-groove method fully constrains the fiber while allowing the fiber to be placed onto the grooves from the top. Light curing adhesive [31] can be used to adhere two V-grooves to hold the incoming fiber. Passive alignment techniques can be used for aligning the fibers with the optical devices such as PD, LD, and waveguide. Image processing of the alignment marks on the Si bench and the LD or PD chip realizes the passive automatic alignment. Coupling loss as low as 5 dB can be achieved using this technique.

13.3.2. TECHNOLOGIES FOR WIRING

Typical Interconnection

For a comparatively lower speed optical module, conventional wire bonds are frequently used for electrical interconnection, as shown in the previous chapter (see Fig. 12.11). However, in higher speed optical modules, especially the multi-gigabit ranges, every signal line is required to be considered as a transmission line, and the carrier with transmission line and interconnection must be designed for minimizing the resonance/reflection, which can degrade the electrical signal and thus the optical signal.

Planar Lightwave Circuit (PLC) Platform

In hybrid-integrated modules where the PLC platform is used, internal wiring is designed based on the speed and the cost of the module. For low-cost applications where the operation bandwidth is less than 1 GHz, the main concern about wiring is the reduction of the parasitic capacitance. The parasitic capacitance can be decreased by inserting silica under cladding between the electrode and the silicon substrate [32]. The electrode on the under cladding still experiences significant capacitance if the wire length is over 1 mm. The capacitance could be minimized more if the electrode is placed over the clad. In the wide bandwidth module application, however, the coplanar line or micro-strip line on the PLC platform is necessary. In that case signal distortion due to propagation loss, as described in Chapter 12 (section 12.4.2), also occurs if the underlying clad thickness is not thick enough. For example, the propagation loss of the co-planar line at 10 GHz can be reduced from 17 dB/cm to 4 dB/cm, if the co-planar line is on the 15-μm-thick silica layers instead of 1.5 μm-thick layers [33]. The higher propagation in 1.5-μm-thick layers is due to the larger loss tangent (tan δ) of the silicon substrate. Depositing a thicker silica layer on silicon, on the other hand, creates large bending of the substrate because of the thermal expansion coefficient difference between the silicon and silica. To avoid any bending, the waveguide (lower clad and upper clad) should be designed with a thickness of around 25 μm. For this, a coplanar line on 15-μm-thick silica is appropriate [33]. The propagation loss for the coplanar line is larger in silica than in the coplanar line formed on alumina substrate.

For longer electrical interconnections on the PLC platform, the micro-strip line can be used, which is composed of a lower electrode on the platform (acting as the ground), polymide insulation layer, and upper electrode for the interconnection. Its propagation loss is dependent on the polymide thickness. Recently, propagation loss as low as 0.9 dB/cm at 10 GHz is reported for the micro-strip line on 29 μm of polymide. It has the advantage of being insensitive to under cladding thickness, because the ground electrode shields the electrical field to the silicon substrate [33]. Separate printed wiring carrier (PWC) with well-designed transmission lines [34], as shown in Fig. 13.3, can also be used in the PLC-based integration. Passive alignment technique could be used for mounting this PWC, simultaneously

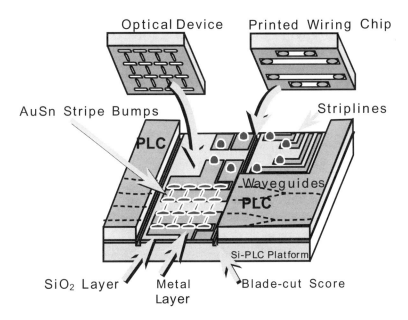

Fig. 13.3 Schematic showing the electrical interconnection in the PLC platform.

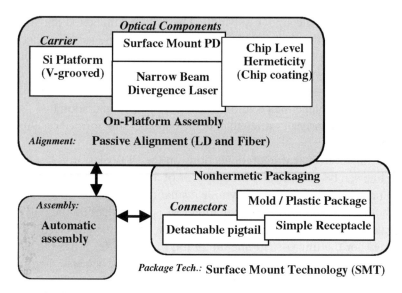

Fig. 13.4   Enabling technologies for low-cost optical module/package.

with other hybrid components, and it also be beneficial in cases where the grooves are used in the PLC platform.

13.3.3.   ENABLING TECHNOLOGIES: LOW-COST PACKAGES

The component costs, as well as the assembly techniques as required for optical packaging, are to be considered seriously, as assembly and packaging account for more than 50% of the total module cost. A few enabling technologies for producing cost-effective optical modules have already been developed, and some are in the development stage. Figure 13.4 shows the block diagrams representing the different enabling technologies. The availability of automatic assembly, low-cost connectors and components basically determine the degree of cost reduction in module manufacturing. In addition, easy alignment technique is also required, and thereby the optical components are to be designed suitable for the respective alignment technique.

Related technologies and components, as shown in Fig. 13.4, are developed to reduce the module cost. Today's optical module assemblies that are performed serially (one at a time and manually by hand) using mostly active alignment mode, are needed as a foundation to change to the parallel mode (automation) using the passive alignment technique [12]. This alignment technique for optical coupling between different optical components is considered to be the least time consuming, and most accurate technique [35–39]. Recently, accuracy of passive alignment as low as 0.5 μm ranges in x-y-z direction [40] has been attained using the V-groove on the Si-substrate.

Further advances in automatic module assembly and reduced module cost can be realized when optical pigtail cords can be replaced by newly developed detachable pigtail cords. Small glass/plastic ferrule can be used inside the package to facilitate the coupling of incoming/outgoing beam for Laser/PD module. In this case, Si-substrate is also used as a motherboard for mounting different optical components. Conventional Si process technology is generally used for defining pits and standoffs on the Si-platform, and a passive self-alignment technique is used for precision aligning of the fiber, LD, and PD in assembly. To achieve low coupling loss by passive alignment, either hemispherical lens and/or modification of the LD/PD are very much essential. Narrow beam divergence laser diode [41–43] and surface mount photodiode [44, 45] are also developed specifically to further increase the coupling efficiency and also the alignment tolerance in assembly. Generally, the coupling efficiency and optical alignment tolerance are in a trade-off relation. Alignment tolerance has to be balanced to actual positioning accuracy. Increasing alignment tolerance helps to increase the yields in manufacturing. The coupling tolerance and coupling efficiency of LD to SMF or silica-waveguide is generally dependent on how closely the mode field of the LD beam matches with that of fiber. A laser structure with an integrated mode size converter, as shown in Fig. 13.5, yields a better overlapping mode to fiber or waveguide from laser output and makes coupling efficiency as high as 70% [41–43]. The PD to be used either for monitoring the laser power or for the receiver module must also be of easy self-alignment structure. Figure 13.6 shows the cross sectional view of self-aligned PD. High sensitivity of more than 0.85 A/W ($\lambda = 1.3$ μm) can be obtained with wide tolerance bands of 25 μm and 65 μm (10% decreases) in the vertical and horizontal direction respectively [46, 47].

Nonhermetic package technology, especially using the injection mold or plastic molding, is also required for further reducing the module cost. For the nonhermetic package, not only the package technology itself, but also the chip technology, especially the chip's coating, has to be developed. Recent studies have shown that high reliability could be achievable with even nonhermetic package due to the improvement of the chip quality.

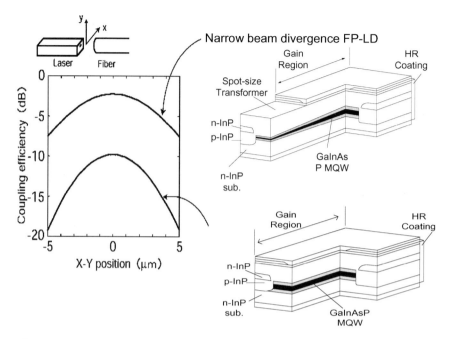

Fig. 13.5 Comparing the coupling efficiency between the LD with spot-size converter and the conventional LD. Tapered-thickness waveguide works as a mode converter and the divergence of the laser beam becomes about 1/3 of the conventional laser.

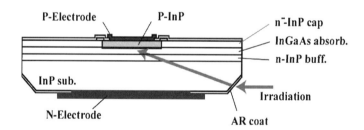

Fig. 13.6 Structure of the self-aligned photodiode.

These are all enabling technologies, some of which are already available, and some under development. The trends toward using a nonhermetic packaging scheme in telecom applications will be widely accepted by further development of laser and detector chips (the fabrication technology along with the passivation technology), and demonstration of long-term reliability. The following section explains the state-of-the-art of different kinds of optical modules including low-cost modules.

13.4.   Different Packages

13.4.1.   *LOW-COST PACKAGES FOR SINGLE FUNCTIONAL DEVICE*

To date, the telecommunication industries mostly use single functional devices in their systems. The package of this device could be hermetic or nonhermetic, which is especially dependent on the area of applications. The packages for single functional device also include similar packages; butterfly and coaxial package, already cited in the previous chapter. This section describes mostly nonhermetic packages for single photonics devices.

Uncooled Mini-DIL Module: DFB-LD

The 14-pin butterfly package for cooled laser module has been the industrial standard for a while. The thermoelectric cooler (TEC) is generally used inside the 14-pin LD module to stabilize LD performance in any operating temperature. Performance improvement of LD chip for high-temperature operation can remove TEC from laser module, and could make the module a smaller size and more cost effective. In contrast with the cooled butterfly package, uncooled coaxial (TO-can) packages (hermetic and nonhermetic) are also considered to be the low-cost package in the industries. However, this package has the disadvantage in placing on the PCB (planner circuit board). To make for easier placement, there are mini dip-in-line (DIL) packages, which are flat and uncooled, and a smaller size than the standard butterfly package (14-pins). This is expected to replace TO-can package and cooled laser module (14 pins butterfly package) in the near future.

   Figure 13.7 shows the schematic structure of a mini-DIL laser module. Silicon submount is used in the mini-DIL package for mounting the laser and the monitor photodiode. V-groove is formed on the Si substrate to hold the first lens positioning. The Si substrate is inexpensive, and has a good thermal conductivity, and is also easy to wafer level volume production using commercially available Si process technologies [1–10]. Anisotropic etching of the substrate for making V-groove is of sufficiently high precision, and easily achieves one-micron-meter range dimensional accuracy. Using a combination of 1st spherical lens on Si-platform and 2nd aspherical lens inside the package window, optical coupling efficiency as high as 40% can be reproducibly achieved. The fiber can be attached with the package using YAG welding.

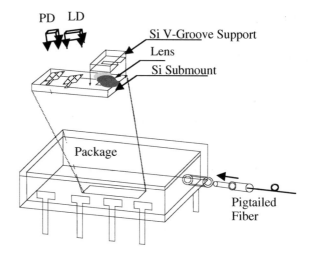

(a)

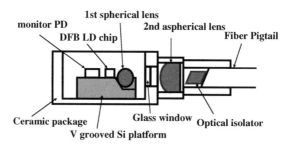

(b)

(c)

Fig. 13.7 Schematic showing the mini-DIL module: (a) assembly technique, (b) cross sectional view, and (c) actual mini-DIL laser module @ 2.5 Gb/s.

Value-added functionality such as micro-strip line and terminating resistor can also be easily added on Si-substrate for good electrical behavior.
The module can perform well at 2.5 Gb/s. Also, more than 40 km single
mode fiber transmission is easily possible using the distributed feedback
laser diode (DFB-LD) of 1310 nm lasing wavelength and also an appropriate electrical connection design. Typical small signal frequency responses
(S_{21} and S_{11}) of the mini-DIL module are shown in Fig. 13.8. The perfect
design of the signal line and also the high frequency feedthrough section,
as discussed in the previous chapter, helps to achieve better frequency

O/E characteristic

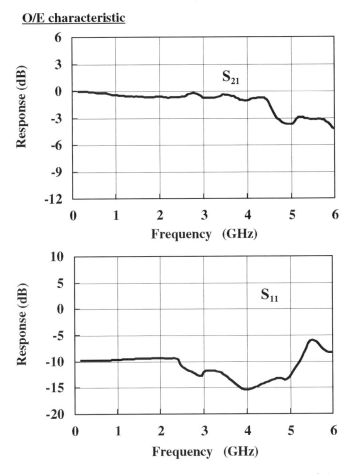

Fig. 13.8 Frequency responses: (a) O/E characteristics, and (b) S_{11}, of the mini-DIL
LD module. The package structure used is shown in Fig. 13.7. The 1310 nm λ/4 shifted
DFB-LD was used in the package.

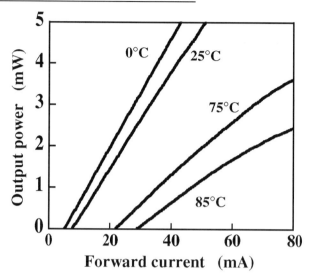

Fig. 13.9 L-I characteristics of the mini-DIL LD module at different case temperatures. The package structure used is shown in Fig. 13.7. The 1310 nm λ/4 shifted DFB-LD was used in the package.

response for the module. An operating temperature range of 0 to 85°C can also be achieved with uncooled operation. Figure 13.9 shows an example of typical light vs. current characteristics of an uncooled mini-DIL LD module at different case temperatures. Appropriate thermal design of the package, as was discussed in the previous chapter, and also design of the LD chip, could help to attain a module for uncooled operation with considerably lower threshold current. The λ/4 shifted DFB-LD [48–50], developed for uncooled operation, is used in the mini-DIL package. Recent developments also indicate that an uncooled LD module can be accomplished at 10 Gb/s [51, 52], if the parasitic effect associated in the LD chip could be minimized.

Plastic/Mold Packages

Flat (Surface Mount) Module: LD/PD

The concept of the surface mount module came from the idea of the re-flow soldering on an electric circuit board. Its package is compact, 12 mm by 7.6 mm, and its flat shape, usually only 3 mm height [52], houses the

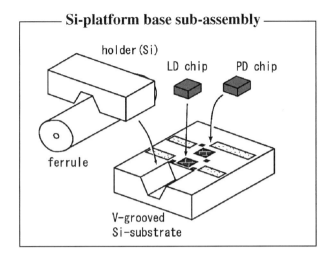

(a)

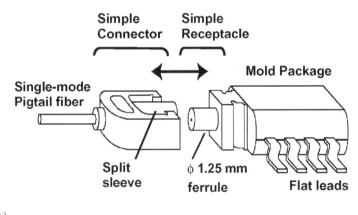

(b)

Fig. 13.10 Schematic diagrams showing the flat (surface mount)-type mold package and its assembly: (a) Si-platform base subassembly, and (b) complete module with detachable connector.

optical units and meets all requirements for the high density mounting on both sides of the electric circuit board. The LD/PD inside the package is usually mounted using the low-cost packaging technique, already explained earlier. Figure 13.10(a) shows the Si-platform base typical subassembly technique. Image processing technique is used to position the LD or PD onto the submount carrier.

Others functionalities such as micro-strip line, solder bumps, and terminating thin film resistors are easily added. To make it more cost effective, the passive self-alignment technique along with Si-V grooves technology can be used for mounting LD on Si substrate, and aligning it with incoming fiber. High precision control of <0.5 um in the x-y-z direction can be achieved in the V-groove technique. For increasing the coupling efficiency compatible for passive alignment, the tapered LD, explained in an earlier section, is now used, with the added benefit of increased alignment tolerance in assembly. For further cost reduction, a detachable connector is usually preferred in the flat package. Figure 13.10(b) shows an example of a flat module with a detachable connector. A simple receptacle structure in the fiber out port can be used, which consists of a glass ferrule with short length fiber and hooking parts. The connector pigtail with a clip and a ferrule in a sleeve is easily mated to the receptacle by using the clip-hooking parts, as shown in Fig. 13.10(b). The two ferrule end faces (receptacle portion and connector cord) physically contact with each other in the sleeve [52]. Based on the high mechanical accuracy of the glass ferrule and the fine polishing of the ferrule end face, low insertion loss and high return loss could be achieved [52, 53].

Plastic Package: VCSEL and Edge-Emitting LD

Cost reduction is the key for the future development of the optoelectronics components, mostly driven by upcoming large photonic network applications. To obtain a low-cost optical module, total assembly techniques are to be seriously considered, as assembly and packaging accounts for a considerable portion of the module cost. In achieving low-cost manufacturing, the photonic devices are to be encapsulated by plastic package, usually a nonhermetic and uncooled type package. Approaches with beam deflections between laser (vertical-cavity or edge-emitting) and optical fiber are generally used. Common for this solution is the submount, mostly made from Si, that carries the active devices but also some mechanical or optical features.

The surface-emitting type devices such as LED and vertical-cavity surface-emitting laser (VCSEL) and also edge-emitting-type LD, can be encapsulated by molded plastic, as shown in Fig. 13.11 [54–60]. The top-emitting device can easily be attached onto the lead frame. However, for the edge-emitting LD, the transformation of the lateral emission to a vertical transmission can be accomplished by using the micro optical 45-degree mirrors in front of the emitting facet. The common base of the LD chip

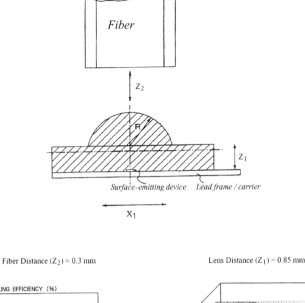

(a)

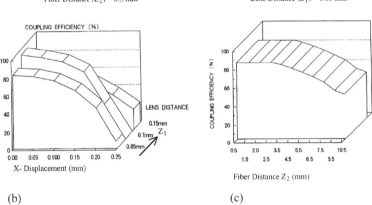

(b)   (c)

Fig. 13.11   Cost-effective molded lens: (a) Cross sectional view, (b) & (c) simulated coupling efficiency with different parameters of molded lens. In the simulation, plastic fiber with large core, and refractive index of 1.56 were used.

and 45-degree micromirror is a Si-submount, which acts also as the heat sink for the LD. Well-designed molded plastic lens, as shown in Fig. 13.11, can make the radiation divergence below <9 degrees, and coupling efficiency with fiber (e.g., plastic optical fiber >90%) [54, 57]. An example of such molded plastic package with surface-emitting device is shown in Fig. 13.12. This package is usually suited for the short-haul data-links application.

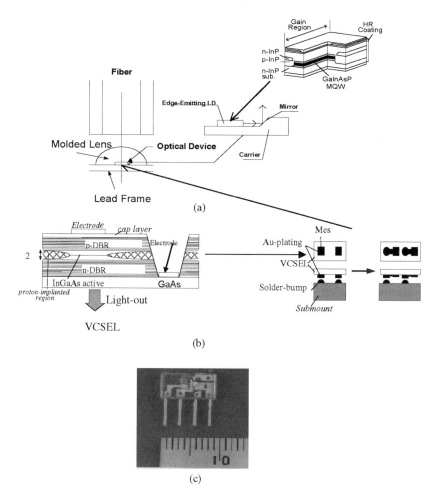

Fig. 13.12 Molded package: (a) Cross sectional view of package with edge-emitting LD scheme, (b) cross-sectional view of VCSEL and its mounting on the carrier using flip-chip bonding technique, and (c) with VCSEL package scheme.

The optical devices to be encapsulated by plastic package are required to have improved performance, and also to satisfy specific performances. As plastic materials have the property of absorbing moisture, and also lower thermal conductivity, the photonics device development must be compatible for plastic package. It requires not only good passivation technology for device fabrication for preventing any moisture ingression, but also high-temperature characteristics. In addition to chip development itself, the development of proper plastic material for packaging and also

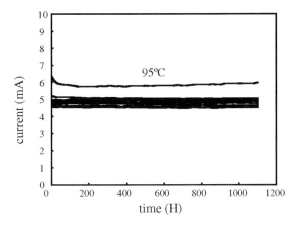

Fig. 13.13 Aging characteristics of plastic molded package with VCSELs @ 95°C and 85% relative humidity.

thermal design is required to achieve good performance. Highly reliable operation of the plastic package must be assured for implementation in the data or telecommunication. Figure 13.13 shows the aging characteristics of the VCSELs, encapsulated by the molded plastic. High temperature and humidity aging in long run operation (>1000 hr) assures compatibility. Recent development of new plastic material for preventing moisture penetration, and also the LD chip, push the molded plastic-based package in new trends of packaging [60].

13.4.2. PACKAGING FOR MULTI-FUNCTIONAL DEVICES

Wavelength-Locker Integrated Laser Module

Dense wavelength-division multiplexing (DWDM) technique is the only technique to increase the transmission capacity, in which several signals at different wavelengths (around 1550 nm) are transmitted in parallel on the single fiber. Newer DWDM systems, with up to almost 100 channels, are used with the channel space of 0.4 nm (50 GHz), and they require the laser module with a highly stable wavelength operation against ambient temperature during full life (20 yrs) cycle operation. It is expected that wavelength variation of 50 pm or less (<6 GHz) is required in a DWDM system with 0.8 nm (100 GHz) channel spacing, and much more less <25 pm (<3 GHz) wavelength variation is expected to maintain for 0.4 nm channel space. As channel space gets down further in the DWDM system, the laser module

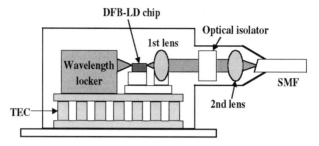

Fig. 13.14 Schematic structure of wavelength-locker integrated module. All components constructing the wavelength-locker are built into the standard 14-pin butterfly package.

having wavelength variation of further lower than <25 pm (3 GHz) is hard to achieve. The main reason is that the wavelength variation isn't only caused by the chip itself, but also due to other components such as thermistor, TEC, etc. used inside the module. Unfortunately, such high-quality chips and components are not readily achieved with today's chip and component fabrication technology. To overcome this, the laser module with wavelength monitoring and controlling capabilities using the passive optical components is necessary for this DWDM application.

Recently, a micro-optics wavelength locker module has been integrated into the conventional high-power CW laser package of 14-pins butterfly package [61]. An example of such a laser module, hybrid integrated with the wavelength locker capability, is shown in Fig. 13.14. A 1550-nm DFB laser diode designed for single mode and high power operation is mounted on a heat sink carrier. Light beam from laser front facet is coupled into a fiber with about 70% efficiency using two aspherical lens systems. Light beam from laser rear facet is introduced into the internal wavelength locker module, whose conceptual illustration is shown in Fig. 13.15. Beam splitter is used at the rear end of the laser to divide the beam into two: one is to be incident to PD1 for laser power monitor and the second is to be incident to Fabry-Perot etalon and PD2 for laser wavelength monitor. Constant wavelength operation (wavelength locking) can be achieved by a feedback to the laser temperature to maintain PD2 monitors current constant, even when the laser oscillation wavelength changes by an aging effect. Fabry-Perot etalon filter is used as the wavelength dispersive elements. Figure 13.16 shows an example of such a filter characteristic. A periodic transmission characteristic of Fabry-Perot etalon filter makes a tunable option to ITU-T's wavelength grids. Its temperature realizes setting the oscillation wavelength of DFB LD to each desired grid. A good wavelength locker should give

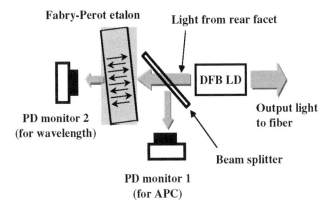

Fig. 13.15   Concept of the internal wavelength-locker.

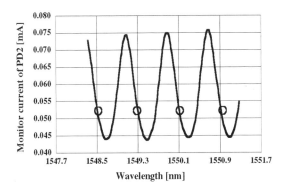

Fig. 13.16   Transmission characteristics of Fabry-Perot etalon against the laser temperature variation from 10 to 55°C. Vertical grids represent ITU-T's grid wavelengths and open circles show the monitor current to retain each setting wavelength.

a wavelength resolution of about 5 pm and have sufficient locking range around each wavelength at which the laser is aimed to operate. This type of integrated wavelength locker showed close to 5 pm wavelength shift with variation of temperature from 0 to 70°C.

The aging of the integrated wavelength locker might be an issue, but because only passive F-P etalon and simple detector are used, the degradation should be much less than that of the laser sections with current injection. The wavelength shift of less than 5 pm in long time accelerating test can be obtained for the type of integrated wavelength-locker, as shown in Fig. 13.14. This confirms the wavelength shift of less than 5 pm in long-run operation.

Receiver Module with Integrated Pre-Amplifier IC

With increasing of the data rate in optical communication, integration is very much necessary. Higher electronics functionalities for high-speed optical system are also necessary for the high-speed optical receiver. In many multielement systems or circuits, the performance is strongly influenced by those elements located near the input system or circuit. This is also true in the digital optical receiver, where the performance of the photodetector and the amplifier elements will have a strong impact on the receiver and system performance. In addition to the individual performance of these elements, the electrical and physical design of the interface between them is equally critical.

Photodiode (PIN or APD) with hybrid integration of the preamplifier is the first step in the high-speed optical receiver. Various interconnection options can be adopted for the high-speed optical receiver package. Figures 13.17 and 13.18 show the internal structure of 10 Gb/s optical receivers as the examples. In the package [62], a high-speed InP/InGaAs avalanche photodiode (APD) and a low noise and wide-bandwidth GaAs heterobipolar transistor (HBT) IC are integrated as shown in the schematic circuit-diagram of Fig. 13.19. The differences between these two are the

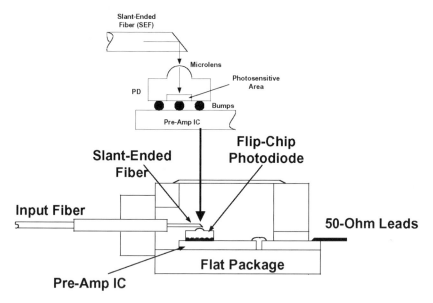

Fig. 13.17 Cross sectional view of the receiver (APD and pre-amplifier) module. Slant-ended fiber and the surface-emitting photodiode are used in the optical module.

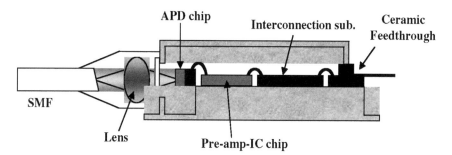

Fig. 13.18 Cross sectional view of the receiver (APD and pre-amplifier) module. Spherical lens is used in this case for the optical coupling.

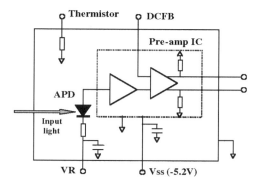

Fig. 13.19 Circuit diagram of 10 Gb/s InP/InGaAs APD and GaAs HBT pre-amplifier IC integrated receiver module.

optical coupling and also the bonding of the photodiode carrier. High optical coupling, and fewer discontinuities in the signal line due to the interconnection, are very much necessary for a high-peformance receiver. Slanted fiber [63] or the combination of the lens can be used for coupling the optical signal to the photodetector. Fabricating a monolithic micro-lens on InP APD helps to increase the optical coupling tolerance, even with a small APD active area of about 10 μm in diameter [62]. The flip-chip bonding or short wire bond technique can be used for interconnecting the PD with the pre-amplifier IC chip. Interconnection between the APD chip and the HBT-IC chip is designed to achieve low stray capacitance and inductance for wide bandwidth operation. Figure 13.20 shows an example of measured small signal frequency response of the APD/Pre-amp integrated receiver. As high as >8 GHz @ $M = 3$ can be attained with wire bond and >10 GHz can be attained if flip-chip bonding is used [63, 64]. Coplanar differential

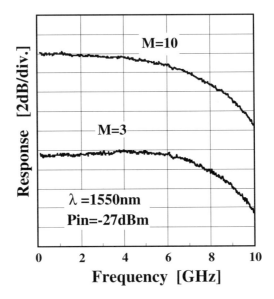

Fig. 13.20   Small signal frequency response of the receiver module at various APD gain ($M = 3$ and 10).

output with ceramic feedthrough and 50 ohms impedance is suitable for connecting the module on a circuit board. Figure 13.21 show the modulation and the BER performances of the 10 Gb/s optical receiver integrated with the APD and pre-amp. A similar package is also used to design the PIN photodiode receiver.

A common standard package is highly desirable for the high-speed receiver module such as the 14-pin butterfly package. However, there has never been a common industrial standard package with common form factor and pin-function assignment, in the case of integrated photodiode and pre-amplifier receiver package for high-bit-rate telecom application. This is mainly due to the differences in pre-amplifier functional design. Establishment of a multi-source agreement would help the development of a standard package style. Figure 13.22 shows an example of a 10 Gb/s receiver module, which is in a multi-source package.

Hybrid Integrated Module on PLC Platform

It is widely recognized that the bandwidth in customer premises will only be possible if the fiber-based network is extended from metro areas to customer premises. Bringing optics closer to the end-users, especially in fiber-to-the-home (FTTH), can be realized based on the availability of the low-cost,

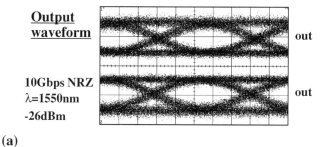

(a)

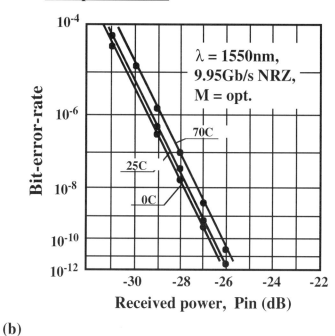

(b)

Fig. 13.21 Modulation performances of the receiver (APD + pre-amplifier): (a) output waveform and (b) bit-error-rate @ 10 Gb/s NRZ.

mass-produced optical modules. As explained in earlier sections, conventional optical module usually consists of many components, including a LD mounted on heat sink, a lens in a metallic holder, and a fiber ferule surrounded by a metallic fiber sleeve in the LD module. These components are assembled using active alignment (by operating the LD), and the maximum fiber output power is adjusted aligning the lens and the fiber. A module

Fig. 13.22   Picture of the 10 G receiver (APD and pre-amplifier) module.

architecture of such kind complicates the assembly process and increases the number of components required, which results in increasing the cost.

Besides FTTH application, it is also essential to increase the module functionality in many applications where the size and shape are the main concerns. In last few years, much work has been going on to avoid complication in the assembly, and also to increase the module's functionality by integrating several functional devices in a single module. Photonics and electronics devices are to be integrated inside the single module, either monolithically or hybrid, to increase functionality. Monolithic integration is unfortunately not readily achieved with today's fabrication technology. Today's hybrid integration using the Si-platform can only allow to have integrated optical module with higher electronics and optical functionalities [65]. The technology for surface-mounting optical devices onto a platform such as silica-based PLC is very attractive [66, 67] due to its potential for cost reduction in the optical module assembly process and increases in the number of optical components required with easy coupling technique.

Bi-Directional Transmitter/Receiver Module

In FTTH application, a bi-directional module is necessary to transmit and receive separate wavelength optical signals over the single fiber. Figure 13.23 shows an example of the bi-directional module consisting of 3 dB Y-branch silica waveguide, an LD transmitter, a PD receiver,

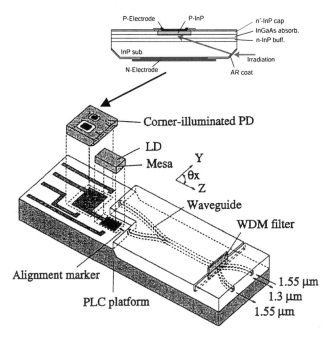

Fig. 13.23 Schematic diagram showing the integrated bi-directional module on the Si-platform for FTTH application. Corner-illuminated PD is used in the module.

a monitor PD to control the LD power automatically, and a 1.3/1.55-μm optical WDM filter [45]. The module can multiplex and demultiplex the 1.3/1.55 μm wavelength optical signals, and transmit and receive the 1.3 μm bi-directional signal. Optical waveguide is used to mux/demux the optical wavelengths and also to guide the optical signal to/from the PD/LD.

Silica-on-silicon technology, as mentioned earlier, is used in fabrication of the bi-directional transmitter/receiver modules. Alignment markers are generally used on the PLC platform for accurate positioning of the wave-guide in the lateral direction, necessary in passive alignment technique for assembling the LD/PD. Flip-chip self-alignment bonding can be used in attachment of LD/PD on the PLC platform. The coupling loss and alignment tolerance can be adjusted by using the LD with spot-size converter, as described in Section 13.3.1, and also by using the refractive index gel around the LD and fiber [38]. Si-resin or per-fluoride liquid having lower refractive index (than optical fiber) can be used as the index matching gels. Of them, Si-resin, having a refractive index of 1.39, is suitable for the module assembly because of its gelatinous and stable chemical and electrical characteristics. Fiber with micro-spherical lens is also used to increase the coupling efficiency and the alignment tolerance. To achieve coupling loss as low as 5 dB and high alignment tolerance ($>12\,\mu$m horizontally, and 2 μm vertically), fiber with spherical lens, and the refractive index matching gels (refractive index of 1.39) must be used [38]. Typical surface-emitting type PD or corner-illuminated PD can be used for this module. In the case of corner-illuminated type PD, as shown in Fig. 13.23, the light beam is illuminated at the back of the substrate, a portion of which is formed into the angle surface, and it is guided to the absorption region. This provides high coupling efficiency (making $>$0.85 A/W) and good uniformity [45]. For the purpose of fiber alignment to couple with waveguide, V-groove approach, as mentioned in the earlier section, can be made on the Si-bench. Sufficient grooves length is necessary to ensure sufficient strength in the fiber mount, and generally 3 mm [38] or more groove length is necessary. In this method, the excess coupling loss between the fiber and the waveguide caused by displacement is as low as 0.1 dB.

Using a similar approach, bi-directional transmitter/receiver module incorporated with receiver IC onto the PLC platform can also be thought about for the subscriber-end device (e.g., FTTH) application. Figure 13.24

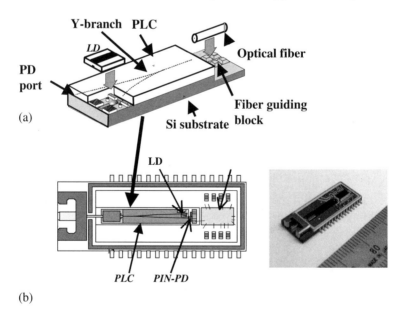

Fig. 13.24 An example of the integrated bi-directional module on the Si-platform for FTTH application; (a) Subassembly technique using the Si platform, and (b) Complete schematic of the module and its picture.

shows the schematic of a bi-directional module encapsulated with ceramic or metal package. A Y-branch splitter/mixer can also be used to split or mix the incoming and outgoing signal to the fiber. The O/E conversion is possible with the PD and receiver IC (pre-amp) directly attaching onto the PLC platform. The fiber alignment and also LD/PD attachment on the PLC can be accomplished using the passive technique, as mentioned earlier. In both packaging schemes, as shown in Figs. 13.23 and 13.24, the module can be connectorized with the pigtail, which is detachable or non-detachable.

Matrix Switch with Optical Amplifier Module

In standard telecom applications, especially in trunks and metro systems, an optical module with higher functionality is required. As the monolithic technology is not available to fabricate the higher functionality optical module, hybrid technique is the most effective way to make such optical modules. An example of such an optical module is the matrix switch with the semiconductor optical amplifier gate [34, 68]. Such an optical module is necessary in high throughput telecommunications, capable to cope with large increases in traffic. This kind of module is promising, because of its high extinction ratio, high optical gain, and fast switching time (sub nanosecond).

Figure 13.25 shows an example of the hybrid integrated optical gate-matrix switch. This module consists of four 1:4 optical splitters, four 4:1 combiners, and four-channel semiconductor optical amplifier gate (SOAG) array chips, which are mounted on the PLC platform. The splitter and combiners are formed on the common PLC platform, on which SOAG array chips are also mounted. A single-mode optical fiber (SMF) array is assembled in the fiber guides at each end of the PLC platform. All hybrid components such as SOAG and SMF can be hybrid integrated on the PLC platform using the passive alignment techniques, mentioned in earlier sections. Figure 13.26 shows the schematic of the configuration of the hybrid integration and module complete picture. Coupling loss between SOAG and waveguides is typically ∼8–9 dB. To reduce such loss, spot-size converter (SSC) integrated SOAG arrays [69] can be used. Radiation divergence of 13–14 degree (FWHM) in both the azimuth and elevation planes can also be achieved, as opposed to ∼40 degree in non-SSC devices. In this case the coupling loss can also be down to 4–5 dB. Fiber-to-silica waveguide coupling can be accomplished using the V-groove on the PLC platform, as mentioned earlier. The coupling loss <0.8 dB can

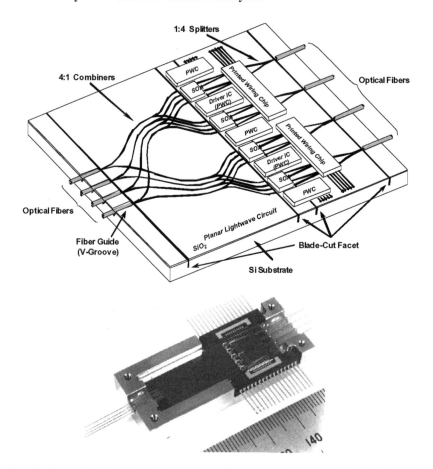

Fig. 13.25 A schematic and a photograph of the hybrid integrated 4×4 optical matrix switch module, composed of SOA, PLC splitters in both ends.

be achieved between single-mode fiber and silica waveguide. The attenuation of silica waveguide is dependent on the silica fabrication technology. APCVD-based technology provides an attenuation loss below 0.1 dB/cm [6]. To have electrical connection for SOAG, printed wiring carrier having the transmission lines in one side can be used, and that PWC can be mounted with SOAG onto the PLC platform using flip-chip bonding. This PWC helps to bridge the electrical connection across the PLC platform, if the grooves are made across the PLC platform. This PWC could also be applicable to the assembly of driver IC chips, if it is required for higher integration. In this module, extinction ratio as high as 40 dB can be

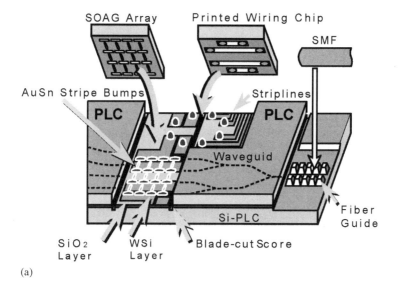

(a)

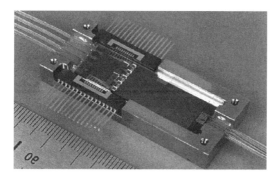

(b)

Fig. 13.26 Optical gate matrix switch: (a) Schematic showing the configuration of the hybrid integration, and (b) its picture.

achieved, large enough for optical gates so as to suppress coherent cross talk [70].

High-speed packet switching from one port to others could be easily possible using this integrated module. Figure 13.27 shows the operation of the hybrid integrated matrix optical gate switch. One $(4 \rightarrow 1')$ of two packet (cell) signals, propagating to the two different ports $(1 \rightarrow 1'$ and $4 \rightarrow 1')$ can be transferred to the same output port $(1')$ selecting the gate driving current. In this example, the input signals are modulated into 10 Gb/s, and

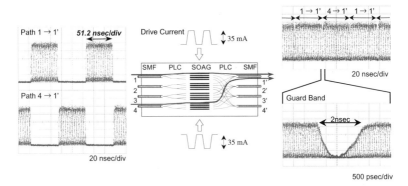

Fig. 13.27 Schematic showing the operation principal of the hybrid integrated optical gate matrix switch.

its power was −5 dBm/channel. The packet length was settled for 512 bits (51.2 ns/packet). The driving current was 35 mA, applied alternately on the two SOA gates. Each packet in the output port is separated by 2 ns guard band. As this kind of module is built-up using the discrete components, fabricated by well-matured technologies, the reliability is also high. Using such hybrid-integration techniques, it is possible to fabricate novel optical modules with higher functionality.

13.5. Conclusions and Future Prospects

The state-of-the-art technologies, as required for integrated optical module, are presented in this chapter. Today's hybrid integration techniques allow achieving of the high functional optical module, eliminating the technology barrier required in monolithic integration. Future developments for the optical module will see further miniaturization, and incorporation of both electrical and optical functional devices inside the single package, reducing the space requirements to cut the system complexities, high volume manufacturability, and yet lower cost.

Bandwidth demands push the optical networks technologies from comparatively low speed to higher speed per channel. For this, the module technology development is moving toward extending module performance to higher speeds, better signal fidelity, less electromagnetic emissions, and hot pluggability for field installable flexibility. This pushes the necessity of high-scale integration of a variety of photonics and electronics devices in a single package. Figure 13.28 summarizes the trends and road maps

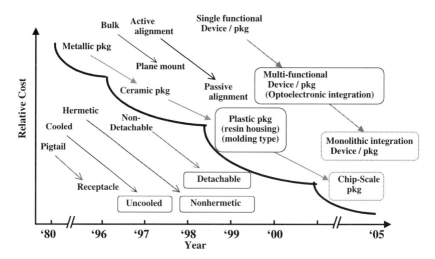

Fig. 13.28 Summary of the trends and road maps of the packaging technologies for optical components.

of the optical package technologies. Past and current experiences suggest that future packaging technologies should move toward more integration, shrinking of the electronics and photonics devices, and lower power consumption. The competitive edge of this integration comes from the increased scale of integration plus the corresponding reduction in packaging costs. In the short run, the combination of III-V and PLC technologies on Si wafer and hybridization by passive alignment techniques will act as a key role to achieve low cost manufacturing and also higher optical functionalities. To offset the cost and complexities, future developments will need a breakthrough in photonics ICs, which includes the monolithic integration of the various photonics device to increase the optical functionality, and their easy coupling with the incoming single-mode fiber.

Fabrication process developments, including the selective area growth, multiple growth, and etch processes, substrate fusions, etc., is another area of trends for fabricating the various optical devices. In packaging, alignment technique covers most of the packaging cost, and this must be improved for highly integrated packaging. Today's optical module assembly, performed serially (one at a time and manually by hand) using mostly active alignment mode, will be required to change to parallel mode (automation) using the passive alignment technique. Passive alignment technique is the best approach making the high-density integration. This requires development of both component and assembly technologies in parallel.

In designing of the highly integrated package in the future, thermal and mechanical designs would be the challenging part as the power consumption inside the package is increased. This pushes the requirement of efficient thermal management consideration in the package design stage. Several simulation tools are available to estimate the behavior separately in the design stage. However, integrated simulation tools coupling with mechanical, thermal, optical, and electrical simulations would be helpful to estimate total performance of the module in the design stage. System-level simulation is also helpful to visualize system performance in the module design level. Several opportunities for coupling all simulation tools are left, which would allow the package designer to estimate the overall performance considering the package and system worst operating conditions.

A key photonics packaging to look for in the near future is totally plastic packages (SMT style) where optical devices must reliably operate in a nonhermatic environment, or uncooled metal/ceramic package. The package could be pigtailed or detachable based on the area of application. This trend also eliminates the TO-can package that contains the optoelectronics devices. These photonics packages will be expected to operate in subscriber network (short-haul) application where data rates are expected to be varied from 622 Mb/s to 10 Gb/s. As an optical device, vertical-cavity surface-emitting laser (VCSEL) in both 850 nm and 1300 nm could meet the packaging requirements.

References

1. H. Kawachi, "Silica waveguides on silicon and their applications to integrated components," Opt. Quantum Electron., 22 (1990) 391–416.
2. Y. Hibino, H. Okazaki, Y. Hida, and Y. Ohmori, "Propagation loss characteristics of long silica-based optical waveguides on 5 inch Si wafers," Electron. Lett., 29 (1993) 1847–1848.
3. Y. Hibino, F. Hanawa, H. Nakagome, M. Ishii, and N. Takato, "High reliability optical splitters composed on silica based planar lightwave circuits," IEEE J. Lightwave Technol., 13 (1995) 1728–1735.
4. Y. P. Li and C. H. Henry, "Silica-based optical integrated circuits," IEE Proc. Optoelectron., 143 (1996) 263–280.
5. Y. Yamada, "Optoelectronic Hybrid Integration Technique using Planar Lightwave Circuit (PLC) Platform," Proceed. of 2001 ICEP, pp. 9–15, 2001.
6. A. K. Dutta, "Annealing effect on interface of APCVD deposited multilayer doped silicon oxides for optical waveguides," J. Electrochemical Soc., 144 (1997) 1073–1081.

7. A. K. Dutta, "Smooth and vertical etching of thick silicon oxide using C_2F_6 based reactive ion beam etching," J. Vacuum Science and Technol. B,13 (1995) 1456–1459.

8. A. K. Dutta, "ESCA study on influence of polymerization on anisotropic etching of deep silicon oxide using C_2F_6 based ECR-RIBE," Japan J. Appl. Phys., 34 (1995) 1663–1667.

9. A. K. Dutta, "Side wall roughness reduction in deep silicon oxide etching using the C_2F_6 based ECR-RIBE," Japan J. Appl. Phys., 34 (1995) 365–369.

10. K. E. Bean, "Änisotropic etching of silicon," IEEE Trans., WD-25 (1978) 1185.

11. K. Miura, S. Sasaki, H. Nobuhara, and M. Yano, "Hybridly integrated six-channel strained MQW semiconductor laser array coupling to a fiber array on a V-grooved Si wafer bond," Tech. Digest of OFC'94, TuC4 (1994).

12. S. Sasaki, G. Nakagawa, K. Miura, and M. Yano, "Passive alignment technique for single channel laser diode coupling to single mode fiber on V-grooved Si wafer board," Tech. Digest of CLEO/Pacific Rim,'95, WO5 (1995).

13. Y. Muramoto, K. Takahata, H. Fukano, K. Kato, A. Kozen, O. Nakajima, and Y. Matsuoka, "46.5 GHz bandwidth monolithic receiver OEIC consisting of the waveguide PIN photodiode and a distributed amplifier," ECOC'97, 5 (1997) 37–40.

14. H. G. Bach, W. Schlaak, G. G. Mekonnen, R. Steingiiber, A. Seeger, T. Engel, W. Passenberg, A. Umbach, C. Schramm, and G. Unetrborsch, "50 Gb/s InP based photoreceiver OEIC with gain flattened transfer characteristics," ECOC'98, 1 (1998) 55–56.

15. K. Takahata, Y. Miyamoto, Y. Muramoto, H. Fukano, and Y. Matsuoka, "50 Gb/s operation of monolithic WGPD/HEMT receiver OEIC module," ECOC'98, 3 (1998) 67–68.

16. H. Soda, K. Nakai, and H. Ishikawa, "Frequency response of an optical intensity modulator monolithically integrated with a DFB laser," Proc. ECOC'88 (1988) 227–230.

17. D. Marcuse, "DFB laser with attached external intensity modulator," IEEE J. Quantum Electron., 26 (1990) 262–269.

18. K. Sato, K. Wakita, I. Kotaka, Y. Kondo, M. Yamamoto, and A. Takada, "Monolithic strained-InGaAsP multi-quantum-well lasers with integrated electroabsorption modulators for active mode locking," Appl. Phys. Lett., 65 (1994) 1–3.

19. D. M. Adams, C. Rolland, A. Fekecs, D. McGhan, A. Somani, S. Bradshaw, M. Poirier, E. Dupont, E. Cremer, and K. Anderson, "1.55 um transmission at 2.5 Gbit/s over 1102 km of NDSF using discrete and monolithically integrated InGaAsP-InP Mach-Zehnder modulator and DFB laser," Electron. Lett., 34 (1998) 771–773.

20. D. M. Adams, C. Rolland, N. Puetz, R. S. Moore, F. R. Shepherd, H. B. Kim, and S. Bradshaw, "Mach-Zehnder modulator integrated with a gain-coupled DFB laser for 10 Gbit/s, 100 km NDSF transmission," Electron. Lett., 32 (1996) 485.
21. N. Putz, D. M. Adams, C. Rolland, R. Moore, and R. Mallard, "Fabrication of an InP/GaInAsP based integrated gain-coupled DFB laser/M-Z phase modulator for 10 Gb/sec fiber optic transmission," Indium Phosphide and Related Materials, 1996. IPRM '96, 8th Internat. Conf. (1996) 152–154.
22. F. Zamkotsian, K. Sato, H. Okamoto, K. Kishi, I. Kotaka, M. Yamamoto, Y. Kondo, H. Yasaka, Y. Yoshikuni, and K. Oe, "An InP-based optical multiplexer integrated with modulators for 100 Gb/s transmission," ECOC'94, post deadline paper (1994) 105–108.
23. F. Zamkotsian, K. Sato, H. Okamoto, K. Kishi, I. Kotaka, M. Yamamoto, Y. Kondo, H. Yasaka, Y. Yoshikuni, and K. Oe, "Monolithic integration of MQW modulators on an optical multiplexer on InP 100 Gb/s transmission," IEEE J. Lightwave Technol., 14 (1996) 2344–2352.
24. Y. Tohmori, "Semiconductor photonic devices for WDM applications," Proc. SPIE, Active and Passive Optical Components for WDM Communications, A. K. Dutta, A. A. S. Awwal, N. K. Dutta, and K. Okamoto, eds., 4532 (2001) 101–113.
25. B. Mason, G. A. Fish, S. Denbaars, and L. A. Coldren, "Widely tunable sampled grating DBR laser with integrated electroabsorption modulator," IEEE Photonic. Technol. Lett., 11 (1999) 638–640.
26. B. Mason, "Electroabsorption modulators," WDM Technologies: Active Optical Components, A. K. Dutta, N. K. Dutta, and M. Fujiwara, eds. Academic Press, San Diego (2002), Chap. 6.
27. D. Yap, Y. K. Brown, R. H. Walden, T. P. E. Broekaert, K. R. Elliott, M. W. Yung, D. L. Persechini, W. W. Ng, and A. R. Kost, "Integrated optoelectrnic circuits with InP-Based HBTs," SPIE Proc., 4290 (2001) 1–11.
28. W. Zhou and M. H. Ervin, "Fabrication and analysis of optoelectronic integrated circuits with one selective area growth," SPIE Proc., 4290 (2001) 21–27.
29. T. Kato, F. Yuuki, K. Tanaka, T. Habu, T. Shimura, A. Takai, K. Mizuhashi, T. Teraoka, and Y. Motegi, "A new assembly architecture for multichannel single mode fiber pigtail LD/PD modules," IEEE 42nd Electron. Components and Technol. Conf. (1992) 853.
30. S. G. Kang, M. K. Song, S. S. Park, S. H. Lee, N. Hwang, H. T. Lee, K. R. Oh, G. C. Joo, and D. Lee, "Fabrication of semiconductor optical switch module using laser welding technique," IEEE Trans. On Advanced Packaging, 23 (2000) 672–680.
31. N. Murata and K. Nakamura, "UV-curable adhesives for optical communication," J. Adhesion, 35 (1991) 251.

32. S. Mino, T. Ohyama, T. Hashimoto, Y. Akahori, K. Yoshino, Y. Yamada, K. Kato, M. Yasu, and K. Moriwaki, "High frequency electrical circuits on a planar lightwave circuit platform," IEEE J. Lightwave Technol., 14 (1996) 860–871.
33. S. Mino, T. Ohyama, Y. Akahori, Y. Yamada, M. Yanagisawa, T. Hashimoto, and Y. Itaya, "10 Gb/s hybrid-integrated laser diode array using a planar lightwave circuit (PLC)-platform," Electron. Lett., 32 (1996) 2232–2233.
34. J. Sasaki, H. Hatakeyama, T. Tamanuki, S. Kitamura, M. Yamaguchi, N. Kitamura, T. Shimoda, M. Kitamura, T. Kato, and M. Itoh, "Hybrid integration 4×4 optical matrix switch using self-aligned semiconductor optical amplifier gate arrays and silica planer lightwave circuit," Electron. Lett., 34 (1998).
35. M. J. Vale and C. Edge, "Self aligned flip-chip assembly of photonic devices with electrical and optical connections," IEEE Trans. Compon., Hybrids Manuf. Technol., CHMT-13 (4) (1990) 780–786.
36. K. Kurata, K. Yamauchi, A. Kawakami, H. Tanaka, H. Honmou, and S. Ishikawa, "A surface mount type single mode laser module using passive alignment," Proc. ECTC'95 (1995) 759–765.
37. K. Kurata, K. Yamauchi, A. Kawakami, H. Tanaka, H. Honmou, and S. Ishikawa, "A surface mount type single mode laser module using passive alignment," VI International Workshop on Optical Access Networks, S3.4 (1994).
38. A. Goto, S. Nakamura, K. Kurata, M. Funabashi, T. Tanabe, K. Komatsu, O. Akiyama, N. Kitamura, T. Tamura, and S. Ishikawa, "Hybrid WDM transmitter/receiver module using alignment free assembly technique," Proc. ECTC'97 (1997) 620–625.
39. J. J. Koh, M. H. Choi, H. J. Song, and J. U. Bu, "On wafer process for mass production of hybridly integrated optical components using passive alignment on silicon motherboard," Proc. ECTC '99 (1999) 216–221.
40. M. Itoh, K. Sasaki, A. Ueda, I. Yoneda, H. Honmou, and K. Fukushima, "Use of solder bumps in three dimensional passive aligned packaging of LD/PD arrays on Si optical benches," Proc. Electron. Components & Technol. Conf. '96 (1996) 1–7.
41. T. Yamamoto, H. Kobayashi, M. Egawa, T. Fujii, H. Soda, and M. Kobayashi, "High temperature operation of 1.3 um narrow beam divergence tapered-thickness waveguide BH MQW lasers," Electron. Lett., 31:25 (1995) 2178–2179.
42. T. Yamamoto, H. Kobayashi, T. Ishikawa, T. Takeuchi, T. Watanabe, T. Fujii, S. Ogita, and M. Kobayashi, "Low threshold current operation of 1.3 um narrow beam divergence tapered-thickness waveguide lasers," Electron. Lett., 33 (1997) 55–56.

43. T. Yamamoto, H. Kobayashi, M. Ekawa, S. Ogita, T. Fujii, and M. Kobayashi, "156 Mbit/s, penalty-free 40 km transmission using a tapered-thickness waveguide BH MQW laser with external optical feedback," Electron. Lett., 33 (1997) 65–67.

44. H. Kobayashi, M. Ekawa, N. Okazaki, O. Aoki, S. Ogita, and H. Soda, "A corner illuminated structure PIN photodiode suitable for planner lightwave circuit," Conf. Proc. LEOS'96, 9th annual meeting, MA5 (1996).

45. G. Nakagawa, S. Sasaki, N. Yamamoto, K. Tanaka, K. Miura, and M. Yano, "High power and high sensitivity PLC module using a novel corner-illuminated PIN photodiode," Proc. ECTC'97 (1997) 607–613.

46. G. Nakagawa, K. Tanaka, T. Yamamoto, N. Yamamoto, M. Norimatsu, M. Kobayashi, K. Miura, and M. Yano, "A corner-illuminated photodiode for receiver and LD-monitor on a planar lightwave circuit platform," Technical Digest CPT'98 (1998) 181–182.

47. G. Nakagawa, T. Yamamoto, T. Sasaki, N. Yamamoto, K. Tanaka, M. Norimatsu, M. Kobayashi, K. Miura, and M. Yano, "Simple assembly scheme of a corner-illuminated PD on a hybridly integrated planar lightwave circuit (PLC) platform," Proc. 2nd IEMT/IMC Symp. (1998) 118–122.

48. S. Ogita, Y. Kotaki, M. Matsuda, Y. Kuwahara, H. Onaka, H. Miyata, and H. Ishikawa, "FM response of narrow-linewidth, multielectrode lambda/4 shift DFB laser," IEEE Photonic. Technol. Lett., 2 (1990) 165–166.

49. Y. Kotaki, S. Ogita, M. Matsude, Y. Kuwahara, and H. Ishikawa, "Tunable, narrow-linewidth and high-power lambda/4-shifted DFB laser," Electron. Lett., 25 (1989) 990–992.

50. H. Ishikawa, K. Kamite, K. Kihara, H. Soda, and H. Imai, "Gigabit/s performance of quarter-wavelength shifted DFB laser," Electron. Lett., 24 (1988) 495–496.

51. A. B. Massara, K. A. Williams, R. V. Penty, I. H. White, F. H. G. M. Wijnands, and P. Crump, "Ridge waveguide 1.55 um lasers with uncooled 10 Gb/s operation at 55 C," Photonics West 2001, Proc. in Plane Semiconductor Lasers IV, 3947 (2001) 203.

52. K. Kurata, K. Yamauchi, A. Kawatani, E. Tanaka, H. Honmou, and S. Ishikawa, "A surface mount single-mode laser module using passive alignment," IEEE Trans. on Components, Packaging, and Manufacturing Technology, Part B: Advanced Packaging, [see also IEEE Trans. Components, Hybrids, and Manufacturing Technolog], 19 (1996) 524–531.

53. N. Takehashi and M. Horii, "A new premolded packaging technology for low cost E/O device applications," 47th ECTC'97, 16–18.

54. A. K. Dutta, "Prospects of highly efficient AlGaInP based surface emitting type ring-LED for 50 and 156 Mb/s POF data link systems," IEEE J. Lightwave Technol., 16 (1998) 106–113.

55. A. K. Dutta, K. Ueda, K. Hara, and K. Kobayashi, "High brightness and reliable AlGaInP-based visible light emitting device for POF data-links," IEEE Photonic. Technol. Lett., 9 (1997) 1567–1569.

56. A. K. Dutta, A. Suzuki, K. Kurihara, F. Miyasaka, H. Hotta, and K. Sugita, "High-brightness, AlGaInP-based visible light-emitting device for efficient coupling," IEEE Photonic. Technol. Lett., 7 (1995) 1134–1136.

57. A. K. Dutta, H. Kosaka, K. Kurihara, Y. Sugimoto, and K. Kasahara, "High speed VCSEL of modulation bandwidth of 7.3 GHz and its application to 100 m PCF transmission," IEEE J. Lightwave Technol., 16 (1998) 870–875.

58. H. Kosaka, A. K. Dutta, K. Kurihara, Y. Sugimoto, and K. Kasahara, "Gigabit-rate optical-signal transmission using vertical-cavity surface-emitting lasers with large-core plastic-cladding fibers," IEEE Photonic. Technol. Lett., 7 (1995) 926–928.

59. A. K. Dutta and K. Kasahara, "Polarization control in VCSELs using uniaxial stress," Solid-State Electron., 42 (1998) 907–910.

60. M. Kobayashi and A. K. Dutta, "Optical packaging/Module technologies: past and future," Proc. SPIE Active and Passive Optical Components for WDM Communications, A. K. Dutta, A. A. S. Awwal, N. K. Dutta, and K. Okamoto, eds., 4532 (2001) 270–280.

61. H. Yonetani, T. Machida, and S. Yamakoshi, "Laser module with wavelength locker for WDM system," Fujitsu Tech. J., 51:3 (2000) 148–151.

62. K. Satoh, I. Hanawa, and M. Kobayashi, "APD receiver module 10 Gb/s high speed data communication," Fujitsu Tech. J., 51:3 (2000) 152–154.

63. Y. Oikawa, H. Kuwatsuka, T. Yamamoto, T. Ihara, H. Hamano, and T. Minami, "Packaging Technology for a 10-Gb/s photoreceiver module," IEEE J. Lightwave Technol., 12 (1994) 343–352.

64. T. Mikawa, M. Kobayashi, and T. Kaneda, "Avalanche photodiodes: present and future," Proc. SPIE Active and Passive Optical Components for WDM Communications, A. K. Dutta, A. A. S. Awwal, N. K. Dutta, and K. Okamoto, eds., 4532 (2001) 139–145.

65. C. H. Henry, G. E. Blonder, and R. F. Kazarinov, "Glass waveguides on silicon for hybrid optical packaging," IEEE J. Lightwave Technol., 7 (1989) 1530–1538.

66. A. Himeno, K. Kato, and T. Miya, "Silica-based planar lightwave circuits," IEEE J. Selective Top. Quant. Elect., 6 (2000) 38–45.

67. K. Kato and Y. Tohmori, "PLC hybrid integration technology and its application to photonic components," IEEE J. Selective Top. Quantum. Elect., 6 (2000) 4–13.

68. S. Nakamura, Y. Ueno, K. Tajima, J. Sasaki, T. Sugimoto, T. Kato, T. Shimoda, M. Itoh, H. Hatakeyama, T. Tamanuki, and T. Sasaki, "Demultiplexing of 168 Gb/s data pulses with a hybrid integrated symmetric Mach-Zehnder all optical switch," IEEE Photonic. Technol. Lett., 12 (2000) 425–427.

69. H. Hatakeyama, S. Kitamura, K. Hamamoto, M. Yamaguchi, and K. Komatsu, "Spot size coneverter integrated semiconductor optical amplifier gates with low power conversion," Tech. Dig. Optoelectronics and Communications Conf. '97, 9C3-2 (1997).
70. S. Takahashi, T. Kato, H. Takeshita, S. Kitamura, and N. Henmi, "10 Gbs/ch space division optical cell switching with 8×8 gate type switch matrix employing gate turn-on-dely compensator," Tech. Digest Photonics in Switching'96, post deadline paper PThC1, 2 (1996) 12–15.

Index

669

Wavelength-division multiplexing
 (WDM) (*continued*)
 photodetectors, 9
  avalanche photodiodes (APDs), 9, 379–434
  P-I-N photodiodes, 9, 317–372
 semiconductor laser sources, 6
  for EDFA pumping, 6, 59, 61–62, 317
  high-power lasers, 6, 13–54, 59–104
  tunable laser diodes, 7, 105–159
  VCSELs, 7–8, 38, 167–197
Wavelength filter, 156
Wavelength-locker integrated laser module,
 647–649
Wavelength-selectable light sources. *See* WSLs
WDM. *See* Wavelength-division multiplexing

Weighted-dose allocation variable pitch EB
 lithography. *See* WAVE
WGR. *See* Waveguide grating router
WG structure. *See* Waveguide structure
Whisker growth, 600
Wire bonding, 607–608
WSLs
 EA-WSL, 506–510
 fabrication, 481–482, 505–513

Z

ZnSe system, surface-emitting lasers,
 188